ELECTRICAL ENGINEER'S PORTABLE HANDBOOK

ROBERT B. HICKEY, P.E.
Electrical Engineer

McGraw-Hill

New York   San Francisco   Washington, D.C.   Auckland   Bogotá
Caracas   Lisbon   London   Madrid   Mexico City   Milan
Montreal   New Delhi   San Juan   Singapore
Sydney   Tokyo   Toronto

Library of Congress Cataloging-in-Publication Data

Hickey, Robert, B.
    Electrical engineer's portable handbook / Robert B. Hickey.
       p.   cm.
    ISBN 0-07-028698-1
    1. Electric engineering Handbooks, manuals, etc. I. Title.
TK151.H53   1999
621.3—dc21   99-22023
                                       CIP

McGraw-Hill

A Division of The McGraw·Hill Companies

ISBN 0-07-028698-1

*The sponsoring editor for this book was Scott Grillo, the editing supervisor was
Tom Laughman, and the production supervisor was Sherri Souffrance. It was set
in Times Ten by North Market Street Graphics.*

Printed and bound by R. R. Donnelley & Sons Company.

This book is printed on recycled, acid-free paper
containing a minimum of 50% recycled, de-inked fiber.

McGraw-Hill books are available at special quantity discounts to use as
premiums and sales promotions, or for use in corporate training programs.
For more information, please write to the Director of Special Sales,
McGraw-Hill, Professional Publishing, Two Penn Plaza, New York, NY
10121-2298. Or contact your local bookstore.

To Vandi Lynn Larson,
my stepdaughter,
in loving memory.

Contents

Chapter 7. Special Systems   437

Chapter 8. Miscellaneous Special Applications   509

Acknowledgments

Thanks to the following people for their professional knowledge and valuable input in helping with this book: John Lloyd, Bruce Dalkowski, Dennis Julian, Dave Woolley, Pat Mitchell, Lenny Baldino, Denise Hart, Kristine Melroy, and the entire electrical staff at vanZelm, Heywood, & Shadford, Inc.

A special thanks to Charles MacDaniel (Mac) Billmyer, my mentor and sponsor in this business.

A much deserved thanks to Scott Grillo and his team at McGraw-Hill and to Tom Laughman and his team at North Market Street Graphics, whose wonderful collaborative spirit and many professional talents have made transforming the raw manuscript into a published reality a very enjoyable experience.

Finally, a very special thanks to my wife Pat, who has supported and encouraged me to write this book, and who has endured seemingly interminable hours of being separated during the preparation of this manuscript, not to mention shouldering my share of the domestic responsibilities.

Credits

Cablec Industrial Cable Co., Caterpillar, Cerberus Pyrotronics, Cooper Bussmann, Cutler-Hammer, Electronic Industries Association/Telecommunications Industry Association, Gould Shawmut, Illuminating Engineering Society of North America, Institute of Electrical and Electronics Engineers, Kerite (Hubbell Power Systems), Leviton Telcom, Okonite Company, Osram Sylvania, Pyrotenax USA, Siemens Energy and Automation, Square D Company, The Siemon Company

Introduction:
How to Use This Book

The concept of this book is that of a *personal tool*, which compacts 20 percent of the data that is needed 80 percent of the time by *electrical design professionals* in the preliminary design of buildings of all types and sizes.

This tool is meant to always be at one's fingertips (open on a drawing board, desk, or computer table; carried in a briefcase; or kept in one's pocket). It is never meant to sit on a bookshelf. It is meant to be used *everyday!*

Because design professionals are individualistic and their practices are so varied, the user is encouraged to *individualize this book* by adding notes or changing data as experience dictates.

Building codes and laws, new technologies, and materials are ever changing in this industry. Therefore, this book should be viewed as a *starter of simple data collection* that must be updated over time. New editions may be published in the future.

Because this book is so broad in scope, yet so compact, information can be presented in only one location, and not repeated. It is expected that the experienced practitioner is generally knowledgeable about the data and knows how to apply it properly. Information is often presented in the form of simple ratios, coefficients, application tips, or rules of thumb that leave the need for commonsense judgment.

This book is unique among handbooks. It provides myriad valuable time-saving data for the experienced practitioner, yet there are enough concept explanations and examples on critical topics to use it as a teaching tool for the fledgling electrical design professional. Also, the topics of Chapters 3 through 7, in particular, are arranged in a sequence that closely approximates the normal design process flow to facilitate speed for the experienced practitioner and learning for the beginner. The Index has been expanded to facilitate quickly locating needed information.

This book is *not a substitute* for professional expertise or other books of a more detailed and specialized nature, but will be a continuing everyday aid that takes the more useful "cream" off the top of other sources.

ELECTRICAL ENGINEER'S
PORTABLE HANDBOOK

FIGURE 1.2 Drawing design checklist (electrical).

Page 1 of 3

Project Status
☐ SD
☐ DD
☐ CD

Project: _____
Proj. No: _____
PM/PE: _____
Date: _____

Items Included
☐ Power Plan
☐ Lighting Plan
☐ Site Plan
☐ Special System Plans
☐ Symbol List
☐ Abbreviation List
☐ One Line - Power Diagram
☐ One Line - Special Systems
☐ Switchboard Schedules
☐ Panelboard Schedules
☐ Fixture Schedules
☐ Site Details
☐ Electrical Details
☐ Building Grounding Plan
☐ Lightning Protection Plan
☐ General Notes
☐ _____
☐ _____
☐ _____

General Items to Check
☐ Title Blocks
    ☐ Firm Logo
    ☐ Job Number
    ☐ Drawing Title
    ☐ Drawing Numbers
    ☐ Date
☐ Plan Titles with Scale
☐ Detail Titles with Scale
☐ Detail Designation Symbols
☐ Symbol List Agrees with Drawing
☐ Abbreviation List Agrees with Drawings

☐ Openings and Floor Plans for Installation and Removal of Electrical and Generator Equipment
☐ Electrical equipment access and clearances
☐ Elevator Size Accommodates All Equipment
☐ Electrical Plans Overlayed on:
    ☐ Architectural Plans
    ☐ Reflected Ceiling Plans
    ☐ Mechanical Plans

One-Line Power Diagram
☐ Primary Distribution
    ☐ Voltage
    ☐ Fault Current Available
    ☐ Cables and Raceways
    ☐ Manholes and Pullboxes
    ☐ Terminations and Splices

☐ Primary Switchgear
    ☐ Enclosure
        ☐ Indoor ☐ Weatherproof ☐ Walk-In
    ☐ Selector Switches
        ☐ Non-fused ☐ Fuse Size
    ☐ Protective Devices
        ☐ Stationary ☐ Drawout
        ☐ Manual ☐ Electrical
        ☐ Active ☐ Space & Busing
        ☐ Breaker ☐ Trip Setting
        ☐ Relay ☐ Trip Setting
        ☐ Circuit Numbering
        ☐ Arresters
        ☐ Interlocks
        ☐ Fault Rating

FIGURE 1.2 Drawing design checklist (electrical). (*Continued*)

Page 2 of 3

Project Status   Project: _____
☐ SD   Proj. No: _____
☐ DD   PM/PE: _____
☐ CD   Date: _____

☐ Primary Metering
     ☐ Owner   ☐ Power Co.
☐ Transformers
   ☐ Primary Voltage
   ☐ Primary Connection
     ☐ Delta ☐ Wye ☐ Double Bushing
   ☐ Secondary Voltage
   ☐ Secondary Connection
     ☐ Delta ☐ Wye
   ☐ Grounding
   ☐ KVA & Percent Impedance (Min.)
   ☐ Type:
   (Oil, Dry, Padmount, Open, WP, etc.)
   ☐ Secondary Compartment C/Bs
   ☐ Surge Arresters
   ☐ Power Company Supplied

☐ Secondary Distribution
   ☐ Voltage
   ☐ Fault Current Available
   ☐ Cables and Raceways
   ☐ Manholes and Pullboxes
   ☐ Termination and Splices
Secondary Switchboard
   ☐ Switchboard (NEMA PB-2 and UL 891)
   ☐ Switchgear (ANSI C37 and UL 1558)
   ☐ Rating   ☐ Current   ☐ Voltage
   ☐ Phase   ☐ Wire
   ☐ Fault Rating
   ☐ Service Entrance?
☐ Enclosure
   ☐ Free-standing   ☐ Non-freestanding
☐ Accessible
   ☐ Front   ☐ Rear   ☐ Side

☐ Main Protective Device
   ☐ Fuse/Sw ☐ Size & Class of Fuse
   ☐ Power Breaker   ☐ Insulated Case
     ☐ Molded Case
     ☐ Indv. Mount ☐ Group Mount
   ☐ Stationary   ☐ Drawout
   ☐ Manual ☐ Electrical
   ☐ Thermal/Magnetic ☐ Solid State
   ☐ Number of Poles & Trip/Frame Amps
   ☐ 100% Duty   ☐ 80% Duty
   ☐ Shunt Trip
   ☐ Interlocks or Ties
☐ Ground Fault Protection
   ☐ Selective ☐ Time Delay
☐ Service Ground
   ☐ Water Service
   ☐ Building Steel
   ☐ Ground Rod
   ☐ Ground Grid - Substation
   ☐ Ground Grid - Building
☐ Revenue Metering
   ☐ Active   ☐ Reactive
   ☐ CT's   ☐ PT's
☐ Owner Metering
   ☐ Volt ☐ Amp ☐ Watt ☐ VA
   ☐ Watt Hr ☐ VARS
   ☐ AMSS ☐ VMSS
   ☐ Electronic
☐ Busing
   ☐ Full Neutral
   ☐ Ground Bus
   ☐ Equipment Ground
   ☐ Grounding Electrode Conductor
   ☐ Connection

FIGURE 1.2   Drawing design checklist (electrical). (*Continued*)

Page 3 of 3

Project Status
- ☐ SD
- ☐ DD
- ☐ CD

Project: _____
Proj. No: _____
PM/PE: _____
Date: _____

☐ Operation

☐ Main Feeder Cable and Raceways

☐ Transfer Switches
- ☐ Type
  - ☐ Automatic   ☐ Manual
- ☐ Current Rating and # Poles
- ☐ Control Connection
- ☐ Load Feeder Cable and Raceway
- ☐ 3 Pole or 4 Pole
- ☐ Neutral and Ground Connection

☐ Standby Generator ☐ Emergency Generator
- ☐ Line Circuit Breaker   ☐ Main Lug
  - ☐ Thermal   ☐ Magnetic
    - ☐ Solid State
- ☐ Number of Poles & Trip/Frame Amps
- ☐ GFP   ☐ Sel.   ☐ Timedelay
- ☐ Load Feeder Cable and Raceway
- ☐ Neutral and Ground Connections

☐ Power Distribution (Panelboard and MCC)
- ☐ Bus Data
  - ☐ Current
  - ☐ Voltage
  - ☐ Phase   ☐ Wire
  - ☐ Fault Current
  - ☐ Full Neutral
  - ☐ Equipment Ground
  - ☐ Insulated
  - ☐ Enclosure
  - ☐ Weatherproof   ☐ Walk-in
- ☐ Mounting
  - ☐ Individual   ☐ Group (Panel Sched.)
  - ☐ Stationary   ☐ Drawout

☐ Manual   ☐ Automatic

☐ Protective Devices
- ☐ Circuit Numbering
- ☐ Fuse/Switch
  - ☐ Fuse Size/Class
- ☐ Combination Starter
  - ☐ Fuse/Switch & Fuses
  - ☐ Circuit Breaker
    - ☐ Mag. Only
  - ☐ Starter Size & Type
  - ☐ Overload Relays

☐ Circuit Breaker
- ☐ Power
- ☐ Insulated Case
- ☐ Molded Case
- ☐ 100% Duty
  - ☐ Mixed Duty
- ☐ Thermal/Magnetic
- ☐ Magnetic
- ☐ Solid State
- ☐ Number of Poles
- ☐ Trip/Frame Amps
- ☐ Ground Fault Protection
- ☐ Selective   ☐ Time Delay
- ☐ Interlocks
- ☐ Key   ☐ Electric

FIGURE 1.3   Site design checklist (electrical).

Page 1 of 2

Project Status   Project: _____
☐ SD   Proj. No: _____
☐ DD   PM/PE: _____
☐ CD   Date: _____

Site Drawings - Plans

☐ Title
☐ Scale
☐ Benchmark
☐ Topo Lines

Top Elevation on:
    ☐ Transformer Pads
    ☐ Switchgear Pads
    ☐ Pole Bases for Site Lighting
    ☐ Standby Generator Pads
    ☐ Manholes
    ☐ Pullboxes

☐ Existing Utility Poles and Numbers
☐ New Utility Poles and Guys (by whom)
☐ Pole Transformers (by whom)
☐ Pad Mount Transformers (by whom)
☐ Revenue Meters
☐ Site Lighting Poles
☐ Generator (Outdoor)
☐ Switchgear (Outdoor)
☐ Manholes
☐ Pullboxes

Check Site Planting, Grades, Fences,
Equipment for Truck Access to:
    ☐ Padmount Transformers
    ☐ Utility Poles
    ☐ Site Lighting Poles

Aerial Distribution
    ☐ Electric Primary
    ☐ Electric Secondary
    ☐ Telephone
    ☐ Site Lighting
    ☐ TV
Underground Distribution
    ☐ Electric Primary
    ☐ Electric Secondary
    ☐ Telephone
    ☐ TV
    ☐ Site Lighting
    ☐ Conduit Sleeves Under Pavement

Fuel Oil Systems
    ☐ Fuel Oil Tank
    ☐ Supply and Return Lines
    ☐ Fill Cap and Fill Lines
    ☐ Vent Cap and Vent Lines
    ☐ Tank Level Gauge Line
    ☐ Soil Condition - Anodes, FG
    ☐ Direction of Line Pitch

Check Truck Wheel Loading Cover:
    ☐ Fuel Oil Tanks
    ☐ Underground Lines
    ☐ Manholes
    ☐ Pullboxes

FIGURE 1.3   Site design checklist (electrical). (*Continued*)

Page 2 of 2

Project Status
☐ SD
☐ DD
☐ CD

Project: _____
Proj. No: _____
PM/PE: _____
Date: _____

Site Drawings - Details

☐ Titles
☐ Scale
☐ Utility Pole Riser
☐ Revenue Meter Riser

Trench Cross Sections
    ☐ Electric, Telephone and TV Lines
    ☐ Duct Banks, Concrete and Grounding

☐ Padmount Transformer, Concrete Pad & Grounding
☐ Exterior Switchgear, Concrete Pad & Grounding
☐ Generator, Concrete Pad & Grounding
☐ Manholes, Concrete, Cable Racks & Grounding
☐ Pullboxes, Concrete, & Grounding
☐ Pole Bases for Site Lighting and Signs

Fuel Oil Systems
    ☐ Fuel Oil Tank, Concrete Pad
    ☐ Trench Cross Sections for Supply & Return Lines
    ☐ Fill, Vent and Level Gage Lines
    ☐ Fuel Fill Cap
    ☐ Fuel Vent Cap

FIGURE 1.4 Existing condition service & distribution checklist.

Page 1 of 3

Project: _____
Proj. No: _____
PM/PE: _____
Date: _____

Power Company Service

Power Company: _____
  Rep Name: _____
  Telephone: _____

Type of Service:
  ☐ Primary ☐ Secondary ☐ Unknown

  ☐ Underground ☐ Overhead
  ☐ Combination ☐ Unknown
  ☐ _____

Transformation
  ☐ Pad ☐ Pole ☐ N/A ☐ Unknown
  KVA: _____ ☐ Unknown
  % Impedance: _____ ☐Unknown
  Primary Voltage _____ ☐ Unknown
  Secondary Voltage: ____ ☐ Unknown

Short Circuit Fault Current Available
  ☐ Power Company   ☐ Sym
  ☐ Primary   ☐ MVA
  ☐ Secondary: _____   ☐ A
  ☐ Unknown

Power Company Pole #: _____ ☐ Unknown

New Poles: ☐ Street Line ☐ Private
    ☐ N/A   ☐ Unknown

Primary Service
  Raceway Size: _____ ☐ Unknown
  Type: ☐ RSC ☐ PVC ☐ PVC/Conc.
    ☐ DB:____ ☐ Unknown
  Cable: _____ ☐ Unknown
  Ground Conductor: _____ ☐ Unknown

Secondary Service
  Raceway Size: _____ ☐ Unknown
  Type: ☐ RSC ☐ PVC ☐ PVC/Conc.
    ☐ DB:____ ☐ Unknown
  Cable: _____ ☐ Unknown

Type of Power Available at Site Line
  Primary ☐ 1PH ☐ 3PH ☐ Unknown
  Sec   ☐ 1PH ☐ 3 PH ☐ Unknown

Has Power Company Been Contacted for
Existing Loads and Requirements for new
services?
    ☐ Yes ☐ No ☐ Not Req.

Comments:
_____
_____
_____

Main Electric Service

Main Entrance Capacity:
  Size _____ A ☐ Unknown
  Total Load
    _____ KW   _____ KVA
    Power Factor _____ ☐ Unknown
  Largest Connected Motor ☐ N/A
    _____ HP ☐ Unknown
  Starter Size & Type ____ ☐Unknown

FIGURE 1.4 Existing condition service & distribution checklist. (*Continued*)

Page 2 of 3
Project: _____
Proj. No: _____
PM/PE: _____
Date: _____

Main Protective Device:

☐ Fuse/Switch ☐ MCCB ☐ ICCB

☐ Power breaker_____ ☐Unknown

Duty: ☐ 80% ☐ 100% ☐Unknown

Type of Trip: ☐ Thermal ☐Magnetic

☐ Solid State

GFP ☐Yes ☐No ☐Unknown

☐ Selective ☐Time Delay

Current Limiting

☐Yes ☐No ☐Unknown

CT's Required: ☐Yes ☐No ☐Unknown

PT's Required: ☐Yes ☐No ☐Unknown

Who Supplies CT's and PT's:

_____ ☐Unknown

Revenue Meters

☐ Active ☐ Reactive ☐Unknown

☐ Inside ☐ Outside ☐Unknown

Type of Construction

☐ Panelboard ☐ Switchboard

☐ Unitized ☐ MCC _____

Grounding Electrode Conductor

Size _____

☐ Ground Rod ☐ Water Service

Rating of Gear

_____ AIC Sym. ☐Unknown

Comments:

_____

_____

_____

Power Distribution System ☐ N/A

Main Distribution Bus _____ A ☐Unknown

Rating _____ AIC Sym ☐Unknown

Distribution Devices

☐ Fuse/Switch ☐ MCCB ☐ ICCB

☐ Power breaker_____ ☐Unknown

Duty: ☐ 80% ☐ 100% ☐Unknown

Type of Trip: ☐ Thermal ☐Magnetic

☐ Solid State

GFP ☐Yes ☐No ☐Unknown

☐ Selective ☐Time Delay

Current Limit ☐Yes ☐No ☐Unknown

Raceways ☐ Aluminum ☐ RSC ☐ ISC

☐ EMT ☐ PVC ☐ Unknown

Conductor Type_____ ☐ Unknown

Voltage Systems #1_____ ☐ Unknown

#2_____

Raceway Location ☐ Exposed ☐ Unknown

Concealed in: ☐ Walls ☐ Ceilings

☐ Floors ☐ Unknown

Busway ☐ Aluminum ☐ Copper ☐ WP

☐ N/A ☐ Unknown

☐ Feeder ☐ Plug-in ☐ Standard

☐ LVD ☐ CL

Plug In Unit: ☐ Fuse/Switch ☐ N/A

☐Circuit Breaker ☐Unknown

Dry Type Transformer

☐ 1 PH ☐ 3 PH ☐ N/A ☐Unknown

Minimum Impedance _____ %

FIGURE 1.4   Existing condition service & distribution checklist. (*Continued*)

Page 3 of 3

Project: _____
Proj. No: _____
PM/PE: _____
Date: _____

Sub Panels:
☐ 1 PH   ☐ 3 PH   ☐ N/A   ☐ Unknown
Rating: _____ AIC sym   ☐ Unknown
Branch Breakers: ☐ Standard
☐ Switching Duty   ☐ Unknown
Comments:
_____
_____
_____

FIGURE 1.5 Design coordination checklist (electrical).

Page 1 of 3

Project Status	Project: _____
☐ SD	Proj. No: _____
☐ DD	PM/PE: _____
☐ CD	Date: _____

Electrical Drawings - Plans   Coord./% Complete

Check that electrical floor plans match architectural and mechanical plans.	Y	N	N/A
Check that the location of floor mounted equipment is consistent between disciplines.	Y	N	N/A
Check that the location of light fixtures matches architectural reflected ceiling plan.	Y	N	N/A
Check that elevator power, telephone and recall systems are shown and coordinated with architectural and fire protection	Y	N	N/A
Check that light fixtures do not conflict with the structure or the mechanical HVAC system.	Y	N	N/A
Check electrical connections to major equipment. Check that horsepower rating, phase, voltage, starter and drive types are consistent with other trade schedules.	Y	N	N/A
Check that locations of panelboards are consistent with architectural floor plans, mechanical floor plans, plumbing & fire protection floor plans.	Y	N	N/A
Check that the panelboards are indicated on the electrical riser diagram.	Y	N	N/A
Check that HVAC control power needs are addressed.			
Check that notes are referenced.	Y	N	N/A
Check that locations of electrical conduit runs, floor trenches, and openings are coordinated with structural plans.	Y	N	N/A
Check that electrical panels are not recessed in fire rated walls.	Y	N	N/A
Check that locations of exterior electrical equipment are coordinated with site paving, grading and landscaping.	Y	N	N/A
Check that structural supports are provided for rooftop electrical equipment.	Y	N	N/A

Food Service Drawings

Check that the equipment layout matches other trade floor plans.	Y	N	N/A
Check that there are no conflicts with columns.	Y	N	N/A
Check that equipment is connected to utility systems.	Y	N	N/A

FIGURE 1.5 Design coordination checklist (electrical). (*Continued*)

Page 2 of 3

Project Status | Project: _____
☐ SD | Proj. No: _____
☐ DD | PM/PE: _____
☐ CD | Date: _____

				Coord./% Complete

	Y	N	N/A
Check that equipment as scheduled on the drawings matches the kitchen floor plans and specifications.	Y	N	N/A
Check that floor depressions and floor troughs are coordinated.	Y	N	N/A
Check that kitchen equipment is schedule and coordinated with floor plans.	Y	N	N/A

Communication Drawings

	Y	N	N/A
Check that equipment layout matches Architect and Consultant Plans.	Y	N	N/A
Check for conflicts between equipment/device spacing, clearances and access.	Y	N	N/A
Check for Architects or Consultant's typical elevations and details showing special device location and mounting heights.	Y	N	N/A
Check empty raceway systems for coordination with Consultant's equipment and wiring.	Y	N	N/A
Check for coordination between Specialty Contractor responsibility and Electrical Contractor responsibility.	Y	N	N/A

A/V Drawings

	Y	N	N/A
Check that equipment layout matches Architect and Consultant Plans.	Y	N	N/A
Check for conflicts between equipment/device spacing, clearances and access.	Y	N	N/A
Check for Architects or Consultant's typical elevations and details showing special device location and mounting heights.	Y	N	N/A
Check empty raceway systems for coordination with Consultant's equipment and wiring.	Y	N	N/A
Check for coordination between Specialty Contractor responsibility and Electrical Contractor responsibility.	Y	N	N/A

Theatre Drawings

	Y	N	N/A
Check that equipment layout matches Architect and Consultant Plans.	Y	N	N/A
Check for conflicts between equipment/device spacing, clearances and access.	Y	N	N/A

FIGURE 1.5   Design coordination checklist (electrical). (*Continued*)

Page 3 of 3

Project Status
☐ SD
☐ DD
☐ CD

Project: _____
Proj. No: _____
PM/PE: _____
Date: _____

<u>Coord./% Complete</u>

Check for Architects or Consultant's typical elevations and details showing special device location and mounting heights.   Y   N   N/A

Check empty raceway systems for coordination with Consultant's equipment and wiring.   Y   N   N/A

Check for coordination between Specialty Contractor responsibility and Electrical Contractor responsibility.   Y   N   N/A

Specifications

Check that bid items explicitly state what is intended.   Y   N   N/A

Check specifications for phasing of construction.   Y   N   N/A

Check that architectural finish schedule agrees with specification index.   Y   N   N/A

Check that major equipment items are coordinated with contract drawings.   Y   N   N/A

Check that items specified "as indicated" and "where indicated" in the specifications are in fact indicated on the contract drawings.   Y   N   N/A

Check that the table of contents matches the sections contained in the body of the specifications.   Y   N   N/A

FIGURE 1.6 Fire alarm system checklist.

Project: _____
Proj. No: _____
PM/PE: _____
Date: _____

Part One - Central Reporting Requirements

Emergency Forces Notification	Y	N	N/A
Auxiliary Alarm System: (Alarms transmitted directly to municipal communication center)	Y	N	N/A
Central Station: (Alarms transmitted to a station location with 24 hour supervision?)	Y	N	N/A
Central Station System: (Alarms automatically transmitted to, recorded in, maintained and supervised from an approved central supervising station)	Y	N	N/A
Proprietary Protective System: (Alarms automatically transmitted to a central supervising station on the Agency property with trained personnel and 24 hour supervision)	Y	N	N/A
Remote Station System: (Alarms transmitted to a location remote from the building where circuits are supervised and appropriate action is taken)	Y	N	N/A

Part Two - Fire Alarm System

Is there a building presently equipped with a Fire Alarm System?   Y   N   N/A
If yes: indicate Make/Model   _____
       Type: _____
       Date Installed: _____

Will this project extend/expand the existing system?   Y   N   N/A

Does the existing system conform to current Codes?		Y	N	N/A
	NFPA	Y	N	N/A
	BOCA	Y	N	N/A
	ADA	Y	N	N/A
	NEC	Y	N	N/A

Is the existing system a conventional or an addressable system?   Y   N   N/A

Is all existing equipment of the same make and manufacturer?   Y   N   N/A

Is the "Fire Alarm Control Panel", located at the Primary Building   Y   N   N/A
Entrance or Main Lobby?

Is the "Fire Alarm Control Panel" and "Annunciator" currently located at a   Y   N   N/A
location approved by the State or local Fire Marshal?

Are system components readily available?   Y   N   N/A

FIGURE 1.6   Fire alarm system checklist. (*Continued*)

Page 2 of 3

Project: _____
Proj. No: _____
PM/PE: _____
Date: _____

Have you inspected the existing Fire Alarm System?	Y	N	N/A
Have you received Agency information on the operational status of the existing system?	Y	N	N/A
Is the building equipped with adequate peripheral devices (i.e., pull stations, back up power, heat and smoke detectors, horn/speaker and strobe lights?)	Y	N	N/A
Is the existing panel and annunciator capable of accommodating the system expansion due to the new renovations?	Y	N	N/A
Have you requested copies of the latest State Fire Marshal citations?	Y	N	N/A
Are there smoke detectors at the elevator lobbies for the elevator recall system where required by Code?	Y	N	N/A
Are there smoke detectors in locations required by the Elevator Code (ASME/ANSI A 17.1)?	Y	N	N/A
Are there adequate quantities of horn/speaker and strobe lights in the corridors?	Y	N	N/A
Is the building equipped with a Fire-Fighter's phone system at each stairwell and elevator lobby?	Y	N	N/A
Have you verified that smoke detectors in residential rooms have been located away from cooking stoves and shower stalls?	Y	N	N/A
Have you specified "single-station", and not "system" detectors in the sleeping residential areas?	Y	N	N/A
Have air handling units been equipped with duct-smoke detectors, as required by NFPA Codes?	Y	N	N/A
Are air handling units annunciated at the building annunciator for easy identification of alarm location?	Y	N	N/A
Is the existing system connected to a Fire Department or other answering service?	Y	N	N/A
If a new building, is the system specified compatible with the existing campus system?	Y	N	N/A
Is the system specified as a "Proprietary" system?	Y	N	N/A
Does the Specification cite three manufacturers of equal quality meetng DPW and Agency requirements?	Y	N	N/A

FIGURE 1.6 Fire alarm system checklist. (*Continued*)

Page 3 of 3

Project: _____
Proj. No: _____
PM/PE: _____
Date: _____

If building is a high-rise, does the fire alarm system conform to BOCA high-rise requirements?	Y	N	N/A
Are stairwells required to have a pressurized smoke ventilation system?	Y	N	N/A
Is the building sprinkler system connected to the Fire Alarm Control Panel and "Annunciator" system?	Y	N	N/A
Is the building equipped with a fire pump?	Y	N	N/A
Is the fire alarm system backed up by a battery and standby generator system?	Y	N	N/A
Is the Fire Pump Electrical Service connected ahead of the Main Service Entrance switch?	Y	N	N/A

Part Three - Elevator Related Questions

Does BOCA or NFPA Code require this building to be fully sprinklered?	Y	N	N/A
When the building is fully sprinklered; are there sprinkler heads in the Elevator Machine Room, and at the top and bottom of each elevator shaft?	Y	N	N/A
Is the power to the elevator automatically shut off by a heat detector and shunt trip breaker; prior to sprinkler discharge?	Y	N	N/A
Are elevator recall smoke detectors isolated from the building's Fire Alarm System?	Y	N	N/A
Do the elevator detectors report to the main Fire Alarm Panel?	Y	N	N/A
Is the proposed elevator room steel fire proofing provided by a material acceptable to the State Elevator Inspector?	Y	N	N/A
Is there a sump pit and duplex outlet in each elevator pit?	Y	N	N/A
Is the elevator pit equipped with a guarded lighting fixture, light switch and duplex outlet?	Y	N	N/A
Does the electrical wiring, equipment, pipes, ducts, etc. in hoistways and machine rooms conform to Section 102 of the Elevator Code (ASME/ANSI A17.1 Code)?	Y	N	N/A
Is there any water piping in the elevator shaft or machine room which does not serve the shaft or machine room?	Y	N	N/A
If there is a standby generator in the building, is any elevator connected to the standby power?	Y	N	N/A
Does the design comply with ADA Section 4.10 requirements for elevators?	Y	N	N/A

1.2   ELECTRICAL SYMBOLS AND MOUNTING HEIGHTS

Electrical Symbols

Electrical symbols can vary widely, but the following closely adhere to industry standards. Industry standard symbols are often modified to meet client and/or project specific requirements.

FIGURE 1.7   Electrical symbols.

SYMBOL	DESCRIPTION
O^X	CEILING MOUNTED LIGHT FIXTURE; SUBLETTER INDICATES FIXTURE TYPE
^X O⊣	WALL MOUNTED LIGHT FIXTURE; SUBLETTER INDICATES FIXTURE TYPE
▭ ^X	2'x4' CEILING MOUNTED FLUORESCENT LIGHT FIXTURE; SUBLETTER INDICATES FIXTURE TYPE
▭ ^X	DUAL BALLAST 2'x4' CEILING MOUNTED FLUORESCENT LIGHT FIXTURE; SUBLETTER INDICATES FIXTURE TYPE
▭ ^X	1'x4' CEILING MOUNTED FLUORESCENT LIGHT FIXTURE; SUBLETTER INDICATES FIXTURE TYPE
▭ ^X	2'x2' CEILING MOUNTED FLUORESCENT LIGHT FIXTURE; SUBLETTER INDICATES FIXTURE TYPE
^X ◣	TYPICAL CEILING MOUNTED FLUORESCENT FIXTURE— NORMAL/EMERGENCY
▭	CONTINUOUS FLUORESCENT LIGHT FIXTURE
◖^X	WALL WASHER LIGHT FIXTURE
● ●⊣ ▮	LIGHT ON EMERGENCY CIRCUIT
⊢—•—⊣ _X	FLUORESCENT STRIP LIGHT FIXTURE; SUBLETTER INDICATES FIXTURE TYPE
▽ ▽	POWER LIGHT TRACK WITH NUMBER OF FIXTURES AS INDICATED ON PLANS; SUBLETTER INDICATES FIXTURE TYPE
Y Y Y	SINGLE OR DUAL HEAD, WALL MOUNTED, REMOTE EMERGENCY LIGHT
⊠ ⊠⊣	DOUBLE FACED CEILING OR WALL—MOUNTED, EXIT SIGN WITH EMERGENCY POWER BACK UP AND DIRECTIONAL ARROWS AS INDICATED ON PLANS
⊗ ⊗⊣	SINGLE FACED CEILING OR WALL—MOUNTED EXIT SIGN WITH EMERGENCY POWER BACK UP AND DIRECTIONAL ARROWS AS INDICATED ON PLANS
⬖	CEILING OR WALL—MOUNTED, SELF—CONTAINED EMERGENCY LIGHT UNIT; FIXTURE SHALL MONITOR LIGHTING CIRCUIT IN AREA.
▥	EMERGENCY LIGHTING BATTERY UNIT

Table titled: **LIGHTING**

FIGURE 1.7   Electrical symbols. (*Continued*)

SWITCHES	
S	SINGLE—POLE SWITCH
S$_2$	DOUBLE—POLE SWITCH
S$_3$	3—WAY SWITCH
S$_4$	4—WAY SWITCH
S$_P$	SINGLE—POLE SWITCH AND PILOT LIGHT
S$_{be}$	BOILER EMERGENCY SWITCH
S$_{DM}$	SINGLE—POLE DIMMER SWITCH
S$_{DM3}$	3—WAY DIMMER SWITCH
S$_T$	SINGLE—POLE SWITCH WITH THERMAL OVERLOAD PROTECTION
S$_K$	SINGLE—POLE KEYED SWITCH
S$_{K3}$	KEYED, 3—WAY SWITCH
S$_{K4}$	KEYED, 4—WAY SWITCH
S$_{MC}$	MOMENTARY CONTACT SWITCH
S$_{PROJ}$	MOTORIZED PROJECTION SCREEN RAISE/LOWER SWITCH
S$_O$	OCCUPANCY SENSOR SWITCH
[S$_O$]	CEILING MOUNTED OCCUPANCY SENSOR
[C]	CONTACTOR, COMPLETE WITH NEMA ENCLOSURE
[TC]	TIME CLOCK, AS INDICATED ON PLANS
[PC]	PHOTOCELL
[■]	PUSHBUTTON SWITCH
E,G ⊂H	EMERGENCY SHUT—OFF SWITCH. SUBLETTER "E" INDICATES ELECTRICAL. SUBLETTER "G" INDICATES GAS
KE,KG ⊂H	MASTER EMERGENCY SHUT—OFF/KEYED RESET SWITCH. SUBLETTER "KE" INDICATES ELECTRICAL. SUBLETTER "KG" INDICATES GAS

FIGURE 1.7 Electrical symbols. (*Continued*)

POWER	
⊜a,b	DUPLEX RECEPTACLE; SUBLETTER "a" INDICATES RECEPTACLE TO BE MOUNTED 6" ABOVE COUNTER TOP OR 48" AFF. SUBLETTER "b" INDICATES MOUNTED IN ARCHITECTURAL MILLWORK. COORDINATE INSTALLATION WITH ARCHITECT.
⊕a,b	DOUBLE DUPLEX RECEPTACLE; SUBLETTER "a" INDICATES RECEPTACLE TO BE MOUNTED 6" ABOVE COUNTER TOP OR 48" AFF. SUBLETTER "b" INDICATES MOUNTED IN ARCHITECTURAL MILLWORK. COORDINATE INSTALLATION WITH ARCHITECT.
⊖	SINGLE RECEPTACLE
⊙R,F,S	FLOOR MOUNTED DUPLEX RECEPTACLE: SUBLETTER "R" INDICATES RECESSED BACKBOX. SUBLETTER "F" INDICATES FLUSH BACKBOX. SUBLETTER "S" INDICATED SURFACE BACKBOX (MONUMENT)
⊜	DUPLEX RECEPTACLE—ONE OUTLET SWITCHED
⊜c	DUPLEX RECEPTACLE. SUBLETTER "C" INDICATES CEILING MOUNTED
⊜TV	DUPLEX RECEPTACLE FOR TELEVISION. MOUNTING HEIGHT AS NOTED ON PLANS
Ⓔ⌐	ELECTRICAL FLOOR MONUMENT WITH LFMC WHIP CONNECTION
◁	SPECIAL—PURPOSE OUTLET. AMPERAGE AND VOLTAGE AS INDICATED ON PLANS. VERIFY NEMA CONFIGURATION WITH EQUIPMENT MANUFACTURER
⬤a,b	DUPLEX RECEPTACLE, EMERGENCY POWER; SUBLETTER "a" INDICATES RECEPTACLE TO BE MOUNTED 6" ABOVE COUNTER TOP OR 48" AFF. SUBLETTER "b" INDICATES MOUNTED IN ARCHITECTURAL MILLWORK. COORDINATE INSTALLATION WITH ARCHITECT.
⬤a,b	DOUBLE DUPLEX RECEPTACLE, EMERG. POWER; SUBLETTER "a" INDICATES RECEPTACLE TO BE MOUNTED 6" ABOVE COUNTER TOP OR 48" AFF. SUBLETTER "b" INDICATES MOUNTED IN ARCHITECTURAL MILLWORK. COORDINATE INSTALLATION WITH ARCHITECT.
⊕⊕⊕	SURFACE RACEWAY WITH OUTLETS AS INDICATED ON PLANS, MOUNTED AT 18" AFF, UNLESS OTHERWISE NOTED
▨	TELEPHONE/POWER POLE
▬	ELECTRICAL PANEL 480/277 VOLT
▭	ELECTRICAL PANEL 120/208 VOLT
▭	SPECIAL—PURPOSE ELECTRICAL PANEL OR EQUIPMENT
Ⓣ	ELECTRICAL POWER TRANSFORMER
☒	MAGNETIC STARTER
F⊐ XX/XX	FUSED DISCONNECT SWITCH WITH SIZE/RATING
☐⊐	NON—FUSED DISCONNECT SWITCH
☒⊐	COMBINATION MAGNETIC STARTER AND DISCONNECT SWITCH
Ⓜ	ELECTRIC MOTOR
VFD	VARIABLE FREQUENCY DRIVE
J	FLOOR OR CEILING MOUNTED JUNCTION BOX
Ⓙ	WALL MOUNTED JUNCTION BOX
▰▰▰	ELECTRIFIED BUS DUCT WITH FUSIBLE, PLUG—IN, BRANCH CIRCUIT DEVICE
▪	HARD—WIRED EQUIPMENT CONNECTION
R	RELAY
EDO	ELECTRIC DOOR OPENER
☐H	ELECTRIC DOOR OPENER ACTUATOR PUSH PLATE

FIGURE 1.7 Electrical symbols. (*Continued*)

SPECIAL SYSTEMS

SYMBOL	DESCRIPTION
⊕$_{R,F,S}$	FLOOR MOUNTED TEL/DATA OUTLET: SUBLETTER "R" INDICATES RECESSED BACKBOX. SUBLETTER "F" INDICATES FLUSH BACKBOX. SUBLETTER "S" INDICATED SURFACE BACKBOX (MONUMENT)
©⌒	COMMUNICATIONS FLOOR MONUMENT WITH LFMC WHIP CONNECTION
▷	COMBINATION DATA/TELEPHONE OUTLET WITH BACKBOX AND EMPTY CONDUIT STUBBED UP TO ABOVE FINISHED CEILING, INCLUDING DRAG LINE
W ►	TELEPHONE OUTLET WITH BACKBOX AND EMPTY CONDUIT, STUBBED UP TO ABOVE FINISHED CEILING, INCLUDING DRAG LINE. SUBLETTER "W" INDICATES WALL–MOUNTED;
►$_{HP}$	HANDICAP PAY TELEPHONE OUTLET WITH BACKBOX AND CONDUIT STUBBED UP TO ABOVE FINISHED CEILING, INCLUDING DRAG LINE
▷	DATA OUTLET WITH BACKBOX AND EMPTY CONDUIT STUBBED UP TO ABOVE FINISHED CEILING, INCLUDING DRAGLINE
⟨TV⟩	TELEVISION CABLE OUTLET; MOUNT AT 18" AFF UNLESS OTHERWISE NOTED.
Ⓢ	CEILING–MOUNTED, SOUND SYSTEM SPEAKER
Ⓢ⊣	WALL–MOUNTED, SOUND SYSTEM SPEAKER
⟦VC⟧	SOUND SYSTEM VOLUME CONTROLLER
Ⓜ$_{F,W}$	SOUND SYSTEM MICROPHONE JACK; SUBLETTER "F" INDICATES FLOOR–MOUNTED. SUBLETTER "W" INDICATES WALL–MOUNTED
⌒	PA/SOUND SYSTEM HANDSET
Ⓢ ⊕	PA/SOUND SYSTEM CLOCK AND SPEAKER MOUNTED IN COMMON ENCLOSURE
⊕	WALL CLOCK WITH HANGER TYPE OUTLET
⟦P⟧○	PROGRAM BELL
⟦E⟧○	EMERGENCY CALL–FOR–AID AUDIO INDICATING UNIT
⟦E⟧$_{PC}$	EMERGENCY CALL–FOR–AID SWITCH
⟦E⟧$_{PB}$	EMERGENCY CALL–FOR–AID PUSHBUTTON
●⊣	EMERGENCY CALL–FOR–AID VISUAL INDICATING UNIT
⟦E⟧●	EMERGENCY CALL–FOR–AID VISUAL/AUDIO INDICATING UNIT
⟦AMP⟧	AMPLIFIER
⟦!⟧$_M$	INTERCOM STATION; SUBLETTER "M" INDICATES MASTER

FIGURE 1.7   Electrical symbols. (*Continued*)

HOSPITAL SYMBOLS

SYMBOL	DESCRIPTION
$\boxed{N}_{1/2}$	NURSE CALL BEDSIDE STATION – SUBNUMBER INDICATES SINGLE OR DOUBLE BED
$\boxed{N}_E$	EMERGENCY NURSE CALL STATION
$\boxed{N}_{M/S}$	NURSE CALL MICROPHONE/SPEAKER UNIT
$\boxed{N}_{SR}$	NURSE CALL STAFF REGISTER
$\boxed{N}_{ACU}$	NURSE CALL AREA CONTROL UNIT
$\boxed{N}_{FCS}$	NURSE CALL FLOOR CONTROL STATION
$\boxed{N}_B$	NURSE CALL CODE BLUE
$\boxed{N}_{SS}$	NURSE CALL STAFF STATION
$\boxed{N}_{DS}$	NURSE CALL DUTY STATION
$\boxed{FM}$	FETAL MONITORING STATION
$\boxed{PM}$	PATIENT MONITORING STATION
$D_N \quad \dashv D_N$	NURSE CALL CORRIDOR DOME LIGHT – CEILING OR WALL MOUNTED.
$D_Z \quad \dashv D_Z$	NURSE CALL CORRIDOR ZONE LIGHT – CEILING OR WALL MOUNTED.
$\boxed{TA}$	TELEMETRY RECEIVER
$\boxed{CTM}$	CENTRAL TELEMETRY UNIT
$\boxed{PU}_{NC/PM}$	PRINTER UNIT, SUBLETTER "NC" INDICATES NURSE CALL; SUBLETTER "PM" INDICATES PATIENT MONITOR
$\boxed{MGAP}$	MEDICAL GAS ALARM PANEL
$\boxed{CMS}$	CENTRAL PATIENT MONITOR STATION

FIGURE 1.7   Electrical symbols. (*Continued*)

	FIRE
M	FIRE ALARM MAGNETIC DOOR HOLD DEVICE
FS	SPRINKLER FIRE ALARM FLOW SWITCH
SS	SPRINKLER FIRE ALARM SUPERVISORY SWITCH
PS	SPRINKLER FIRE ALARM PRESSURE SWITCH
F	MASTER FIRE ALARM PULL BOX
S$_E$	SMOKE DETECTOR FOR ELEVATOR RECALL CONTROLS
✖	EXTERIOR REMOTE FIRE ALARM FLASHING STROBE LIGHT
FACP	FIRE ALARM CONTROL PANEL
RAP	REMOTE ANNUNCIATOR PANEL
F	MANUAL FIRE ALARM PULL STATION
◼	FIRE ALARM VISUAL INDICATING UNIT
◼◁	FIRE ALARM AUDIO/VISUAL INDICATING UNIT
◼◁$_S$	FIRE ALARM SPEAKER/VISUAL INDICATING UNIT (VOICE EVAC. SYSTEM)
F	FIRE ALARM CEILING—MOUNTED SPEAKER
◼◁$_M$	FIRE ALARM MINI SPEAKER
H$_B$	AUTOMATIC FIRE ALARM HEAT DETECTOR. SUBLETTER "B" INDICATES 200 DEGREES F. HEAT DETECTOR
▶^{FP}	FIREFIGHTERS TELEPHONE OUTLET
▶^{ADR}	AREA OF REFUGE TELEPHONE OUTLET
▶^{EM}	EMERGENCY TELEPHONE OUTLET
S	AUTOMATIC FIRE ALARM SMOKE DETECTOR
S$_S$	AUTOMATIC FIRE ALARM SMOKE DETECTOR WITH SOUNDER BASE
DS	DUCT SMOKE FIRE ALARM DETECTOR
DH	DUCT HEAT FIRE ALARM DETECTOR
TS	SMOKE DETECTOR TEST SWITCH

FIGURE 1.7 Electrical symbols. (*Continued*)

	SECURITY
ES	DOOR STRIKE
⬢	DOOR/WINDOW CONTACT
◼◀	VIDEO CAMERA, WITH MOUNTING HARDWARE
VM	VIDEO MONITOR
VR	VIDEO RECORDER
CR	CARD READER
Ⓜ Ⓜ⊣	CEILING OR WALL-MOUNTED MOTION DETECTOR

	WIRING
———	BRANCH CIRCUIT POWER WIRING
— — —	BRANCH CIRCUIT SWITCHED WIRING
— - —	BRANCH CIRCUIT AC OR DC CONTROL WIRING
—EM—	BRANCH CIRCUIT EMERGENCY AC OR DC WIRING. 3/4" CONDUIT, 2#10 AND 1#10 GROUND, UNLESS OTHERWISE NOTED
—CT—	CABLETRAY
○———	CONDUIT DOWN
⊂———	CONDUIT UP
——▶	HOME RUN. 3/4" CONDUIT, 2#12 AND 1#12 GROUND, UNLESS OTHERWISE NOTED. NOTE: HOME RUN SHALL BE FROM FIRST ELECTRICAL DEVICE BACKBOX IN CIRCUIT TO ELECTRICAL PANEL

FIGURE 1.7 Electrical symbols. (*Continued*)

ONE–LINE	
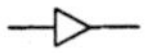 —▷—	POTHEAD
 —▶—	STRESSCONE
₣	CURRENT TRANSFORMER
 ∃₣	POTENTIAL TRANSFORMER
 —⌒—	FUSE
 —ᵒ—	FUSE CUT OUT
 —⌒—	FUSE & SWITCH
—⟍|	SWITCH
 —⌒—	CIRCUIT BREAKER
 ≺←⌒→≻	DRAWOUT CIRCUIT BREAKER
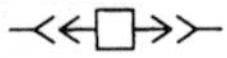 ≺←▢→≻	MEDIUM VOLTAGE DRAWOUT CIRCUIT BREAKER
 ←⌒—	BUSPLUG CIRCUIT BREAKER
 ←⌒—	BUSPLUG FUSE & SWITCH
⏚	GROUND
 —ᨏ—	THERMAL OVERLOAD
Ⓒ	RELAY/COIL
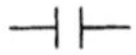 —‖—	N/O CONTACT
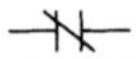 —⊬—	N.C. CONTACT
 —⋀—	PROTECTIVE RELAY
Ⓐ	AMMETER
AS	AMMETER SWITCH
V	VOLTMETER
VS	VOLTMETER SWITCH
 WHM⟩	WATTHOUR METER
 WM⟩	WATTMETER

FIGURE 1.7   Electrical symbols. (*Continued*)

ONE—LINE	
 (WHD)→	WATTHOUR DEMAND METER
	TRANSFORMER
	SHIELDED TRANSFORMER
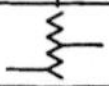	AUTO TRANSFORMER
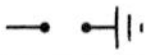 →• ⊣I·	LIGHTNING ARRESTER
 Ⓖ	GENERATOR
Δ	DELTA
Y	WYE
Ⓚ#	KEY INTERLOCK
 N•⟍ •E ⟍•L	AUTOMATIC TRANSFER SWITCH (A.T.S.)
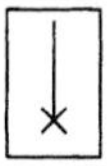 ╳	MAIN LUG ONLY PANELBOARD
 ⌇	MAIN CIRCUIT BREAKER PANELBOARD
) XXXAF XXXAT	CIRCUIT BREAKER WITH AMP FRAME OVER AMP TRIP
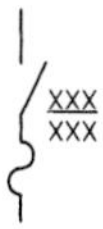 ⌇ XXX XXX	FUSED DISCONNECT SWITCH, WITH SWITCH SIZE OVER FUSE SIZE

FIGURE 1.7   Electrical symbols. (*Continued*)

ABBREVIATIONS	
SYMBOL	DESCRIPTION
A	AMPERE
C	CONDUIT
P	POLE
W	WIRE
T	TELEPHONE SERVICE
FA	FIRE ALARM
NF	NON–FUSED
WP	WEATHERPROOF
C/B	CIRCUIT BREAKER
AFF	ABOVE FINISHED FLOOR
AFG	ABOVE FINISHED GRADE
CIR	CIRCUIT
TX	TRANSFORMER
MD	MOTORIZED DAMPER
PE	PRIMARY ELECTRIC SERVICE
SE	SECONDARY ELECTRIC SERVICE
RTU	ROOFTOP UNIT
TCP	TEMPERATURE CONTROL PANEL
SD	SMOKE DAMPER
IG	ISOLATED GROUND
RMC	RIGID METALLIC CONDUIT
EMT	ELECTRIC METALLIC TUBING

FIGURE 1.7   Electrical symbols. (*Continued*)

ABBREVIATIONS	
SYMBOL	DESCRIPTION
FMC	FLEXIBLE METALLIC TUBING
TV	TELEVISION
PVC	POLYVINYL CHLORIDE CONDUIT
EF	EXHAUST FAN
REF	ROOF EXHAUST FAN
AHU	AIR HANDLING UNIT
CUH	CABINET UNIT HEATER
EWC	ELECTRIC WATER COOLER
EWH	ELECTRIC WATER HEATER
GFI	GROUND FAULT INTERRUPTER
MAU	MAKE-UP AIR UNIT
WG	WIRE GUARD
S&P	SPACE AND PROVISION
E	EXISTING TO REMAIN
RE	REMOVE EXISTING
RL	RELOCATE EXISTING
NL	NEW LOCATION OF EXISTING RELOCATED
NR	NEW TO REPLACE EXISTING
RR	REMOVE AND REPLACE ON NEW SURFACE

Mounting Heights

Mounting heights of electrical devices are influenced by and must be closely coordinated with the architectural design. However, there are industry standard practices followed by architects as well as code and legal requirements, such as Americans with Disabilities Act (ADA) guidelines. The following recommended mounting heights for electrical devices provide a good guideline in the absence of any specific information and are ADA compliant.

TABLE 1.1   Mounting Heights for Electrical Devices

	DEVICE	MOUNTING HEIGHTS
1.	Light switches, wall mounted occupancy sensors	48″ to centerline of box Exception: 44" maximum to top above counters which are 20"-25"D.
2.	Wall mounted exit signs	90″ to centerline of sign or centered in wall area between top of door and ceiling.
2A.	Ceiling mounted exit signs and pendant mounted fixtures.	80" to bottom of fixture.
3.	Receptacles	16″ to bottom of box Exception: 44" maximum to top above counters which are 20"-25"D.
4.	Special outlets or receptacles	16″ to bottom of box or as noted on drawings Exception: 44" maximum to top above counters which are 20"-25"D.
5.	Plugmold or Wiremold	As noted on drawings. Exception: 44" maximum to top above counters which are 20"-25"D.

TABLE 1.1 Mounting Heights for Electrical Devices (*Continued*)

6.	Clock outlets	12″ from ceiling to centerline or 7'-0" to centerline if ceiling is over 8'-0"
7.	Data/communication or telephone outlets	16″ to bottom of box Exception: 44" maximum to top above counters which are 20"-25"D.
8.	Telephone outlets - wall type	54″ to Dial Center (non-accessible) 48″ to highest operable part (accessible)
9.	Pay type telephone outlets	48″ maximum to coin slot
10.	Fire alarm manual pull stations	48″ to centerline of box - not more than 5' - 0″ from exit
11.	Fire alarm horns, bells, flashing lights, etc.	80″ to centerline of device or 6″ below ceiling, whichever is lower so that no point is more than 50' away without obstruction
12.	Wall mounted remote indicator light	80″ to centerline of device or 6″ below ceiling, whichever is lower
13.	Area of Refuge Telephone	Same as telephone - accessible
14.	Call-For-Aid switch with pull chain to floor	48″ to centerline of box minimum (toilets) 66" to centerline of box maximum (showers - located out of spray area)
15.	Card reader	48″ to highest operable part (side or forward access)
16.	Intercom station	54″ to highest operable part (side access) 48″ highest operable part (forward access)
17.	Sound system volume control	54″ to highest operable part (side access) 48″ to highest operable part (forward access)

TABLE 1.1 Mounting Heights for Electrical Devices (*Continued*)

18.	Microphone outlets	16″ to bottom of box
19.	Thermostats	54″ to highest operable part (side access) 48″ to highest operable part (forward access)
20.	Temperature/Humidity Sensors	60″ to center line of box

NOTES:
1. All dimensions are considered from finished floor and, unless noted otherwise, shall not vary.

2. All dimensions shall be coordinated with architectural details and may be adjusted to conform with architectural requirements as long as no code restriction is violated.

3. Outlets installed lower than 15″ AFF (forward reach) and 9″ AFF (side reach) are in violation of ADA.

SPECIAL NOTES:

1. Exit signs shall NOT be installed so that it blocks fire alarm visual devices.

2. Wall mounted light fixtures:

   a. Bottom of fixture at 80″ AFF. or greater.

   b. Bottom of fixture at less than 80″ AFF. protrusion into space shall be no more than 4″.

1.3 NEMA DEVICE CONFIGURATIONS

Nonlocking

FIGURE 1.8 Configuration chart for general-purpose nonlocking plugs and receptacles.

		15 AMPERE		20 AMPERE		30 AMPERE		50 AMPERE		60 AMPERE	
		RECEPTACLE	PLUG	RECEPTACLE	PLUG	RECEPTACLE	PLUG	RECEPTACLE	PLUG	RECEPTACLE	PLUG
2 POLE 2 WIRE	1 — 125 V	1-15R	1-15P		1-20P				1-20P		
	2 — 250 V		2-15P	2-20R	2-20P	2-30R	2-30P				
	3 — 277 V	(RESERVED FOR FUTURE CONFIGURATIONS)									
	4 — 600 V	(RESERVED FOR FUTURE CONFIGURATIONS)									
2 POLE 3 WIRE GROUNDING	5 — 125 V	5-15R	5-15P	5-20R	5-20P	5-30R	5-30P	5-50R	5-50P		
	6 — 250 V	6-15R	6-15P	6-20R	6-20P	6-30R	6-30P	6-50R	6-50P		
	7 — 277 V AC	7-15R	7-15P	7-20R	7-20P	7-30R	7-30P	7-50R	7-50P		
	24 — 347 V AC	24-15R	24-15P	24-20R	24-20P	24-30R	24-30P	24-50R	24-50P		
	8 — 480 V AC	(RESERVED FOR FUTURE CONFIGURATIONS)									
	9 — 600 V AC	(RESERVED FOR FUTURE CONFIGURATIONS)									
3 POLE 3 WIRE	10 — 125/250 V			10-20R	10-20P	10-30R	10-30P	10-50R	10-50P		
	11 — 3ø 250 V	11-15R	11-15P	11-20R	11-20P	11-30R	11-30P	11-50R	11-50P		
	12 — 3ø 480 V	(RESERVED FOR FUTURE CONFIGURATIONS)									
	13 — 3ø 600 V	(RESERVED FOR FUTURE CONFIGURATIONS)									
3 POLE 4 WIRE GROUNDING	14 — 125/250 V	14-15R	14-15P	14-20R	14-20P	14-30R	14-30P	14-50R	14-50P	14-60R	14-60P
	15 — 3ø 250 V	15-15R	15-15P	15-20R	15-20P	15-30R	15-30P	15-50R	15-50P	15-60R	15-60P
	16 — 3ø 480 V	(RESERVED FOR FUTURE CONFIGURATIONS)									
	17 — 3ø 600 V	(RESERVED FOR FUTURE CONFIGURATIONS)									
4 POLE 4 WIRE	18 — 3ø 208 Y 120 V	18-15R	18-15P	18-20R	18-20P	18-30R	18-30P	18-50R	18-50P	18-60R	18-60P
	19 — 3ø 408 Y 277 V	(RESERVED FOR FUTURE CONFIGURATIONS)									
	20 — 3ø 600 Y 347 V	(RESERVED FOR FUTURE CONFIGURATIONS)									
4 POLE 5 WIRE GROUNDING	21 — 3ø 208 Y 120 V	(RESERVED FOR FUTURE CONFIGURATIONS)									
	22 — 3ø 408 Y 277 V	(RESERVED FOR FUTURE CONFIGURATIONS)									
	23 — 3ø 600 Y 347 V	(RESERVED FOR FUTURE CONFIGURATIONS)									

Locking

FIGURE 1.9 Configuration chart for specific-purpose locking plugs and receptacles.

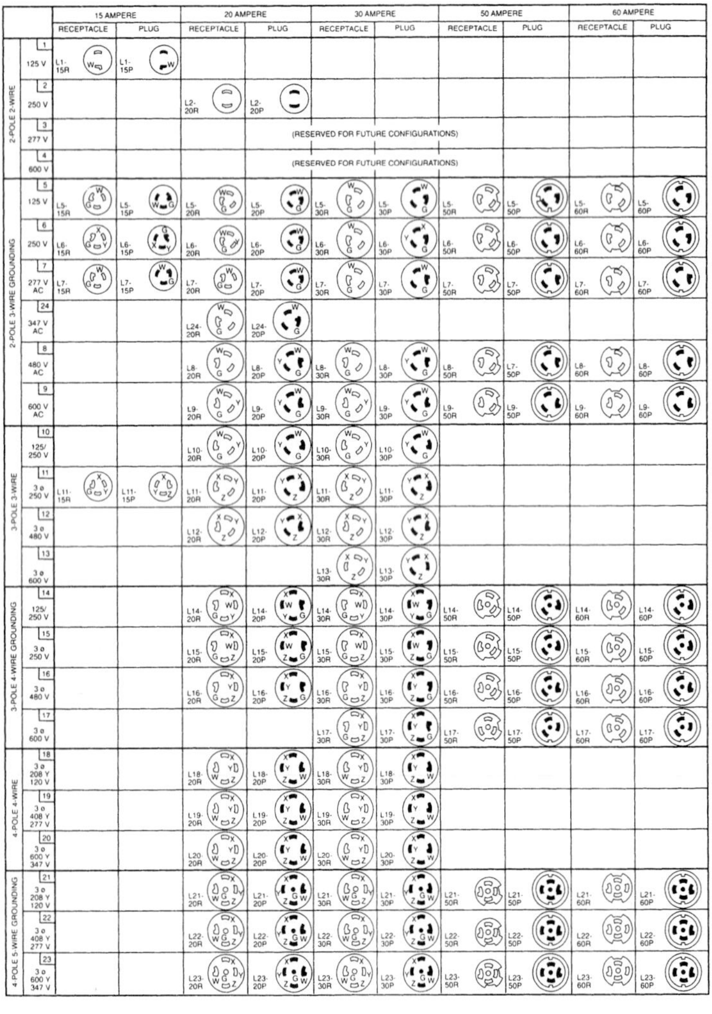

1.4 IEEE PROTECTIVE DEVICE NUMBERS

TABLE 1.2 IEEE Standard Protective Device Numbers

device number	definition and function	device number	definition and function
1	**master element** is the initiating device, such as a control switch, voltage relay, float switch, etc., which serves either directly, or through such permissive devices as protective and time-delay relays to place an equipment in or out of operation.	13	**synchronous-speed device,** such as a centrifugal-speed switch, a slip-frequency relay, a voltage relay, an undercurrent relay or any type of device, operates at approximately synchronous speed of a machine.
2	**time-delay starting, or closing, relay** is a device which functions to give a desired amount of time delay before or after any point or operation in a switching sequence or protective relay system, except as specifically provided by device functions 62 and 79 described later.	14	**under-speed device** functions when the speed of a machine falls below a predetermined value.
		15	**speed or frequency, matching device** functions to match and hold the speed or the frequency of a machine or of a system equal to, or approximately equal to, that of another machine, source or system.
3	**checking or interlocking relay** is a device which operates in response to the position of a number of other devices, or to a number of predetermined conditions in an equipment to allow an operating sequence to proceed, to stop, or to provide a check of the position of these devices or of these conditions for any purpose.	16	Reserved for future application.
		17	**shunting, or discharge, switch** serves to open or to close a shunting circuit around any piece of apparatus (except a resistor), such as a machine field, a machine armature, a capacitor or a reactor. **note:** This excludes devices which perform such shunting operations as may be necessary in the process of starting a machine by devices 6 or 42, or their equivalent, and also excludes device 73 function which serves for the switching of resistors.
4	**master contactor** is a device, generally controlled by device No. 1 or equivalent, and the necessary permissive and protective devices, which serves to make and break the necessary control circuits to place an equipment into operation under the desired conditions and to take it out of operation under other or abnormal conditions.		
		18	**accelerating or decelerating device** is used to close or to cause the closing of circuits which are used to increase or to decrease the speed of a machine.
5	**stopping device** functions to place and hold an equipment out of operation.	19	**starting-to-running transition contactor** is a device which operates to initiate or cause the automatic transfer of a machine from the starting to the running power connection.
6	**starting circuit breaker** is a device whose principal function is to connect a machine to its source of starting voltage.		
7	**anode circuit breaker** is one used in the anode circuits of a power rectifier for the primary purpose of interrupting the rectifier circuit if an arc back should occur.	20	**electrically operated valve** is a solenoid- or motor-operated valve which is used in a vacuum, air, gas, oil, water, or similar, lines. **note:** The function of the valve may be indicated by the insertion of descriptive words such as "Brake" or "Pressure Reducing" in the function name, such as "Electrically Operated *Brake* Valve".
8	**control power disconnecting device** is a disconnecting device—such as a knife switch, circuit breaker or pullout fuse block—used for the purpose of connecting and disconnecting, respectively, the source of control power to and from the control bus or equipment. **note:** Control power is considered to include auxiliary power which supplies such apparatus as small motors and heaters.		
		21	**distance relay** is a device which functions when the circuit admittance, impedance or reactance increases or decreases beyond predetermined limits.
9	**reversing device** is used for the purpose of reversing a machine field or for performing any other reversing functions.	22	**equalizer circuit breaker** is a breaker which serves to control or to make and break the equalizer or the current-balancing connections for a machine field, or for regulating equipment, in a multiple-unit installation.
10	**unit sequence switch** is used to change the sequence in which units may be placed in and out of service in multiple-unit equipments.	23	**temperature control device** functions to raise or to lower the temperature of a machine or other apparatus, or of any medium, when its temperature falls below, or rises above, a predetermined value. **note:** An example is a thermostat which switches on a space heater in a switchgear assembly when the temperature falls to a desired value as distinguished from a device which is used to provide automatic temperature regulation between close limits and would be designated as 90T.
11	Reserved for future application.		
12	**over-speed device** is usually a direct-connected speed switch which functions on machine overspeed.		

TABLE 1.2 IEEE Standard Protective Device Numbers (*Continued*)

24 Reserved for future application.

25 **synchronizing, or synchronism-check, device** operates when two a-c circuits are within the desired limits of frequency, phase angle or voltage, to permit or to cause the paralleling of these two circuits.

26 **apparatus thermal device** functions when the temperature of the shunt field or the armortisseur winding of a machine, or that of a load limiting or load shifting resistor or of a liquid or other medium exceeds a predetermined value; or if the temperature of the protected apparatus, such as a power rectifier, or of any medium decreases below a predetermined value.

27 **undervoltage relay** is a device which functions on a given value of undervoltage.

28 Reserved for future application.

29 **isolating contactor** is used expressly for disconnecting one circuit from another for the purposes of emergency operation, maintenance, or test.

30 **annunciator relay** is a nonautomatically reset device which gives a number of separate visual indications upon the functioning of protective devices, and which may also be arranged to perform a lockout function.

31 **separate excitation device** connects a circuit such as the shunt field of a synchronous converter to a source of separate excitation during the starting sequence; or one which energizes the excitation and ignition circuits of a power rectifier.

32 **directional power relay** is one which functions on a desired value of power flow in a given direction, or upon reverse power resulting from arc back in the anode or cathode circuits of a power rectifier.

33 **position switch** makes or breaks contact when the main device or piece of apparatus, which has no device function number, reaches a given position.

34 **motor-operated sequence switch** is a multicontact switch which fixes the operating sequence of the major devices during starting and stopping, or during other sequential switching operations.

35 **brush-operating, or slip-ring short-circuiting, device** is for raising, lowering, or shifting the brushes of a machine, or for short-circuiting its slip rings, or for engaging or disengaging the contacts of a mechanical rectifier.

36 **polarity device** operates or permits the operation of another device on a predetermined polarity only.

37 **undercurrent or underpower relay** is a device which functions when the current or power flow decreases below a predetermined value.

38 **bearing protective device** is one which functions on excessive bearing temperature, or on other abnormal mechanical conditions, such as undue wear, which may eventually result in excessive bearing temperature.

39 Reserved for future application.

40 **field relay** is a device that functions on a given or abnormally low value or failure of machine field current, or on an excessive value of the reactive component of armature current in an a-c machine indicating abnormally low field excitation.

41 **field circuit breaker** is a device which functions to apply, or to remove, the field excitation of a machine.

42 **running circuit breaker** is a device whose principal function is to connect a machine to its source of running voltage after having been brought up to the desired speed on the starting connection.

43 **manual transfer or selector device** transfers the control circuits so as to modify the plan of operation of the switching equipment or of some of the devices.

44 **unit sequence starting relay** is a device which functions to start the next available unit in a multiple-unit equipment on the failure or on the non-availability of the normally preceding unit.

45 Reserved for future application.

46 **reverse-phase, or phase-balance, current relay** is a device which functions when the polyphase currents are of reverse-phase sequence, or when the polyphase currents are unbalanced or contain negative phase-sequence components above a given amount.

47 **phase-sequence voltage relay** is a device which functions upon a predetermined value of polyphase voltage in the desired phase sequence.

48 **incomplete sequence relay** is a device which returns the equipment to the normal, or off, position and locks it out if the normal starting, operating or stopping sequence is not properly completed within a predetermined time.

TABLE 1.2   IEEE Standard Protective Device Numbers (*Continued*)

49	**machine, or transformer, thermal relay** is a device which functions when the temperature of an a-c machine armature, or of the armature or other load carrying winding or element of a d-c machine, or converter or power rectifier or power transformer (including a power rectifier transformer) exceeds a predetermined value.
50	**instantaneous overcurrent, or rate-of-rise relay** is a device which functions instantaneously on an excessive value of current, or on an excessive rate of current rise, thus indicating a fault in the apparatus or circuit being protected.
51	**a-c time overcurrent relay** is a device with either a definite or inverse time characteristic which functions when the current in an a-c circuit exceeds a predetermined value.
52	**a-c circuit breaker** is a device which is used to close and interrupt an a-c power circuit under normal conditions or to interrupt this circuit under fault or emergency conditions.
53	**exciter or d-c generator relay** is a device which forces the d-c machine field excitation to build up during starting or which functions when the machine voltage has built up to a given value.
54	**high-speed d-c circuit breaker** is a circuit breaker which starts to reduce the current in the main circuit in 0.01 second or less, after the occurrence of the d-c overcurrent or the excessive rate of current rise.
55	**power factor relay** is a device which operates when the power factor in an a-c circuit becomes above or below a predetermined value.
56	**field application relay** is a device which automatically controls the application of the field excitation to an a-c motor at some predetermined point in the slip cycle.
57	**short-circuiting or grounding device** is a power or stored energy operated device which functions to short-circuit or to ground a circuit in response to automatic or manual means.
58	**power rectifier misfire relay** is a device which functions if one or more of the power rectifier anodes fails to fire.
59	**overvoltage relay** is a device which functions on a given value of overvoltage.
60	**voltage balance relay** is a device which operates on a given difference in voltage between two circuits.

61	**current balance relay** is a device which operates on a given difference in current input or output of two circuits.
62	**time-delay stopping, or opening, relay** is a time-delay device which serves in conjunction with the device which initiates the shutdown, stopping, or opening operation in an automatic sequence.
63	**liquid or gas pressure, level, or flow relay** is a device which operates on given values of liquid or gas pressure, flow or level, or on a given rate of change of these values.
64	**ground protective relay** is a device which functions on failure of the insulation of a machine, transformer or of other apparatus to ground, or on flashover of a d-c machine to ground. **note:** This function is assigned only to a relay which detects the flow of current from the frame of a machine or enclosing case or structure of a piece of apparatus to ground, or detects a ground on a normally ungrounded winding or circuit. It is not applied to a device connected in the secondary circuit or secondary neutral of a current transformer, or current transformers, connected in the power circuit of a normally grounded system.
65	**governor** is the equipment which controls the gate or valve opening of a prime mover.
66	**notching, or jogging, device** functions to allow only a specified number of operations of a given device, or equipment, or a specified number of successive operations within a given time of each other. It also functions to energize a circuit periodically, or which is used to permit intermittent acceleration or jogging of a machine at low speeds for mechanical positioning.
67	**a-c directional overcurrent relay** is a device which functions on a desired value of a-c overcurrent flowing in a predetermined direction.
68	**blocking relay** is a device which initiates a pilot signal for blocking of tripping on external faults in a transmission line or in other apparatus under predetermined conditions, or co-operates with other devices to block tripping or to block reclosing on an out-of-step condition or on power swings.
69	**permissive control device** is generally a two-position, manually operated switch which in one position permits the closing of a circuit breaker, or the placing of an equipment into operation, and in the other position prevents the circuit breaker or the equipment from being operated.
70	**electrically operated rheostat** is a rheostat which is used to vary the resistance of a circuit in response to some means of electrical control.

TABLE 1.2 IEEE Standard Protective Device Numbers (*Continued*)

71 Reserved for future application.

72 **d-c circuit breaker** is used to close and interrupt a d-c power circuit under normal conditions or to interrupt this circuit under fault or emergency conditions.

73 **load-resistor contactor** is used to shunt or insert a step of load limiting, shifting, or indicating resistance in a power circuit, or to switch a space heater in circuit, or to switch a light, or regenerative, load resistor of a power rectifier or other machine in and out of circuit.

74 **alarm relay** is a device other than an annunciator, as covered under device No. 30, which is used to operate, or to operate in connection with, a visual or audible alarm.

75 **position changing mechanism** is the mechanism which is used for moving a removable circuit breaker unit to and from the connected, disconnected, and test positions.

76 **d-c overcurrent relay** is a device which functions when the current in a d-c circuit exceeds a given value.

77 **pulse transmitter** is used to generate and transmit pulses over a telemetering or pilot-wire circuit to the remote indicating or receiving device.

78 **phase angle measuring, or out-of-step protective relay** is a device which functions at a predetermined phase angle between two voltages or between two currents or between voltage and current.

79 **a-c reclosing relay** is a device which controls the automatic reclosing and locking out of an a-c circuit interrupter.

80 Reserved for future application.

81 **frequency relay** is a device which functions on a predetermined value of frequency—either under or over or on normal system frequency—or rate of change of frequency.

82 **d-c reclosing relay** is a device which controls the automatic closing and reclosing of a d-c circuit interrupter, generally in response to load circuit conditions.

83 **automatic selective control, or transfer, relay** is a device which operates to select automatically between certain sources or conditions in an equipment, or performs a transfer operation automatically.

84 **operating mechanism** is the complete electrical mechanism or servo-mechanism, including the operating motor, solenoids, position switches, etc., for a tap changer, induction regulator or any piece of apparatus which has no device function number.

85 **carrier, or pilot-wire, receiver relay** is a device which is operated or restrained by a signal used in connection with carrier-current or d-c pilot-wire fault directional relaying.

86 **locking-out relay** is an electrically operated hand or electrically reset device which functions to shut down and hold an equipment out of service on the occurrence of abnormal conditions.

87 **differential protective relay** is a protective device which functions on a percentage or phase angle or other quantitative difference of two currents or of some other electrical quantities.

88 **auxiliary motor, or motor generator** is one used for operating auxiliary equipment such as pumps, blowers, exciters, rotating magnetic amplifiers, etc.

89 **line switch** is used as a disconnecting or isolating switch in an a-c or d-c power circuit, when this device is electrically operated or has electrical accessories, such as an auxiliary switch, magnetic lock, etc.

90 **regulating device** functions to regulate a quantity, or quantities, such as voltage, current, power, speed, frequency, temperature, and load, at a certain value or between certain limits for machines, tie lines or other apparatus.

91 **voltage directional relay** is a device which operates when the voltage across an open circuit breaker or contactor exceeds a given value in a given direction.

92 **voltage and power directional relay** is a device which permits or causes the connection of two circuits when the voltage difference between them exceeds a given value in a predetermined direction and causes these two circuits to be disconnected from each other when the power flowing between them exceeds a given value in the opposite direction.

93 **field changing contactor** functions to increase or decrease in one step the value of field excitation on a machine.

94 **tripping, or trip-free, relay** is a device which functions to trip a circuit breaker, contactor, or equipment, or to permit immediate tripping by other devices; or to prevent immediate reclosure of a circuit interrupter, in case it should open automatically even though its closing circuit is maintained closed.

95
to
99 Used only for specific applications on individual installations where none of the assigned numbered functions from 1 to 94 is suitable.

note: A similar series of numbers, starting with 201 instead of 1, shall be used for those device functions in a machine, feeder or other equipment when these are controlled directly from the supervisory system. Typical examples of such device functions are 201, 205, and 294.

TABLE 1.2 IEEE Standard Protective Device Numbers (*Continued*)

suffix letters

Suffix letters are used with device function numbers for various purposes. In order to prevent possible conflict, any suffix letter used singly, or any combination of letters, denotes only one word or meaning in an individual equipment. All other words should use the abbreviations as contained in American Standard Z32.13-1950, or latest revision thereof, or should use some other distinctive abbreviation, or be written out in full each time they are used. Furthermore, the meaning of each single suffix letter, or combination of letters, should be clearly designated in the legend on the drawings or publications applying to the equipment.

The following suffix letters generally form part of the device function designation and thus are written directly behind the device number, such as 23X, 90V, or 52BT.

These letters denote **separate auxiliary devices,** such as

$\left.\begin{array}{l} X \\ Y \\ Z \end{array}\right\}$ —auxiliary relay*

R   —raising relay
L   —lowering relay
O   —opening relay
C   —closing relay
CS   —control switch
CL   —"a" auxiliary-switch relay
OP —"b" auxiliary-switch relay
U   —"up" position-switch relay
D   —"down" position-switch relay
PB   —push button

*note: In the control of a circuit breaker with so-called X-Y relay control scheme, the X relay is the device whose main contacts are used to energize the closing coil and the contacts of the Y relay provide the anti-pump feature for the circuit breaker.

These letters indicate the **condition or electrical** quantity to which the device responds, or the medium in which it is located, such as:

A   —air, or amperes
C   —current
E   —electrolyte
F   —frequency, or flow
L   —level, or liquid
P   —power, or pressure
PF   —power factor
Q   —oil
S   —speed
T   —temperature
V   —voltage, volts, or vacuum
VAR—reactive power
W   —water, or watts

These letters denote the **location of the main device in the circuit,** or the type of circuit in which the device is used or the type of circuit or apparatus with which it is associated, when this is necessary, such as:

A   —alarm or auxiliary power
AC   —alternating current
AN   —anode
B   —battery, or blower, or bus
BK   —brake
BP   —bypass
BT   —bus tie
C   —capacitor, or condenser, compensator, or carrier current
CA   —cathode
DC   —direct current
E   —exciter
F   —feeder, or field, or filament
G   —generator, or ground**
H   —heater, or housing
L   —line
M   —motor, or metering
N   —network, or neutral**
P   —pump
R   —reactor, or rectifier
S   —synchronizing
T   —transformer, or test, or thyratron
TH   —transformer (high-voltage side)
TL   —transformer (low-voltage side)
TM   —telemeter
U   —unit

**Suffix "N" is generally used in preference to "G" for devices connected in the secondary neutral of current transformers, or in the secondary of a current transformer whose primary winding is located in the neutral of a machine or power transformer, except in the case of transmission line relaying, where the suffix "G" is more commonly used for those relays which operate on ground faults.

These letters denote **parts of the main device,** divided in the two following categories:

all parts, except auxiliary contacts and limit switches as covered later.

Many of these do not form part of the device number, and should be written directly below the device number, such as $\frac{20}{IS}$ or $\frac{43}{A}$.

BB   —bucking bar (for high speed d-c circuit breaker)
BK   —brake
C   —coil, or condenser, or capacitor
CC   —closing coil
HC   —holding coil
IS   —inductive shunt
L   —lower operating coil
M   —operating motor
MF   —fly-ball motor
ML   —load-limit motor
MS   —speed adjusting, or synchronizing, motor
S   —solenoid
TC   —trip coil
U   —upper operating coil
V   —valve

TABLE 1.2   IEEE Standard Protective Device Numbers (*Continued*)

All auxiliary contacts and limit switches for such devices and equipment as circuit breakers, contactors, valves and rheostats. These are designated as follows:

a --Auxiliary switch, open when the main device is in the de-energized or non-operated position.

b —Auxiliary switch, closed when the main device is in the de-energized or non-operated position.

aa—Auxiliary switch, open when the operating mechanism of the main device is in the de-energized or non-operated position.

bb--Auxiliary switch, closed when the operating mechanism of the main device is in the de-energized or non-operated position.

e, f, etc., ab, ac, ad, etc., or ba, bc, bd, etc., are special auxiliary switches other than a, b, aa, and bb. Lower-case (small) letters are to be used for the above auxiliary switches.

note: If several similar auxiliary switches are present on the same device, they should be designated numerically 1, 2, 3, etc. when necessary.

LC—Latch-checking switch, closed when the circuit breaker-mechanism linkage is relatched after an opening operation of the circuit breaker.

LS—limit switch

These letters cover **all other distinguishing features or characteristics or conditions,** not specifically described in 2-9.4.1 to 2-9.4.4, which serve to describe the use of the device or its contacts in the equipment such as:

A   —accelerating, or automatic
B   —blocking, or backup
C   —close, or cold
D   —decelerating, detonate, or down
E   —emergency
F   —failure, or forward

H   —hot, or high
HR --hand reset
HS ---high speed
IT ---inverse time
L   ---left, or local, or low, or lower, or leading
M   —manual
OFF —off
ON —on
O   —open
P   --polarizing
R   —right, or raise, or reclosing, or receiving, or remote, or reverse
S   ---sending, or swing
T   —test, or trip, or trailing
TDC---time-delay closing
TDO---time-delay opening
U   --up

suffix numbers

If two or more devices with the same function number and suffix letter (if used) are present in the same equipment, they may be distinguished by numbered suffixes as for example, 52X-1, 52X-2 and 52X-3, when necessary.

devices performing more than one function

If one device performs two relatively important functions in an equipment so that it is desirable to identify both of these functions, this may be done by using a double function number and name such as:

27-59 undervoltage and overvoltage relay.

1.5   NEMA STANDARD ENCLOSURES

Indoor Nonhazardous Locations (Table 1.3)

Outdoor Nonhazardous Locations (Table 1.4)

Indoor Hazardous Locations (Table 1.5)

Knockout Dimensions (Table 1.6)

TABLE 1.3 Comparison of Specific Applications of Enclosures for Indoor Nonhazardous Locations

Provides a Degree of Protection Against the Following Environmental Conditions	Type of Enclosures									
	1[1]	2[1]	4	4X	5	6	6P	12	12K	13
Incidental contact with the enclosed equipment	X	X	X	X	X	X	X	X	X	X
Falling dirt	X	X	X	X	X	X	X	X	X	X
Falling liquids and light splashing	...	...	X	X	X	X	X	X	X	X
Circulating dust, lint, fibers, and flyings[2]	...	...	X	X	...	X	X	X	X	X
Settling airborne dust, lint, fibers, and flyings[2]	...	...	X	X	X	X	X	X	X	X
Hosedown and splashing water	...	...	X	X	...	...	X	...	...	...
Oil and coolant seepage	...	...	...	...	...	...	...	X	X	X
Oil or coolant spraying and splashing	...	...	...	...	...	...	...	...	...	X
Corrosive agents	...	...	...	X	...	...	X	...	...	...
Occasional temporary submersion	...	...	...	...	...	X	X	...	...	...
Occasional prolonged submersion	...	...	...	...	...	...	X	...	...	...

TABLE 1.4 Comparison of Specific Applications of Enclosures for Outdoor Nonhazardous Locations

Provides a Degree of Protection Against the Following Environmental Conditions	Type of Enclosures						
	3	3R[3]	3S	4	4X	6	6P
Incidental contact with the enclosed equipment	X	X	X	X	X	X	X
Rain, snow, and sleet[4]	X	X	X	X	X	X	X
Sleet[5]	...	...	X	...	...	...	...
Windblown dust	X	...	X	X	X	X	X
Hosedown	...	...	...	X	X	X	X
Corrosive agents	...	...	...	...	X	...	X
Occasional temporary submersion	...	...	...	...	...	X	X
Occasional prolonged submersion	...	...	...	...	...	...	X

[1] These enclosures may be ventilated. However, Type 1 may not provide protection against small particles of falling dirt when ventilation is provided in the enclosure top. Consult the manufacturer.

[2] These fibers and flying are nonhazardous materials and are not considered the Class III type ignitable fibers or combustible flyings. For Class III type ignitable fibers or combustible flyings see the National Electrical Code, Article 500.

[3] External operating mechanisms are not required to be operable when the enclosure is ice covered.

[4] External operating mechanisms are operable when the enclosure is ice covered.

[5] These enclosures may be ventilated.

TABLE 1.5 Comparison of Specific Applications for Indoor Hazardous Locations

Provides a Degree of Protection Against Atmospheres Typically Containing (For Complete Listing. See NFPA 497M-1986, Classification of Gases, Vapors and Dusts for Electrical Equipment in Hazardous (Classified) Locations)	Class	Type of Enclosure 7 and 8, Class I Groups⑦				Type of Enclosure 9, Class II Groups⑦			10
		A	B	C	D	E	F	G	
Acetylene	I	X	…	…	…	…	…	…	…
Hydrogen, manufactured gas	I	…	X	…	…	…	…	…	…
Diethel ether, ethylene, cyclopropane	I	…	…	X	…	…	…	…	…
Gasoline, hexane, butane, naphtha, propane, acetone, toluene, isoprene	I	…	…	…	X	…	…	…	…
Metal dust	II	…	…	…	…	X	…	…	…
Carbon black, coal dust, coke dust	II	…	…	…	…	…	X	…	…
Flour, starch, grain dust	II	…	…	…	…	…	…	X	…
Fibers, flyings⑥	III	…	…	…	…	…	…	X	…
Methane with or without coal dust	MSHA	…	…	…	…	…	…	…	X

⑥ For Class III type ignitable fibers or combustible flyings see the National Electrical Code, Article 500.

⑦ Due to the characteristics of the gas, vapor, or dust, a product suitable for one Class or Group may not be suitable for another Class or Group unless so marked on the product.

TABLE 1.6 Knockout Dimensions

Conduit Trade Size, Inches	Knockout Diameter, Inches		
	Minimum	Nominal	Maximum
1/2	0.859	0.875	0.906
3/4	1.094	1.109	1.141
1	1.359	1.375	1.406
1 1/4	1.719	1.734	1.766
1 1/2	1.958	1.984	2.016
2	2.433	2.469	2.500
2 1/2	2.938	2.969	3.000
3	3.563	3.594	3.625
3 1/2	4.063	4.125	4.156
4	4.563	4.641	4.672
5	5.625	5.719	5.750
6	6.700	6.813	6.844

1.6 FORMULAS AND TERMS

FIGURE 1.10 Formulas and terms.

Formulas for Determining Amperes, hp, kW, and kVA

To Find	Direct Current	Alternating Current		
		Single-Phase	Two-Phase — 4-Wire①	Three-Phase
Amperes (I) When Horsepower is Known	$\dfrac{hp \times 746}{E \times \% \text{ eff}}$	$\dfrac{hp \times 746}{E \times \% \text{ eff} \times pf}$	$\dfrac{hp \times 746}{2 \times E \times \% \text{ eff} \times pf}$	$\dfrac{hp \times 746}{\sqrt{3} \times E \times \% \text{ eff} \times pf}$
Amperes (I) When Kilowatts is Known	$\dfrac{kW \times 1000}{E}$	$\dfrac{kW \times 1000}{E \times pf}$	$\dfrac{kW \times 1000}{2 \times E \times pf}$	$\dfrac{kW \times 1000}{\sqrt{3} \times E \times pf}$
Amperes (I) When kVA is Known		$\dfrac{kVA \times 1000}{E}$	$\dfrac{kVA \times 1000}{2 \times E}$	$\dfrac{kVA \times 1000}{\sqrt{3} \times E}$
Kilowatts	$\dfrac{I \times E}{1000}$	$\dfrac{I \times E \times pf}{1000}$	$\dfrac{I \times E \times 2 \times pf}{1000}$	$\dfrac{I \times E \times \sqrt{3} \times pf}{1000}$
kVA		$\dfrac{I \times E}{1000}$	$\dfrac{I \times E \times 2}{1000}$	$\dfrac{I \times E \times \sqrt{3}}{1000}$
Horsepower (Output)	$\dfrac{I \times E \times \% \text{ eff}}{746}$	$\dfrac{I \times E \times \% \text{ eff} \times pf}{746}$	$\dfrac{I \times E \times 2 \times \% \text{ eff} \times pf}{746}$	$\dfrac{I \times E \times \sqrt{3} \times \% \text{ eff} \times pf}{746}$

Common Electrical Terms

Ampere (I)	= unit of current or rate of flow of electricity
Volt (E)	= unit of electromotive force
Ohm (R)	= unit of resistance
	Ohms law: $I = \dfrac{E}{R}$ (DC or 100% pf)
Megohm	= 1,000,000 ohms
Volt Amperes (VA)	= unit of apparent power = $E \times I$ (single-phase) = $E \times I \times \sqrt{3}$
Kilovolt Amperes (kVA)	= 1000 volt-amperes
Watt (W)	= unit of true power = $VA \times pf$ = .00134 hp
Kilowatt (kW)	= 1000 watts
Power Factor (pf)	= ratio of true to apparent power = $\dfrac{W}{VA} = \dfrac{kW}{kVA}$
Watt-hour (Wh)	= unit of electrical work = one watt for one hour = 3,413 Btu = 2,655 ft. lbs.
Kilowatt-hour (kWh)	= 1000 watt-hours
Horsepower (hp)	= measure of time rate of doing work = equivalent of raising 33,000 lbs. one ft. in one minute = 746 watts
Demand Factor	= ratio of maximum demand to the total connected load
Diversity Factor	= ratio of the sum of individual maximum demands of the various subdivisions of a system to the maximum demand of the whole system
Load Factor	= ratio of the average load over a designated period of time to the peak load occurring in that period

① For 2-phase, 3-wire circuits the current in the common conductor is $\sqrt{2}$ times that in either of the two other conductors.

How to Compute Power Factor

Determining watts: $pf = \dfrac{watts}{volts \times amperes}$

1. From watt-hour meter.
   Watts = rpm of disc × 60 × Kh

   Where Kh is meter constant printed on face or nameplate of meter.

   If metering transformers are used, above must be multiplied by the transformer ratios.

2. Directly from wattmeter reading.
   Where:

   Volts = line-to-line voltage as measured by voltmeter.

   Amps = current measured in line wire (not neutral) by ammeter.

Temperature Conversion

(F° to C°)				$C^\circ = 5/9 \ (F^\circ \text{-} 32^\circ)$				
(C° to F°)				$F^\circ = 9/5 (C^\circ) + 32^\circ$				
C°	-15	-10	-5	0	5	10	15	20
F°	5	14	23	32	41	50	59	68
C°	25	30	35	40	45	50	55	60
F°	77	86	95	104	113	122	131	140
C°	65	70	75	80	85	90	95	100
F°	149	158	167	176	185	194	203	212

1 inch	= 2.54 centimeters
1 kilogram	= 2.20 lbs.
1 square inch	= 1,273,200 circular mills
1 circular mill	= .785 square mil
1 Btu	= 778 ft. lbs. = 252 calories
1 year	= 8,760 hours

1.7 TYPICAL EQUIPMENT SIZES AND WEIGHTS

Tables 1.7 to 1.12 provide typical equipment sizes and weights to assist in the preliminary design and layout of an electrical distribution system. The reader is cautioned that this data is only representative of industry manufacturers and should consult specific vendors for detailed information. This information could prove useful in determining initial space requirements and weight impacts for structural purposes.

1.8 SEISMIC REQUIREMENTS

The design of seismic restraint systems for electrical distribution equipment and raceways is usually done by a structural engineer through performance specifications by the electrical design professional. It is therefore necessary for the electrical designer to be generally familiar with the seismic code requirements and the seismic zone that are applicable to a project. Figure 1.11 will serve as an introduction.

TABLE 1.7 Typical Equipment Sizes—600-Volt Class

Equipment	KVA Rating	Dimensions (inches)			Weight Lbs. (CU)	Weight Lbs. (AL)
		H	W	D		
Switchboards (per Section)	N/A	90	26 - 45	24 - 60	Varies	Varies
Motor Control Centers (per Section)	N/A	90	20	16 - 22	Varies	Varies
Power Panel	N/A	To 80	30 - 48	6 - 12	Varies	Varies
Lighting/Small Appliance Panels	N/A	30 - 50	22	6	Varies	Varies
Transformers 3-phase, Dry Type, General Purpose	30	30	20	15	300	230
	45	30	20	15	370	310
	75	40	26	20	550	480
	112.5	40	26	20	675	600
	150	46	26	21	850	760
	300	56	32	24	1750	1300
	500	75	45	36	3100	2400
Transformers 3-phase, Dry Type, K-Rated	30	31	21	15	370	310
	45	40	26	20	575	480
	75	40	26	20	675	600
	112.5	56	31	24	850	760
	150	56	31	24	1200	1100
	300	75	45	36	3100	2400
	500	90	69	42	see mfg.	4500

TABLE 1.8 Transformer Weight (lbs) by KVA

Oil Filled 3 Phase 5/15 KV To 480/277			
KVA	Lbs.	KVA	Lbs.
150	1800	1000	6200
300	2900	1500	8400
500	4700	2000	9700
750	5300	3000	15000

Dry 240/480 To 120/240 Volt			
1 Phase		3 Phase	
KVA	Lbs.	KVA	Lbs.
1	23	3	90
2	36	6	135
3	59	9	170
5	73	15	220
7.5	131	30	310
10	149	45	400
15	205	75	600
25	255	112.5	950
37.5	295	150	1140
50	340	225	1575
75	550	300	1870
100	670	500	2850
167	900	750	4300

TABLE 1.9 Generator Weight (lbs) by KW

3 Phase 4 Wire 277/480 Volt			
Gas		Diesel	
KW	Lbs.	KW	Lbs.
7.5	600	30	1800
10	630	50	2230
15	960	75	2250
30	1500	100	3840
65	2350	125	4030
85	2570	150	5500
115	4310	175	5650
170	6530	200	5930
		250	6320
		300	7840
		350	8220
		400	10750
		500	11900

TABLE 1.10 Weight (lbs/lf) of Four-Pole Aluminum and Copper Bus Duct by Ampere Load

Amperes	Aluminum Feeder	Copper Feeder	Aluminum Plug–In	Copper Plug–In
225			7	7
400			8	13
600	10	10	11	14
800	10	19	13	18
1000	11	19	16	22
1350	14	24	20	30
1600	17	26	25	39
2000	19	30	29	46
2500	27	43	36	56
3000	30	48	42	73
4000	39	67		
5000		78		

TABLE 1.11 Conduit Weight Comparisons (lbs per 100 ft) Empty

Type	1/2"	3/4"	1"	1-1/4"	1-1/2"	2"	2-1/2"	3"	3-1/2"	4"	5"	6"
Rigid Aluminum	28	37	55	72	89	119	188	246	296	350	479	630
Rigid Steel	79	105	153	201	249	332	527	683	831	972	1314	1745
Intermediate Steel (IMC)	60	82	116	150	182	242	401	493	573	638		
Electrical Metallic Tubing (EMT)	29	45	65	96	111	141	215	260	365	390		
Polyvinyl Chloride, Schedule 40	16	22	32	43	52	69	109	142	170	202	271	350
Polyvinyl Chloride Encased Burial						38		67	88	105	149	202
Fibre Duct Encased Burial						127		164	180	206	400	511
Fibre Duct Direct Burial						150		251	300	354		
Transite Encased Burial						160		240		330	450	550
Transite Direct Burial						220		310	290	400	540	640

TABLE 1.12 Conduit Weight Comparisons (lbs per 100 ft) with Maximum Cable Fill

Type	1/2"	3/4"	1"	1-1/4"	1-1/2"	2"	2-1/2"	3"	3-1/2"	4"	5"	6"
Rigid Galvanized Steel (RGS)	104	140	235	358	455	721	1022	1451	1749	2148	3083	4343
Intermediate Steel (IMC)	84	113	186	293	379	611	883	1263	1501	1830		
Electrical Metallic Tubing (EMT)	54	116	183	296	368	445	641	930	1215	1540		

*Conduit & Heaviest Conductor Combination

FIGURE 1.11 Seismic requirements. (a) Seismic zone map of the United States. (b) Normalized response spectra shapes.

Seismic Requirements

Uniform Building Code (UBC)

The 1994 Uniform Building Code (UBC) includes Volume 2 for earthquake design requirements. Sections 1624-1633 of this reference specifically require that structures and portions of structures shall be designed to withstand the seismic ground motion specified in the code. The design engineer must evaluate the effect of lateral forces not only on the building structure but also on the equipment in determining whether the design will withstand those forces. In the code electrical equipment such as control panels, motors, switchgear, transformers, and associated conduit are specifically identified.

The criteria for selecting the seismic requirements are defined in Section 1627 of the code. Panel a of the code includes a seismic zone map of the United States. Panel b of the code includes the normalized response spectra shapes for different soil conditions. The damping value is 5% of the critical damping.

(a)

FIGURE 1.11 Seismic requirements. (a) Seismic zone map of the United States. (b) Normalized response spectra shapes. (Continued)

The seismic requirements in the UBC can be completely defined as the Zero Period Acceleration (ZPA) and Spectrum Accelerations are computed. In a test program, these values are computed conservatively to envelop the requirements of all seismic zones. The lateral force on elements of structures and nonstructural components are defined in Section 1630. The dynamic lateral forces are defined in Section 1629. These loads are converted to seismic accelerations according to the normalized response spectra shown in Panel b of the UBC.

The total design lateral force required is:

Force F_p = Z I_p C_p W_p

Dividing both sides by W_p, the acceleration requirement in g's is equal to:

Acceleration = F_p/W_p = Z I_p C_p

Where:

Z: is the seismic zone factor and is taken equal to 0.4. This is the maximum value provided in Table 16-I of the code.

I_p: is the importance factor and is taken equal to 1.5. This is the maximum value provided in Table 16-K of the code.

C_p: is the horizontal force factor and is taken equal to 0.75 for rigid equipment as defined in Table 16-O. For flexible equipment, this value is equal to twice the value for the rigid equipment: 2 x 0.75 = 1.5. This is the maximum value provided in the code.

W_p: is the weight of the equipment.

Therefore, the maximum acceleration for rigid equipment is:

**Acceleration = F_p/W_p
= Z I_p C_p
= 0.4 x 1.5 x 0.75
= 0.45g**

The maximum acceleration for flexible equipment is:

**Acceleration = F_p/W_p
= Z I_p C_p
= 0.4 x 1.5 x 1.5
= 0.9g**

Flexible equipment is defined in the UBC as equipment with a period of vibration equal to or greater than 0.06 seconds. This period of vibration corresponds to a dominant frequency of vibration equal to 16.7 Hz.

Equipment must be designed and tested to the UBC requirements to determine that it will be functional following a seismic event. In addition, a structural or civil engineer must perform calculations based on data received from the equipment manufacturer specifying the size, weight, center of gravity, and mounting provisions of the equipment to determine its method of attachment so it will remain attached to its foundation during a seismic event. Finally, the contractor must properly install the equipment in accordance with the anchorage design.

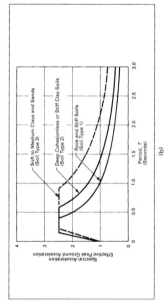

(b)

NOTES

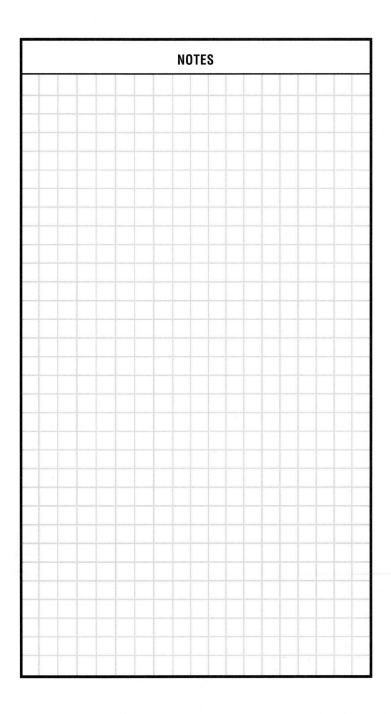

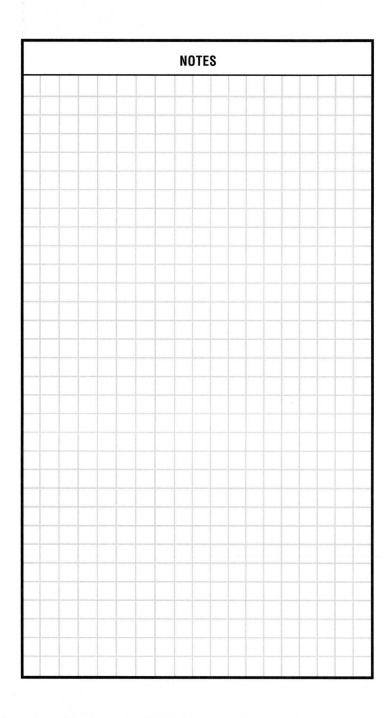

NOTES

CHAPTER TWO
National Electrical Code® (NEC®)*
Articles, Tables, and Data

2.0 WORKING SPACE AROUND ELECTRICAL EQUIPMENT

Introduction

The National Electrical Code (NEC), produced by the National Fire Protection Association® (NFPA®),[†] is known as NFPA-70, and is the "bible" of electrical design and construction. It is developed and written by a committee of some of the best electrical professionals who are knowledgeable in the safe and effective design, construction, operation, and maintenance of electrical systems, with input from the industry at large. It sets forth the *minimum* standards by which electrical systems should be designed and constructed.

Although complying with the NEC minimum requirements will ensure safe and effective electrical system design and operation, good design practice often dictates that more stringent requirements be met, or more stringent requirements may be mandated by the local electrical inspector. Keep in mind that the authority having final jurisdiction for acceptance of an electrical system design and installation is the local electrical inspector for the project. It may be prudent, therefore, to involve the local electrical inspector in the early stages of design and from time to time throughout the design process, to help him or her become familiar with the project and your design intent, to see if there are any special requirements or possible differences in interpretation of the NEC, and thus to facilitate a design that will not only be safe and effective, but will be accepted with no costly surprises once construction has begun.

Interpretations of the NEC can be obtained from the NFPA, both formally and informally, with the latter being the quickest. This is sometimes needed for clarification of *Code* articles, which may be subject to broad interpretation of the *Code's* intent.

---

*National Electrical Code® and NEC® are registered trademarks of the National Fire Protection Association, Inc., Quincy, MA 02269.

[†] National Fire Alarm Code® is a registered trademark of the National Fire Protection Association, Inc., Quincy, MA 02269.

This part of the handbook brings together, in one convenient location, the most frequently used NEC articles, tables, and data used by electrical design professionals. For the most part, NEC articles are only referenced for the applicable topic, or abstracted, highlighted, or abbreviated, without the full text. Tables and data from the NEC are given in their entirety. The user is encouraged to read the complete text of the NEC article under consideration for more comprehensive understanding, cross-references to related NEC articles, and total context.

The following article, NEC Article 110-16, is repeated in its entirety.

NEC Article 110-16: Working Space About Electric Equipment (600 Volts, Nominal, or Less)

Sufficient access and working clearance shall be provided and maintained about all electric equipment to permit ready and safe operation and maintenance of such equipment.

(A) WORKING CLEARANCES

Except as elsewhere required or permitted in this *Code,* the dimension of the working space in the direction of access to live parts operating at 600 volts, nominal, or less to ground and likely to require examination, adjustment, servicing, or maintenance while energized shall not be less than indicated in Table 2.1 (NEC Table 110-16a). Distances shall be measured from the live parts if such are exposed or from the enclosure front or opening if such are enclosed. Concrete, brick, or tile walls shall be considered as grounded.

The "Conditions" [as mentioned in Table 2.1] are as follows:

1. Exposed live parts on one side and no live or grounded parts on the other side of the working space, or exposed live parts on both sides effectively guarded by suitable wood or other insulating materials. Insulated wire or insulated busbars operating at not over 300 volts shall not be considered live parts.

TABLE 2.1 NEC Table 110-16(a): Working Clearances

Nominal Voltage to Ground	Conditions Minimum Clear Distance (feet)		
	1	2	3
	(Feet)	(Feet)	(Feet)
0-150	3	3	3
151-600	3	$3^1/2$	4

For SI units: 1 ft = 0.3048 m.

FIGURE 2.1   Examples of conditions 1, 2, and 3.

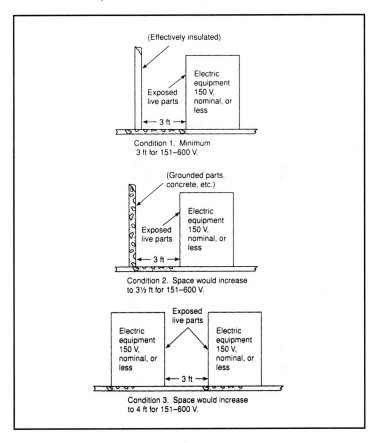

(Effectively insulated)

Exposed live parts

Electric equipment 150 V, nominal, or less

← 3 ft →

Condition 1. Minimum 3 ft for 151–600 V.

(Grounded parts, concrete, etc.)

Exposed live parts

Electric equipment 150 V, nominal, or less

← 3 ft →

Condition 2. Space would increase to 3½ ft for 151–600 V.

Exposed live parts

Electric equipment 150 V, nominal, or less

Electric equipment 150 V, nominal, or less

← 3 ft →

Condition 3. Space would increase to 4 ft for 151–600 V.

2. Exposed live parts on one side and grounded parts on the other side.

3. Exposed live parts on both sides of the work space (not guarded as provided in Condition 1) with the operator between.

Examples of Conditions 1, 2, and 3 are shown in Figure 2.1.

In addition to the dimensions shown in Table 2.1, the work space shall not be less than 30 in. (762 mm) wide in front of the electric equipment. The work space shall be clear and extend from the floor or platform to the height required by this section. In all cases, the work space shall permit at least a 90-degree opening of equipment doors or hinged panels. See Figures 2.2 and 2.3. Equipment of equal depth shall be permitted within the height requirements of this section.

FIGURE 2.2 90° clearance.

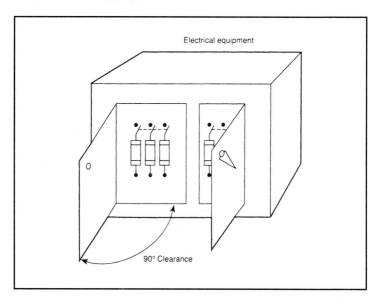

FIGURE 2.3 30-inch-wide working space.

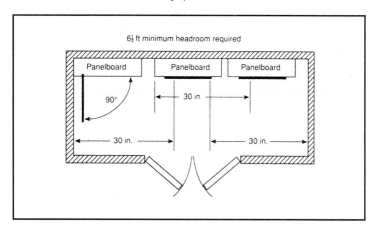

Exception No. 1: Working space shall not be required in back of assemblies such as dead-front switchboards or motor control centers where there are no renewable or adjustable parts such as fuses or switches on the back and where all the connections are accessible from locations other than the back. Where rear access is required to work on de-energized parts on the back of enclosed equipment, a minimum working space of 30 in. (762 mm) horizontally shall be provided. See Figure 2.4 for an example of this exception.

Exception No. 2: By special permission, smaller spaces shall be permitted where all uninsulated parts are at a voltage no greater than 30 volts RMS, 42 volts peak, or 60 volts DC.

Exception No. 3: In existing buildings where electrical equipment is being replaced, Condition 2 working clearance shall be permitted between dead-front switchboards, panelboards, or motor control centers located across the aisle from each other where conditions of maintenance and supervision ensure that written procedures have been adopted to prohibit equipment on both sides of the aisle from being open at the same time and qualified persons who are authorized will service the installation. See Figure 2.5 for an example of this exception.

FIGURE 2.4   30-inch minimum work space—rear of equipment.

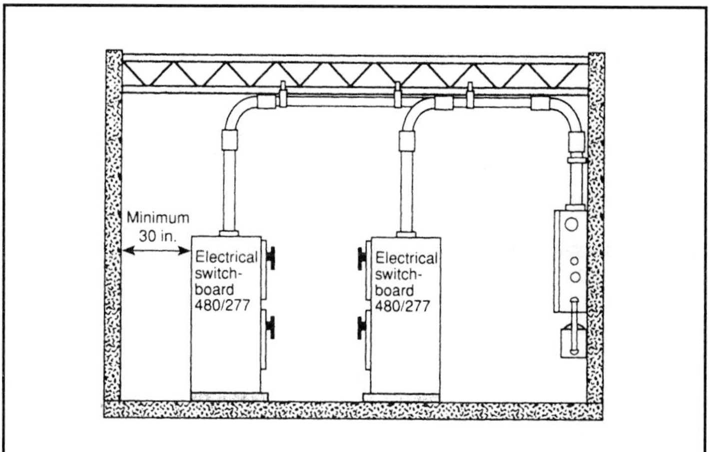

FIGURE 2.5   Example of NEC Section 110-16(a), Exception No. 3.

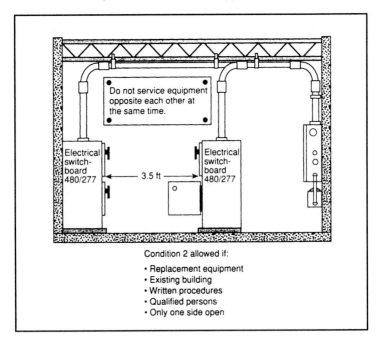

Condition 2 allowed if:

• Replacement equipment
• Existing building
• Written procedures
• Qualified persons
• Only one side open

(B) CLEAR SPACES

Working space required by this section shall not be used for storage. When normally enclosed live parts are exposed for inspection or servicing, the working space, if in a passageway or general open space, shall be suitably guarded.

(C) ACCESS AND ENTRANCE TO WORKING SPACE

At least one entrance of sufficient area shall be provided to give access to the working space about electric equipment.

For equipment rated 1200 amperes or more and over 6 ft (1.83 m) wide, containing overcurrent devices, switching devices, or control devices, there shall be one entrance not less than 24 in. (610 mm) wide and 6½ ft. (1.98 m) high at each end. For examples of the preceding two paragraphs, refer to Figures 2.6 and 2.7, respectively.

An example of an unacceptable arrangement of a large switchboard is shown in Figure 2.8.

FIGURE 2.6   NEC Section 110-16(c), basic rule, first paragraph.

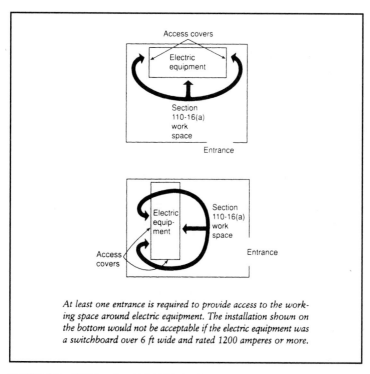

At least one entrance is required to provide access to the working space around electric equipment. The installation shown on the bottom would not be acceptable if the electric equipment was a switchboard over 6 ft wide and rated 1200 amperes or more.

FIGURE 2.7   NEC Section 110-16(c), basic rule, second paragraph.

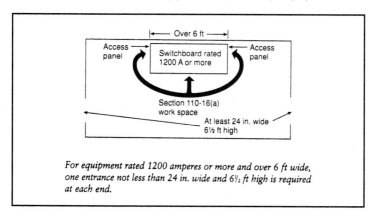

For equipment rated 1200 amperes or more and over 6 ft wide, one entrance not less than 24 in. wide and 6½ ft high is required at each end.

FIGURE 2.8 Example of an unacceptable arrangement of a large switchboard.

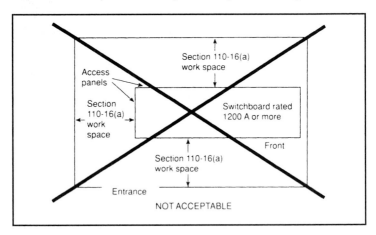

Exception No. 1: Where the location permits a continuous and unob-
structed way of exit travel, one means of exit shall be permitted.
See Figure 2.9.

Exception No. 2: Where the work space required by Section 110-16(a)
is doubled, only one entrance to the working space is required, and
it shall be located so that the edge of the entrance nearest the equipment is the minimum clear distance given in Table 2.1 [NEC Table
110-16(a)] away from such equipment. See Figure 2.10.

FIGURE 2.9 Example of Exception No. 1.

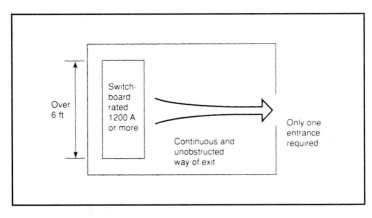

FIGURE 2.10   Example of Exception No. 2.

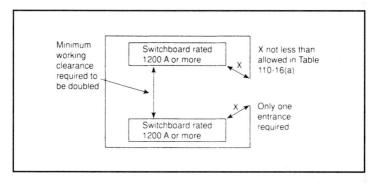

(D) ILLUMINATION

Illumination shall be provided for all working spaces about service equipment, switchboards, panelboards, or motor control centers installed indoors. Additional lighting fixtures shall not be required where the work space is illuminated by an adjacent light source. In electrical equipment rooms, the illumination shall not be controlled by automatic means only.

(E) HEADROOM

The minimum headroom of working spaces about service equipment, switchboards, panelboards, or motor control centers shall be 6½ ft (1.98 m). Where the electrical equipment exceeds 6½ ft (1.98 m) in height, the minimum headroom shall not be less than the height of the equipment.

> *Exception:* Service equipment or panelboards, in existing dwelling units, that do not exceed 200 amperes.

NOTE   For higher voltages, see Article 710.

Other applicable articles for working space about electric equipment (600 volts, nominal, or less), are articles 110-17 through 110-22, inclusive, including articles referenced therein.

2.1   OVER 600 VOLTS, NOMINAL

For working space over 600 volts, nominal, refer to NEC articles 110-30 through 110-40, inclusive, which supplement or modify the preceding articles that also apply.

In no case do the provisions of this part apply to the equipment on the supply side of the service point. Equipment on the supply side of the service point is outside the scope of the NEC. Such equipment is covered by the *National Electrical Safety Code* (ANSI C2), published by the Institute of Electrical and Electronics Engineers (IEEE).

Generally speaking, in most applications involving electrical equipment over 600 volts, nominal, encountered by electrical design professionals in the building industry, the equipment is in metal-enclosed switchgears located in secure rooms or vaults accessible to qualified persons only.

NEC Article 110-34. Work Space and Guarding

NEC Article 110-34 covers work space and guarding. In particular, Table 2.2 [NEC Table 110-34(a)] covers the minimum depth of clear working space at electrical equipment.

The "Conditions" [as mentioned in Table 2.2] are as follows:

1. Exposed live parts on one side and no live or grounded parts on the other side of the working space or exposed live parts on both sides effectively guarded by suitable wood or other insulating materials. Insulated wire or insulated busbars operating at not over 300 volts shall not be considered live parts.

2. Exposed live parts on one side and grounded parts on the other side. Concrete, brick, or tile walls will be considered as grounded surfaces.

3. Exposed live parts on both sides of the work space (not guarded as provided in Condition 1) with the operator between.

TABLE 2.2 NEC Table 110-34(a): Minimum Depth of Clear Working Space at Electrical Equipment

Nominal Voltage to Ground	Conditions (feet)		
	1	2	3
	(Feet)	(Feet)	(Feet)
601-2500	3	4	5
2501-9000	4	5	6
9001-25,000	5	6	9
25,001-75 kV	6	8	10
Above 75 kV	8	10	12

For SI units: 1 ft = 0.3048 m

TABLE 2.3   NEC Table 110-34(e): Elevation of Unguarded Live Parts Above Working Space

Nominal Voltage Between Phases	Elevation
601-7500	8 ft, 6 in.
7501-35,000	9 ft
Over 35kV	9 ft + 0.37 in. per kV above 35

For SI units: 1 in. = 25.4 mm; 1 ft = 0.3048 m.

Exception: Working space shall not be required in back of equipment such as dead-front switchboards or control assemblies where there are no renewable or adjustable parts (such as fuses or switches) on the back and where all connections are accessible from locations other than the back. Where rear access is required to work on de-energized parts on the back of enclosed equipment, a minimum working space of 30 in. (762 mm) horizontally shall be provided.

Elevation of Unguarded Live Parts Above Working Space

Table 2.3 [NEC Table 110-34(e)], gives the elevation of unguarded live parts above working space.

2.2   OVERCURRENT-PROTECTION STANDARD AMPERE RATINGS

NEC Article 240-6, Standard Ampere Ratings, is repeated here in its entirety.

NEC Article 240-6: Standard Ampere Ratings
(A) FUSES AND FIXED-TRIP CIRCUIT BREAKERS

The standard ampere ratings for fuses and inverse-time circuit breakers shall be considered 15, 20, 25, 30, 35, 40, 45, 50, 60, 70, 80, 90, 100, 110, 125, 150, 175, 200, 225, 250, 300, 350, 400, 450, 500, 600, 700, 800, 1000, 1200, 1600, 2000, 2500, 3000, 4000, 5000, and 6000 amperes.

Exception: Additional standard ratings for fuses shall be considered 1, 3, 6, 10, and 601.

(B) ADJUSTABLE-TRIP CIRCUIT BREAKERS

The rating of an adjustable-trip circuit breaker having external means for adjusting the long-time pickup (ampere rating or overload) setting shall be the maximum setting possible.

Exception: Circuit breakers that have removable and sealable covers over the adjusting means, or are located behind bolted equipment enclosure doors, or are located behind locked doors accessible only to qualified personnel, shall be permitted to have ampere ratings equal to the adjusted (set) long-time pickup settings.

NOTE It is not the intent to prohibit the use of nonstandard ampere ratings for fuses and inverse-time circuit breakers.

2.3 NEC ARTICLE 240-21. FEEDER TAP RULES

This article, dealing with location in circuit (feeder tap rules), is repeated in its entirety.

NEC Article 240-21: Location in Circuit

An overcurrent device shall be connected in each ungrounded circuit conductor as follows:

(A) FEEDER AND BRANCH-CIRCUIT CONDUCTORS

Feeder and branch-circuit conductors shall be protected by overcurrent-protective devices connected at the point at which the conductors receive their supply, unless otherwise permitted in (b) through (m), listed hereafter. See Figure 2.11.

(B) FEEDER TAPS NOT OVER 10 FEET (3.05 M) LONG

Conductors shall be permitted to be tapped, without overcurrent protection at the tap, to a feeder or transformer secondary, where all the following conditions are met:

1. The length of the tap conductors does not exceed 10 ft (3.05 m).
2. The ampacity of the tap conductors is:
  a. Not less than the combined computed loads on the circuits supplied by the tap conductors, and
  b. Not less than the rating of the device supplied by the tap conductors, or not less than the rating of the overcurrent-protective device at the termination of the tap conductors.

FIGURE 2.11   NEC Section 240-3 permits the circuit breaker protecting the feeder conductors to protect the tap conductors to the cabinet.

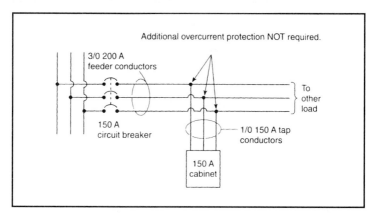

3. The tap conductors do not extend beyond the switchboard, panel-board, disconnecting means, or control devices they supply.
4. Except at the point of connection to the feeder, the tap conductors are enclosed in a raceway, which shall extend from the tap to the enclosure of an enclosed switchboard, panelboard, or control devices, or to the back of an open switchboard.
5. For field installations where the tap conductors leave the enclosure or vault in which the tap is made, the rating of the overcurrent device on the line side of the tap conductors shall not exceed 1000 percent of the tap conductor's ampacity.

NOTE   See Sections 384-16(a) and (d) for lighting and appliance branch-circuit panelboards.

(C) FEEDER TAPS NOT OVER 25 FT (7.62 M) LONG

Conductors shall be permitted to be tapped, without overcurrent protection at the tap, to a feeder where all the following conditions are met:

1. The length of the tap conductors does not exceed 25 ft (7.62 m).
2. The ampacity of the tap conductors is not less than $\frac{1}{3}$ of the rating of the overcurrent device protecting the feeder conductors.
3. The tap conductors terminate in a single circuit breaker or a single set of fuses that will limit the load to the ampacity of the tap conductors. This device shall be permitted to supply any number of additional overcurrent devices on its load side.

FIGURE 2.12   Feeder taps terminating in a single circuit breaker.

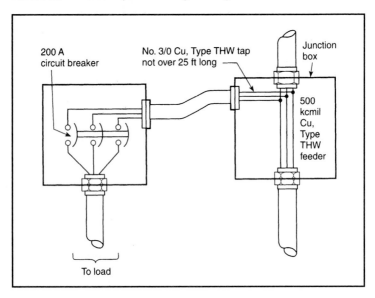

4. The tap conductors are suitably protected from physical damage or are enclosed in a raceway.

See Figure 2.12.

(D) FEEDER TAPS SUPPLYING A TRANSFORMER (PRIMARY PLUS SECONDARY) NOT OVER 25 FT (7.62 M) LONG

Conductors supplying a transformer shall be permitted to be tapped, without overcurrent protection at the tap, from a feeder where all the following conditions are met:

1. The conductors supplying the primary of a transformer have an ampacity at least one-third of the rating of the overcurrent device protecting the feeder conductors.
2. The conductors supplied by the secondary of the transformer shall have an ampacity that, when multiplied by the ratio of the secondary-to-primary voltage, is at least one-third of the rating of the overcurrent device protecting the feeder conductors.
3. The total length of one primary plus one secondary conductor, excluding any portion of the primary conductor that is protected at its ampacity, is not over 25 ft (7.62 m).

FIGURE 2.13 Example of transformer feeder taps not over 25 ft long.

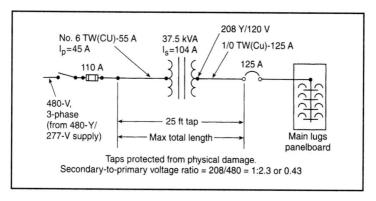

Taps protected from physical damage.
Secondary-to-primary voltage ratio = 208/480 = 1:2.3 or 0.43

4. The primary and secondary conductors are suitably protected from physical damage.
5. The secondary conductors terminate in a single circuit breaker or set of fuses that will limit the load current to not more than the conductor ampacity that is permitted by Section 310-15.

See Figure 2.13.

(E) FEEDER TAPS OVER 25 FT (7.62 M) LONG

Conductors over 25 ft (7.62 m) long shall be permitted to be tapped from feeders in high-bay manufacturing buildings over 35 ft (10.67 m) high at walls, where conditions of maintenance and supervision ensure that only qualified persons will service the systems. Conductors tapped, without overcurrent protection at the tap, to a feeder shall be permitted to be not over 25 ft (7.62 m) long horizontally and not over 100 ft (30.5 m) total length where all the following conditions are met:

1. The ampacity of the tap conductors is not less than one-third of the rating of the overcurrent device protecting the feeder conductors.
2. The tap conductors terminate at a single circuit breaker or a single set of fuses that will limit the load to the ampacity of the tap conductors. This single overcurrent device shall be permitted to supply any number of additional overcurrent devices on its load side.
3. The tap conductors are suitably protected from physical damage or are enclosed in a raceway.
4. The tap conductors are continuous from end to end and contain no splices.

5. The tap conductors are sized No. 6 copper or No. 4 aluminum or larger.

6. The tap conductors do not penetrate walls, floors, or ceilings.

7. The tap is made no less than 30 ft (9.14 m) from the floor.

See Figure 2.14.

(F) BRANCH-CIRCUIT TAPS

Taps to individual outlets and circuit conductors supplying a single electric household range shall be permitted to be protected by the branch-circuit overcurrent devices where in accordance with the requirements of Sections 210-19, 210-20, and 210-24.

(G) BUSWAY TAPS

Busways and busway taps shall be permitted to be protected against overcurrent in accordance with Sections 364-10 through 364-13.

(H) MOTOR CIRCUIT TAPS

Motor-feeder and branch-circuit conductors shall be permitted to be protected against overcurrent in accordance with Sections 430-28 and 430-53, respectively.

FIGURE 2.14   Example of NEC Section 240-21(e)(7) tap rule.

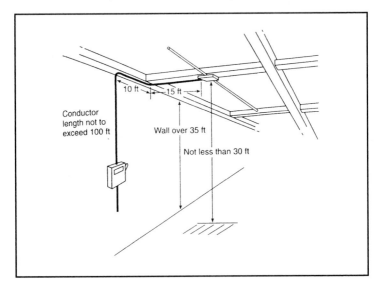

(I) CONDUCTORS FROM GENERATOR TERMINALS

Conductors from generator terminals shall be permitted to be protected against overcurrent in accordance with Section 445-5.

(J) TRANSFORMER SECONDARY CONDUCTORS OF SEPARATELY DERIVED SYSTEMS FOR INDUSTRIAL INSTALLATIONS

Conductors shall be permitted to be connected to a transformer secondary of a separately derived system for industrial installations, without overcurrent protection at the connection, where all the following conditions are met:

1. The length of the secondary conductors does not exceed 25 ft (7.62 m).
2. The ampacity of the secondary conductors is not less than the secondary current rating of the transformer, and the sum of the ratings of the overcurrent devices does not exceed the ampacity of the secondary conductors.
3. All overcurrent devices are grouped.
4. The secondary conductors are suitably protected from physical damage.

(M) OUTSIDE FEEDER TAPS

Outside conductors shall be permitted to be tapped to a feeder or to be connected at the transformer secondary, without overcurrent protection at the tap or connection, where all the following conditions are met:

1. The conductors are suitably protected from physical damage.
2. The conductors terminate at a single circuit breaker or a single set of fuses that will limit the load current to the ampacity of the conductors. This single overcurrent device shall be permitted to supply any number of additional overcurrent devices on its load side.
3. The tap conductors are installed outdoors, except at the point of termination.
4. The overcurrent device for the conductors is an integral part of a disconnecting means or shall be located immediately adjacent thereto.
5. The disconnecting means for the conductors are installed at a readily accessible location either outside of a building or structure, or inside nearest the point of entrance of the conductors.

(N) SERVICE CONDUCTORS

Service-entrance conductors shall be permitted to be protected by overcurrent devices in accordance with Section 230-91.

2.4   NEC ARTICLE 310: CONDUCTORS FOR GENERAL WIRING

Introduction

This article covers conductors for general wiring and includes Articles 310-1 through 310-15. Only Articles 310-3, 310-4, 310-5, 310-13, and 310-15 are included here in their entirety. The user of this handbook is encouraged to refer to the NEC for the complete text of the *Code*.

NEC Article 310-3: Stranded Conductors

Where installed in raceways, conductors of size No. 8 and larger shall be stranded.

Exception: As permitted or required elsewhere in this *Code*.

NEC Article 310-4: Conductors in Parallel

Aluminum, copper-clad aluminum, or copper conductors of size No. 1/0 and larger, comprising each phase, neutral, or grounded circuit conductor, shall be permitted to be connected in parallel (electrically joined at both ends to form a single conductor).

Exception No. 1: As permitted in Section 620-12(a)(1).

Exception No. 2: Conductors in sizes smaller than No. 1/0 shall be permitted to be run in parallel to supply control power to indicating instruments, contactors, relays, solenoids, and similar control devices provided (a) they are contained in the same raceway or cable; (b) the ampacity of each individual conductor is sufficient to carry the entire load current shared by the parallel conductors; and (c) the overcurrent protection is such that the ampacity of each individual conductor will not be exceeded if one or more of the parallel conductors become inadvertently disconnected.

Exception No. 3: Conductors in sizes smaller than No. 1/0 shall be permitted to be run in parallel for frequencies of 360 hertz and higher where conditions (a), (b), and (c) of Exception No. 2 are met.

Exception No. 4: Under engineering supervision, grounded neutral conductors in sizes No. 2 and larger shall be permitted to be run in parallel for existing installations.

NOTE   Exception No. 4 can be used to alleviate overheating of neutral conductors in existing installations due to high content of triplen harmonic currents.

The paralleled conductors in each phase, neutral, or grounded circuit conductor shall:

1. Be the same length
2. Have the same conductor material
3. Be the same size in circular mil area
4. Have the same insulation type
5. Be terminated in the same manner

Where run in separate raceways or cables, the raceways or cables shall have the same physical characteristics.

NOTE   Differences in inductive reactance and unequal division of current can be minimized by choice of materials, methods of construction, and orientation of conductors. It is not necessary for conductors of one phase, neutral, or grounded circuit conductor to be the same as those of another phase, neutral, or grounded circuit conductor to achieve balance.

Where equipment grounding conductors are used with conductors in parallel, they shall comply with the requirements of this section, except that they shall be sized in accordance with Section 250-95.

Where conductors are used in parallel, space in enclosures shall be given consideration (see Articles 370 and 373).

Conductors installed in parallel shall comply with the provisions of Article 310, Note 8(a), Notes to Ampacity Tables of 0 to 2000 Volts.

NEC Article 310-5: Minimum Size of Conductors

The minimum size of conductors shall be as shown in Table 2.4 (NEC Table 310-5).

Exception No. 1: For flexible cords as permitted by Section 400-12.

Exception No. 2: For fixture wire as permitted by Section 410-24.

TABLE 2.4   NEC Table 310-5: Minimum Size of Conductors

Voltage Rating of Conductor—Volts	Minimum Conductor Size—AWG
0 through 2000	14 Copper 12 Aluminum or Copper-Clad Aluminum
2001 through 8000	8
8001 through 15000	2
15001 through 28000	1
28001 through 35000	1/0

Exception No. 3: For motors rated 1 horsepower or less as permitted by Section 430-22(b).

Exception No. 4: For cranes and hoists as permitted by Section 610-14.

Exception No. 5: For elevator control and signaling circuits as permitted by Section 620-12.

Exception No. 6: For Class 1, Class 2, and Class 3 circuits as permitted by Sections 725-27 and 725-51.

Exception No. 7: Fire alarm circuits as permitted by Sections 760-27, 760-51, and 760-71.

Exception No. 8: For motor-control circuits as permitted by Section 430-72.

NEC Article 310-13: Conductor Constructions and Applications

Insulated conductors shall comply with the applicable provisions of one or more of the following: Tables 310-13, 310-61, 310-62, 310-63, and 310-64.

These conductors shall be permitted for use in any of the wiring methods recognized in Chapter 3 and as specified in their respective tables.

NOTE   Thermoplastic insulation may stiffen at temperatures colder than −10°C (+14°F). Thermoplastic insulation may also be deformed at normal temperatures where subjected to pressure, such as points of support. Thermoplastic insulation, where used on DC circuits in wet locations, may result in electroendosmosis between conductor and insulation.

Table 2.5, which is not a part of the NEC, but is a part of the NEC Handbook, is included for your convenience.

For conductor applications and insulations, see Table 2.6 (NEC Table 310-13).

NEC Article 310-15: Ampacities

Ampacities for conductors shall be permitted to be determined by (a) or (b), which follow.

NOTE   Ampacities provided by this section do not take voltage drop into consideration. See Section 210-19(a) Note No. 4 for branch circuits and Section 215-2(b) Note No. 2 for feeders.

(A) GENERAL

Ampacities for conductors rated 0 to 2,000 volts shall be as specified in the Allowable Ampacity Tables 310-16 through 310-19 and their accom-

TABLE 2.5   Conductor Characteristics

Conductor Characteristics

	Copper	Cu/Al	Aluminum
Density lbs/in^3	0.323	0.121	0.098
Density gm/cm^3	8.91	3.34	2.71
Resistivity ohms/CMF	10.37	16.08	16.78
Resistivity Microhm-CM	1.724	2.673	2.790
Conductivity (IACS%)	100	61–63	61
Weight % Copper	100	26.8	. . .
Tensile K psi-Hard	65.0	30.0	27.0
Tensile kg/mm^2-Hard	45.7	21.1	19.0
Tensile K psi-Annealed	35.0	17.0	17.0*
Tensile kg/mm^2-Annealed	24.6	12.0	12.0
Specific Gravity	8.91	3.34	2.71

*Semi-annealed

panying notes. Ampacities for solid dielectric insulated conductors rated 2,001 through 35,000 volts shall be as specified in Tables 310-67 through 310-86 and their accompanying notes.

NOTE   Tables 310-16 through 310-19 are application tables for determining conductor sizes on loads calculated in accordance with Article 220. Allowable ampacities result from consideration of one or more of the following:

1. Temperature compatibility with connected equipment, especially at the connection points.
2. Coordination with circuit and system overcurrent protection.
3. Compliance with the requirements of product listings or certifications. See Section 110-3(b).
4. Preservation of the safety benefits of established industry practices and standardized procedures.

(B) ENGINEERING SUPERVISION

Under engineering supervision, conductor ampacities shall be permitted to be calculated by means of the following general formula (shown in Figure 2.15):

FIGURE 2.15   Conductor ampacity calculation.

$$I = \sqrt{\frac{TC - (TA + DELTA\ TD)}{RDC\,(1 + YC)\,RCA}}$$

Where: TC = Conductor in degrees C.

                          TA = Ambient temperature in degrees C.

              DELTA TD = Dielectric loss temperature rise.

                          RDC = DC resistance of a conductor at temperature TC.

                          YC = Component AC resistance resulting from skin effect and proximity effect.

                          RCA = Effective thermal resistance between conductor and surrounding ambient.

NOTE See Appendix B of NEC for examples of formula applications.

(C) SELECTION OF AMPACITY

Where more than one calculated or tabulated ampacity could apply for a given circuit length, the lowest value shall be used.

> *Exception:* Where two different ampacities apply to adjacent portions of a circuit, the higher ampacity shall be permitted to be used beyond the point of transition, a distance equal to 10 ft (3.05 m) or 10 percent of the circuit length figured at the higher ampacity, whichever is less.

NOTE See Section 110-14(c) for conductor temperature limitations due to termination provisions.

(D) ELECTRICAL DUCTS

Electrical ducts as used in Article 310 shall include any of the electrical conduits recognized in Chapter 3 as suitable for use underground, and other raceways round in cross-section, listed for underground use, embedded in earth or concrete.

Ampacities for conductors of 0 to 2000 volts are given in Tables 2.7 through 2.10 (NEC Tables 310-16 through 310-19).

Notes to Ampacity Tables of 0 to 2000 Volts

1. *Explanation of tables:* For explanation of type letters and for recognized size of conductors for the various conductor insulations, see Section 310-13. For installation requirements, see Sections 310-1 through 310-10 and the various articles of this *Code.* For flexible cords, see Tables 400-4, 400-5(A), and 400-5(B).

3. *120/240-volt, three-wire, single-phase dwelling services and feeders:* For dwelling units, conductors, as listed below, shall be permitted to be utilized as 120/240-volt, three-wire, single-phase service-entrance conductors, service lateral conductors, and

TABLE 2.6 NEC Table 310-13: Conductor Application and Insulations

Trade Name	Type Letter	Max. Operating Temp.	Application Provisions	Insulation	AWG or kcmil	Thickness of Insulation	Mils	Outer Covering****
Fluorinated Ethylene Propylene	FEP or FEPB	90° C 194° F 200° C 392° F	Dry and damp locations. Dry locations—special applications.†	Fluorinated Ethylene Propylene	14-10 8-2		20 30	None
				Fluorinated Ethylene Propylene	14-8		14	Glass braid
					6-2		14	Asbestos or other suitable braid material
Mineral Insulation (Metal Sheathed)	MI	90° F 194° F 250° C 482° F	Dry and wet locations. For special application.†	Magnesium Oxide	18-16***** 16-10 9-4 3-500		23 36 50 55	Copper or Alloy Steel
Moisture-, Heat-, and Oil-Resistant Thermoplastic	MTW	60° C 140° F 90° C 194° F	Machine tool wiring in wet locations as permitted in NFPA 79. (See Article 670.) Machine tool wiring in dry locations as permitted in NFPA 79. (See Article 670.)	Flame-Retardant, Moisture-, Heat-, and Oil-Resistant Thermoplastic	22-12 10 8 6 4-2 1-4/0 213-500 501-1000	(A) 30 30 45 60 60 80 95 110	(B) 15 20 30 30 40 50 60 70	(A) None (B) Nylon jacket or equivalent
Paper		85° C 185° F	For underground service conductors, or by special permission.	Paper				Lead sheath

TABLE 2.6 NEC Table 310-13: Conductor Application and Insulations (*Continued*)

Trade Name	Type Letter	Max. Operating Temp.	Application Provisions	Insulation	AWG or kcmil	Thickness of Insulation Mils	Outer Covering****
Perfluoroalkoxy	PFA	90° C 194° F 200° C 392° F	Dry and damp locations. Dry locations—special applications.†	Perfluoroalkoxy	14-10 8-2 1-4/0	20 30 45	None
Perfluoroalkoxy	PFAH	250° C 482° F	Dry locations only. Only for leads within apparatus or within raceways connected to apparatus. (Nickel or nickel-coated copper only.)	Perfluoroalkoxy	14-10 8-2 1-4/0	20 30 45	None
Thermoset	RH	75° C 167° F	Dry and damp locations.	Flame-Retardant Thermoset	14-12** 10 8-2 1-4/0	30 45 60 80	Moisture-resistant, flame-retardant, non-metallic covering*
Thermoset	RHH	90° C 194° F	Dry and damp locations.		213-500 501-1000 1001-2000 For 601-2000 volts, see Table 310-62.	95 110 125	

TABLE 2.6 NEC Table 310-13: Conductor Application and Insulations (*Continued*)

Trade Name	Type Letter	Max. Operating Temp.	Application Provisions	Insulation	AWG or kcmil	Thickness of Insulation Mils	Outer Covering****
Moisture-Resistant Thermoset	RHW †††	75° C 167° F	Dry and wet locations. Where over 2000 volts insulation, shall be ozone-resistant.	Flame-Retardant, Moisture-Resistant Thermoset	14-10 .... 8-2 .... 1-4/0 .... 213-500 .... 501-1000 .... 1001-2000 .... For 601-2000 volts, see Table 310-62.	45 60 80 95 110 125	Moisture-resistant, flame-retardant, non-metallic covering*
Moisture-Resistant Thermoset	RHW-2	90° C 194° F	Dry and wet locations.	Flame-Retardant, Moisture-Resistant Thermoset	14-10 .... 8-2 .... 1-4/0 .... 213-500 .... 501-1000 .... 1001-2000 .... For 601-2000 volts, see Table 310-62.	45 60 80 95 110 125	Moisture-resistant, flame-retardant, nonmetallic covering*
Silicone	SA	90° C 194° F / 200° C 392° F	Dry and damp locations. / For special application.†	Silicone Rubber	14-10 .... 8-2 .... 1-4/0 .... 213-500 .... 501-1000 .... 1001-2000	45 60 80 95 110 125	Glass or other suitable braid material
Thermoset	SIS	90° C 194° F	Switchboard wiring only.	Flame-Retardant Thermoset	14-10 .... 8-2 .... 6-2 .... 1-4/0	30 45 60 55	None

TABLE 2.6 NEC Table 310-13: Conductor Application and Insulations (*Continued*)

Trade Name	Type Letter	Max. Operating Temp.	Application Provisions	Insulation	AWG or kcmil	Thickness of Insulation Mils	Outer Covering****
Thermoplastic and Fibrous Outer Braid	TBS	90° C 194° F	Switchboard wiring only.	Thermo-plastic	14-10 8 6-2 1-4/0	30 45 60 80	Flame-retardant, nonmetallic covering
Extended Polytetra-fluoroethylene	TFE	250° C 482° F	Dry locations only. Only for leads within apparatus or within raceways connected to apparatus, or as open wiring. (Nickel or nickel-coated copper only.)	Extruded Polytetra-fluoro-ethylene	14-10 8-2 1-4/0	20 30 45	None
Heat-Resistant Thermoplastic	THHN	90° C 194° F	Dry and damp locations.	Flame-Retardant, Heat-Resistant Thermo-plastic	14-12 10 8-6 4-2 1-4/0 250-500 501-1000	15 20 30 40 50 60 70	Nylon jacket or equivalent
Moisture- and Heat-Resistant Thermoplastic	THHW	75° F 167° F 90° C 194° F	Wet location. Dry location.	Flame-Retardant, Moisture- and Heat-Resistant Thermo-plastic	14-10 8-2 1-4/0 213-500 501-1000	45 60 80 95 110	None
Moisture- and Heat-Resistant Thermoplastic	THW †††	75° F 167° F 90° C 194° F	Dry and wet locations. Special applications within electric discharge lighting equipment. Limited to 1000 open-circuit volts or less. (Size 14-8 only as permitted in Section 410-31.)	Flame-Retardant, Moisture- and Heat-Resistant Thermo-plastic	14-10 8-2 1-4/0 213-500 501-1000 1001-2000	45 60 80 95 110 125	None

TABLE 2.6 NEC Table 310-13: Conductor Application and Insulations (*Continued*)

Trade Name	Type Letter	Max. Operating Temp.	Application Provisions	Insulation	AWG or kcmil	Thickness of Insulation Mils	Outer Covering****
Moisture- and Heat-Resistant Thermoplastic	THWN††	75° C 167° F	Dry and wet locations.	Flame-Retardant, Moisture- and Heat-Resistant Thermoplastic	14-12 10 8-6 4-2 1-4/0 250-500 501-1000	15 20 30 40 50 60 70	Nylon jacket or equivalent
Moisture-Resistant Thermoplastic	TW	60° C 140° F	Dry and wet locations.	Flame-Retardant, Moisture-Resistant Thermoplastic	14-10 8 6-2 1-4/0 213-500 501-1000 1001-2000	30 45 60 80 95 110 125	None
Underground Feeder and Branch-Circuit Cable—Single Conductor. (For Type UF cable employing more than one conductor, see Article 339.)	UF	60° C 140° F 75° C 167° F††††	See Article 339.	Moisture-Resistant Moisture- and Heat-Resistant	14-10 8-2 1-4/0	60§ 80§ 95§	Integral with insulation
Underground Service-Entrance Cable—Single Conductor (For Type USE cable employing more than one conduc-	USE†††	75° C 167° F	See Article 338.	Heat- and Moisture-Resistant	12-10 8-2 1-4/0 213-500 501-1000 1001-2000	45 60 80 95§§ 110 125	Moisture-resistant nonmetallic covering [See Section 338-1(b).]

TABLE 2.6 NEC Table 310-13: Conductor Application and Insulations (*Continued*)

Trade Name	Type Letter	Max. Operating Temp.	Application Provisions	Insulation	AWG or kcmil	Thickness of Insulation — Mils	Outer Covering****
Thermoset	XHH	90° C 194° F	Dry and damp locations.	Flame-Retardant Thermoset	14-10 8-2 1-4/0 213-500 501-1000 1001-2000	30 45 55 65 80 95	None
Moisture-Resistant Thermoset	XHHW†††	90° C 194° F 75° C 167° F	Dry and damp locations. Wet locations.	Flame-Retardant, Moisture-Resistant Thermoset	14-10 8-2 1-4/0 213-500 501-1000 1001-2000	30 45 55 65 80 95	None
Moisture-Resistant Thermoset	XHHW-2	90° C 194° F	Dry and wet locations.	Flame-Retardant, Moisture-Resistant Thermoset	14-10 8-2 1-4/0 213-500 501-1000 1001-2000	30 45 55 65 80 95	None
Modified Ethylene Tetrafluoroethylene	Z	90° C 194° F 150° C 302° F	Dry and damp locations. Dry locations—special applications.†	Modified Ethylene Tetrafluoroethylene	14-12 10 8-4 3-1 1/0-4/0	15 20 25 35 45	None
Modified Ethylene Tetrafluoroethylene	ZW†††	75° C 167° F 90° C 194° F 150° C 302° F	Wet locations. Dry and damp locations. Dry locations—special applications.†	Modified Ethylene Tetrafluoroethylene	14-10 8-2	30 45	None

****Some insulations do not require an outer covering.
†Where design conditions require maximum conductor operating temperatures above 90°C.
*****For signaling circuits permitting 300-volt insulation.
*For size Nos. 1-12, RHH shall be 45 mils-thickness insulation.
**Some rubber insulations do not require an outer covering.
†††Listed wire types designated with the suffix "-2," such as RHW-2, shall be permitted to be used at a continuous 90°C operating temperature, wet or dry.
§ Includes integral jacket.
†††† For ampacity limitation, see Section 339-5.
§§ Insulation thickness shall be permitted to be 80 mils for listed Type USE conductors that have been subjected to special investigations.
The nonmetallic covering over individual rubber-covered conductors of aluminum-sheathed cable and of lead-sheathed or multiconductor cable shall not be required to the flame-retardant. For Type MC cable, see Section 334-20. For nonmetallic-sheathed cable, see Section 336-25. For Type UF cable, see Section 339-1.

TABLE 2.7 NEC Table 310-16: Allowable Ampacities of Insulated Conductors Rated 0 Through 2000 Volts, 60° to 90°C (140° to 194°F), Not More Than Three Current-Carrying Conductors in Raceway or Cable or Earth (Directly Buried), Based on Ambient Temperature of 30°C (86°F)

Size	Temperature Rating of Conductor. See Table 310-13.						Size
	60°C (140°F)	75°C (167°F)	90°C (194°F)	60°C (140°F)	75°C (167°F)	90°C (194°F)	
AWG kcmil	TYPES TW†, UF†	TYPES FEPW†, RH†, RHW†, THHW†, THW†, THWN†, XHHW† USE†, ZW†	TYPES TBS, SA SIS, FEP†, FEPB†, MI RHH†, RHW-2, THHN†, THHW†, THW-2†, THWN-2†, USE-2, XHH, XHHW† XHHW-2, ZW-2	TYPES TW†, UF†	TYPES RH†, RHW†, THHW†, THW†, THWN†, XHHW†, USE†	TYPES TBS, SA, SIS, THHN†, THHW†, THW-2, THWN-2, RHH†, RHW-2, USE-2, XHH, XHHW, XHHW-2, ZW-2	AWG kcmil
	COPPER			ALUMINUM OR COPPER-CLAD ALUMINUM			
18	....	....	14	....	....	....	....
16	....	....	18	....	....	....	....
14	20†	20†	25†	....	....	....	....
12	25†	25†	30†	20†	20†	25†	12
10	30	35†	40†	25	30†	35†	10
8	40	50	55	30	40	45	8
6	55	65	75	40	50	60	6
4	70	85	95	55	65	75	4
3	85	100	110	65	75	85	3
2	95	115	130	75	90	100	2
1	110	130	150	85	100	115	1
1/0	125	150	170	100	120	135	1/0
2/0	145	175	195	115	135	150	2/0
3/0	165	200	225	130	155	175	3/0
4/0	195	230	260	150	180	205	4/0
250	215	255	290	170	205	230	250
300	240	285	320	190	230	255	300
350	260	310	350	210	250	280	350
400	280	335	380	225	270	305	400
500	320	380	430	260	310	350	500
600	355	420	475	285	340	385	600
700	385	460	520	310	375	420	700
750	400	475	535	320	385	435	750
800	410	490	555	330	395	450	800
900	435	520	585	355	425	480	900
1000	455	545	615	375	445	500	1000
1250	495	590	665	405	485	545	1250
1500	520	625	705	435	520	585	1500
1750	545	650	735	455	545	615	1750
2000	560	665	750	470	560	630	2000

CORRECTION FACTORS							
Ambient Temp.°C	For ambient temperatures other than 30°C (86°F), multiply the allowable ampacities shown above by the appropriate factor shown below.						Ambient Temp. °F
21-25	1.08	1.05	1.04	1.08	1.05	1.04	70-77
26-30	1.00	1.00	1.00	1.00	1.00	1.00	78-86
31-35	.91	.94	.96	.91	.94	.96	87-95
36-40	.82	.88	.91	.82	.88	.91	96-104
41-45	.71	.82	.87	.71	.82	.87	105-113
46-50	.58	.75	.82	.58	.75	.82	114-122
51-55	.41	.67	.76	.41	.67	.76	123-131
56-60	....	.58	.71	....	.58	.71	132-140
61-70	....	.33	.58	....	.33	.58	141-158
71-80	....	....	.41	....	....	.41	159-176

†Unless otherwise specifically permitted elsewhere in this *Code*, the overcurrent protection for conductor types marked with an obelisk (†) shall not exceed 15 amperes for No. 14, 20 amperes for No. 12, and 30 amperes for No. 10 copper; or 15 amperes for No. 12 and 25 amperes for No. 10 aluminum and copper-clad aluminum after any correction factors for ambient temperature and number of conductors have been applied.

TABLE 2.8 NEC Table 310-17: Allowable Ampacities of Single Insulated Conductors, Rated 0 Through 2000 Volts, in Free Air Based on Ambient Air Temperature of 30°C (86°F)

Size	Temperature Rating of Conductor. See Table 310-13.						Size
	60°C (140°F)	75°C (167°F)	90°C (194°F)	60°C (140°F)	75°C (167°F)	90°C (194°F)	
AWG kcmil	TYPES TW†, UF†	TYPES FEPW†, RH†, RHW†, THHW†, THW†, THWN†, XHHW† ZW†	TYPES TBS, SA, SIS, FEP†, F E P B †, MI, RHH†, RHW-2, THHN†, THHW†, THW-2†, THWN-2†, USE-2, XHH, XHHW†, XHHW-2, ZW-2	TYPES TW†, UF†	TYPES RH†, RHW†, THHW†, THW†, THWN†, XHHW†	TYPES TBS, SA, SIS, THHN†, THHW†, THW-2, THWN-2, RHH†, RHW-2, USE-2, XHH, XHHW†, XHHW-2, ZW-2	AWG kcmil
		COPPER		ALUMINUM OR COPPER-CLAD ALUMINUM			
18	....	....	18	....	....	....	....
16	....	....	24	....	....	....	....
14	25†	30†	35†	....	....	....	....
12	30†	35†	40†	25†	30†	35†	12
10	40†	50†	55†	35†	40†	40†	10
8	60	70	80	45	55	60	8
6	80	95	105	60	75	80	6
4	105	125	140	80	100	110	4
3	120	145	165	95	115	130	3
2	140	170	190	110	135	150	2
1	165	195	220	130	155	175	1
1/0	195	230	260	150	180	205	1/0
2/0	225	265	300	175	210	235	2/0
3/0	260	310	350	200	240	275	3/0
4/0	300	360	405	235	280	315	4/0
250	340	405	455	265	315	355	250
300	375	445	505	290	350	395	300
350	420	505	570	330	395	445	350
400	455	545	615	355	425	480	400
500	515	620	700	405	485	545	500
600	575	690	780	455	540	615	600
700	630	755	855	500	595	675	700
750	655	785	885	515	620	700	750
800	680	815	920	535	645	725	800
900	730	870	985	580	700	785	900
1000	780	935	1055	625	750	845	1000
1250	890	1065	1200	710	855	960	1250
1500	980	1175	1325	795	950	1075	1500
1750	1070	1280	1445	875	1050	1185	1750
2000	1155	1385	1560	960	1150	1335	2000

CORRECTION FACTORS

Ambient Temp. °C	For ambient temperatures other than 30°C (86°F), multiply the allowable ampacities shown above by the appropriate factor shown below.						Ambient Temp. °F
21-25	1.08	1.05	1.04	1.08	1.05	1.04	70-77
26-30	1.00	1.00	1.00	1.00	1.00	1.00	78-86
31-35	.91	.94	.96	.91	.94	.96	87-95
36-40	.82	.88	.91	.82	.88	.91	96-104
41-45	.71	.82	.87	.71	.82	.87	105-113
46-50	.58	.75	.82	.58	.75	.82	114-122
51-55	.41	.67	.76	.41	.67	.76	123-131
56-60	....	.58	.71	....	.58	.71	132-140
61-70	....	.33	.58	....	.33	.58	141-158
71-80	....	....	.41	....	....	.41	159-176

†Unless otherwise specifically permitted elsewhere in this *Code*, the overcurrent protection for conductor types marked with an obelisk (†) shall not exceed 15 amperes for No. 14, 20 amperes for No. 12, and 30 amperes for No. 10 copper; or 15 amperes for No. 12 and 25 amperes for No. 10 aluminum and copper-clad aluminum.

TABLE 2.9   NEC Table 310-18: Allowable Ampacities of Three Single Insulated Conductors Rated 0 Through 2000 Volts, 150° to 250°C (302° to 482°F), in Raceway or Cable Based on Ambient Air Temperature of 40°C (104°F)

Size	Temperature Rating of Conductor. See Table 310-13.				Size
	150°C (302°F)	200°C (392°F)	250°C (482°F)	150°C (302°F)	
AWG kcmil	TYPE Z	TYPES FEP, FEPB, PFA	TYPES PFAH, TFE	TYPE Z	AWG kcmil
		COPPER	NICKEL OR NICKEL-COATED COPPER	ALUMINUM OR COPPER-CLAD ALUMINUM	
14	34	36	39	. . . .	14
12	43	45	54	30	12
10	55	60	73	44	10
8	76	83	93	57	8
6	96	110	117	75	6
4	120	125	148	94	4
3	143	152	166	109	3
2	160	171	191	124	2
1	186	197	215	145	1
1/0	215	229	244	169	1/0
2/0	251	260	273	198	2/0
3/0	288	297	308	227	3/0
4/0	332	346	361	260	4/0
250	. . . .	. . . .	. . . .	. . . .	250
300	. . . .	. . . .	. . . .	. . . .	300
350	. . . .	. . . .	. . . .	. . . .	350
400	. . . .	. . . .	. . . .	. . . .	400
500	. . . .	. . . .	. . . .	. . . .	500
600	. . . .	. . . .	. . . .	. . . .	600
700	. . . .	. . . .	. . . .	. . . .	700
750	. . . .	. . . .	. . . .	. . . .	750
800	. . . .	. . . .	. . . .	. . . .	800
1000	. . . .	. . . .	. . . .	. . . .	1000
1500	. . . .	. . . .	. . . .	. . . .	1500
2000	. . . .	. . . .	. . . .	. . . .	2000

CORRECTION FACTORS

Ambient Temp. °C	For ambient temperatures other than 40°C (104°F), multiply the allowable ampacities shown above by the appropriate factor shown below.				Ambient Temp. °F
41-50	.95	.97	.98	.95	105-122
51-60	.90	.94	.95	.90	123-140
61-70	.85	.90	.93	.85	141-158
71-80	.80	.87	.90	.80	159-176
81-90	.74	.83	.87	.74	177-194
91-100	.67	.79	.85	.67	195-212
101-120	.52	.71	.79	.52	213-248
121-140	.30	.61	.72	.30	249-284
141-160	. . . .	.50	.65	. . . .	285-320
161-180	. . . .	.35	.58	. . . .	321-356
181-200	. . . .	. . . .	.49	. . . .	357-392
201-225	. . . .	. . . .	.35	. . . .	393-437

TABLE 2.10   NEC Table 310-19: Allowable Ampacities for Single Insulated Conductors Rated 0 Through 2000 Volts, 150° to 250°C (302° to 482°F), in Free Air Based on Ambient Air Temperature of 40°C (104°F)

Size	Temperature Rating of Conductor. See Table 310-13.					Size
AWG kcmil	150°C (302°F) TYPE Z	200°C (392°F) TYPES FEP, FEPB, PFA	Bare or covered conductors	250°C (482°F) TYPES PFAH, TFE	150°C (302°F) TYPE Z	AWG kcmil
	COPPER	COPPER	COPPER	NICKEL OR NICKEL-COATED COPPER	ALUMINUM OR COPPER-CLAD ALUMINUM	
14	46	54	30	59	....	14
12	60	68	35	78	47	12
10	80	90	50	107	63	10
8	106	124	70	142	83	8
6	155	165	95	205	112	6
4	190	220	125	278	148	4
3	214	252	150	327	170	3
2	255	293	175	381	198	2
1	293	344	200	440	228	1
1/0	339	399	235	532	263	1/0
2/0	390	467	275	591	305	2/0
3/0	451	546	320	708	351	3/0
4/0	529	629	370	830	411	4/0
250	....	....	415	....	....	250
300	....	....	460	....	....	300
350	....	....	520	....	....	350
400	....	....	560	....	....	400
500	....	....	635	....	....	500
600	....	....	710	....	....	600
700	....	....	780	....	....	700
750	....	....	805	....	....	750
800	....	....	835	....	....	800
900	....	....	865	....	....	900
1000	....	....	895	....	....	1000
1500	....	....	1205	....	....	1500
2000	....	....	1420	....	....	2000

CORRECTION FACTORS

Ambient Temp. °C	For ambient temperatures other than 40°C (104°F), multiply the allowable ampacities shown above by the appropriate factor shown below.					Ambient Temp. °F
41-50	.95	.97	....	.98	.95	105-122
51-60	.90	.94	....	.95	.90	123-140
61-70	.85	.90	....	.93	.85	141-158
71-80	.80	.87	....	.90	.80	159-176
81-90	.74	.83	....	.87	.74	177-194
91-100	.67	.79	....	.85	.67	195-212
101-120	.52	.71	....	.79	.52	213-248
121-140	.30	.61	....	.72	.30	249-284
141-160	....	.50	....	.65	....	285-320
161-180	....	.35	....	.58	....	321-356
181-200	....	....	....	.49	....	357-392
201-225	....	....	....	.35	....	393-437

feeder conductors that serve as the main power feeder to a dwelling unit and are installed in a raceway or cable with or without an equipment-grounding conductor. For applications of this note, the feeder conductors to a dwelling unit shall not be required to be larger than its service-entrance conductors. The grounded conductor shall be permitted to be smaller than the ungrounded conductors, provided the requirements of Sections 215-2, 220-22, and 230-42 are met (see Table 2.11).

5. *Bare or covered conductors:* Where bare or covered conductors are used with insulated conductors, their allowable ampacities shall be limited to those permitted for the adjacent insulated conductors.

6. *Mineral-insulated, metal-sheathed cable:* The temperature limitation on which the ampacities of mineral-insulated, metal-sheathed cable are based is determined by the insulating materials used in the end seal. Termination fittings incorporating unimpregnated, organic, insulating materials are limited to 90°C (194°F) operation.

7. *Type MTW (machine tool wire):* For allowable ampacities of Type MTW wire, see Table 11 in the *Electrical Standard for Industrial Machinery,* NFPA 79-1994.

8. *Adjustment factors:*
   a. *More than three current-carrying conductors in a raceway:* Where the number of current-carrying conductors in a raceway or cable exceeds three, the allowable ampacities shall be reduced, as shown in Table 2.12.

TABLE 2.11   Dwelling Units, Conductor Types, and Sizes

Copper	Aluminum or Copper-Clad Aluminum	Service or Feeder Rating in Amps
AWG	AWG	
4	2	100
3	1	110
2	1/0	125
1	2/0	150
1/0	3/0	175
2/0	4/0	200
3/0	250 kcmil	225
4/0	300 kcmil	250
250 kcmil	350 kcmil	300
350 kcmil	500 kcmil	350
400 kcmil	600 kcmil	400

TABLE 2.12 Adjustment Factors (Derating)

Number of Current-Carrying Conductors	Percent of Values in Tables as Adjusted for Ambient Temperature if Necessary
4 through 6	80
7 through 9	70
10 through 20	50
21 through 30	45
31 through 40	40
41 and above	35

Where single conductors or multiconductor cables are stacked or bundled longer than 24 in. (610 mm) without maintaining spacing and are not installed in raceways, the allowable ampacity of each conductor shall be reduced as shown in Table 2.12.

Exception No. 1: Where conductors of different systems, as provided in Section 300-3, are installed in a common raceway or cable, the derating factors shown above shall apply to the number of power and lighting (Articles 210, 215, 220, and 230) conductors only.

Exception No. 2: For conductors installed in cable trays, the provisions of Section 318-11 shall apply.

Exception No. 3: Derating factors shall not apply to conductors in nipples having a length not exceeding 24 in. (610 mm).

Exception No. 4: Derating factors shall not apply to underground conductors entering or leaving an outdoor trench if those conductors have physical protection in the form of a rigid-metal conduit, intermediate-metal conduit, or rigid nonmetallic conduit having a length not exceeding 10 ft (3.05 m) and the number of conductors does not exceed four.

Exception No. 5: For other loading conditions, adjustment factors and ampacities shall be permitted to be calculated under Section 310-15(b).

NOTE See NEC Appendix B, Table B-310-11, for adjustment factors for more than three current-carrying conductors in a raceway or cable with diversity.

    b. More than one conduit, tube, or raceway: Spacing between conduits, tubing, or raceways shall be maintained.

9. *Overcurrent protection:* Where the standard ratings and settings of overcurrent devices do not correspond with the ratings and settings allowed for conductors, the next higher standard rating and setting shall be permitted.

Exception: As limited in Section 240-3.

10. *Neutral conductor:*

a. A neutral conductor that carries only the unbalanced current from other conductors of the same circuit need not be counted when applying the provisions of Note 8.

b. In a three-wire circuit consisting of two phase wires and the neutral of a four-wire, three-phase wye-connected system, a common conductor carries approximately the same current as the line-to-neutral load currents of the other conductors and shall be counted when applying the provisions of Note 8.

c. On a four-wire, three-phase wye circuit where the major portion of the load consists of nonlinear loads, there are harmonic currents present in the neutral conductor, and the neutral shall be considered to be a current-carrying conductor.

11. *Grounding or bonding conductor:* A grounding or bonding conductor shall not be counted when applying the provisions of Note 8.

NOTE Tables 2.13 (NEC Table 310-61) and 2.14 (NEC Table 310-62) are included here for convenient reference. NEC Tables 310-63 through 310-86, which cover ampacities for conductors rated 2001 volts and higher, are not included in this handbook.

TABLE 2.13 NEC Table 310-61: Conductor Application and Insulation

Trade Name	Type Letter	Maximum Operating Temperature	Application Provision	Insulation	Outer Covering
Medium voltage solid dielectric	MV-90 MV-105†	90° C 105° C	Dry or wet locations rated 2001 volts and higher	Thermoplastic or Thermosetting	Jacket, Sheath, or Armor

† Where design conditions require maximum conductor temperatures above 90°C.

TABLE 2.14 NEC Table 310-62: Thickness of Insulation for 601–2000-Volt Nonshielded Types RHH and RHW, in Mils

Conductor Size AWG-kcmil	A	B
14-10	80	60
8	80	70
6-2	95	70
1-2/0	110	90
3/0-4/0	110	90
213-500	125	105
501-1000	140	120

Note: 1. Column A insulations are limited to natural, SBR, and butyl rubbers.
Note: 2. Column B insulations are materials such as cross-linked polyethylene, ethylene propylene rubber, and composites thereof.

2.5 NEC CHAPTER 9: TABLES (PARTIAL)

Introduction

Although the percentage fill allowed for conduit and tubing has not changed in the 1996 NEC, there have been substantial changes in this edition with the addition of many new conduit tables for 12 types of conduit. In lieu of old NEC Tables 3A, 3B, or 3C, the user is now offered Appendix C. These tables have been expanded to provide specific and accurate raceway fill information, with separate tables for both metallic- and nonmetallic-type conduit and tubing raceways, as well as flexible raceways. NEC Appendix C follows in Section 2.6.

Included here are Tables 2.15 through 2.20, inclusive, which are NEC Chapter 9, Tables 1, 4, 5, 8, 9, and 10, respectively.

NOTE Table 1 is based on common conditions of proper cabling and alignment of conductors where the length of the pull and the number of bends are within reasonable limits. It should be recognized that, for certain conditions, a larger-size conduit or a lesser conduit fill should be considered.

TABLE 2.15 Percent of Cross-Section of Conduit and Tubing for Conductors

Number of Conductors	1	2	Over 2
All Conductor Types	53	31	40

TABLE 2.16 NEC Chapter 9, Table 4: Dimensions and Percent Area of Conduit and Tubing

Electrical Metallic Tubing

Trade Size Inches	Internal Diameter Inches	Total Area 100% Sq. In.	2 Wires 31% Sq. In.	Over 2 Wires 40% Sq. In.	1 Wire 53% Sq. In.
½	0.622	0.304	0.094	0.122	0.161
¾	0.824	0.533	0.165	0.213	0.283
1	1.049	0.864	0.268	0.346	0.458
1¼	1.380	1.496	0.464	0.598	0.793
1½	1.610	2.036	0.631	0.814	1.079
2	2.067	3.356	1.040	1.342	1.778
2½	2.731	5.858	1.816	2.343	3.105
3	3.356	8.846	2.742	3.538	4.688
3½	3.834	11.545	3.579	4.618	6.119
4	4.334	14.753	4.573	5.901	7.819

Electrical Nonmetallic Tubing

Trade Size Inches	Internal Diameter Inches	Total Area 100% Sq. In.	2 Wires 31% In Sq. In.	Over 2 Wires 40% Sq. In.	1 Wire 53% Sq. In.
½	0.560	0.246	0.076	0.099	0.131
¾	0.760	0.454	0.141	0.181	0.240
1	1.000	0.785	0.243	0.314	0.416
1¼	1.340	1.410	0.437	0.564	0.747
1½	1.570	1.936	0.600	0.774	1.026
2	2.020	3.205	0.994	1.282	1.699
2½	—	—	—	—	—
3	—	—	—	—	—
3½	—	—	—	—	—
4	—	—	—	—	—

Flexible Metal Conduit

Trade Size Inches	Internal Diameter Inches	Total Area 100% Sq. In.	2 Wires 31% Sq. In.	Over 2 Wires 40% Sq. In.	1 Wire 53% Sq. In.
⅜	0.384	0.116	0.036	0.046	0.061
½	0.635	0.317	0.098	0.127	0.168
¾	0.824	0.533	0.165	0.213	0.282
1	1.020	0.817	0.253	0.327	0.433
1¼	1.275	1.277	0.396	0.511	0.677
1½	1.538	1.857	0.576	0.743	0.984
2	2.040	3.269	1.013	1.307	1.732
2½	2.500	4.909	1.522	1.964	2.602
3	3.000	7.069	2.191	2.827	3.746
3½	3.500	9.621	2.983	3.848	5.099
4	4.000	12.566	3.896	5.027	6.660

Intermediate Metal Conduit

Trade Size Inches	Internal Diameter Inches	Total Area 100% Sq. In.	2 Wires 31% In Sq. In.	Over 2 Wires 40% Sq. In.	1 Wire 53% Sq. In.
⅜	—	—	—	—	—
½	0.660	0.342	0.106	0.137	0.181
¾	0.864	0.586	0.182	0.235	0.311
1	1.105	0.959	0.297	0.384	0.508
1¼	1.448	1.646	0.510	0.658	0.872
1½	1.683	2.223	0.689	0.889	1.178
2	2.150	3.629	1.125	1.452	1.923
2½	2.557	5.135	1.592	2.054	2.722
3	3.176	7.922	2.456	3.169	4.199
3½	3.671	10.584	3.281	4.234	5.610
4	4.166	13.631	4.226	5.452	7.224

TABLE 2.16 NEC Chapter 9, Table 4: Dimensions and Percent Area of Conduit and Tubing (*Continued*)

Trade Size Inches	Internal Diameter Inches	Rigid PVC Conduit, Schedule 80 Total Area 100% Sq. In.	2 Wires 31% Sq. In.	Over 2 Wires 40% Sq. In.	1 Wire 53% Sq. In.	Internal Diameter Inches	Rigid PVC Conduit Schedule 40 and HDPE Conduit Total Area 100% Sq. In.	2 Wires 31% Sq. In.	Over 2 Wires 40% Sq. In.	1 Wire 53% Sq. In.
½	0.526	0.217	0.067	0.087	0.115	0.602	0.285	0.088	0.114	0.151
¾	0.722	0.409	0.127	0.164	0.217	0.804	0.508	0.157	0.203	0.269
1	0.936	0.688	0.213	0.275	0.365	1.029	0.832	0.258	0.333	0.441
1¼	1.255	1.237	0.383	0.495	0.656	1.360	1.453	0.450	0.581	0.770
1½	1.476	1.711	0.530	0.684	0.907	1.590	1.986	0.616	0.794	1.052
2	1.913	2.874	0.891	1.150	1.1523	2.047	3.291	1.020	1.316	1.744
2½	2.290	4.119	1.277	1.647	2.183	2.445	4.695	1.455	1.878	2.488
3	2.864	6.442	1.997	2.577	3.414	3.042	7.268	2.253	2.907	3.852
3½	3.326	8.688	2.693	3.475	4.605	3.521	9.737	3.018	3.895	5.161
4	3.786	11.258	3.490	4.503	5.967	3.998	12.554	3.892	5.022	6.654
5	4.768	17.855	5.535	7.142	9.463	5.016	19.761	6.126	7.904	10.473
6	5.709	25.598	7.935	10.239	13.567	6.031	28.567	8.856	11.427	15.141

Trade Size Inches	Internal Diameter Inches	Type A, Rigid PVC Conduit Total Area 100% Sq. In.	2 Wires 31% Sq. In.	Over 2 Wires 40% Sq. In.	1 Wire 53% Sq. In.	Internal Diameter Inches	Type EB, PVC Conduit Total Area 100% Sq. In.	2 Wires 31% Sq. In.	Over 2 Wires 40% Sq. In.	1 Wire 53% Sq. In.
½	0.700	0.385	0.119	0.154	0.204	—	—	—	—	—
¾	0.910	0.650	0.202	0.260	0.345	—	—	—	—	—
1	1.175	1.084	0.336	0.434	0.575	—	—	—	—	—
1¼	1.500	1.767	0.548	0.707	0.937	—	—	—	—	—
1½	1.720	2.324	0.720	0.929	1.231	—	—	—	—	—
2	2.155	3.647	1.131	1.459	1.933	2.221	3.874	1.201	1.550	2.053
2½	2.635	5.453	1.690	2.181	2.890	—	—	—	—	—
3	3.230	8.194	2.540	3.278	4.343	3.330	8.709	2.700	3.484	4.616
3½	3.690	10.694	3.315	4.278	5.668	3.804	11.365	3.523	4.546	6.024
4	4.180	13.723	4.254	5.489	7.273	4.289	14.448	4.479	5.779	7.657
5	—	—	—	—	—	5.316	22.195	6.881	8.878	11.764
6	—	—	—	—	—	6.336	31.530	9.774	12.612	16.711

TABLE 2.16 NEC Chapter 9, Table 4: Dimensions and Percent Area of Conduit and Tubing *(Continued)*

Trade Size Inches	Liquidtight Flexible Nonmetallic Conduit (Type FNMC-B*)					Liquidtight Flexible Nonmetallic Conduit (Type FNMC-A*)				
	Internal Diameter Inches	Total Area 100% Sq. In.	2 Wires 31% Sq. In.	Over 2 Wires 40% Sq. In.	1 Wire 53% Sq. In.	Internal Diameter Inches	Total Area 100% Sq. In.	2 Wires 31% Sq. In.	Over 2 Wires 40% Sq. In.	1 Wire 53% Sq. In.
⅜	0.494	0.192	0.059	0.077	0.102	0.495	0.192	0.060	0.077	0.102
½	0.632	0.314	0.097	0.125	0.166	0.630	0.312	0.097	0.125	0.165
¾	0.830	0.541	0.168	0.216	0.287	0.825	0.535	0.166	0.214	0.283
1	1.054	0.872	0.270	0.349	0.462	1.043	0.854	0.265	0.341	0.452
1¼	1.395	1.528	0.474	0.611	0.810	1.383	1.501	0.465	0.600	0.796
1½	1.588	1.979	0.614	0.792	1.049	1.603	2.017	0.625	0.807	1.069
2	2.033	3.245	1.006	1.298	1.720	2.063	3.341	1.036	1.336	1.771

*Corresponds to Section 351-22(2).   *Corresponds to Section 351-22(1)

Trade Size Inches	Liquidtight Flexible Metal Conduit					Rigid Metal Conduit				
	Internal Diameter Inches	Total Area 100% Sq. In.	2 Wires 31% Sq. In.	Over 2 Wires 40% Sq. In.	1 Wire 53% Sq. In.	Internal Diameter Inches	Total Area 100% Sq. In.	2 Wires 31% Sq. In.	Over 2 Wires 40% Sq. In.	1 Wire 53% Sq. In.
⅜	0.494	0.192	0.059	0.077	0.102	—				
½	0.632	0.314	0.097	0.125	0.166	0.632	0.314	0.097	0.125	0.166
¾	0.830	0.541	0.168	0.216	0.287	0.836	0.549	0.170	0.220	0.291
1	1.054	0.872	0.270	0.349	0.462	1.063	0.888	0.275	0.355	0.470
1¼	1.395	1.528	0.474	0.611	0.810	1.394	1.526	0.473	0.610	0.809
1½	1.588	1.979	0.614	0.792	1.049	1.624	2.071	0.642	0.829	1.098
2	2.033	3.245	1.006	1.298	1.720	2.083	3.408	1.056	1.363	1.806
2½	2.493	4.879	1.513	1.952	2.586	2.489	4.866	1.508	1.946	2.579
3	3.085	7.475	2.317	2.990	3.962	3.090	7.499	2.325	3.000	3.975
3½	3.520	9.731	3.017	3.893	5.158	3.570	10.010	3.103	4.004	5.305
4	4.020	12.692	3.935	5.077	6.727	4.050	12.883	3.994	5.153	6.828
5	—	—	—	—	—	5.073	20.213	6.266	8.085	10.713
6	—	—	—	—	—	6.093	29.158	9.039	11.663	15.454

TABLE 2.17 NEC Chapter 9, Table 5: Dimensions of Insulated Conductors and Fixture Wires

Type: AF, FFH-2, RFH-1, RFH-2, RH, RHH*, RHW*, RHW-2*, RHH, RHW, RHW-2, SF-1, SF-2, SFF-1, SFF-2, TF, TFF, THHW, THW, THW-2, TW, XF, XFF

Type	Size	Approx. Diam. In.	Approx. Area Sq. In.
RFH-2	18	0.136	0.0145
FFH-2	16	0.148	0.0172
RH	14	0.163	0.0209
	12	0.182	0.026
RHW-2, RHH	14	0.193	0.0293
RHW	12	0.212	0.0353
RH, RHH	10	0.236	0.0437
RHW	8	0.326	0.0835
RHW-2	6	0.364	0.1041
	4	0.412	0.1333
	3	0.44	0.1521
	2	0.472	0.175
	1	0.582	0.266
	1/0	0.622	0.3039
	2/0	0.668	0.3505
	3/0	0.72	0.4072
	4/0	0.778	0.4754
	250	0.895	0.6291
	300	0.95	0.7088
	350	1.001	0.787
	400	1.048	0.8626
	500	1.133	1.0082
	600	1.243	1.2135
	700	1.314	1.3561
	750	1.348	1.4272
	800	1.38	1.4957
	900	1.444	1.6377
	1000	1.502	1.7719
	1250	1.729	2.3479
	1500	1.852	2.6938
	1750	1.966	3.0357
	2000	2.072	3.3719
SF-2, SFF-2	18	0.121	0.0115
	16	0.133	0.0139
	14	0.148	0.0172
SF-1, SFF-1	18	0.091	0.0065
RFH-1, AF, XF, XFF	18	0.106	0.008
AF, TF, TFF, XF, XFF	16	0.118	0.0109
AF, TW, XF, XFF	14	0.133	0.0139
TW	12	0.152	0.0181
	10	0.176	0.0243
	8	0.236	0.0437
RHH*, RHW*, RHW-2*, THHW, THW, THW-2	14	0.163	0.0209

*Types RHH, RHW, and RHW-2 without outer covering.

TABLE 2.17 NEC Chapter 9, Table 5: Dimensions of Insulated Conductors and Fixture Wires (*Continued*)

Type: AF, RHH*, RHW*, RHW-2*, THHN, THHW, THW, THW-2, TFN, TFFN, THWN, THWN-2, XF, XFF

Type	Size	Approx. Diam. In.	Approx. Area Sq. In.
RHH*, RHW*, RHW-2*	12	0.182	0.026
THHW, THW, AF, XF, XFF	10	0.206	0.0333
RHH*, RHW*, RHW-2*	8	0.266	0.0556
THHW, THW, THW-2			
TW, THW	6	0.304	0.0726
THHW	4	0.352	0.0973
THW-2	3	0.38	0.1134
RHH*	2	0.412	0.1333
RHW*	1	0.492	0.1901
RHW-2*	1/0	0.532	0.2223
	2/0	0.578	0.2624
	3/0	0.63	0.3117
	4/0	0.688	0.3718
	250	0.765	0.4596
	300	0.82	0.5281
	350	0.871	0.5958
	400	0.918	0.6619
	500	1.003	0.7901
	600	1.113	0.9729
	700	1.184	1.101
	750	1.218	1.1652
	800	1.25	1.2272
	900	1.314	1.3561
	1000	1.372	1.4784
	1250	1.539	1.8602
	1500	1.662	2.1695
	1750	1.776	2.4773
	2000	1.882	2.7818
TFN	18	0.084	0.0055
TFFN	16	0.096	0.0133
THHN	14	0.111	0.0097
THWN	12	0.13	0.0133
THWN-2	10	0.164	0.0211
	8	0.216	0.0366
	6	0.254	0.0507
	4	0.324	0.0824
	3	0.352	0.0973
	2	0.384	0.1158
	1	0.446	0.1562
	1/0	0.486	0.1855
	2/0	0.532	0.2223
	3/0	0.584	0.2679
	4/0	0.642	0.3237
	250	0.711	0.397
	300	0.766	0.4608

*Types RHH, RHW, and RHW-2 without outer covering.

TABLE 2.17 NEC Chapter 9, Table 5: Dimensions of Insulated Conductors and Fixture Wires (*Continued*)

Type: FEP, FEPB, PAF, PAFF, PF, PFA, PFAH, PFF, PGF, PGFF, PTF, PTFF, TFE, THHN, THWN, THWN-2, Z, ZF, ZFF

Type	Size	Approx. Diam. In.	Approx. Area Sq. In.
THHN	350	0.817	0.5242
THWN	400	0.864	0.5863
THWN-2	500	0.949	0.7073
	600	1.051	0.8676
	700	1.122	0.9887
	750	1.156	1.0496
	800	1.188	1.1085
	900	1.252	1.2311
	1000	1.31	1.3478
PF, PGFF, PGF, PFF	18	0.086	0.0058
PTF, PAF, PTFF, PAFF	16	0.098	0.0075
PF, PGFF, PGF, PFF PTF, PAF, PTFF PAFF, TFE FEP, PFA FEPB, PFAH	14	0.113	0.01
TFE, FEP	12	0.132	0.0137
PFA, FEPB	10	0.156	0.0191
PFAH	8	0.206	0.0333
	6	0.244	0.0468
	4	0.292	0.067
	3	0.32	0.0804
	2	0.352	0.0973
TFE, PFAH	1	0.422	0.1399
TFE	1/0	0.462	0.1676
PFA	2/0	0.508	0.2027
PFAH, Z	3/0	0.56	0.2463
	4/0	0.618	0.3
ZF, ZFF	18	0.076	0.0045
	16	0.088	0.0061
Z, ZF, ZFF	14	0.103	0.0083
Z	12	0.122	0.0117
	10	0.156	0.0191
	8	0.196	0.0302
	6	0.234	0.043
	4	0.282	0.0625
	3	0.33	0.0855
	2	0.362	0.1029
	1	0.402	0.1269

TABLE 2.17 NEC Chapter 9, Table 5: Dimensions of Insulated Conductors and Fixture Wires (*Continued*)

Type: KF-1, KF-2, KFF-1, KFF-2, XHH, XHHW, XHHW-2, ZW			
Type	Size	Approx. Diam. In.	Approx. Area Sq. In.
XHHW, ZW	14	0.133	0.0139
XHHW-2	12	0.152	0.0181
XHH	10	0.176	0.0243
	8	0.236	0.0437
	6	0.274	0.059
	4	0.322	0.0814
	3	0.35	0.0962
	2	0.382	0.1146
XHHW	1	0.442	0.1534
XHHW-2	1/0	0.482	0.1825
XHH	2/0	0.528	0.219
	3/0	0.58	0.2642
	4/0	0.638	0.3197
	250	0.705	0.3904
	300	0.76	0.4536
	350	0.811	0.5166
	400	0.858	0.5782
	500	0.943	0.6984
	600	1.053	0.8709
	700	1.124	0.9923
	750	1.158	1.0532
	800	1.19	1.1122
	900	1.254	1.2351
	1000	1.312	1.3519
	1250	1.479	1.718
	1500	1.602	2.0157
	1750	1.716	2.3127
	2000	1.822	2.6073
KF-2	18	0.063	0.0031
KFF-2	16	0.075	0.0044
	14	0.09	0.0064
	12	0.109	0.0093
	10	0.133	0.0139
KF-1	18	0.057	0.0026
KFF-1	16	0.069	0.0037
	14	0.084	0.0055
	12	0.103	0.0083
	10	0.127	0.0127

TABLE 2.18 NEC Chapter 9, Table 8: Conductor Properties

Size AWG/ kcmil	Area Cir. Mils	Conductors				DC Resistance at 75°C (167°F)		
		Stranding		Overall		Copper		Aluminum
		Quan-tity	Diam. In.	Diam. In.	Area In.²	Uncoated ohm/kFT	Coated ohm/kFT	ohm/ kFT
18	1620	1	—	0.040	0.001	7.77	8.08	12.8
18	1620	7	0.015	0.046	0.002	7.95	8.45	13.1
16	2580	1	—	0.051	0.002	4.89	5.08	8.05
16	2580	7	0.019	0.058	0.003	4.99	5.29	8.21
14	4110	1	--	0.064	0.003	3.07	3.19	5.06
14	4110	7	0.024	0.073	0.004	3.14	3.26	5.17
12	6530	1	—	0.081	0.005	1.93	2.01	3.18
12	6530	7	0.030	0.092	0.006	1.98	2.05	3.25
10	10380	1	—	0.102	0.008	1.21	1.26	2.00
10	10380	7	0.038	0.116	0.011	1.24	1.29	2.04
8	16510	1	---	0.128	0.013	0.764	0.786	1.26
8	16510	7	0.049	0.146	0.017	0.778	0.809	1.28
6	26240	7	0.061	0.184	0.027	0.491	0.510	0.808
4	41740	7	0.077	0.232	0.042	0.308	0.321	0.508
3	52620	7	0.087	0.260	0.053	0.245	0.254	0.403
2	66360	7	0.097	0.292	0.067	0.194	0.201	0.319
1	83690	19	0.066	0.332	0.087	0.154	0.160	0.253
1/0	105600	19	0.074	0.372	0.109	0.122	0.127	0.201
2/0	133100	19	0.084	0.418	0.137	0.0967	0.101	0.159
3/0	167800	19	0.094	0.470	0.173	0.0766	0.0797	0.126
4/0	211600	19	0.106	0.528	0.219	0.0608	0.0626	0.100
250	—	37	0.082	0.575	0.260	0.0515	0.0535	0.0847
300	—	37	0.090	0.630	0.312	0.0429	0.0446	0.0707
350	--	37	0.097	0.681	0.364	0.0367	0.0382	0.0605
400	—	37	0.104	0.728	0.416	0.0321	0.0331	0.0529
500	—	37	0.116	0.813	0.519	0.0258	0.0265	0.0424
600	—	61	0.099	0.893	0.626	0.0214	0.0223	0.0353
700	—	61	0.107	0.964	0.730	0.0184	0.0189	0.0303
750	—	61	0.111	0.998	0.782	0.0171	0.0176	0.0282
800	—	61	0.114	1.030	0.834	0.0161	0.0166	0.0265
900	—	61	0.122	1.094	0.940	0.0143	0.0147	0.0235
1000	--	61	0.128	1.152	1.042	0.0129	0.0132	0.0212
1250	—.	91	0.117	1.289	1.305	0.0103	0.0106	0.0169
1500	—.	91	0.128	1.412	1.566	0.00858	0.00883	0.0141
1750	-··	127	0.117	1.526	1.829	0.00735	0.00756	0.0121
2000	....	127	0.126	1.632	2.092	0.00643	0.00662	0.0106

These resistance values are valid ONLY for the parameters as given. Using conductors having coated strands, different stranding type, and, especially, other temperatures changes the resistance.
Formula for temperature change: $R_2 = R_1 [1 + \alpha(T_2 - 75)]$ where: $\alpha_{cu} = 0.00323$, $\alpha_{AL} = 0.00330$.
Conductors with compact and compressed stranding have about 9 percent and 3 percent, respectively, smaller bare conductor diameters than those shown. See Table 5A for actual compact cable dimensions.
The IACS conductivities used: bare copper = 100%, aluminum = 61%.
Class B stranding is listed as well as solid for some sizes. Its overall diameter and area is that of its circumscribing circle.
(FPN): The construction information is per NEMA WC8-1988. The resistance is calculated per National Bureau of Standards Handbook 100, dated 1966, and Handbook 109, dated 1972.

TABLE 2.19 NEC Chapter 9, Table 9: AC Resistance and Reactance for 600-Volt Cables

Size AWG/ kcmil	X_L (Reactance) for All Wires		AC Resistance for Uncoated Copper Wires			AC Resistance for Aluminum Wires			Effective Z at 0.85 PF for Uncoated Copper Wires			Effective Z at 0.85 PF for Aluminum Wires			Size AWG/ kcmil
	PVC, Al. Conduits	Steel Conduit	PVC Conduit	Al. Conduit	Steel Conduit	PVC Conduit	Al. Conduit	Steel Conduit	PVC Conduit	Al. Conduit	Steel Conduit	PVC Conduit	Al. Conduit	Steel Conduit	
14	0.058	0.073	3.1	3.1	3.1	—	—	—	2.7	2.7	2.7	—	—	—	14
12	0.054	0.068	2.0	2.0	2.0	3.2	3.2	3.2	1.7	1.7	1.7	2.8	2.8	2.8	12
10	0.050	0.063	1.2	1.2	1.2	2.0	2.0	2.0	1.1	1.1	1.1	1.8	1.8	1.8	10
8	0.052	0.065	0.78	0.78	0.78	1.3	1.3	1.3	0.69	0.69	0.70	1.1	1.1	1.1	8
6	0.051	0.064	0.49	0.49	0.49	0.81	0.81	0.81	0.44	0.45	0.45	0.71	0.72	0.72	6
4	0.048	0.060	0.31	0.31	0.31	0.51	0.51	0.51	0.29	0.29	0.30	0.46	0.46	0.46	4
3	0.047	0.059	0.25	0.25	0.25	0.40	0.41	0.40	0.23	0.24	0.24	0.37	0.37	0.37	3
2	0.045	0.057	0.19	0.20	0.20	0.32	0.32	0.32	0.19	0.19	0.20	0.30	0.30	0.30	2
1	0.046	0.057	0.15	0.16	0.16	0.25	0.26	0.25	0.16	0.16	0.16	0.24	0.24	0.25	1
1/0	0.044	0.055	0.12	0.13	0.12	0.20	0.21	0.20	0.13	0.13	0.13	0.19	0.20	0.20	1/0
2/0	0.043	0.054	0.10	0.10	0.10	0.16	0.16	0.16	0.11	0.11	0.11	0.16	0.16	0.16	2/0
3/0	0.042	0.052	0.077	0.082	0.079	0.13	0.13	0.13	0.088	0.092	0.094	0.13	0.13	0.14	3/0
4/0	0.041	0.051	0.062	0.067	0.063	0.10	0.11	0.10	0.074	0.078	0.080	0.11	0.11	0.11	4/0
250	0.041	0.052	0.052	0.057	0.054	0.085	0.090	0.086	0.066	0.070	0.073	0.094	0.098	0.10	250
300	0.041	0.051	0.044	0.049	0.045	0.071	0.076	0.072	0.059	0.063	0.065	0.082	0.086	0.088	300
350	0.040	0.050	0.038	0.043	0.039	0.061	0.066	0.063	0.053	0.058	0.060	0.073	0.077	0.080	350
400	0.040	0.049	0.033	0.038	0.035	0.054	0.059	0.055	0.049	0.053	0.056	0.066	0.071	0.073	400
500	0.039	0.048	0.027	0.032	0.029	0.043	0.048	0.045	0.043	0.048	0.050	0.057	0.061	0.064	500
600	0.039	0.048	0.023	0.028	0.025	0.036	0.041	0.038	0.040	0.044	0.047	0.051	0.055	0.058	600
750	0.038	0.048	0.019	0.024	0.021	0.029	0.034	0.031	0.036	0.040	0.043	0.045	0.049	0.052	750
1000	0.037	0.046	0.015	0.019	0.018	0.023	0.027	0.025	0.032	0.036	0.040	0.039	0.042	0.046	1000

Notes:
1. *These values are based on the following constants:* UL-type RHH wires with Class B stranding, in cradled configuration. Wire conductivities are 100 percent IACS copper and 61 percent IACS aluminum, and aluminum conduit is 45 percent IACS. Capacitive reactance is ignored, since it is negligible at these voltages.
2. These resistance values are valid only at 75°C (167°F) and for the parameters as given, but are representative for 600-volt wire types operating at 60 Hz.

"Effective Z" is defined as R cos(theta) + X sin(theta), where "theta" is the power factor angle of the circuit. Multiplying current by effective impedance gives a good approximation for line-to-neutral voltage drop. Effective impedance values shown in this table are valid only at 0.85 power factor. For another circuit power factor (PF), effective impedance (Ze) can be calculated from R and X_L values given in this table as follows:
$Z_e = R \times PF + X_L \sin[\arccos(PF)]$.

TABLE 2.20 NEC Chapter 9, Table 10: Expansion Characteristics of PVC Nonmetallic Conduit

Table 10. Expansion Characteristics of PVC Rigid Nonmetallic Conduit Coefficient of Thermal Expansion = 3.38 × 10⁻⁵ In./In./°F

Temperature Change in Degrees F	Length Change in Inches per 100 ft. of PVC Conduit	Temperature Change in Degrees F	Length Change in Inches per 100 ft. of PVC Conduit	Temperature Change in Degrees F	Length Change in Inches per 100 ft. of PVC Conduit	Temperature Change in Degrees F	Length Change in Inches per 100 ft. of PVC Conduit
5	0.2	55	2.2	105	4.2	155	6.3
10	0.4	60	2.4	110	4.5	160	6.5
15	0.6	65	2.6	115	4.7	165	6.7
20	0.8	70	2.8	120	4.9	170	6.9
25	1.0	75	3.0	125	5.1	175	7.1
30	1.2	80	3.2	130	5.3	180	7.3
35	1.4	85	3.4	135	5.5	185	7.5
40	1.6	90	3.6	140	5.7	190	7.7
45	1.8	95	3.8	145	5.9	195	7.9
50	2.0	100	4.1	150	6.1	200	8.1

Notes to Tables

Note 1: See NEC Appendix C for the maximum number of conductors and fixture wires, all of the same size (total cross-sectional area including insulation), permitted in trade sizes of the applicable conduit or tubing.

Note 2: NEC Chapter 9, Table 1 applies only to complete conduit or tubing systems and is not intended to apply to sections of conduit or tubing used to protect exposed wiring from physical damage.

Note 3: Equipment grounding or bonding conductors, where installed, shall be included when calculating conduit or tubing fill. The actual dimensions of the equipment-grounding or -bonding conductor (insulated or bare) shall be used in the calculation.

Note 4: Where conduit or tubing nipples, having a maximum length not to exceed 24 in. (610 mm), are installed between boxes, cabinets, and similar enclosures, the nipples shall be permitted to be filled to 60 percent of their total cross-sectional area, and Article 310, Note 8(a) of Notes to Ampacity Tables of 0 to 2000 Volts, need not apply to this condition.

Note 5: For conductors not included in NEC Chapter 9, such as multiconductor cables, the actual dimensions shall be used.

Note 6: For combinations of conductors of different sizes, use Tables 5 and 5A in NEC Chapter 9 for dimensions of conductors and Table 4 in Chapter 9 for the applicable conduit or tubing dimensions.

Note 7: When calculating the maximum number of conductors permitted in a conduit or tubing, all of the same size (total cross-sectional area including insulation), the next higher whole number shall be used to determine the maximum number of conductors permitted when the calculation results in a decimal of 0.8 or larger.

Note 8: Where bare conductors are permitted by other sections of this *Code,* the dimensions for bare conductors in Table 8 of NEC Chapter 9 shall be permitted.

Note 9: A multiconductor cable of two or more conductors shall be treated as a single conductor for calculating percentage conduit fill area. For cables that have elliptical cross-sections, the cross-sectional area calculation shall be based on using the major diameter of the ellipse as a circle diameter.

Note 10: When pulling three conductors or cables into a raceway, if the ratio of the raceway (inside diameter) to the conductor or cable (outside diameter) is between 2.8 and 3.2, jamming can occur and the next larger size raceway should be used. Although jamming can occur when pulling four or more conductors or cables into a raceway, the probability is very low.

2.6 NEC APPENDIX C (PARTIAL)

Introduction

This appendix is not a part of the requirements of the NEC and is included for information only. However, by using the tables in this appendix, one is afforded very accurate calculations without having to perform the calculations according to NEC Chapter 9, Table 1.

Tables 2.21 through 2.32 (NEC Tables C1 through C12), inclusive, are included. NEC Tables C1A through C12A are not included here, because they cover fill for compact conductors, which are rarely used in the building industry. If you need these fill requirements, please refer to Appendix C of the NEC.

TABLE 2.21   NEC Appendix C, Table C1: Maximum Number of Conductors and Fixture Wires in Electric Metallic Tubing (Based on Table 1, Chapter 9)

Type Letters	Conductor Size AWG/kcmil	Trade Sizes in Inches									
		½	¾	1	1¼	1½	2	2½	3	3½	4
RH	14	6	10	16	28	39	64	112	169	221	282
	12	4	8	13	23	31	51	90	136	177	227
RHH, RHW,	14	4	7	11	20	27	46	80	120	157	201
RHW-2	12	3	6	9	17	23	38	66	100	131	167
RH, RHH,	10	2	5	8	13	18	30	53	81	105	135
RHW,	8	1	2	4	7	9	16	28	42	55	70
RHW-2	6	1	1	3	5	8	13	22	34	44	56
	4	1	1	2	4	6	10	17	26	34	44
	3	1	1	1	4	5	9	15	23	30	38
	2	1	1	1	3	4	7	13	20	26	33
	1	0	1	1	1	3	5	9	13	17	22
	1/0	0	1	1	1	2	4	7	11	15	19
	2/0	0	1	1	1	2	4	6	10	13	17
	3/0	0	0	1	1	1	3	5	8	11	14
	4/0	0	0	1	1	1	3	5	7	9	12
	250	0	0	0	1	1	1	3	5	7	9
	300	0	0	0	1	1	1	3	5	6	8
	350	0	0	0	1	1	1	3	4	6	7
	400	0	0	0	1	1	1	2	4	5	7
	500	0	0	0	0	1	1	2	3	4	6
	600	0	0	0	0	1	1	1	3	4	5
	700	0	0	0	0	0	1	1	2	3	4
	750	0	0	0	0	0	1	1	2	3	4
	800	0	0	0	0	0	1	1	2	3	4
	900	0	0	0	0	0	1	1	1	3	3
	1000	0	0	0	0	0	1	1	1	2	3
	1250	0	0	0	0	0	0	1	1	1	2
	1500	0	0	0	0	0	0	1	1	1	1
	1750	0	0	0	0	0	0	1	1	1	1
	2000	0	0	0	0	0	0	1	1	1	1

TABLE 2.21 NEC Appendix C, Table C1: Maximum Number of Conductors and Fixture Wires in Electric Metallic Tubing (Based on Table 1, Chapter 9) (*Continued*)

Type Letters	Conductor Size AWG/kcmil	Trade Sizes in Inches									
		½	¾	1	1¼	1½	2	2½	3	3½	4
TW	14	8	15	25	43	58	96	168	254	332	424
	12	6	11	19	33	45	74	129	195	255	326
	10	5	8	14	24	33	55	96	145	190	243
	8	2	5	8	13	18	30	53	81	105	135
RHH*, RHW*, RHW-2*, THHW, THW, THW-2	14	6	10	16	28	39	64	112	169	221	282
RHH*, RHW*, RHW-2*, THHW, THW	12	4	8	13	23	31	51	90	136	177	227
	10	3	6	10	18	24	40	70	106	138	177
RHH*, RHW*, RHW-2*, THHW, THW, THW-2	8	1	4	6	10	14	24	42	63	83	106
RHH*, RHW*, RHW-2*, TW, THW, THHW, THW-2	6	1	3	4	8	11	18	32	48	63	81
	4	1	1	3	6	8	13	24	36	47	60
	3	1	1	3	5	7	12	20	31	40	52
	2	1	1	2	4	6	10	17	26	34	44
	1	1	1	1	3	4	7	12	18	24	31
	1/0	0	1	1	2	3	6	10	16	20	26
	2/0	0	1	1	1	3	5	9	13	17	22
	3/0	0	1	1	1	2	4	7	11	15	19
	4/0	0	0	1	1	1	3	6	9	12	16
	250	0	0	1	1	1	3	5	7	10	13
	300	0	0	1	1	1	2	4	6	8	11
	350	0	0	0	1	1	1	4	6	7	10
	400	0	0	0	1	1	1	3	5	7	9
	500	0	0	0	1	1	1	3	4	6	7
	600	0	0	0	1	1	1	2	3	4	6
	700	0	0	0	0	1	1	1	3	4	5
	750	0	0	0	0	1	1	1	3	4	5
	800	0	0	0	0	1	1	1	3	3	5
	900	0	0	0	0	0	1	1	2	3	4
	1000	0	0	0	0	0	1	1	2	3	4
	1250	0	0	0	0	0	1	1	1	2	3
	1500	0	0	0	0	0	1	1	1	1	2
	1750	0	0	0	0	0	0	1	1	1	2
	2000	0	0	0	0	0	0	1	1	1	1

*Types RHH, RHW, and RHW-2 without outer covering.

TABLE 2.21 NEC Appendix C, Table C1: Maximum Number of Conductors and Fixture Wires in Electric Metallic Tubing (Based on Table 1, Chapter 9) (*Continued*)

Type Letters	Conductor Size AWG/kcmil	Trade Sizes in Inches									
		½	¾	1	1¼	1½	2	2½	3	3½	4
THHN,	14	12	22	35	61	84	138	241	364	476	608
THWN,	12	9	16	26	45	61	101	176	266	347	443
THWN-2	10	5	10	16	28	38	63	111	167	219	279
	8	3	6	9	16	22	36	64	96	126	161
	6	2	4	7	12	16	26	46	69	91	116
	4	1	2	4	7	10	16	28	43	56	71
	3	1	1	3	6	8	13	24	36	47	60
	2	1	1	3	5	7	11	20	30	40	51
	1	1	1	1	4	5	8	15	22	29	37
	1/0	1	1	1	3	4	7	12	19	25	32
	2/0	0	1	1	2	3	6	10	16	20	26
	3/0	0	1	1	1	3	5	8	13	17	22
	4/0	0	1	1	1	2	4	7	11	14	18
	250	0	0	1	1	1	3	6	9	11	15
	300	0	0	1	1	1	3	5	7	10	13
	350	0	0	1	1	1	2	4	6	9	11
	400	0	0	0	1	1	1	4	6	8	10
	500	0	0	0	1	1	1	3	5	6	8
	600	0	0	0	1	1	1	2	4	5	7
	700	0	0	0	1	1	1	2	3	4	6
	750	0	0	0	0	1	1	1	3	4	5
	800	0	0	0	0	1	1	1	3	4	5
	900	0	0	0	0	1	1	1	3	3	4
	1000	0	0	0	0	1	1	1	2	3	4
FEP, FEPB,	14	12	21	34	60	81	134	234	354	462	590
PFA, PFAH,	12	9	15	25	43	59	98	171	258	337	430
TFE	10	6	11	18	31	42	70	122	185	241	309
	8	3	6	10	18	24	40	70	106	138	177
	6	2	4	7	12	17	28	50	75	98	126
	4	1	3	5	9	12	20	35	53	69	88
	3	1	2	4	7	10	16	29	44	57	73
	2	1	1	3	6	8	13	24	36	47	60
PFA, PFAH, TFE	1	1	1	2	4	6	9	16	25	33	42
PFA, PFAH,	1/0	1	1	1	3	5	8	14	21	27	35
TFE, Z	2/0	0	1	1	3	4	6	11	17	22	29
	3/0	0	1	1	2	3	5	9	14	18	24
	4/0	0	1	1	1	2	4	8	11	15	19

TABLE 2.21   NEC Appendix C, Table C1: Maximum Number of Conductors and Fixture Wires in Electric Metallic Tubing (Based on Table 1, Chapter 9) (*Continued*)

Type Letters	Conductor Size AWG/kcmil	Trade Sizes in Inches									
		½	¾	1	1¼	1½	2	2½	3	3½	4
Z	14	14	25	41	72	98	161	282	426	556	711
	12	10	18	29	51	69	114	200	302	394	504
	10	6	11	18	31	42	70	122	185	241	309
	8	4	7	11	20	27	44	77	117	153	195
	6	3	5	8	14	19	31	54	82	107	137
	4	1	3	5	9	13	21	37	56	74	94
	3	1	2	4	7	9	15	27	41	54	69
	2	1	1	3	6	8	13	22	34	45	57
	1	1	1	2	4	6	10	18	28	36	46
XHH,	14	8	15	25	43	58	96	168	254	332	424
XHHW,	12	6	11	19	33	45	74	129	195	255	326
XHHW-2,	10	5	8	14	24	33	55	96	145	190	243
ZW	8	2	5	8	13	18	30	53	81	105	135
	6	1	3	6	10	14	22	39	60	78	100
	4	1	2	4	7	10	16	28	43	56	72
	3	1	1	3	6	8	14	24	36	48	61
	2	1	1	3	5	7	11	20	31	40	51
XHH, XHHW,	1	1	1	1	4	5	8	15	23	30	38
XHHW-2	1/0	1	1	1	3	4	7	13	19	25	32
	2/0	0	1	1	2	3	6	10	16	21	27
	3/0	0	1	1	1	3	5	9	13	17	22
	4/0	0	1	1	1	2	4	7	11	14	18
	250	0	0	1	1	1	3	6	9	12	15
	300	0	0	1	1	1	3	5	8	10	13
	350	0	0	1	1	1	2	4	7	9	11
	400	0	0	0	1	1	1	4	6	8	10
	500	0	0	0	1	1	1	3	5	6	8
	600	0	0	0	1	1	1	2	4	5	6
	700	0	0	0	0	1	1	2	3	4	6
	750	0	0	0	0	1	1	1	3	4	5
	800	0	0	0	0	1	1	1	3	4	5
	900	0	0	0	0	1	1	1	3	3	4
	1000	0	0	0	0	0	1	1	2	3	4
	1250	0	0	0	0	0	1	1	1	2	3
	1500	0	0	0	0	0	1	1	1	1	3
	1750	0	0	0	0	0	0	1	1	1	2
	2000	0	0	0	0	0	0	1	1	1	1

TABLE 2.21 NEC Appendix C, Table C1: Maximum Number of Conductors and Fixture Wires in Electric Metallic Tubing (Based on Table 1, Chapter 9) (*Continued*)

	Fixture Wires						
Type Letters	Conductor Size AWG/kcmil	Trade Sizes in Inches					
		½	¾	1	1¼	1½	2
FFH-2, RFH-2, RFHH-3	18	8	14	24	41	56	92
	16	7	12	20	34	47	78
SF-2, SFF-2	18	10	18	30	52	71	116
	16	8	15	25	43	58	96
	14	7	12	20	34	47	78
SF-1, SFF-1	18	18	33	53	92	125	206
AF, RFH-1, RFHH-2, TF, TFF, XF, XFF	18	14	24	39	68	92	152
AF, RFHH-2, TF, TFF, XF, XFF	16	11	19	31	55	74	123
AF, XF, XFF	14	8	15	25	43	58	96
TFN, TFFN	18	22	38	63	108	148	244
	16	17	29	48	83	113	186
PF, PFF, PGF, PGFF, PAF, PTF, PTFF, PAFF	18	21	36	59	103	140	231
	16	16	28	46	79	108	179
	14	12	21	34	60	81	134
ZF, ZFF, ZHF, HF, HFF	18	27	47	77	133	181	298
	16	20	35	56	98	133	220
	14	14	25	41	72	98	161
KF-2, KFF-2	18	39	69	111	193	262	433
	16	27	48	78	136	185	305
	14	19	33	54	93	127	209
	12	13	23	37	64	87	144
	10	8	15	25	43	58	96
KF-1, KFF-1	18	46	82	133	230	313	516
	16	33	57	93	161	220	362
	14	22	38	63	108	148	244
	12	14	25	41	72	98	161
	10	9	16	27	47	64	105
AX, XF, XFF	12	4	8	13	23	31	51
	10	3	6	10	18	24	40

Note: This table is for concentric stranded conductors only. For compact stranded conductors, Table C1A should be used.

TABLE 2.22 NEC Appendix C, Table C2: Maximum Number of Conductors and Fixture Wires in Electrical Nonmetallic Tubing (Based on Table 1, Chapter 9)

Type Letters	Conductor Size AWG/kcmil	Trade Size in Inches					
		1/2	3/4	1	1 1/4	1 1/2	2
RH	14	4	8	15	27	37	61
	12	3	7	12	21	29	49
RHH, RHW,	14	3	6	10	19	26	43
RHW-2	12	2	5	9	16	22	36
RH, RHH,	10	1	4	7	13	17	29
RHW,	8	1	1	3	6	9	15
RHW-2	6	1	1	3	5	7	12
	4	1	1	2	4	6	9
	3	1	1	1	3	5	8
	2	0	1	1	3	4	7
	1	0	1	1	1	3	5
	1/0	0	0	1	1	2	4
	2/0	0	0	1	1	1	3
	3/0	0	0	1	1	1	3
	4/0	0	0	1	1	1	2
	250	0	0	0	1	1	1
	300	0	0	0	1	1	1
	350	0	0	0	1	1	1
	400	0	0	0	1	1	1
	500	0	0	0	0	1	1
	600	0	0	0	0	1	1
	700	0	0	0	0	0	1
	750	0	0	0	0	0	1
	800	0	0	0	0	0	1
	900	0	0	0	0	0	1
	1000	0	0	0	0	0	1
	1250	0	0	0	0	0	0
	1500	0	0	0	0	0	0
	1750	0	0	0	0	0	0
	2000	0	0	0	0	0	0
TW	14	7	13	22	40	55	92
	12	5	10	17	31	42	71
	10	4	7	13	23	32	52
	8	1	4	7	13	17	29

TABLE 2.22 NEC Appendix C, Table C2: Maximum Number of Conductors and Fixture Wires in Electrical Nonmetallic Tubing (Based on Table 1, Chapter 9) (*Continued*)

Type Letters	Conductor Size AWG/kcmil	Trade Size in Inches					
		½	¾	1	1¼	1½	2
RHH*, RHW*, RHW-2*, THHW, THW, THW-2	14	4	8	15	27	37	61
RHH*, RHW*, RHW-2*, THHW, THW	12	3	7	12	21	29	49
	10	3	5	9	17	23	38
RHH*, RHW*, RHW-2*, THHW, THW, THW-2	8	1	3	5	10	14	23
RHH*, RHW*, RHW-2*, TW, THW, THHW, THW-2	6	1	2	4	7	10	17
	4	1	1	3	5	8	13
	3	1	1	2	5	7	11
	2	1	1	2	4	6	9
	1	0	1	1	3	4	6
	1/0	0	1	1	2	3	5
	2/0	0	1	1	1	3	5
	3/0	0	0	1	1	2	4
	4/0	0	0	1	1	1	3
	250	0	0	1	1	1	2
	300	0	0	0	1	1	2
	350	0	0	0	1	1	1
	400	0	0	0	1	1	1
	500	0	0	0	1	1	1
	600	0	0	0	0	1	1
	700	0	0	0	0	1	1
	750	0	0	0	0	1	1
	800	0	0	0	0	1	1
	900	0	0	0	0	0	1
	1000	0	0	0	0	0	1
	1250	0	0	0	0	0	1
	1500	0	0	0	0	0	0
	1750	0	0	0	0	0	0
	2000	0	0	0	0	0	0

*Types RHH, RHW, and RHW-2 without outer covering.

TABLE 2.22 NEC Appendix C, Table C2: Maximum Number of Conductors and Fixture Wires in Electrical Nonmetallic Tubing (Based on Table 1, Chapter 9) (*Continued*)

Type Letters	Conductor Size AWG/kcmil	Trade Sizes in Inches					
		½	¾	1	1¼	1½	2
THHN,	14	10	18	32	58	80	132
THWN,	12	7	13	23	42	58	96
THWN-2	10	4	8	15	26	36	60
	8	2	5	8	15	21	35
	6	1	3	6	11	15	25
	4	1	1	4	7	9	15
	3	1	1	3	5	8	13
	2	1	1	2	5	6	11
	1	1	1	1	3	5	8
	1/0	0	1	1	3	4	7
	2/0	0	1	1	2	3	5
	3/0	0	1	1	1	3	4
	4/0	0	0	1	1	2	4
	250	0	0	1	1	1	3
	300	0	0	1	1	1	2
	350	0	0	0	1	1	2
	400	0	0	0	1	1	1
	500	0	0	0	1	1	1
	600	0	0	0	1	1	1
	700	0	0	0	0	1	1
	750	0	0	0	0	1	1
	800	0	0	0	0	1	1
	900	0	0	0	0	1	1
	1000	0	0	0	0	0	1
FEP, FEPB,	14	10	18	31	56	77	128
PFA, PFAH,	12	7	13	23	41	56	93
TFE	10	5	9	16	29	40	67
	8	3	5	9	17	23	38
	6	1	4	6	12	16	27
	4	1	2	4	8	11	19
	3	1	1	4	7	9	16
	2	1	1	3	5	8	13
PFA, PFAH, TFE	1	1	1	1	4	5	9

TABLE 2.22 NEC Appendix C, Table C2: Maximum Number of Conductors and Fixture Wires in Electrical Nonmetallic Tubing (Based on Table 1, Chapter 9) (*Continued*)

Type Letters	Conductor Size AWG/kcmil	Trade Size in Inches					
		½	¾	1	1¼	1½	2
PFA, PFAH,	1/0	0	1	1	3	4	7
TFE, Z	2/0	0	1	1	2	4	6
	3/0	0	1	1	1	3	5
	4/0	0	1	1	1	2	4
Z	14	12	22	38	68	93	154
	12	8	15	27	48	66	109
	10	5	9	16	29	40	67
	8	3	6	10	18	25	42
	6	1	4	7	13	18	30
	4	1	3	5	9	12	20
	3	1	1	3	6	9	15
	2	1	1	3	5	7	12
	1	1	1	2	4	6	10
XHH,	14	7	13	22	40	55	92
XHHW,	12	5	10	17	31	42	71
XHHW-2,	10	4	7	13	23	32	52
ZW	8	1	4	7	13	17	29
	6	1	3	5	9	13	21
	4	1	1	4	7	9	15
	3	1	1	3	6	8	13
	2	1	1	2	5	6	11
XHH, XHHW,	1	1	1	1	3	5	8
XHHW-2	1/0	0	1	1	3	4	7
	2/0	0	1	1	2	3	6
	3/0	0	1	1	1	3	5
	4/0	0	0	1	1	2	4
	250	0	0	1	1	1	3
	300	0	0	1	1	1	3
	350	0	0	1	1	1	2
	400	0	0	0	1	1	1
	500	0	0	0	1	1	1
	600	0	0	0	1	1	1
	700	0	0	0	0	1	1
	750	0	0	0	0	1	1
	800	0	0	0	0	1	1
	900	0	0	0	0	1	1
	1000	0	0	0	0	0	1
	1250	0	0	0	0	0	1
	1500	0	0	0	0	0	1
	1750	0	0	0	0	0	0
	2000	0	0	0	0	0	0

TABLE 2.22   NEC Appendix C, Table C2: Maximum Number of Conductors and Fixture Wires in Electrical Nonmetallic Tubing (Based on Table 1, Chapter 9) (*Continued*)

	Fixture Wires						
Type Letters	Conductor Size AWG/kcmil	Trade Size in Inches					
		1/2	3/4	1	1 1/4	1 1/2	2
FFH-2, RFH-2, RFHH-3	18	6	12	21	39	53	88
	16	5	10	18	32	45	74
SF-2, SFF-2	18	8	15	27	49	67	111
	16	7	13	22	40	55	92
	14	5	10	18	32	45	74
SF-1, SFF-1	18	15	28	48	86	119	197
AF, RFH-1, RFHH-2, TF,TFF, XF, XFF	18	11	20	35	64	88	145
AF, RFHH-2, TF,TFF, XF, XFF	16	9	16	29	51	71	117
AF, XF, XFF	14	7	13	22	40	55	92
TFN, TFFN	18	18	33	57	102	141	233
	16	13	25	43	78	107	178
PF, PFF, PGF, PGFF, PAF, PTF, PTFF, PAFF	18	17	31	54	97	133	221
	16	13	24	42	75	103	171
	14	10	18	31	56	77	128
ZF, ZFF, ZHF, HF, HFF	18	22	40	70	125	172	285
	16	16	29	51	92	127	210
	14	12	22	38	68	93	154
KF-2, KFF-2	18	31	58	101	182	250	413
	16	22	41	71	128	176	291
	14	15	28	49	88	121	200
	12	10	19	33	60	83	138
	10	7	13	22	40	55	92
KF-1, KFF-1	18	38	69	121	217	298	493
	16	26	49	85	152	209	346
	14	18	33	57	102	141	233
	12	12	22	38	68	93	154
	10	7	14	24	44	61	101
AF, XF, XFF	12	3	7	12	21	29	49
	10	3	5	9	17	23	38

Note: This table is for concentric stranded conductors only. For compact stranded conductors, Table C2A should be used.

TABLE 2.23   NEC Appendix C, Table C3: Maximum Number of Conductors and Fixture Wires in Flexible Metallic Conduit (Based on Table 1, Chapter 9)

Type Letters	Conductor Size AWG/kcmil	Trade Sizes in Inches									
		$\frac{1}{2}$	$\frac{3}{4}$	1	$1\frac{1}{4}$	$1\frac{1}{2}$	2	$2\frac{1}{2}$	3	$3\frac{1}{2}$	4
RH	14	6	10	15	24	35	62	94	135	184	240
	12	5	8	12	19	28	50	75	108	148	193
RHH, RHW,	14	4	7	11	17	25	44	67	96	131	171
RHW-2	12	3	6	9	14	21	37	55	80	109	142
RH, RHH,	10	3	5	7	11	17	30	45	64	88	115
RHW,	8	1	2	4	6	9	15	23	34	46	60
RHW-2	6	1	1	3	5	7	12	19	27	37	48
	4	1	1	2	4	5	10	14	21	29	37
	3	1	1	1	3	5	8	13	18	25	33
	2	1	1	1	3	4	7	11	16	22	28
	1	0	1	1	1	2	5	7	10	14	19
	1/0	0	1	1	1	2	4	6	9	12	16
	2/0	0	1	1	1	1	3	5	8	11	14
	3/0	0	0	1	1	1	3	5	7	9	12
	4/0	0	0	1	1	1	2	4	6	8	10
	250	0	0	0	1	1	1	3	4	6	8
	300	0	0	0	1	1	1	2	4	5	7
	350	0	0	0	1	1	1	2	3	5	6
	400	0	0	0	0	1	1	1	3	4	6
	500	0	0	0	0	1	1	1	3	4	5
	600	0	0	0	0	1	1	1	2	3	4
	700	0	0	0	0	0	1	1	1	3	3
	750	0	0	0	0	0	1	1	1	2	3
	800	0	0	0	0	0	1	1	1	2	3
	900	0	0	0	0	0	1	1	1	2	3
	1000	0	0	0	0	0	1	1	1	1	3
	1250	0	0	0	0	0	0	1	1	1	1
	1500	0	0	0	0	0	0	1	1	1	1
	1750	0	0	0	0	0	0	1	1	1	1
	2000	0	0	0	0	0	0	0	1	1	1
TW	14	9	15	23	36	53	94	141	203	277	361
	12	7	11	18	28	41	72	108	156	212	277
	10	5	8	13	21	30	54	81	116	158	207
	8	3	5	7	11	17	30	45	64	88	115

TABLE 2.23   NEC Appendix C, Table C3: Maximum Number of Conductors and Fixture Wires in Flexible Metallic Conduit (Based on Table 1, Chapter 9) (*Continued*)

Type Letters	Conductor Size AWG/kcmil	Trade Sizes in Inches									
		1/2	3/4	1	1 1/4	1 1/2	2	2 1/2	3	3 1/2	4
RHH*, RHW*, RHW-2*, THHW, THW, THW-2	14	6	10	15	24	35	62	94	135	184	240
RHH*, RHW*, RHW-2*, THHW, THW	12	5	8	12	19	28	50	75	108	148	193
	10	4	6	10	15	22	39	59	85	115	151
RHH*, RHW*, RHW-2*, THHW, THW, THW-2	8	1	4	6	9	13	23	35	51	69	90
RHH*, RHW*, RHW-2*, TW, THW, THHW, THW-2	6	1	3	4	7	10	18	27	39	53	69
	4	1	1	3	5	7	13	20	29	39	51
	3	1	1	3	4	6	11	17	25	34	44
	2	1	1	2	4	5	10	14	21	29	37
	1	1	1	1	2	4	7	10	15	20	26
	1/0	0	1	1	1	3	6	9	12	17	22
	2/0	0	1	1	1	3	5	7	10	14	19
	3/0	0	1	1	1	2	4	6	9	12	16
	4/0	0	0	1	1	1	3	5	7	10	13
	250	0	0	1	1	1	3	4	6	8	11
	300	0	0	1	1	1	2	3	5	7	9
	350	0	0	0	1	1	1	3	4	6	8
	400	0	0	0	1	1	1	3	4	6	7
	500	0	0	0	1	1	1	2	3	5	6
	600	0	0	0	0	1	1	1	3	4	5
	700	0	0	0	0	1	1	1	2	3	4
	750	0	0	0	0	1	1	1	2	3	4
	800	0	0	0	0	1	1	1	1	3	4
	900	0	0	0	0	0	1	1	1	3	3
	1000	0	0	0	0	0	1	1	1	2	3
	1250	0	0	0	0	0	1	1	1	1	2
	1500	0	0	0	0	0	0	1	1	1	1
	1750	0	0	0	0	0	0	1	1	1	1
	2000	0	0	0	0	0	0	1	1	1	1

*Types RHH, RHW, and RHW-2 without outer covering.

TABLE 2.23 NEC Appendix C, Table C3: Maximum Number of Conductors and Fixture Wires in Flexible Metallic Conduit (Based on Table 1, Chapter 9) (*Continued*)

Type Letters	Conductor Size AWG/kcmil	Trade Sizes in Inches									
		½	¾	1	1¼	1½	2	2½	3	3½	4
THHN,	14	13	22	33	52	76	134	202	291	396	518
THWN,	12	9	16	24	38	56	98	147	212	289	378
THWN-2	10	6	10	15	24	35	62	93	134	182	238
	8	3	6	9	14	20	35	53	77	105	137
	6	2	4	6	10	14	25	38	55	76	99
	4	1	2	4	6	9	16	24	34	46	61
	3	1	1	3	5	7	13	20	29	39	51
	2	1	1	3	4	6	11	17	24	33	43
	1	1	1	1	3	4	8	12	18	24	32
	1/0	1	1	1	2	4	7	10	15	20	27
	2/0	0	1	1	1	3	6	9	12	17	22
	3/0	0	1	1	1	2	5	7	10	14	18
	4/0	0	1	1	1	1	4	6	8	12	15
	250	0	0	1	1	1	3	5	7	9	12
	300	0	0	1	1	1	3	4	6	8	11
	350	0	0	1	1	1	2	3	5	7	9
	400	0	0	0	1	1	1	3	5	6	8
	500	0	0	0	1	1	1	2	4	5	7
	600	0	0	0	0	1	1	1	3	4	5
	700	0	0	0	0	1	1	1	3	4	5
	750	0	0	0	0	1	1	1	2	3	4
	800	0	0	0	0	1	1	1	2	3	4
	900	0	0	0	0	0	1	1	1	3	4
	1000	0	0	0	0	0	1	1	1	3	3
FEP, FEPB,	14	12	21	32	51	74	130	196	282	385	502
PFA, PFAH,	12	9	15	24	37	54	95	143	206	281	367
TFE	10	6	11	17	26	39	68	103	148	201	263
	8	4	6	10	15	22	39	59	85	115	151
	6	2	4	7	11	16	28	42	60	82	107
	4	1	3	5	7	11	19	29	42	57	75
	3	1	2	4	6	9	16	24	35	48	62
	2	1	1	3	5	7	13	20	29	39	51
PFA, PFAH, TFE	1	1	1	2	3	5	9	14	20	27	36

TABLE 2.23 NEC Appendix C, Table C3: Maximum Number of Conductors and Fixture Wires in Flexible Metallic Conduit (Based on Table 1, Chapter 9) (*Continued*)

Type Letters	Conductor Size AWG/kcmil	Trade Sizes in Inches									
		½	¾	1	1¼	1½	2	2½	3	3½	4
PFA, PFAH,	1/0	1	1	1	3	4	8	11	17	23	30
TFE, Z	2/0	1	1	1	2	3	6	9	14	19	24
	3/0	0	1	1	1	3	5	8	11	15	20
	4/0	0	1	1	1	2	4	6	9	13	16
Z	14	15	25	39	61	89	157	236	340	463	605
	12	11	18	28	43	63	111	168	241	329	429
	10	6	11	17	26	39	68	103	148	201	263
	8	4	7	11	17	24	43	65	93	127	166
	6	3	5	7	12	17	30	45	65	89	117
	4	1	3	5	8	12	21	31	45	61	80
	3	1	2	4	6	8	15	23	33	45	58
	2	1	1	3	5	7	12	19	27	37	49
	1	1	1	2	4	6	10	15	22	30	39
XHH,	14	9	15	23	36	53	94	141	203	277	361
XHHW,	12	7	11	18	28	41	72	108	156	212	277
XHHW-2,	10	5	8	13	21	30	54	81	116	158	207
ZW	8	3	5	7	11	17	30	45	64	88	115
	6	1	3	5	8	12	22	33	48	65	85
	4	1	2	4	6	9	16	24	34	47	61
	3	1	1	3	5	7	13	20	29	40	52
	2	1	1	3	4	6	11	17	24	33	44
XHH, XHHW,	1	1	1	1	3	5	8	13	18	25	32
XHHW-2	1/0	1	1	1	2	4	7	10	15	21	27
	2/0	0	1	1	2	3	6	9	13	17	23
	3/0	0	1	1	1	3	5	7	10	14	19
	4/0	0	1	1	1	2	4	6	9	12	15
	250	0	0	1	1	1	3	5	7	10	13
	300	0	0	1	1	1	3	4	6	8	11
	350	0	0	1	1	1	2	4	5	7	9
	400	0	0	0	1	1	1	3	5	6	8
	500	0	0	0	1	1	1	3	4	5	7
	600	0	0	0	0	1	1	1	3	4	5
	700	0	0	0	0	1	1	1	3	4	5
	750	0	0	0	0	1	1	1	2	3	4
	800	0	0	0	0	1	1	1	2	3	4
	900	0	0	0	0	0	1	1	1	3	4
	1000	0	0	0	0	0	1	1	1	3	3
	1250	0	0	0	0	0	1	1	1	1	3
	1500	0	0	0	0	0	1	1	1	1	2
	1750	0	0	0	0	0	0	1	1	1	1
	2000	0	0	0	0	0	0	1	1	1	1

TABLE 2.23   NEC Appendix C, Table C3: Maximum Number of Conductors and Fixture Wires in Flexible Metallic Conduit (Based on Table 1, Chapter 9) (*Continued*)

	Fixture Wires					
Type Letters	Conductor Size AWG/kcmil	Trade Sizes in Inches				
		½	¾	1	1¼	1½	2
FFH-2, RFH-2,	18	8	14	22	35	51	90
RFHH-3,	16	7	12	19	29	43	76
SF-2, SFF-2	18	11	18	28	44	64	113
	16	9	15	23	36	53	94
	14	7	12	19	29	43	76
SF-1, SFF-1	18	19	32	50	78	114	201
AF,RFH-1, RFHH-2, TF, TFF, XF, XFF	18	14	24	37	58	84	148
AF, RFHH-2, TF, TFF, XF, XFF	16	11	19	30	47	68	120
AF, XF, XFF	14	9	15	23	36	53	94
TFN, TFFN	18	23	38	59	93	135	237
	16	17	29	45	71	103	181
PF, PFF,	18	22	36	56	88	128	225
PGF, PGFF,	16	17	28	43	68	99	174
PAF, PTF, PTFF, PAFF	14	12	21	32	51	74	130
ZF, ZFF,	18	28	47	72	113	165	290
ZHF, HF,	16	20	35	53	83	121	214
HFF	14	15	25	39	61	89	157
KF-2, KFF-2	18	41	68	105	164	239	421
	16	28	48	74	116	168	297
	14	19	33	51	80	116	204
	12	13	23	35	55	80	140
	10	9	15	23	36	53	94
KF-1, KFF-1	18	48	82	125	196	285	503
	16	34	57	88	138	200	353
	14	23	38	59	93	135	237
	12	15	25	39	61	89	157
	10	10	16	25	40	58	103
AF, XF, XFF	12	5	8	12	19	28	50
	10	4	6	10	15	22	39

Note: This table is for concentric stranded conductors only. For compact stranded conductors, Table C3A should be used.

TABLE 2.24 NEC Appendix C, Table C4: Maximum Number of Conductors and Fixture Wires in Intermediate Metallic Conduit (Based on Table 1, Chapter 9)

Type Letters	Conductor Size AWG/kcmil	Trade Sizes in Inches									
		½	¾	1	1¼	1½	2	2½	3	3½	4
RH	14	6	11	18	31	42	69	98	151	202	261
	12	5	9	14	25	34	56	79	122	163	209
RHH, RHW,	14	4	8	13	22	30	49	70	108	144	186
RHW-2	12	4	6	11	18	25	41	58	89	120	154
RH, RHH,	10	3	5	8	15	20	33	47	72	97	124
RHW,	8	1	3	4	8	10	17	24	38	50	65
RHW-2	6	1	1	3	6	8	14	19	30	40	52
	4	1	1	3	5	6	11	15	23	31	41
	3	1	1	2	4	6	9	13	21	28	36
	2	1	1	1	3	5	8	11	18	24	31
	1	0	1	1	2	3	5	7	12	16	20
	1/0	0	1	1	1	3	4	6	10	14	18
	2/0	0	1	1	1	2	4	6	9	12	15
	3/0	0	0	1	1	1	3	5	7	10	13
	4/0	0	0	1	1	1	3	4	6	9	11
	250	0	0	1	1	1	1	3	5	6	8
	300	0	0	0	1	1	1	3	4	6	7
	350	0	0	0	1	1	1	2	4	5	7
	400	0	0	0	1	1	1	2	3	5	6
	500	0	0	0	1	1	1	1	3	4	5
	600	0	0	0	0	1	1	1	2	3	4
	700	0	0	0	0	1	1	1	2	3	4
	750	0	0	0	0	1	1	1	1	3	4
	800	0	0	0	0	0	1	1	1	3	3
	900	0	0	0	0	0	1	1	1	2	3
	1000	0	0	0	0	0	1	1	1	2	3
	1250	0	0	0	0	0	1	1	1	1	2
	1500	0	0	0	0	0	0	1	1	1	1
	1750	0	0	0	0	0	0	1	1	1	1
	2000	0	0	0	0	0	0	1	1	1	1
TW	14	10	17	27	47	64	104	147	228	304	392
	12	7	13	21	36	49	80	113	175	234	301
	10	5	9	15	27	36	59	84	130	174	224
	8	3	5	8	15	20	33	47	72	97	124

TABLE 2.24 NEC Appendix C, Table C4: Maximum Number of Conductors and Fixture Wires in Intermediate Metallic Conduit (Based on Table 1, Chapter 9) (*Continued*)

Type Letters	Conductor Size AWG/kcmil	Trade Sizes in Inches									
		½	¾	1	1¼	1½	2	2½	3	3½	4
RHH*, RHW*, RHW-2*, THHW, THW, THW-2	14	6	11	18	31	42	69	98	151	202	261
RHH*, RHW*, RHW-2*, THHW, THW	12	5	9	14	25	34	56	79	122	163	209
	10	4	7	11	19	26	43	61	95	127	163
RHH*, RHW*, RHW-2*, THHW, THW, THW-2	8	2	4	7	12	16	26	37	57	76	98
RHH*, RHW*, RHW-2*, TW, THW, THHW, THW-2	6	1	3	5	9	12	20	28	43	58	75
	4	1	2	4	6	9	15	21	32	43	56
	3	1	1	3	6	8	13	18	28	37	48
	2	1	1	3	5	6	11	15	23	31	41
	1	1	1	1	3	4	7	11	16	22	28
	1/0	1	1	1	3	4	6	9	14	19	24
	2/0	0	1	1	2	3	5	8	12	16	20
	3/0	0	1	1	1	3	4	6	10	13	17
	4/0	0	1	1	1	2	4	5	8	11	14
	250	0	0	1	1	1	3	4	7	9	12
	300	0	0	1	1	1	2	4	6	8	10
	350	0	0	1	1	1	2	3	5	7	9
	400	0	0	0	1	1	1	3	4	6	8
	500	0	0	0	1	1	1	2	4	5	7
	600	0	0	0	1	1	1	1	3	4	5
	700	0	0	0	0	1	1	1	3	4	5
	750	0	0	0	0	1	1	1	2	3	4
	800	0	0	0	0	1	1	1	2	3	4
	900	0	0	0	0	1	1	1	2	3	4
	1000	0	0	0	0	0	1	1	1	3	3
	1250	0	0	0	0	0	1	1	1	1	3
	1500	0	0	0	0	0	1	1	1	1	2
	1750	0	0	0	0	0	0	1	1	1	1
	2000	0	0	0	0	0	0	1	1	1	1

*Types RHH, RHW, and RHW-2 without outer covering.

TABLE 2.24   NEC Appendix C, Table C4: Maximum Number of Conductors and Fixture Wires in Intermediate Metallic Conduit (Based on Table 1, Chapter 9) (*Continued*)

Type Letters	Conductor Size AWG/kcmil	Trade Sizes in Inches									
		½	¾	1	1¼	1½	2	2½	3	3½	4
THHN,	14	14	24	39	68	91	149	211	326	436	562
THWN,	12	10	17	29	49	67	109	154	238	318	410
THWN-2	10	6	11	18	31	42	68	97	150	200	258
	8	3	6	10	18	24	39	56	86	115	149
	6	2	4	7	13	17	28	40	62	83	107
	4	1	3	4	8	10	17	25	38	51	66
	3	1	2	4	6	9	15	21	32	43	56
	2	1	1	3	5	7	12	17	27	36	47
	1	1	1	2	4	5	9	13	20	27	35
	1/0	1	1	1	3	4	8	11	17	23	29
	2/0	1	1	1	3	4	6	9	14	19	24
	3/0	0	1	1	2	3	5	7	12	16	20
	4/0	0	1	1	1	2	4	6	9	13	17
	250	0	0	1	1	1	3	5	8	10	13
	300	0	0	1	1	1	3	4	7	9	12
	350	0	0	1	1	1	2	4	6	8	10
	400	0	0	1	1	1	2	3	5	7	9
	500	0	0	0	1	1	1	3	4	6	7
	600	0	0	0	1	1	1	2	3	5	6
	700	0	0	0	1	1	1	1	3	4	5
	750	0	0	0	1	1	1	1	3	4	5
	800	0	0	0	0	1	1	1	3	4	5
	900	0	0	0	0	1	1	1	2	3	4
	1000	0	0	0	0	1	1	1	2	3	4
FEP, FEPB,	14	13	23	38	66	89	145	205	317	423	545
PFA, PFAH,	12	10	17	28	48	65	106	150	231	309	398
TFE	10	7	12	20	34	46	76	107	166	221	285
	8	4	7	11	19	26	43	61	95	127	163
	6	3	5	8	14	19	31	44	67	90	116
	4	1	3	5	10	13	21	30	47	63	81
	3	1	3	4	8	11	18	25	39	52	68
	2	1	2	4	6	9	15	21	32	43	56
PFA, PFAH, TFE	1	1	1	2	4	6	10	14	22	30	39

TABLE 2.24 NEC Appendix C, Table C4: Maximum Number of Conductors and Fixture Wires in Intermediate Metallic Conduit (Based on Table 1, Chapter 9) (*Continued*)

Type Letters	Conductor Size AWG/kcmil	Trade Sizes in Inches									
		½	¾	1	1¼	1½	2	2½	3	3½	4
PFA, PFAH,	1/0	1	1	1	4	5	8	12	19	25	32
TFE, Z	2/0	1	1	1	3	4	7	10	15	21	27
	3/0	0	1	1	2	3	6	8	13	17	22
	4/0	0	1	1	1	3	5	7	10	14	18
Z	14	16	28	46	79	107	175	247	381	510	657
	12	11	20	32	56	76	124	175	271	362	466
	10	7	12	20	34	46	76	107	166	221	285
	8	4	7	12	21	29	48	68	105	140	180
	6	3	5	9	15	20	33	47	73	98	127
	4	1	3	6	10	14	23	33	50	67	87
	3	1	2	4	7	10	17	24	37	49	63
	2	1	1	3	6	8	14	20	30	41	53
	1	1	1	3	5	7	11	16	25	33	43
XHH,	14	10	17	27	47	64	104	147	228	304	392
XHHW,	12	7	13	21	36	49	80	113	175	234	301
XHHW-2,	10	5	9	15	27	36	59	84	130	174	224
ZW	8	3	5	8	15	20	33	47	72	97	124
	6	1	4	6	11	15	24	35	53	71	92
	4	1	3	4	8	11	18	25	39	52	67
	3	1	2	4	7	9	15	21	33	44	56
	2	1	1	3	5	7	12	18	27	37	47
XHH, XHHW,	1	1	1	2	4	5	9	13	20	27	35
XHHW-2	1/0	1	1	1	3	5	8	11	17	23	30
	2/0	1	1	1	3	4	6	9	14	19	25
	3/0	0	1	1	2	3	5	7	12	16	20
	4/0	0	1	1	1	2	4	6	10	13	17
	250	0	0	1	1	1	3	5	8	11	14
	300	0	0	1	1	1	3	4	7	9	12
	350	0	0	1	1	1	3	4	6	8	10
	400	0	0	1	1	1	2	3	5	7	9
	500	0	0	0	1	1	1	3	4	6	8
	600	0	0	0	1	1	1	2	3	5	6
	700	0	0	0	1	1	1	1	3	4	5
	750	0	0	0	1	1	1	1	3	4	5
	800	0	0	0	0	1	1	1	3	4	5
	900	0	0	0	0	1	1	1	2	3	4
	1000	0	0	0	0	1	1	1	2	3	4
	1250	0	0	0	0	0	1	1	1	2	3
	1500	0	0	0	0	0	1	1	1	1	2
	1750	0	0	0	0	0	1	1	1	1	2
	2000	0	0	0	0	0	0	1	1	1	1

TABLE 2.24 NEC Appendix C, Table C4: Maximum Number of Conductors and Fixture Wires in Intermediate Metallic Conduit (Based on Table 1, Chapter 9) (*Continued*)

	Fixture Wires						
Type Letters	**Conductor Size AWG/kcmil**	**Trade Sizes in Inches**					
		½	**¾**	**1**	**1¼**	**1½**	**2**
FHH-2, RFH-2,	18	9	16	26	45	61	100
RFHH-3	16	8	13	22	38	51	84
SF-2, SFF-2	18	12	20	33	57	77	126
	16	10	17	27	47	64	104
	14	8	13	22	38	51	84
SF-1, SFF-1	18	21	36	59	101	137	223
AF, RFH-1, RFHH-2, TF, TFF, XF, XFF	18	15	26	43	75	101	165
AF, RFH-2, TF, TFF, XF, XFF	16	12	21	35	60	81	133
AF, XF, XFF	14	10	17	27	47	64	104
TFN, TFFN	18	25	42	69	119	161	264
	16	19	32	53	91	123	201
PF, PFF, PGF, PGFF, PAF, PTF, PTFF, PAFF	18	23	40	66	113	153	250
	16	18	31	51	87	118	193
	14	13	23	38	66	89	145
ZF, ZFF, ZHF, HF, HFF	18	30	52	85	146	197	322
	16	22	38	63	108	145	238
	14	16	28	46	79	107	175
KF-2, KFF-2	18	44	75	123	212	287	468
	16	31	53	87	149	202	330
	14	21	36	60	103	139	227
	12	14	25	41	70	95	156
	10	10	17	27	47	64	104
KF-1, KFF-1	18	52	90	147	253	342	558
	16	37	63	103	178	240	392
	14	25	42	69	119	161	264
	12	16	28	46	79	107	175
	10	10	18	30	52	70	114
AX, XF, XFF	12	5	9	14	25	34	56
	10	4	7	11	19	26	43

Note: This table is for concentric stranded conductors only. For compact stranded conductors, Table C4A should be used.

TABLE 2.25 NEC Appendix C, Table C5: Maximum Number of Conductors and Fixture Wires in Liquidtight Flexible Nonmetallic Conduit (Type FNMC-B) (Based on Table 1, Chapter 9)

Type Letters	Conductor Size AWG/kcmil	Trade Sizes in Inches						
		⅜	½	¾	1	1¼	1½	2
RH	14	3	6	10	16	29	38	62
	12	3	5	8	13	23	30	50
RHH, RHW,	14	2	4	7	12	21	27	44
RHW-2	12	1	3	6	10	17	22	36
RH, RHH,	10	1	3	5	8	14	18	29
RHW,	8	1	1	2	4	7	9	15
RHW-2	6	1	1	1	3	6	7	12
	4	0	1	1	2	4	6	9
	3	0	1	1	1	4	5	8
	2	0	1	1	1	3	4	7
	1	0	0	1	1	1	3	5
	1/0	0	0	1	1	1	2	4
	2/0	0	0	1	1	1	1	3
	3/0	0	0	0	1	1	1	3
	4/0	0	0	0	1	1	1	2
	250	0	0	0	0	1	1	1
	300	0	0	0	0	1	1	1
	350	0	0	0	0	1	1	1
	400	0	0	0	0	1	1	1
	500	0	0	0	0	1	1	1
	600	0	0	0	0	0	1	1
	700	0	0	0	0	0	0	1
	750	0	0	0	0	0	0	1
	800	0	0	0	0	0	0	1
	900	0	0	0	0	0	0	1
	1000	0	0	0	0	0	0	1
	1250	0	0	0	0	0	0	0
	1500	0	0	0	0	0	0	0
	1750	0	0	0	0	0	0	0
	2000	0	0	0	0	0	0	0

TABLE 2.25   NEC Appendix C, Table C5: Maximum Number of Conductors and Fixture Wires in Liquidtight Flexible Nonmetallic Conduit (Type FNMC-B) (Based on Table 1, Chapter 9) (*Continued*)

Type Letters	Conductor Size AWG/kcmil	Trade Sizes in Inches						
		⅜	½	¾	1	1¼	1½	2
TW	14	5	9	15	25	44	57	93
	12	4	7	12	19	33	43	71
	10	3	5	9	14	25	32	53
	8	1	3	5	8	14	18	29
RHH*, RHW*, RHW-2*, THHW, THW, THW-2	14	3	6	10	16	29	38	62
RHH*, RHW*, RHW-2*, THHW, THW	12	3	5	8	13	23	30	50
	10	1	3	6	10	18	23	39
RHH*, RHW*, RHW-2*, THHW, THW, THW-2	8	1	1	4	6	11	14	23
RHH*, RHW*, RHW-2*, TW, THW, THHW, THW-2	6	1	1	3	5	8	11	18
	4	1	1	1	3	6	8	13
	3	1	1	1	3	5	7	11
	2	0	1	1	2	4	6	9
	1	0	1	1	1	3	4	7
	1/0	0	0	1	1	2	3	6
	2/0	0	0	1	1	2	3	5
	3/0	0	0	1	1	1	2	4
	4/0	0	0	0	1	1	1	3
	250	0	0	0	1	1	1	3
	300	0	0	0	1	1	1	2
	350	0	0	0	0	1	1	1
	400	0	0	0	0	1	1	1
	500	0	0	0	0	1	1	1
	600	0	0	0	0	1	1	1
	700	0	0	0	0	0	1	1
	750	0	0	0	0	0	1	1
	800	0	0	0	0	0	1	1
	900	0	0	0	0	0	0	1
	1000	0	0	0	0	0	0	1
	1250	0	0	0	0	0	0	1
	1500	0	0	0	0	0	0	0
	1750	0	0	0	0	0	0	0
	2000	0	0	0	0	0	0	0

*Types RHH, RHW, and RHW-2 without outer covering.
**Corresponds to Section 351-22(2).

TABLE 2.25　NEC Appendix C, Table C5: Maximum Number of Conductors and Fixture Wires in Liquidtight Flexible Nonmetallic Conduit (Type FNMC-B) (Based on Table 1, Chapter 9) (*Continued*)

Types Letters	Conductor Size AWG/kcmil	Trade Sizes in Inches						
		⅜	½	¾	1	1¼	1½	2
THHN,	14	8	13	22	36	63	81	133
THWN,	12	5	9	16	26	46	59	97
THWN-2	10	3	6	10	16	29	37	61
	8	1	3	6	9	16	21	35
	6	1	2	4	7	12	15	25
	4	1	1	2	4	7	9	15
	3	1	1	1	3	6	8	13
	2	1	1	1	3	5	7	11
	1	0	1	1	1	4	5	8
	1/0	0	1	1	1	3	4	7
	2/0	0	0	1	1	2	3	6
	3/0	0	0	1	1	1	3	5
	4/0	0	0	1	1	1	2	4
	250	0	0	0	1	1	1	3
	300	0	0	0	1	1	1	3
	350	0	0	0	1	1	1	2
	400	0	0	0	0	1	1	1
	500	0	0	0	0	1	1	1
	600	0	0	0	0	1	1	1
	700	0	0	0	0	1	1	1
	750	0	0	0	0	0	1	1
	800	0	0	0	0	0	1	1
	900	0	0	0	0	0	1	1
	1000	0	0	0	0	0	0	1
FEP, FEPB,	14	7	12	21	35	61	79	129
PFA, PFAH,	12	5	9	15	25	44	57	94
TFE	10	4	6	11	18	32	41	68
	8	1	3	6	10	18	23	39
	6	1	2	4	7	13	17	27
	4	1	1	3	5	9	12	19
	3	1	1	2	4	7	10	16
	2	1	1	1	3	6	8	13
PFA, PFAH, TFE	1	0	1	1	2	4	5	9

TABLE 2.25   NEC Appendix C, Table C5: Maximum Number of Conductors and Fixture Wires in Liquidtight Flexible Nonmetallic Conduit (Type FNMC-B) (Based on Table 1, Chapter 9) (*Continued*)

Type Letters	Conductor Size AWG/kcmil	Trade Sizes in Inches						
		⅜	½	¾	1	1¼	1½	2
PFA, PFAH,	1/0	0	1	1	1	3	4	7
TFE, Z	2/0	0	1	1	1	3	4	6
	3/0	0	0	1	1	2	3	5
	4/0	0	0	1	1	1	2	4
Z	14	9	15	26	42	73	95	156
	12	6	10	18	30	52	67	111
	10	4	6	11	18	32	41	68
	8	2	4	7	11	20	26	43
	6	1	3	5	8	14	18	30
	4	1	1	3	5	9	12	20
	3	1	1	2	4	7	9	15
	2	0	1	1	3	6	7	12
	1	0	1	1	2	5	6	10
XHH,	14	5	9	15	25	44	57	93
XHHW,	12	4	7	12	19	33	43	71
XHHW-2,	10	3	5	9	14	25	32	53
ZW	8	1	3	5	8	14	18	29
	6	1	1	3	6	10	13	22
	4	1	1	2	4	7	9	16
	3	1	1	1	3	6	8	13
	2	1	1	1	3	5	7	11
XHH, XHHW,	1	0	1	1	1	4	5	8
XHHW-2	1/0	0	1	1	1	3	4	7
	2/0	0	0	1	1	2	3	6
	3/0	0	0	1	1	1	3	5
	4/0	0	0	1	1	1	2	4
	250	0	0	0	1	1	1	3
	300	0	0	0	1	1	1	3
	350	0	0	0	1	1	1	2
	400	0	0	0	0	1	1	1
	500	0	0	0	0	1	1	1
	600	0	0	0	0	1	1	1
	700	0	0	0	0	1	1	1
	750	0	0	0	0	0	1	1
	800	0	0	0	0	0	1	1
	900	0	0	0	0	0	1	1
	1000	0	0	0	0	0	0	1
	1250	0	0	0	0	0	0	1
	1500	0	0	0	0	0	0	1
	1750	0	0	0	0	0	0	0
	2000	0	0	0	0	0	0	0

**Corresponds to Section 351-22(2).

TABLE 2.25 NEC Appendix C, Table C5: Maximum Number of Conductors and Fixture Wires in Liquidtight Flexible Nonmetallic Conduit (Type FNMC-B) (Based on Table 1, Chapter 9) (*Continued*)

		Fixtures Wires						
Type Letters	Conductor Size AWG/kcmil	Trade Sizes in Inches						
		³⁄₈	¹⁄₂	³⁄₄	1	1¹⁄₄	1¹⁄₂	2
FFH-2, RFH-2	18	5	8	15	24	42	54	89
	16	4	7	12	20	35	46	75
SF-2, SFF-2	18	6	11	19	30	53	69	113
	16	5	9	15	25	44	57	93
	14	4	7	12	20	35	46	75
SF-1, SFF-1	18	11	19	33	53	94	122	199
AF, RFH-1, RFHH-2, TF, TFF, XF, XFF	18	8	14	24	39	69	90	147
AF, RFHH-2, TF, TFF, XF, XFF	16	7	11	20	32	56	72	119
AF, XF, XFF	14	5	9	15	25	44	57	93
TFN, TFFN	18	14	23	39	63	111	144	236
	16	10	17	30	48	85	110	180
PF, PFF, PGF, PGFF, PAF, PTF, PTFF, PAFF	18	13	21	37	60	105	136	223
	16	10	16	29	46	81	105	173
	14	7	12	21	35	61	79	129
HF, HFF, ZF, ZFF, ZHF	18	17	28	48	77	136	176	288
	16	12	20	35	57	100	129	212
	14	9	15	26	42	73	95	156
KF-2, KFF-2	18	24	40	70	112	197	255	418
	16	17	28	49	79	139	180	295
	14	12	19	34	54	95	123	202
	12	8	13	23	37	65	85	139
	10	5	9	15	25	44	57	93
KF-1, KFF-1	18	29	48	83	134	235	304	499
	16	20	34	58	94	165	214	350
	!4	14	23	39	63	111	144	236
	12	9	15	26	42	73	95	156
	10	6	10	17	27	48	62	102
AF, XF, XFF	12	3	5	8	13	23	30	50
	10	1	3	6	10	18	23	39

**Corresponds to Section 351-22(2).
Note: This table is for concentric stranded conductors only. For compact stranded conductors, Table C5A should be used.

TABLE 2.26   NEC Appendix C, Table C6: Maximum Number of Conductors and Fixture Wires in Liquidtight Flexible Nonmetallic Conduit (Type FNMC-A) (Based on Table 1, Chapter 9)

Type Letters	Conductor Size AWG/kcmil	Trade Sizes in Inches						
		⅜	½	¾	1	1¼	1½	2
RH	14	3	6	10	16	28	38	64
	12	3	4	8	13	23	31	51
RHH, RHW,	14	2	4	7	11	20	27	45
RHW-2	12	1	3	6	9	17	23	38
RH, RHH,	10	1	3	5	8	13	18	30
RHW,	8	1	1	2	4	7	9	16
RHW-2	6	1	1	1	3	5	7	13
	4	0	1	1	2	4	6	10
	3	0	1	1	1	4	5	8
	2	0	1	1	1	3	4	7
	1	0	0	1	1	1	3	5
	1/0	0	0	1	1	1	2	4
	2/0	0	0	1	1	1	1	4
	3/0	0	0	0	1	1	1	3
	4/0	0	0	0	1	1	1	3
	250	0	0	0	0	1	1	1
	300	0	0	0	0	1	1	1
	350	0	0	0	0	1	1	1
	400	0	0	0	0	1	1	1
	500	0	0	0	0	0	1	1
	600	0	0	0	0	0	1	1
	700	0	0	0	0	0	0	1
	750	0	0	0	0	0	0	1
	800	0	0	0	0	0	0	1
	900	0	0	0	0	0	0	1
	1000	0	0	0	0	0	0	1
	1250	0	0	0	0	0	0	0
	1500	0	0	0	0	0	0	0
	1750	0	0	0	0	0	0	0
	2000	0	0	0	0	0	0	0

TABLE 2.26 NEC Appendix C, Table C6: Maximum Number of Conductors and Fixture Wires in Liquidtight Flexible Nonmetallic Conduit (Type FNMC-A) (Based on Table 1, Chapter 9) (*Continued*)

Type Letters	Conductor Size AWG/kcmil	Trade Sizes in Inches						
		³⁄₈	½	¾	1	1¼	1½	2
TW	14	5	9	15	24	43	58	96
	12	4	7	12	19	33	44	74
	10	3	5	9	14	24	33	55
	8	1	3	5	8	13	18	30
RHH*, RHW*, RHW-2*, THHW, THW, THW-2	14	3	6	10	16	28	38	64
RHH*, RHW*, RHW-2*, THHW, THW	12	3	4	8	13	23	31	51
	10	1	3	6	10	18	24	40
RHH*, RHW*, RHW-2*, THHW, THW, THW-2	8	1	1	4	6	10	14	24
RHH*, RHW*, RHW-2*, TW, THW, THHW, THW-2	6	1	1	3	4	8	11	18
	4	1	1	1	3	6	8	13
	3	1	1	1	3	5	7	11
	2	0	1	1	2	4	6	10
	1	0	1	1	1	3	4	7
	1/0	0	0	1	1	2	3	6
	2/0	0	0	1	1	1	3	5
	3/0	0	0	1	1	1	2	4
	4/0	0	0	0	1	1	1	3
	250	0	0	0	1	1	1	3
	300	0	0	0	1	1	1	2
	350	0	0	0	0	1	1	1
	400	0	0	0	0	1	1	1
	500	0	0	0	0	1	1	1
	600	0	0	0	0	1	1	1
	700	0	0	0	0	0	1	1
	750	0	0	0	0	0	1	1
	800	0	0	0	0	0	1	1
	900	0	0	0	0	0	0	1
	1000	0	0	0	0	0	0	1
	1250	0	0	0	0	0	0	1
	1500	0	0	0	0	0	0	1
	1750	0	0	0	0	0	0	0
	2000	0	0	0	0	0	0	0

*Types RHH, RHW, and RHW-2 without outer covering.
** Corresponds to Section 351-22(1).

TABLE 2.26   NEC Appendix C, Table C6: Maximum Number of Conductors and Fixture Wires in Liquidtight Flexible Nonmetallic Conduit (Type FNMC-A) (Based on Table 1, Chapter 9) (*Continued*)

Type Letters	Conductor Size AWG/kcmil	Trade Sizes in Inches						
		³⁄₈	¹⁄₂	³⁄₄	1	1¹⁄₄	1¹⁄₂	2
THHN,	14	8	13	22	35	62	83	137
THWN,	12	5	9	16	25	45	60	100
THWN-2	10	3	6	10	16	28	38	63
	8	1	3	6	9	16	22	36
	6	1	2	4	6	12	16	26
	4	1	1	2	4	7	9	16
	3	1	1	1	3	6	8	13
	2	1	1	1	3	5	7	11
	1	0	1	1	1	4	5	8
	1/0	0	1	1	1	3	4	7
	2/0	0	0	1	1	2	3	6
	3/0	0	0	1	1	1	3	5
	4/0	0	0	1	1	1	2	4
	250	0	0	0	1	1	1	3
	300	0	0	0	1	1	1	3
	350	0	0	0	1	1	1	2
	400	0	0	0	0	1	1	1
	500	0	0	0	0	1	1	1
	600	0	0	0	0	1	1	1
	700	0	0	0	0	1	1	1
	750	0	0	0	0	0	1	1
	800	0	0	0	0	0	1	1
	900	0	0	0	0	0	1	1
	1000	0	0	0	0	0	0	1
FEP, FEPB,	14	7	12	21	34	60	80	133
PFA, PFAH,	12	5	9	15	25	44	59	97
TFE	10	4	6	11	18	31	42	70
	8	1	3	6	10	18	24	40
	6	1	2	4	7	13	17	28
	4	1	1	3	5	9	12	20
	3	1	1	2	4	7	10	16
	2	1	1	1	3	6	8	13
PFA, PFAH, TFE	1	0	1	1	2	4	5	9

TABLE 2.26 NEC Appendix C, Table C6: Maximum Number of Conductors and Fixture Wires in Liquidtight Flexible Nonmetallic Conduit (Type FNMC-A) (Based on Table 1, Chapter 9) (*Continued*)

Type Letters	Conductor Size AWG/kcmil	Trade Sizes in Inches						
		⅜	½	¾	1	1¼	1½	2
PFA, PFAH,	1/0	0	1	1	1	3	5	8
TFE, Z	2/0	0	1	1	1	3	4	6
	3/0	0	0	1	1	2	3	5
	4/0	0	0	1	1	1	2	4
Z	14	9	15	25	41	72	97	161
	12	6	10	18	29	51	69	114
	10	4	6	11	18	31	42	70
	8	2	4	7	11	20	26	44
	6	1	3	5	8	14	18	31
	4	1	1	3	5	9	13	21
	3	1	1	2	4	7	9	15
	2	1	1	1	3	6	8	13
	1	1	1	1	2	4	6	10
XHH,	14	5	9	15	24	43	58	96
XHHW,	12	4	7	12	19	33	44	74
XHHW-2,	10	3	5	9	14	24	33	55
ZW	8	1	3	5	8	13	18	30
	6	1	1	3	5	10	13	22
	4	1	1	2	4	7	10	16
	3	1	1	1	3	6	8	14
	2	1	1	1	3	5	7	11
XHH, XHHW,	1	0	1	1	1	4	5	8
XHHW-2	1/0	0	1	1	1	3	4	7
	2/0	0	0	1	1	2	3	6
	3/0	0	0	1	1	1	3	5
	4/0	0	0	1	1	1	2	4
	250	0	0	0	1	1	1	3
	300	0	0	0	1	1	1	3
	350	0	0	0	1	1	1	2
	400	0	0	0	0	1	1	1
	500	0	0	0	0	1	1	1
	600	0	0	0	0	1	1	1
	700	0	0	0	0	1	1	1
	750	0	0	0	0	0	1	1
	800	0	0	0	0	0	1	1
	900	0	0	0	0	0	1	1
	1000	0	0	0	0	0	0	1
	1250	0	0	0	0	0	0	1
	1500	0	0	0	0	0	0	1
	1750	0	0	0	0	0	0	0
	2000	0	0	0	0	0	0	0

**Corresponds to Section 351-22(1).

TABLE 2.26   NEC Appendix C, Table C6: Maximum Number of Conductors and Fixture Wires in Liquidtight Flexible Nonmetallic Conduit (Type FNMC-A) (Based on Table 1, Chapter 9) (*Continued*)

Type Letters	Conductor Size AWG/kcmil	Trade Sizes in Inches						
		⅜	½	¾	1	1¼	1½	2
FFH-2, RFH-2,	18	5	8	14	23	41	55	92
RFHH-3	16	4	7	12	20	35	47	77
SF-2, SFF-2	18	6	11	18	29	52	70	116
	16	5	9	15	24	43	58	96
	14	4	7	12	20	35	47	77
SF-1, SFF-1	18	12	19	33	52	92	124	205
AF, RFH-1, RFHH-2, TF, TFF, XF, XFF	18	8	14	24	39	68	91	152
AF, RFHH-2, TF, TFF, XF, XFF	16	7	11	19	31	55	74	122
AF, XF, XFF	14	5	9	15	24	43	58	96
TFN, TFFN	18	14	22	39	62	109	146	243
	16	10	17	29	47	83	112	185
PF, PFF, PGF, PGFF, PAF, PTF, PTFF, PAFF	18	13	21	37	59	103	139	230
	16	10	16	28	45	80	107	178
	14	7	12	21	34	60	80	133
HF, HFF, ZF, ZFF, ZHF	18	17	27	47	76	133	179	297
	16	12	20	35	56	98	132	219
	14	9	15	25	41	72	97	161
KF-2, KFF-2	18	25	40	69	110	193	260	431
	16	17	28	48	77	136	183	303
	14	12	19	33	53	94	126	209
	12	8	13	23	36	64	86	143
	10	5	9	15	24	43	58	96
KF-1, KFF-1	18	29	48	82	131	231	310	514
	16	21	33	57	92	162	218	361
	14	14	22	39	62	109	146	243
	12	9	15	25	41	72	97	161
	10	6	10	17	27	47	63	105
AF, XF, XFF	12	3	4	8	13	23	31	51
	10	1	3	6	10	18	24	40

**Corresponds to Section 351-22(1).
Note: This table is for concentric stranded conductors only. For compact stranded conductors, Table C6A should be used.

TABLE 2.27 NEC Appendix C, Table C7: Maximum Number of Conductors and Fixture Wires in Liquidtight Flexible Metallic Conduit (Based on Table 1, Chapter 9)

Type Letters	Conductor Size AWG/kcmil	Trade Sizes in Inches									
		1/2	3/4	1	1 1/4	1 1/2	2	2 1/2	3	3 1/2	4
RH	14	6	10	16	29	38	62	93	143	186	243
	12	5	8	13	23	30	50	75	115	149	195
RHH, RHW,	14	4	7	12	21	27	44	66	102	133	173
RHW-2	12	3	6	10	17	22	36	55	84	110	144
RH, RHH,	10	3	5	8	14	18	29	44	68	89	116
RHW,	8	1	2	4	7	9	15	23	36	46	61
RHW-2	6	1	1	3	6	7	12	18	28	37	48
	4	1	1	2	4	6	9	14	22	29	38
	3	1	1	1	4	5	8	13	19	25	33
	2	1	1	1	3	4	7	11	17	22	29
	1	0	1	1	1	3	5	7	11	14	19
	1/0	0	1	1	1	2	4	6	10	13	16
	2/0	0	1	1	1	1	3	5	8	11	14
	3/0	0	0	1	1	1	3	4	7	9	12
	4/0	0	0	1	1	1	2	4	6	8	10
	250	0	0	0	1	1	1	3	4	6	8
	300	0	0	0	1	1	1	2	4	5	7
	350	0	0	0	1	1	1	2	3	5	6
	400	0	0	0	1	1	1	1	3	4	6
	500	0	0	0	1	1	1	1	3	4	5
	600	0	0	0	0	1	1	1	2	3	4
	700	0	0	0	0	0	1	1	1	3	3
	750	0	0	0	0	0	1	1	1	2	3
	800	0	0	0	0	0	1	1	1	2	3
	900	0	0	0	0	0	1	1	1	2	3
	1000	0	0	0	0	0	1	1	1	1	3
	1250	0	0	0	0	0	0	1	1	1	1
	1500	0	0	0	0	0	0	1	1	1	1
	1750	0	0	0	0	0	0	1	1	1	1
	2000	0	0	0	0	0	0	0	1	1	1

TABLE 2.27 NEC Appendix C, Table C7: Maximum Number of Conductors and Fixture Wires in Liquidtight Flexible Metallic Conduit (Based on Table 1, Chapter 9) (*Continued*)

Type Letters	Conductor Size AWG/kcmil	Trade Sizes in Inches									
		1/2	3/4	1	1 1/4	1 1/2	2	2 1/2	3	3 1/2	4
TW	14	9	15	25	44	57	93	140	215	280	365
	12	7	12	19	33	43	71	108	165	215	280
	10	5	9	14	25	32	53	80	123	160	209
	8	3	5	8	14	18	29	44	68	89	116
RHH*, RHW*, RHW-2*, THHW, THW, THW-2	14	6	10	16	29	38	62	93	143	186	243
RHH*, RHW*, RHW-2*, THHW, THW	12	5	8	13	23	30	50	75	115	149	195
	10	3	6	10	18	23	39	58	89	117	152
RHH*, RHW*, RHW-2*, THHW, THW, THW-2	8	1	4	6	11	14	23	35	53	70	91
RHH*, RHW*, RHW-2*, TW, THW, THHW, THW-2	6	1	3	5	8	11	18	27	41	53	70
	4	1	1	3	6	8	13	20	30	40	52
	3	1	1	3	5	7	11	17	26	34	44
	2	1	1	2	4	6	9	14	22	29	38
	1	1	1	1	3	4	7	10	15	20	26
	1/0	0	1	1	2	3	6	8	13	17	23
	2/0	0	1	1	2	3	5	7	11	15	19
	3/0	0	1	1	1	2	4	6	9	12	16
	4/0	0	0	1	1	1	3	5	8	10	13
	250	0	0	1	1	1	3	4	6	8	11
	300	0	0	1	1	1	2	3	5	7	9
	350	0	0	0	1	1	1	3	5	6	8
	400	0	0	0	1	1	1	3	4	6	7
	500	0	0	0	1	1	1	2	3	5	6
	600	0	0	0	1	1	1	1	3	4	5
	700	0	0	0	0	1	1	1	2	3	4
	750	0	0	0	0	1	1	1	2	3	4
	800	0	0	0	0	1	1	1	2	3	4
	900	0	0	0	0	0	1	1	1	3	3
	1000	0	0	0	0	0	1	1	1	2	3
	1250	0	0	0	0	0	1	1	1	1	2
	1500	0	0	0	0	0	0	1	1	1	2
	1750	0	0	0	0	0	0	1	1	1	1
	2000	0	0	0	0	0	0	1	1	1	1

*Types RHH, RHW, and RHW-2 without outer covering.

TABLE 2.27 NEC Appendix C, Table C7: Maximum Number of Conductors and Fixture Wires in Liquidtight Flexible Metallic Conduit (Based on Table 1, Chapter 9) (*Continued*)

Type Letters	Conductor Size AWG/kcmil	Trade Sizes in Inches									
		$\frac{1}{2}$	$\frac{3}{4}$	1	$1\frac{1}{4}$	$1\frac{1}{2}$	2	$2\frac{1}{2}$	3	$3\frac{1}{2}$	4
THHN,	14	13	22	36	63	81	133	201	308	401	523
THWN,	12	9	16	26	46	59	97	146	225	292	381
THWN-2	10	6	10	16	29	37	61	92	141	184	240
	8	3	6	9	16	21	35	53	81	106	138
	6	2	4	7	12	15	25	38	59	76	100
	4	1	2	4	7	9	15	23	36	47	61
	3	1	1	3	6	8	13	20	30	40	52
	2	1	1	3	5	7	11	17	26	33	44
	1	1	1	1	4	5	8	12	19	25	32
	1/0	1	1	1	3	4	7	10	16	21	27
	2/0	0	1	1	2	3	6	8	13	17	23
	3/0	0	1	1	1	3	5	7	11	14	19
	4/0	0	1	1	1	2	4	6	9	12	15
	250	0	0	1	1	1	3	5	7	10	12
	300	0	0	1	1	1	3	4	6	8	11
	350	0	0	1	1	1	2	3	5	7	9
	400	0	0	0	1	1	1	3	5	6	8
	500	0	0	0	1	1	1	2	4	5	7
	600	0	0	0	1	1	1	1	3	4	6
	700	0	0	0	1	1	1	1	3	4	5
	750	0	0	0	0	1	1	1	3	3	5
	800	0	0	0	0	1	1	1	2	3	4
	900	0	0	0	0	1	1	1	2	3	4
	1000	0	0	0	0	0	1	1	1	3	3
FEP, FEPB,	14	12	21	35	61	79	129	195	299	389	507
PFA, PFAH,	12	9	15	25	44	57	94	142	218	284	370
TFE	10	6	11	18	32	41	68	102	156	203	266
	8	3	6	10	18	23	39	58	89	117	152
	6	2	4	7	13	17	27	41	64	83	108
	4	1	3	5	9	12	19	29	44	58	75
	3	1	2	4	7	10	16	24	37	48	63
	2	1	1	3	6	8	13	20	30	40	52
PFA, PFAH, TFE	1	1	1	2	4	5	9	14	21	28	36

TABLE 2.27   NEC Appendix C, Table C7: Maximum Number of Conductors and Fixture Wires in Liquidtight Flexible Metallic Conduit (Based on Table 1, Chapter 9) (*Continued*)

Type Letters	Conductor Size AWG/kcmil	Trade Sizes in Inches									
		1/2	3/4	1	1 1/4	1 1/2	2	2 1/2	3	3 1/2	4
PFA, PFAH,	1/0	1	1	1	3	4	7	11	18	23	30
TFE, Z	2/0	1	1	1	3	4	6	9	14	19	25
	3/0	0	1	1	2	3	5	8	12	16	20
	4/0	0	1	1	1	2	4	6	10	13	17
Z	14	20	26	42	73	95	156	235	360	469	611
	12	14	18	30	52	67	111	167	255	332	434
	10	8	11	18	32	41	68	102	156	203	266
	8	5	7	11	20	26	43	64	99	129	168
	6	4	5	8	14	18	30	45	69	90	118
	4	2	3	5	9	12	20	31	48	62	81
	3	2	2	4	7	9	15	23	35	45	59
	2	1	1	3	6	7	12	19	29	38	49
	1	1	1	2	5	6	10	15	23	30	40
XHH,	14	9	15	25	44	57	93	140	215	280	365
XHHW,	12	7	12	19	33	43	71	108	165	215	280
XHHW-2,	10	5	9	14	25	32	53	80	123	160	209
ZW	8	3	5	8	14	18	29	44	68	89	116
	6	1	3	6	10	13	22	33	50	66	86
	4	1	2	4	7	9	16	24	36	48	62
	3	1	1	3	6	8	13	20	31	40	52
	2	1	1	3	5	7	11	17	26	34	44
XHH, XHHW,	1	1	1	1	4	5	8	12	19	25	33
XHHW-2	1/0	1	1	1	3	4	7	10	16	21	28
	2/0	0	1	1	2	3	6	9	13	17	23
	3/0	0	1	1	1	3	5	7	11	14	19
	4/0	0	1	1	1	2	4	6	9	12	16
	250	0	0	1	1	1	3	5	7	10	13
	300	0	0	1	1	1	3	4	6	8	11
	350	0	0	1	1	1	2	3	5	7	10
	400	0	0	0	1	1	1	3	5	6	8
	500	0	0	0	1	1	1	2	4	5	7
	600	0	0	0	1	1	1	1	3	4	6
	700	0	0	0	1	1	1	1	3	4	5
	750	0	0	0	0	1	1	1	3	3	5
	800	0	0	0	0	1	1	1	2	3	4
	900	0	0	0	0	1	1	1	2	3	4
	1000	0	0	0	0	0	1	1	1	3	3
	1250	0	0	0	0	0	1	1	1	1	3
	1500	0	0	0	0	0	1	1	1	1	2
	1750	0	0	0	0	0	0	1	1	1	2
	2000	0	0	0	0	0	0	1	1	1	2

TABLE 2.27 NEC Appendix C, Table C7: Maximum Number of Conductors and Fixture Wires in Liquidtight Flexible Metallic Conduit (Based on Table 1, Chapter 9) (*Continued*)

				Fixture Wires			
Type Letters	Conductor Size AWG/kcmil	Trade Sizes in Inches					
		1/2	3/4	1	1 1/4	1 1/2	2
FFH-2, RFH-2,	18	8	15	24	42	54	89
RFHH-3	16	7	12	20	35	46	75
SF-2, SFF-2	18	11	19	30	53	69	113
	16	9	15	25	44	57	93
	14	7	12	20	35	46	75
SF-1, SFF-1	18	19	33	53	94	122	199
AF, RFH-1, RFHH-2, TF, TFF, XF, XFF	18	14	24	39	69	90	147
AF, RFHH-2, TF, TFF, XF, XFF	16	11	20	32	56	72	119
AF, XF, XFF	14	9	15	25	44	57	93
TFN, TFFN	18	23	39	63	111	144	236
	16	17	30	48	85	110	180
PF, PFF, PGF, PGFF, PAF, PTF, PTFF, PAFF	18	21	37	60	105	136	223
	16	16	29	46	81	105	173
	14	12	21	35	61	79	129
HF, HFF, ZF, ZFF, ZHF	18	28	48	77	136	176	288
	16	20	35	57	100	129	212
	14	15	26	42	73	95	156
KF-2, KFF-2	18	40	70	112	197	255	418
	16	28	49	79	139	180	295
	14	19	34	54	95	123	202
	12	13	23	37	65	85	139
	10	9	15	25	44	57	93
KF-1, KFF-1	18	48	83	134	235	304	499
	16	34	58	94	165	214	350
	14	23	39	63	111	144	236
	12	15	26	42	73	95	156
	10	10	17	27	48	62	102
AF, XF, XFF	12	5	8	13	23	30	50
	10	3	6	10	18	23	39

Note: This table is for concentric stranded conductors only. For compact stranded conductors, Table C7A should be used.

TABLE 2.28   NEC Appendix C, Table C8: Maximum Number of Conductors and Fixture Wires in Rigid Metallic Conduit (Based on Table 1, Chapter 9)

Type Letters	Conductor Size AWG/kcmil	Trade Sizes in Inches											
		½	¾	1	1¼	1½	2	2½	3	3½	4	5	6
RH	14	6	10	17	29	39	65	93	143	191	246	387	558
	12	5	8	13	23	32	52	75	115	154	198	311	448
RHH, RHW, RHW-2	14	4	7	12	21	28	46	66	102	136	176	276	398
	12	3	6	10	17	23	38	55	85	113	146	229	330
RH, RHH, RHW, RHW-2	10	3	5	8	14	19	31	44	68	91	118	185	267
	8	1	2	4	7	10	16	23	36	48	61	97	139
	6	1	1	3	6	8	13	18	29	38	49	77	112
	4	1	1	2	4	6	10	14	22	30	38	60	87
	3	1	1	2	4	5	9	12	19	26	34	53	76
	2	1	1	1	3	4	7	11	17	23	29	46	66
	1	0	1	1	1	3	5	7	11	15	19	30	44
	1/0	0	1	1	1	2	4	6	10	13	17	26	38
	2/0	0	1	1	1	2	4	5	8	11	14	23	33
	3/0	0	0	1	1	1	3	4	7	10	12	20	28
	4/0	0	0	1	1	1	3	4	6	8	11	17	24
	250	0	0	0	1	1	1	3	4	6	8	13	18
	300	0	0	0	1	1	1	2	4	5	7	11	16
	350	0	0	0	1	1	1	2	4	5	6	10	15
	400	0	0	0	1	1	1	1	3	4	6	9	13
	500	0	0	0	1	1	1	1	3	4	5	8	11
	600	0	0	0	0	1	1	1	2	3	4	6	9
	700	0	0	0	0	1	1	1	1	3	4	6	8
	750	0	0	0	0	0	1	1	1	3	3	5	8
	800	0	0	0	0	0	1	1	1	2	3	5	7
	900	0	0	0	0	0	1	1	1	2	3	5	7
	1000	0	0	0	0	0	1	1	1	1	3	4	6
	1250	0	0	0	0	0	0	1	1	1	1	3	5
	1500	0	0	0	0	0	0	1	1	1	1	3	4
	1750	0	0	0	0	0	0	1	1	1	1	2	4
	2000	0	0	0	0	0	0	0	1	1	1	2	3

TABLE 2.28 NEC Appendix C, Table C8: Maximum Number of Conductors and Fixture Wires in Rigid Metallic Conduit (Based on Table 1, Chapter 9) (*Continued*)

Type Letters	Conductor Size AWG/kcmil	Trade Sizes in Inches											
		1/2	3/4	1	1 1/4	1 1/2	2	2 1/2	3	3 1/2	4	5	6
TW	14	9	15	25	44	59	98	140	216	288	370	581	839
	12	7	12	19	33	45	75	107	165	221	284	446	644
	10	5	9	14	25	34	56	80	123	164	212	332	480
	8	3	5	8	14	19	31	44	68	91	118	185	267
RHH*, RHW*, RHW-2*, THHW, THW, THW-2	14	6	10	17	29	39	65	93	143	191	246	387	558
RHH*, RHW*, RHW-2*, THHW, THW	12	5	8	13	23	32	52	75	115	154	198	311	448
	10	3	6	10	18	25	41	58	90	120	154	242	350
RHH*, RHW*, RHW-2*, THHW, THW, THW-2	8	1	4	6	11	15	24	35	54	72	92	145	209
RHH*, RHW*, RHW-2*, TW, THW, THHW, THW-2	6	1	3	5	8	11	18	27	41	55	71	111	160
	4	1	1	3	6	8	14	20	31	41	53	83	120
	3	1	1	3	5	7	12	17	26	35	45	71	103
	2	1	1	2	4	6	10	14	22	30	38	60	87
	1	1	1	1	3	4	7	10	15	21	27	42	61
	1/0	0	1	1	2	3	6	8	13	18	23	36	52
	2/0	0	1	1	2	3	5	7	11	15	19	31	44
	3/0	0	1	1	1	2	4	6	9	13	16	26	37
	4/0	0	0	1	1	1	3	5	8	10	14	21	31
	250	0	0	1	1	1	3	4	6	8	11	17	25
	300	0	0	1	1	1	2	3	5	7	9	15	22
	350	0	0	0	1	1	1	3	5	6	8	13	19
	400	0	0	0	1	1	1	3	4	6	7	12	17
	500	0	0	0	1	1	1	2	3	5	6	10	14
	600	0	0	0	1	1	1	1	3	4	5	8	12
	700	0	0	0	0	1	1	1	2	3	4	7	10
	750	0	0	0	0	1	1	1	2	3	4	7	10
	800	0	0	0	0	1	1	1	2	3	4	6	9
	900	0	0	0	0	1	1	1	1	3	4	6	8
	1000	0	0	0	0	0	1	1	1	2	3	5	8
	1250	0	0	0	0	0	1	1	1	1	2	4	6
	1500	0	0	0	0	0	1	1	1	1	2	3	5
	1750	0	0	0	0	0	0	1	1	1	1	3	4
	2000	0	0	0	0	0	0	1	1	1	1	3	4

*Types RHH, RHW, and RHW-2 without outer covering.

TABLE 2.28   NEC Appendix C, Table C8: Maximum Number of Conductors and Fixture Wires in Rigid Metallic Conduit (Based on Table 1, Chapter 9) (*Continued*)

Type Letters	Conductor Size AWG/kcmil	Trade Sizes in Inches ½	¾	1	1¼	1½	2	2½	3	3½	4	5	6
THHN,	14	13	22	36	63	85	140	200	309	412	531	833	1202
THWN,	12	9	16	26	46	62	102	146	225	301	387	608	877
THWN-2	10	6	10	17	29	39	64	92	142	189	244	383	552
	8	3	6	9	16	22	37	53	82	109	140	221	318
	6	2	4	7	12	16	27	38	59	79	101	159	230
	4	1	2	4	7	10	16	23	36	48	62	98	141
	3	1	1	3	6	8	14	20	31	41	53	83	120
	2	1	1	3	5	7	11	17	26	34	44	70	100
	1	1	1	1	4	5	8	12	19	25	33	51	74
	1/0	1	1	1	3	4	7	10	16	21	27	43	63
	2/0	0	1	1	2	3	6	8	13	18	23	36	52
	3/0	0	1	1	1	3	5	7	11	15	19	30	43
	4/0	0	1	1	1	2	4	6	9	12	16	25	36
	250	0	0	1	1	1	3	5	7	10	13	20	29
	300	0	0	1	1	1	3	4	6	8	11	17	25
	350	0	0	1	1	1	2	3	5	7	10	15	22
	400	0	0	1	1	1	2	3	5	7	8	13	20
	500	0	0	0	1	1	1	2	4	5	7	11	16
	600	0	0	0	1	1	1	1	3	4	6	9	13
	700	0	0	0	1	1	1	1	3	4	5	8	11
	750	0	0	0	0	1	1	1	3	4	5	7	11
	800	0	0	0	0	1	1	1	2	3	4	7	10
	900	0	0	0	0	1	1	1	2	3	4	6	9
	1000	0	0	0	0	1	1	1	1	3	4	6	8
FEP, FEPB,	14	12	22	35	61	83	136	194	300	400	515	808	1166
PFA, PFAH,	12	9	16	26	44	60	99	142	219	292	376	590	851
TFE	10	6	11	18	32	43	71	102	157	209	269	423	610
	8	3	6	10	18	25	41	58	90	120	154	242	350
	6	2	4	7	13	17	29	41	64	85	110	172	249
	4	1	3	5	9	12	20	29	44	59	77	120	174
	3	1	2	4	7	10	17	24	37	50	64	100	145
	2	1	1	3	6	8	14	20	31	41	53	83	120
PFA, PFAH, TFE	1	1	1	2	4	6	9	14	21	28	37	57	83

TABLE 2.28 NEC Appendix C, Table C8: Maximum Number of Conductors and Fixture Wires in Rigid Metallic Conduit (Based on Table 1, Chapter 9) (*Continued*)

Type Letters	Conductor Size AWG/kcmil	Trade Sizes in Inches											
		1/2	3/4	1	1 1/4	1 1/2	2	2 1/2	3	3 1/2	4	5	6
PFA, PFAH,	1/0	1	1	1	3	5	8	11	18	24	30	48	69
TFE, Z	2/0	1	1	1	3	4	6	9	14	19	25	40	57
	3/0	0	1	1	2	3	5	8	12	16	21	33	47
	4/0	0	1	1	1	2	4	6	10	13	17	27	39
Z	14	15	26	42	73	100	164	234	361	482	621	974	1405
	12	10	18	30	52	71	116	166	256	342	440	691	997
	10	6	11	18	32	43	71	102	157	209	269	423	610
	8	4	7	11	20	27	45	64	99	132	170	267	386
	6	3	5	8	14	19	31	45	69	93	120	188	271
	4	1	3	5	9	13	22	31	48	64	82	129	186
	3	1	2	4	7	9	16	22	35	47	60	94	136
	2	1	1	3	6	8	13	19	29	39	50	78	113
	1	1	1	2	5	6	10	15	23	31	40	63	92
XHH,	14	9	15	25	44	59	98	140	216	288	370	581	839
XHHW,	12	7	12	19	33	45	75	107	165	221	284	446	644
XHHW-2,	10	5	9	14	25	34	56	80	123	164	212	332	480
ZW	8	3	5	8	14	19	31	44	68	91	118	185	267
	6	1	3	6	10	14	23	33	51	68	87	137	197
	4	1	2	4	7	10	16	24	37	49	63	99	143
	3	1	1	3	6	8	14	20	31	41	53	84	121
	2	1	1	3	5	7	12	17	26	35	45	70	101
XHH, XHHW,	1	1	1	1	4	5	9	12	19	26	33	52	76
XHHW-2	1/0	1	1	1	3	4	7	10	16	22	28	44	64
	2/0	0	1	1	2	3	6	9	13	18	23	37	53
	3/0	0	1	1	1	3	5	7	11	15	19	30	44
	4/0	0	1	1	1	2	4	6	9	12	16	25	36
	250	0	0	1	1	1	3	5	7	10	13	20	30
	300	0	0	1	1	1	3	4	6	9	11	18	25
	350	0	0	1	1	1	2	3	6	7	10	15	22
	400	0	0	1	1	1	2	3	5	7	9	14	20
	500	0	0	0	1	1	1	2	4	5	7	11	16
	600	0	0	0	1	1	1	1	3	4	6	9	13
	700	0	0	0	1	1	1	1	3	4	5	8	11
	750	0	0	0	0	1	1	1	3	4	5	7	11
	800	0	0	0	0	1	1	1	2	3	4	7	10
	900	0	0	0	0	1	1	1	2	3	4	6	9
	1000	0	0	0	0	1	1	1	1	3	4	6	8
	1250	0	0	0	0	0	1	1	1	2	3	4	6
	1500	0	0	0	0	0	1	1	1	1	2	4	5
	1750	0	0	0	0	0	0	1	1	1	1	3	5
	2000	0	0	0	0	0	0	1	1	1	1	3	4

TABLE 2.28   NEC Appendix C, Table C8: Maximum Number of Conductors and Fixture Wires in Rigid Metallic Conduit (Based on Table 1, Chapter 9) (*Continued*)

Type Letters	Conductor Size AWG/kcmil	Trade Sizes in Inches					
		½	¾	1	1¼	1½	2
FFH-2, RFH-2,	18	8	15	24	42	57	94
RFHH-3	16	7	12	20	35	48	79
SF-2, SFF-2	18	11	19	31	53	72	118
	16	9	15	25	44	59	98
	14	7	12	20	35	48	79
SF-1, SFF-1	18	19	33	54	94	127	209
AF, RFH-1, RFHH-2, TF, TFF, XF, XFF	18	14	25	40	69	94	155
AF, RFHH-2, TF, TFF, XF, XFF	16	11	20	32	56	76	125
AF, XF, XFF	14	9	15	25	44	59	98
TFN, TFFN	18	23	40	64	111	150	248
	16	17	30	49	84	115	189
PF, PFF, PGF, PGFF, PAF, PTF, PTFF, PAFF	18	21	38	61	105	143	235
	16	16	29	47	81	110	181
	14	12	22	35	61	83	136
HF, HFF, ZF, ZFF, ZHF	18	28	48	79	135	184	303
	16	20	36	58	100	136	223
	14	15	26	42	73	100	164
KF-2, KFF-2	18	40	71	114	197	267	439
	16	28	50	80	138	188	310
	14	19	34	55	95	129	213
	12	13	23	38	65	89	146
	10	9	15	25	44	59	98
KF-1, KFF-1	18	48	84	136	235	318	524
	16	34	59	96	165	224	368
	14	23	40	64	111	150	248
	12	15	26	42	73	100	164
	10	10	17	28	48	65	107
AF, XF, XFF	12	5	8	13	23	32	52
	10	3	6	10	18	25	41

Note: This table is for concentric stranded conductors only. For compact stranded conductors, Table C8A should be used.

TABLE 2.29 NEC Appendix C, Table C9: Maximum Number of Conductors and Fixture Wires in Rigid PVC Conduit, Schedule 80 (Based on Table 1, Chapter 9)

Type Letters	Conductor Size AWG/kcmil	Trade Sizes in Inches											
		1/2	3/4	1	1 1/4	1 1/2	2	2 1/2	3	3 1/2	4	5	6
RH	14	4	8	13	23	32	55	79	123	166	215	341	490
	12	3	6	10	19	26	44	63	99	133	173	274	394
RHH, RHW,	14	3	5	9	17	23	39	56	88	118	153	243	349
RHW-2	12	2	4	7	14	19	32	46	73	98	127	202	290
RH, RHH,	10	1	3	6	11	15	26	37	59	79	103	163	234
RHW,	8	1	1	3	6	8	13	19	31	41	54	85	122
RHW-2,	6	1	1	2	4	6	11	16	24	33	43	68	98
	4	1	1	1	3	5	8	12	19	26	33	53	77
	3	0	1	1	3	4	7	11	17	23	29	47	67
	2	0	1	1	3	4	6	9	14	20	25	41	58
	1	0	1	1	1	2	4	6	9	13	17	27	38
	1/0	0	0	1	1	1	3	5	8	11	15	23	33
	2/0	0	0	1	1	1	3	4	7	10	13	20	29
	3/0	0	0	1	1	1	3	4	6	8	11	17	25
	4/0	0	0	0	1	1	2	3	5	7	9	15	21
	250	0	0	0	1	1	1	2	4	5	7	11	16
	300	0	0	0	1	1	1	2	3	5	6	10	14
	350	0	0	0	1	1	1	1	3	4	5	9	13
	400	0	0	0	0	1	1	1	3	4	5	8	12
	500	0	0	0	0	1	1	1	2	3	4	7	10
	600	0	0	0	0	0	1	1	1	3	3	6	8
	700	0	0	0	0	0	1	1	1	2	3	5	7
	750	0	0	0	0	0	1	1	1	2	3	5	7
	800	0	0	0	0	0	1	1	1	2	3	4	7
	1000	0	0	0	0	0	1	1	1	1	2	4	5
	1250	0	0	0	0	0	0	1	1	1	1	3	4
	1500	0	0	0	0	0	0	1	1	1	1	2	4
	1750	0	0	0	0	0	0	0	1	1	1	2	3
	2000	0	0	0	0	0	0	0	1	1	1	1	3

TABLE 2.29   NEC Appendix C, Table C9: Maximum Number of Conductors and Fixture Wires in Rigid PVC Conduit, Schedule 80 (Based on Table 1, Chapter 9) (*Continued*)

Type Letters	Conductor Size AWG/kcmil	Trade Sizes in Inches											
		½	¾	1	1¼	1½	2	2½	3	3½	4	5	6
TW	14	6	11	20	35	49	82	118	185	250	324	514	736
	12	5	9	15	27	38	63	91	142	192	248	394	565
	10	3	6	11	20	28	47	67	106	143	185	294	421
	8	1	3	6	11	15	26	37	59	79	103	163	234
RHH*, RHW*, RHW-2*, THHW, THW, THW-2	14	4	8	13	23	32	55	79	123	166	215	341	490
RHH*, RHW*, RHW-2*, THHW, THW	12	3	6	10	19	26	44	63	99	133	173	274	394
	10	2	5	8	15	20	34	49	77	104	135	214	307
RHH*, RHW*, RHW-2*, THHW, THW, THW-2	8	1	3	5	9	12	20	29	46	62	81	128	184
RHH*, RHW*, RHW-2*, TW, THW, THHW, THW-2	6	1	1	3	7	9	16	22	35	48	62	98	141
	4	1	1	3	5	7	12	17	26	35	46	73	105
	3	1	1	2	4	6	10	14	22	30	39	63	90
	2	1	1	1	3	5	8	12	19	26	33	53	77
	1	0	1	1	2	3	6	8	13	18	23	37	54
	1/0	0	1	1	1	3	5	7	11	15	20	32	46
	2/0	0	1	1	1	2	4	6	10	13	17	27	39
	3/0	0	0	1	1	1	3	5	8	11	14	23	33
	4/0	0	0	1	1	1	3	4	7	9	12	19	27
	250	0	0	0	1	1	2	3	5	7	9	15	22
	300	0	0	0	1	1	1	3	5	6	8	13	19
	350	0	0	0	1	1	1	2	4	6	7	12	17
	400	0	0	0	1	1	1	2	4	5	7	10	15
	500	0	0	0	1	1	1	1	3	4	5	9	13
	600	0	0	0	0	1	1	1	2	3	4	7	10
	700	0	0	0	0	1	1	1	2	3	4	6	9
	750	0	0	0	0	0	1	1	1	3	4	6	8
	800	0	0	0	0	0	1	1	1	3	3	6	8
	900	0	0	0	0	0	1	1	1	2	3	5	7
	1000	0	0	0	0	0	1	1	1	2	3	5	7
	1250	0	0	0	0	0	1	1	1	1	2	4	5
	1500	0	0	0	0	0	0	1	1	1	1	3	4
	1750	0	0	0	0	0	0	1	1	1	1	3	4
	2000	0	0	0	0	0	0	0	1	1	1	2	3

*Types RHH, RHW, and RHW-2 without outer covering.

TABLE 2.29 NEC Appendix C, Table C9: Maximum Number of Conductors and Fixture Wires in Rigid PVC Conduit, Schedule 80 (Based on Table 1, Chapter 9) (*Continued*)

Type Letters	Conductor Size AWG/kcmil	Trade Sizes in Inches											
		½	¾	1	1¼	1½	2	2½	3	3½	4	5	6
THHN,	14	9	17	28	51	70	118	170	265	358	464	736	1055
THWN,	12	6	12	20	37	51	86	124	193	261	338	537	770
THWN-2	10	4	7	13	23	32	54	78	122	164	213	338	485
	8	2	4	7	13	18	31	45	70	95	123	195	279
	6	1	3	5	9	13	22	32	51	68	89	141	202
	4	1	1	3	6	8	14	20	31	42	54	86	124
	3	1	1	3	5	7	12	17	26	35	46	73	105
	2	1	1	2	4	6	10	14	22	30	39	61	88
	1	0	1	1	3	4	7	10	16	22	29	45	65
	1/0	0	1	1	2	3	6	9	14	18	24	38	55
	2/0	0	1	1	1	3	5	7	11	15	20	32	46
	3/0	0	1	1	1	2	4	6	9	13	17	26	38
	4/0	0	0	1	1	1	3	5	8	10	14	22	31
	250	0	0	1	1	1	3	4	6	8	11	18	25
	300	0	0	0	1	1	2	3	5	7	9	15	22
	350	0	0	0	1	1	1	3	5	6	8	13	19
	400	0	0	0	1	1	1	3	4	6	7	12	17
	500	0	0	0	1	1	1	2	3	5	6	10	14
	600	0	0	0	0	1	1	1	3	4	5	8	12
	700	0	0	0	0	1	1	1	2	3	4	7	10
	750	0	0	0	0	1	1	1	2	3	4	7	9
	800	0	0	0	0	1	1	1	2	3	4	6	9
	900	0	0	0	0	0	1	1	1	3	3	6	8
	1000	0	0	0	0	0	1	1	1	2	3	5	7
FEP, FEPB,	14	8	16	27	49	68	115	164	257	347	450	714	1024
PFA, PFAH,	12	6	12	20	36	50	84	120	188	253	328	521	747
TFE	10	4	8	14	26	36	60	86	135	182	235	374	536
	8	2	5	8	15	20	34	49	77	104	135	214	307
	6	1	3	6	10	14	24	35	55	74	96	152	218
	4	1	2	4	7	10	17	24	38	52	67	106	153
	3	1	1	3	6	8	14	20	32	43	56	89	127
	2	1	1	3	5	7	12	17	26	35	46	73	105
PFA, PFAH, TFE	1	1	1	1	3	5	8	11	18	25	32	51	73
PFA, PFAH, TFE, Z	1/0	0	1	1	3	4	7	10	15	20	27	42	61
	2/0	0	1	1	2	3	5	8	12	17	22	35	50
	3/0	0	1	1	1	2	4	6	10	14	18	29	41
	4/0	0	0	1	1	1	4	5	8	11	15	24	34

TABLE 2.29   NEC Appendix C, Table C9: Maximum Number of Conductors and Fixture Wires in Rigid PVC Conduit, Schedule 80 (Based on Table 1, Chapter 9) (*Continued*)

Type Letters	Conductor Size AWG/kcmil	Trade Sizes in Inches											
		$1/2$	$3/4$	1	$1^1/4$	$1^1/2$	2	$2^1/2$	3	$3^1/2$	4	5	6
Z	14	10	19	33	59	82	138	198	310	418	542	860	1233
	12	7	14	23	42	58	98	141	220	297	385	610	875
	10	4	8	14	26	36	60	86	135	182	235	374	536
	8	3	5	9	16	22	38	54	85	115	149	236	339
	6	2	4	6	11	16	26	38	60	81	104	166	238
	4	1	2	4	8	11	18	26	41	55	72	114	164
	3	1	2	3	5	8	13	19	30	40	52	83	119
	2	1	1	2	5	6	11	16	25	33	43	69	99
	1	0	1	2	4	5	9	13	20	27	35	56	80
XHH,	14	6	11	20	35	49	82	118	185	250	324	514	736
XHHW,	12	5	9	15	27	38	63	91	142	192	248	394	565
XHHW-2,	10	3	6	11	20	28	47	67	106	143	185	294	421
ZW	8	1	3	6	11	15	26	37	59	79	103	163	234
	6	1	2	4	8	11	19	28	43	59	76	121	173
	4	1	1	3	6	8	14	20	31	42	55	87	125
	3	1	1	3	5	7	12	17	26	36	47	74	106
	2	1	1	2	4	6	10	14	22	30	39	62	89
XHH, XHHW,	1	0	1	1	3	4	7	10	16	22	29	46	66
XHHW-2	1/0	0	1	1	2	3	6	9	14	19	24	39	56
	2/0	0	1	1	1	3	5	7	11	16	20	32	46
	3/0	0	1	1	1	2	4	6	9	13	17	27	38
	4/0	0	0	1	1	1	3	5	8	11	14	22	32
	250	0	0	1	1	1	3	4	6	9	11	18	26
	300	0	0	1	1	1	2	3	5	7	10	15	22
	350	0	0	0	1	1	1	3	5	6	8	14	20
	400	0	0	0	1	1	1	3	4	6	7	12	17
	500	0	0	0	1	1	1	2	3	5	6	10	14
	600	0	0	0	0	1	1	1	3	4	5	8	11
	700	0	0	0	0	1	1	1	2	3	4	7	10
	750	0	0	0	0	1	1	1	2	3	4	6	9
	800	0	0	0	0	1	1	1	1	3	4	6	9
	900	0	0	0	0	0	1	1		3	3	5	8
	1000	0	0	0	0	0	1	1	1	2	3	5	7
	1250	0	0	0	0	0	1	1	1	1	2	4	6
	1500	0	0	0	0	0	0	1	1	1	1	3	5
	1750	0	0	0	0	0	0	1	1	1	1	3	4
	2000	0	0	0	0	0	0	1	1	1	1	2	4

TABLE 2.29 NEC Appendix C, Table C9: Maximum Number of Conductors and Fixture Wires in Rigid PVC Conduit, Schedule 80 (Based on Table 1, Chapter 9) (*Continued*)

		Fixture Wires					
Type Letters	Conductor Size AWG/kcmil	Trade Sizes in Inches					
		1/2	3/4	1	1 1/4	1 1/2	2
FFH-2, RFH-2,	18	6	11	19	34	47	79
RFHH-3	16	5	9	16	28	39	67
SF-2, SFF-2	18	7	14	24	43	59	100
	16	6	11	20	35	49	82
	14	5	9	16	28	39	67
SF-1, SFF-1	18	13	25	42	76	105	177
AF, RFH-1, RFHH-2, TF, TFF, XF, XFF	18	10	18	31	56	77	130
AF, RFHH-2, TF, TFF, XF, XFF	16	8	15	25	45	62	105
AF, XF, XFF	14	6	11	20	35	49	82
TFN, TFFN	18	16	29	50	90	124	209
	16	12	22	38	68	95	159
PF, PFF, PGF, PGFF, PAF, PTF, PTFF, PAFF	18	15	28	47	85	118	198
	16	11	22	36	66	91	153
	14	8	16	27	49	68	115
HF, HFF, ZF, ZFF, ZHF	18	19	36	61	110	152	255
	16	14	27	45	81	112	188
	14	10	19	33	59	82	138
KF-2, KFF-2	18	28	53	88	159	220	371
	16	19	37	62	112	155	261
	14	13	25	43	77	107	179
	12	9	17	29	53	73	123
	10	6	11	20	35	49	82
KF-1, KFF-1	18	33	63	106	190	263	442
	16	23	44	74	133	185	310
	14	16	29	50	90	124	209
	12	10	19	33	59	82	138
	10	7	13	21	39	54	90
AF, XF, XFF	12	3	6	10	19	26	44
	10	2	5	8	15	20	34

Note: This table is for concentric stranded conductors only. For compact stranded conductors, Table C9A should be used.

TABLE 2.30   NEC Appendix C, Table C10: Maximum Number of Conductors and Fixture Wires in Rigid PVC Conduit, Schedule 40 and HDPE Conduit (Based on Table 1, Chapter 9)

Type Letters	Conductor Size AWG/kcmil	Trade Sizes in Inches											
		½	¾	1	1¼	1½	2	2½	3	3½	4	5	6
RH	14	5	9	16	28	38	63	90	139	186	240	378	546
	12	4	8	12	22	30	50	72	112	150	193	304	439
RHH, RHW,	14	4	7	11	20	27	45	64	99	133	171	269	390
RHW-2	12	3	5	9	16	22	37	53	82	110	142	224	323
RH, RHH,	10	2	4	7	13	18	30	43	66	89	115	181	261
RHW,	8	1	2	4	7	9	15	22	35	46	60	94	137
RHW-2	6	1	1	3	5	7	12	18	28	37	48	76	109
	4	1	1	2	4	6	10	14	22	29	37	59	85
	3	1	1	1	4	5	8	12	19	25	33	52	75
	2	1	1	1	3	4	7	10	16	22	28	45	65
	1	0	1	1	1	3	5	7	11	14	19	29	43
	1/0	0	1	1	1	2	4	6	9	13	16	26	37
	2/0	0	0	1	1	1	3	5	8	11	14	22	32
	3/0	0	0	1	1	1	3	4	7	9	12	19	28
	4/0	0	0	1	1	1	2	4	6	8	10	16	24
	250	0	0	0	1	1	1	3	4	6	8	12	18
	300	0	0	0	1	1	1	2	4	5	7	11	16
	350	0	0	0	1	1	1	2	3	5	6	10	14
	400	0	0	0	1	1	1	1	3	4	6	9	13
	500	0	0	0	0	1	1	1	3	4	5	8	11
	600	0	0	0	0	1	1	1	2	3	4	6	9
	700	0	0	0	0	0	1	1	1	3	3	6	8
	750	0	0	0	0	0	1	1	1	2	3	5	8
	800	0	0	0	0	0	1	1	1	2	3	5	7
	900	0	0	0	0	0	1	1	1	2	3	5	7
	1000	0	0	0	0	0	1	1	1	1	3	4	6
	1250	0	0	0	0	0	0	1	1	1	1	3	5
	1500	0	0	0	0	0	0	1	1	1	1	3	4
	1750	0	0	0	0	0	0	1	1	1	1	2	3
	2000	0	0	0	0	0	0	0	1	1	1	2	3

TABLE 2.30 NEC Appendix C, Table C10: Maximum Number of Conductors and Fixture Wires in Rigid PVC Conduit, Schedule 40 and HDPE Conduit (Based on Table 1, Chapter 9) (*Continued*)

Type Letters	Conductor Size AWG/kcmil	Trade Sizes in Inches											
		½	¾	1	1¼	1½	2	2½	3	3½	4	5	6
TW	14	8	14	24	42	57	94	135	209	280	361	568	822
	12	6	11	18	32	44	72	103	160	215	277	436	631
	10	4	8	13	24	32	54	77	119	160	206	325	470
	8	2	4	7	13	18	30	43	66	89	115	181	261
RHH*, RHW*, RHW-2*, THHW, THW, THW-2	14	5	9	16	28	38	63	90	139	186	240	378	546
RHH*, RHW*, RHW-2*, THHW, THW*	12	4	8	12	22	30	50	72	112	150	193	304	439
	10	3	6	10	17	24	39	56	87	117	150	237	343
RHH*, RHW*, RHW-2*, THHW, THW, THW-2	8	1	3	6	10	14	23	33	52	70	90	142	205
RHH*, RHW*, RHW-2*, TW, THW, THHW, THW-2	6	1	2	4	8	11	18	26	40	53	69	109	157
	4	1	1	3	6	8	13	19	30	40	51	81	117
	3	1	1	3	5	7	11	16	25	34	44	69	100
	2	1	1	2	4	6	10	14	22	29	37	59	85
	1	0	1	1	3	4	7	10	15	20	26	41	60
	1/0	0	1	1	2	3	6	8	13	17	22	35	51
	2/0	0	1	1	1	3	5	7	11	15	19	30	43
	3/0	0	1	1	1	2	4	6	9	12	16	25	36
	4/0	0	0	1	1	1	3	5	8	10	13	21	30
	250	0	0	1	1	1	3	4	6	8	11	17	25
	300	0	0	1	1	1	2	3	5	7	9	15	21
	350	0	0	0	1	1	1	3	5	6	8	13	19
	400	0	0	0	1	1	1	3	4	6	7	12	17
	500	0	0	0	1	1	1	2	3	5	6	10	14
	600	0	0	0	0	1	1	1	3	4	5	8	11
	700	0	0	0	0	1	1	1	2	3	4	7	10
	750	0	0	0	0	1	1	1	2	3	4	6	10
	800	0	0	0	0	1	1	1	2	3	4	6	9
	900	0	0	0	0	0	1	1	1	3	3	6	8
	1000	0	0	0	0	0	1	1	1	2	3	5	7
	1250	0	0	0	0	0	1	1	1	1	2	4	6
	1500	0	0	0	0	0	1	1	1	1	1	3	5
	1750	0	0	0	0	0	0	1	1	1	1	3	4
	2000	0	0	0	0	0	0	1	1	1	1	3	4

*Types RHH, RHW, and RHW-2 without outer covering.

TABLE 2.30   NEC Appendix C, Table C10: Maximum Number of Conductors and Fixture Wires in Rigid PVC Conduit, Schedule 40 and HDPE Conduit (Based on Table 1, Chapter 9) (*Continued*)

Type Letters	Conductor Size AWG/kcmil	Trade Sizes in Inches											
		$\frac{1}{2}$	$\frac{3}{4}$	1	$1\frac{1}{4}$	$1\frac{1}{2}$	2	$2\frac{1}{2}$	3	$3\frac{1}{2}$	4	5	6
THHN,	14	11	21	34	60	82	135	193	299	401	517	815	1178
THWN,	12	8	15	25	43	59	99	141	218	293	377	594	859
THWN-2	10	5	9	15	27	37	62	89	137	184	238	374	541
	8	3	5	9	16	21	36	51	79	106	137	216	312
	6	1	4	6	11	15	26	37	57	77	99	156	225
	4	1	2	4	7	9	16	22	35	47	61	96	138
	3	1	1	3	6	8	13	19	30	40	51	81	117
	2	1	1	3	5	7	11	16	25	33	43	68	98
	1	1	1	1	3	5	8	12	18	25	32	50	73
	1/0	1	1	1	3	4	7	10	15	21	27	42	61
	2/0	0	1	1	2	3	6	8	13	17	22	35	51
	3/0	0	1	1	1	3	5	7	11	14	18	29	42
	4/0	0	1	1	1	2	4	6	9	12	15	24	35
	250	0	0	1	1	1	3	4	7	10	12	20	28
	300	0	0	1	1	1	3	4	6	8	11	17	24
	350	0	0	1	1	1	2	3	5	7	9	15	21
	400	0	0	0	1	1	1	3	5	6	8	13	19
	500	0	0	0	1	1	1	2	4	5	7	11	16
	600	0	0	0	1	1	1	1	3	4	5	9	13
	700	0	0	0	0	1	1	1	3	4	5	8	11
	750	0	0	0	0	1	1	1	2	3	4	7	11
	800	0	0	0	0	1	1	1	2	3	4	7	10
	900	0	0	0	0	1	1	1	2	3	4	6	9
	1000	0	0	0	0	0	1	1	1	3	3	6	8
FEP, FEPB,	14	11	20	33	58	79	131	188	290	389	502	790	1142
PFA, PFAH,	12	8	15	24	42	58	96	137	212	284	366	577	834
TFE	10	6	10	17	30	41	69	98	152	204	263	414	598
	8	3	6	10	17	24	39	56	87	117	150	237	343
	6	2	4	7	12	17	28	40	62	83	107	169	244
	4	1	3	5	8	12	19	28	43	58	75	118	170
	3	1	2	4	7	10	16	23	36	48	62	98	142
	2	1	1	3	6	8	13	19	30	40	51	81	117
PFA, PFAH, TFE	1	1	1	2	4	5	9	13	20	28	36	56	81

TABLE 2.30 NEC Appendix C, Table C10: Maximum Number of Conductors and Fixture Wires in Rigid PVC Conduit, Schedule 40 and HDPE Conduit (Based on Table 1, Chapter 9) (*Continued*)

Type Letters	Conductor Size AWG/kcmil	Trade Sizes in Inches											
		½	¾	1	1¼	1½	2	2½	3	3½	4	5	6
PFA, PFAH,	1/0	1	1	1	3	4	8	11	17	23	30	47	68
TFE, Z	2/0	0	1	1	3	4	6	9	14	19	24	39	56
	3/0	0	1	1	2	3	5	7	12	16	20	32	46
	4/0	0	1	1	1	2	4	6	9	13	16	26	38
Z	14	13	24	40	70	95	158	226	350	469	605	952	1376
	12	9	17	28	49	68	112	160	248	333	429	675	976
	10	6	10	17	30	41	69	98	152	204	263	414	598
	8	3	6	11	19	26	43	62	96	129	166	261	378
	6	2	4	7	13	18	30	43	67	90	116	184	265
	4	1	3	5	9	12	21	30	46	62	80	126	183
	3	1	2	4	6	9	15	22	34	45	58	92	133
	2	1	1	3	5	7	12	18	28	38	49	77	111
	1	1	1	2	4	6	10	14	23	30	39	62	90
XHH,	14	8	14	24	42	57	94	135	209	280	361	568	822
XHHW,	12	6	11	18	32	44	72	103	160	215	277	436	631
XHHW-2,	10	4	8	13	24	32	54	77	119	160	206	325	470
ZW	8	2	4	7	13	18	30	43	66	89	115	181	261
	6	1	3	5	10	13	22	32	49	66	85	134	193
	4	1	2	4	7	9	16	23	35	48	61	97	140
	3	1	1	3	6	8	13	19	30	40	52	82	118
	2	1	1	3	5	7	11	16	25	34	44	69	99
XHH, XHHW,	1	1	1	1	3	5	8	12	19	25	32	51	74
XHHW-2	1/0	1	1	1	3	4	7	10	16	21	27	43	62
	2/0	0	1	1	2	3	6	8	13	17	23	36	52
	3/0	0	1	1	1	3	5	7	11	14	19	30	43
	4/0	0	1	1	1	2	4	6	9	12	15	24	35
	250	0	0	1	1	1	3	5	7	10	13	20	29
	300	0	0	1	1	1	3	4	6	8	11	17	25
	350	0	0	1	1	1	2	3	5	7	9	15	22
	400	0	0	0	1	1	1	3	5	6	8	13	19
	500	0	0	0	1	1	1	2	4	5	7	11	16
	600	0	0	0	1	1	1	1	3	4	5	9	13
	700	0	0	0	1	1	1	1	3	4	5	8	11
	750	0	0	0	0	1	1	1	2	3	4	7	11
	800	0	0	0	0	1	1	1	2	3	4	7	10
	900	0	0	0	0	1	1	1	2	3	4	6	9
	1000	0	0	0	0	1	1	1	1	3	3	6	8
	1250	0	0	0	0	0	1	1	1	1	3	4	6
	1500	0	0	0	0	0	1	1	1	1	2	4	5
	1750	0	0	0	0	0	0	1	1	1	1	3	5
	2000	0	0	0	0	0	0	1	1	1	1	3	4

TABLE 2.30   NEC Appendix C, Table C10: Maximum Number of Conductors and Fixture Wires in Rigid PVC Conduit, Schedule 40 and HDPE Conduit (Based on Table 1, Chapter 9) (*Continued*)

Type Letters	Conductor Size AWG/kcmil	Trade Sizes in Inches					
		½	¾	1	1¼	1½	2
FFH-2, RFH-2,	18	8	14	23	40	54	90
RFHH-3	16	6	12	19	33	46	76
SF-2, SFF-2	18	10	17	29	50	69	114
	16	8	14	24	42	57	94
	14	6	12	19	33	46	76
SF-1, SFF-1	18	17	31	51	89	122	202
AF, RFH-1, RFHH-2, TF, TFF, XF, XFF	18	13	23	38	66	90	149
AF, RFHH-2, TF,TFF, XF, XFF	16	10	18	30	53	73	120
AF, XF, XFF	14	8	14	24	42	57	94
TFN, TFFN	18	20	37	60	105	144	239
	16	16	28	46	80	110	183
PF, PFF, PGF, PGFF, PAF, PTF, PTFF, PAFF	18	19	35	57	100	137	227
	16	15	27	44	77	106	175
	14	11	20	33	58	79	131
HF, HFF, ZF, ZFF, ZHF	18	25	45	74	129	176	292
	16	18	33	54	95	130	216
	14	13	24	40	70	95	158
KF-2, KFF-2	18	36	65	107	187	256	424
	16	26	46	75	132	180	299
	14	17	31	52	90	124	205
	12	12	22	35	62	85	141
	10	8	14	24	42	57	94
KF-1, KFF-1,	18	43	78	128	223	305	506
	16	30	55	90	157	214	355
	14	20	37	60	105	144	239
	12	13	24	40	70	95	158
	10	9	16	26	45	62	103
AF, XF, XFF	12	4	8	12	22	30	50
	10	3	6	10	17	24	39

Note: This table is for concentric stranded conductors only. For compact stranded conductors, Table C10A should be used.

TABLE 2.31 NEC Appendix C, Table C11: Maximum Number of Conductors and Fixture Wires in Type A, Rigid PVC Conduit (Based on Table 1, Chapter 9)

Type Letters	Conductor Size AWG/kcmil	Trade Sizes in Inches									
		1/2	3/4	1	1 1/4	1 1/2	2	2 1/2	3	3 1/2	4
RH	14	7	12	20	34	44	70	104	157	204	262
	12	6	10	16	27	35	56	84	126	164	211
RHH, RHW,	14	5	9	15	24	31	49	74	112	146	187
RHW-2	12	4	7	12	20	26	41	61	93	121	155
RH, RHH,	10	3	6	10	16	21	33	50	75	98	125
RHW,	8	1	3	5	8	11	17	26	39	51	65
RHW-2	6	1	2	4	6	9	14	21	31	41	52
	4	1	1	3	5	7	11	16	24	32	41
	3	1	1	3	4	6	9	14	21	28	36
	2	1	1	2	4	5	8	12	18	24	31
	1	0	1	1	2	3	5	8	12	16	20
	1/0	0	1	1	2	3	5	7	10	14	18
	2/0	0	1	1	1	2	4	6	9	12	15
	3/0	0	1	1	1	1	3	5	8	10	13
	4/0	0	0	1	1	1	3	4	7	9	11
	250	0	0	1	1	1	1	3	5	7	8
	300	0	0	1	1	1	1	3	4	6	7
	350	0	0	0	1	1	1	2	4	5	7
	400	0	0	0	1	1	1	2	4	5	6
	500	0	0	0	1	1	1	1	3	4	5
	600	0	0	0	0	1	1	1	2	3	4
	700	0	0	0	0	1	1	1	2	3	4
	750	0	0	0	0	1	1	1	1	3	4
	800	0	0	0	0	1	1	1	1	3	3
	900	0	0	0	0	0	1	1	1	2	3
	1000	0	0	0	0	0	1	1	1	2	3
	1250	0	0	0	0	0	1	1	1	1	2
	1500	0	0	0	0	0	0	1	1	1	1
	1750	0	0	0	0	0	0	1	1	1	1
	2000	0	0	0	0	0	0	1	1	1	1

TABLE 2.31   NEC Appendix C, Table C11: Maximum Number of Conductors and Fixture Wires in Type A, Rigid PVC Conduit (Based on Table 1, Chapter 9) (*Continued*)

Type Letters	Conductor Size AWG/kcmil	Trade Sizes in Inches									
		½	¾	1	1¼	1½	2	2½	3	3½	4
TW	14	11	18	31	51	67	105	157	235	307	395
	12	8	14	24	39	51	80	120	181	236	303
	10	6	10	18	29	38	60	89	135	176	226
	8	3	6	10	16	21	33	50	75	98	125
RHH*, RHW*, RHW-2*, THHW, THW, THW-2	14	7	12	20	34	44	70	104	157	204	262
RHH*, RHW*, RHW-2*, THHW, THW	12	6	10	16	27	35	56	84	126	164	211
	10	4	8	13	21	28	44	65	98	128	165
RHH*, RHW*, RHW-2*, THHW, THW, THW-2	8	2	4	8	12	16	26	39	59	77	98
RHH*, RHW*, RHW-2*, TW, THW, THHW, THW-2	6	1	3	6	9	13	20	30	45	59	75
	4	1	2	4	7	9	15	22	33	44	56
	3	1	1	4	6	8	13	19	29	37	48
	2	1	1	3	5	7	11	16	24	32	41
	1	1	1	1	3	5	7	11	17	22	29
	1/0	1	1	1	3	4	6	10	14	19	24
	2/0	0	1	1	2	3	5	8	12	16	21
	3/0	0	1	1	1	3	4	7	10	13	17
	4/0	0	1	1	1	2	4	6	9	11	14
	250	0	0	1	1	1	3	4	7	9	12
	300	0	0	1	1	1	2	4	6	8	10
	350	0	0	1	1	1	2	3	5	7	9
	400	0	0	1	1	1	1	3	5	6	8
	500	0	0	1	1	1	1	2	4	5	7
	600	0	0	0	1	1	1	1	3	4	5
	700	0	0	0	1	1	1	1	3	4	5
	750	0	0	0	1	1	1	1	3	3	4
	800	0	0	0	0	1	1	1	2	3	4
	900	0	0	0	0	1	1	1	2	3	4
	1000	0	0	0	0	1	1	1	1	3	3
	1250	0	0	0	0	0	1	1	1	1	3
	1500	0	0	0	0	0	1	1	1	1	2
	1750	0	0	0	0	0	0	1	1	1	1
	2000	0	0	0	0	0	0	1	1	1	1

*Types RHH, RHW, and RHW-2 without outer covering.

TABLE 2.31 NEC Appendix C, Table C11: Maximum Number of Conductors and Fixture Wires in Type A, Rigid PVC Conduit (Based on Table 1, Chapter 9) (*Continued*)

Type Letters	Conductor Size AWG/kcmil	Trade Sizes in Inches									
		½	¾	1	1¼	1½	2	2½	3	3½	4
THHN,	14	16	27	44	73	96	150	225	338	441	566
THWN,	12	11	19	32	53	70	109	164	246	321	412
THWN-2	10	7	12	20	33	44	69	103	155	202	260
	8	4	7	12	19	25	40	59	89	117	150
	6	3	5	8	14	18	28	43	64	84	108
	4	1	3	5	8	11	17	26	39	52	66
	3	1	2	4	7	9	15	22	33	44	56
	2	1	1	3	6	8	12	19	28	37	47
	1	1	1	2	4	6	9	14	21	27	35
	1/0	1	1	2	4	5	8	11	17	23	29
	2/0	1	1	1	3	4	6	10	14	19	24
	3/0	0	1	1	2	3	5	8	12	16	20
	4/0	0	1	1	1	3	4	6	10	13	17
	250	0	1	1	1	2	3	5	8	10	14
	300	0	0	1	1	1	3	4	7	9	12
	350	0	0	1	1	1	2	4	6	8	10
	400	0	0	1	1	1	2	3	5	7	9
	500	0	0	1	1	1	1	3	4	6	7
	600	0	0	0	1	1	1	2	3	5	6
	700	0	0	0	1	1	1	1	3	4	5
	750	0	0	0	1	1	1	1	3	4	5
	800	0	0	0	1	1	1	1	3	4	5
	900	0	0	0	0	1	1	1	2	3	4
	1000	0	0	0	0	1	1	1	2	3	4
FEP, FEPB,	14	15	26	43	70	93	146	218	327	427	549
PFA, PFAH,	12	11	19	31	51	68	106	159	239	312	400
TFE	10	8	13	22	37	48	76	114	171	224	287
	8	4	8	13	21	28	44	65	98	128	165
	6	3	5	9	15	20	31	46	70	91	117
	4	1	4	6	10	14	21	32	49	64	82
	3	1	3	5	8	11	18	27	40	53	68
	2	1	2	4	7	9	15	22	33	44	56
PFA, PFAH, TFE	1	1	1	3	5	6	10	15	23	30	39

TABLE 2.31   NEC Appendix C, Table C11: Maximum Number of Conductors and Fixture Wires in Type A, Rigid PVC Conduit (Based on Table 1, Chapter 9) (*Continued*)

Type Letters	Conductor Size AWG/kcmil	Trade Sizes in Inches									
		½	¾	1	1¼	1½	2	2½	3	3½	4
PFA, PFAH,	1/0	1	1	2	4	5	8	13	19	25	32
TFE, Z	2/0	1	1	1	3	4	7	10	16	21	27
	3/0	1	1	1	3	3	6	9	13	17	22
	4/0	0	1	1	2	3	5	7	11	14	18
Z	14	18	31	52	85	112	175	263	395	515	661
	12	13	22	37	60	79	124	186	280	365	469
	10	8	13	22	37	48	76	114	171	224	287
	8	5	8	14	23	30	48	72	108	141	181
	6	3	6	10	16	21	34	50	76	99	127
	4	2	4	7	11	15	23	35	52	68	88
	3	1	3	5	8	11	17	25	38	50	64
	2	1	2	4	7	9	14	21	32	41	53
	1	1	1	3	5	7	11	17	26	33	43
XHH,	14	11	18	31	51	67	105	157	235	307	395
XHHW,	12	8	14	24	39	51	80	120	181	236	303
XHHW-2,	10	6	10	18	29	38	60	89	135	176	226
ZW	8	3	6	10	16	21	33	50	75	98	125
	6	2	4	7	12	15	24	37	55	72	93
	4	1	3	5	8	11	18	26	40	52	67
	3	1	2	4	7	9	15	22	34	44	57
	2	1	1	3	6	8	12	19	28	37	48
XHH, XHHW,	1	1	1	3	4	6	9	14	21	28	35
XHHW-2	1/0	1	1	2	4	5	8	12	18	23	30
	2/0	1	1	1	3	4	6	10	15	19	25
	3/0	0	1	1	2	3	5	8	12	16	20
	4/0	0	1	1	1	3	4	7	10	13	17
	250	0	1	1	1	2	3	5	8	11	14
	300	0	0	1	1	1	3	5	7	9	12
	350	0	0	1	1	1	3	4	6	8	10
	400	0	0	1	1	1	2	3	5	7	9
	500	0	0	1	1	1	1	3	4	6	8
	600	0	0	0	1	1	1	2	3	5	6
	700	0	0	0	1	1	1	1	3	4	5
	750	0	0	0	1	1	1	1	3	4	5
	800	0	0	0	1	1	1	1	3	4	5
	900	0	0	0	0	1	1	1	2	3	4
	1000	0	0	0	0	1	1	1	2	3	4
	1250	0	0	0	0	0	1	1	1	2	3
	1500	0	0	0	0	0	1	1	1	1	2
	1750	0	0	0	0	0	1	1	1	1	2
	2000	0	0	0	0	0	0	1	1	1	1

TABLE 2.31 NEC Appendix C, Table C11: Maximum Number of Conductors and Fixture Wires in Type A, Rigid PVC Conduit (Based on Table 1, Chapter 9) (*Continued*)

Type Letters	Conductor Size AWG/kcmil	Trade Sizes in Inches					
		$\frac{1}{2}$	$\frac{3}{4}$	1	$1\frac{1}{4}$	$1\frac{1}{2}$	2
FFH-2, RFH-2,	18	10	18	30	48	64	100
RFHH-3	16	9	15	25	41	54	85
SF-2, SFF-2	18	13	22	37	61	81	127
	16	11	18	31	51	67	105
	14	9	15	25	41	54	85
SF-1, SFF-1	18	23	40	66	108	143	224
AF, RFH-1, RFHH-2, TF, TFF, XF, XFF	18	17	29	49	80	105	165
AF, RFHH-2, TF, TFF, XF, XFF	16	14	24	39	65	85	134
AF, XF, XFF	14	11	18	31	51	67	105
TFN, TFFN	18	28	47	79	128	169	265
	16	21	36	60	98	129	202
PF, PFF, PGF, PGFF, PAF, PTF, PTFF, PAFF	18	26	45	74	122	160	251
	16	20	34	58	94	124	194
	14	15	26	43	70	93	146
HF, HFF, ZF, ZFF, ZHF	18	34	58	96	157	206	324
	16	25	42	71	116	152	239
	14	18	31	52	85	112	175
KF-2, KFF-2	18	49	84	140	228	300	470
	16	35	59	98	160	211	331
	14	24	40	67	110	145	228
	12	16	28	46	76	100	157
	10	11	18	31	51	67	105
KF-1, KFF-1	18	59	100	167	272	357	561
	16	41	70	117	191	251	394
	14	28	47	79	128	169	265
	12	18	31	52	85	112	175
	10	12	20	34	55	73	115
AF, XF, XFF	12	6	10	16	27	35	56
	10	4	8	13	21	28	44

Note: This table is for concentric stranded conductors only. For compact stranded conductors, Table C11A should be used.

TABLE 2.32   NEC Appendix C, Table C12: Maximum Number of Conductors in Type EB, PVC Conduit (Based on Table 1, Chapter 9)

Type Letters	Conductor Size AWG/kcmil	Trade Sizes in Inches					
		2	3	3½	4	5	6
RH	14	74	166	217	276	424	603
	12	59	134	175	222	341	485
RHH, RHW, RHW-2	14	53	119	155	197	303	430
	12	44	98	128	163	251	357
RH, RHH, RHW, RHW-2	10	35	79	104	132	203	288
	8	18	41	54	69	106	151
	6	15	33	43	55	85	121
	4	11	26	34	43	66	94
	3	10	23	30	38	58	83
	2	9	20	26	33	50	72
	1	6	13	17	21	33	47
	1/0	5	11	15	19	29	41
	2/0	4	10	13	16	25	36
	3/0	4	8	11	14	22	31
	4/0	3	7	9	12	18	26
	250	2	5	7	9	14	20
	300	1	5	6	8	12	17
	350	1	4	5	7	11	16
	400	1	4	5	6	10	14
	500	1	3	4	5	9	12
	600	1	3	3	4	7	10
	700	1	2	3	4	6	9
	750	1	2	3	4	6	9
	800	1	2	3	4	6	8
	900	1	1	2	3	5	7
	1000	1	1	2	3	5	7
	1250	1	1	1	2	3	5
	1500	0	1	1	1	3	4
	1750	0	1	1	1	3	4
	2000	0	1	1	1	2	3

TABLE 2.32 NEC Appendix C, Table C12: Maximum Number of Conductors in Type EB, PVC Conduit (Based on Table 1, Chapter 9) (*Continued*)

Type Letters	Conductor Size AWG/kcmil	Trade Sizes in Inches					
		2	3	3½	4	5	6
TW	14	111	250	327	415	638	907
	12	85	192	251	319	490	696
	10	63	143	187	238	365	519
	8	35	79	104	132	203	288
RHH*, RHW*, RHW-2*, THHW, THW, THW-2	14	74	166	217	276	424	603
RHH*, RHW*, RHW-2*, THHW, THW	12	59	134	175	222	341	485
	10	46	104	136	173	266	378
RHH*, RHW*, RHW-2*, THHW, THW, THW-2	8	28	62	81	104	159	227
RHH*, RHW*, RHW-2*, TW, THW, THHW, THW-2	6	21	48	62	79	122	173
	4	16	36	46	59	91	129
	3	13	30	40	51	78	111
	2	11	26	34	43	66	94
	1	8	18	24	30	46	66
	1/0	7	15	20	26	40	56
	2/0	6	13	17	22	34	48
	3/0	5	11	14	18	28	40
	4/0	4	9	12	15	24	34
	250	3	7	10	12	19	27
	300	3	6	8	11	17	24
	350	2	6	7	9	15	21
	400	2	5	7	8	13	19
	500	1	4	5	7	11	16
	600	1	3	4	6	9	13
	700	1	3	4	5	8	11
	750	1	3	4	5	7	11
	800	1	3	3	4	7	10
	900	1	2	3	4	6	9
	1000	1	2	3	4	6	8
	1250	1	1	2	3	4	6
	1500	1	1	1	2	4	6
	1750	1	1	1	2	3	5
	2000	0	1	1	1	3	4

*Types RHH, RHW, and RHW-2 without outer covering.

TABLE 2.32   NEC Appendix C, Table C12: Maximum Number of Conductors in Type EB, PVC Conduit (Based on Table 1, Chapter 9) (*Continued*)

Type Letters	Conductor Size AWG/kcmil	Trade Sizes in Inches					
		2	3	3½	4	5	6
THHN,	14	159	359	468	595	915	1300
THWN,	12	116	262	342	434	667	948
THWN-2	10	73	165	215	274	420	597
	8	42	95	124	158	242	344
	6	30	68	89	114	175	248
	4	19	42	55	70	107	153
	3	16	36	46	59	91	129
	2	13	30	39	50	76	109
	1	10	22	29	37	57	80
	1/0	8	18	24	31	48	68
	2/0	7	15	20	26	40	56
	3/0	5	13	17	21	33	47
	4/0	4	10	14	18	27	39
	250	4	8	11	14	22	31
	300	3	7	10	12	19	27
	350	3	6	8	11	17	24
	400	2	6	7	10	15	21
	500	1	5	6	8	12	18
	600	1	4	5	6	10	14
	700	1	3	4	6	9	12
	750	1	3	4	5	8	12
	800	1	3	4	5	8	11
	900	1	3	3	4	7	10
	1000	1	2	3	4	6	9
FEP, FEPB,	14	155	348	454	578	888	1261
PFA, PFAH,	12	113	254	332	422	648	920
TFE	10	81	182	238	302	465	660
	8	46	104	136	173	266	378
	6	33	74	97	123	189	269
	4	23	52	68	86	132	188
	3	19	43	56	72	110	157
	2	16	36	46	59	91	129
PFA, PFAH, TFE	1	11	25	32	41	63	90

TABLE 2.32 NEC Appendix C, Table C12: Maximum Number of Conductors in Type EB, PVC Conduit (Based on Table 1, Chapter 9) (*Continued*)

Type Letters	Conductor Size AWG/kcmil	Trade Sizes in Inches					
		2	3	3½	4	5	6
PFA, PFAH,	1/0	9	20	27	34	53	75
TFE, Z	2/0	7	17	22	28	43	62
	3/0	6	14	18	23	36	51
	4/0	5	11	15	19	29	42
Z	14	186	419	547	696	1069	1519
	12	132	297	388	494	759	1078
	10	81	182	238	302	465	660
	8	51	115	150	191	294	417
	6	36	81	105	134	206	293
	4	24	55	72	92	142	201
	3	18	40	53	67	104	147
	2	15	34	44	56	86	122
	1	12	27	36	45	70	99
XHH,	14	111	250	327	415	638	907
XHHW,	12	85	192	251	319	490	696
XHHW-2,	10	63	143	187	238	365	519
ZW	8	35	79	104	132	203	288
	6	26	59	77	98	150	213
	4	19	42	56	71	109	155
	3	16	36	47	60	92	131
	2	13	30	39	50	77	110
XHH, XHHW,	1	10	22	29	37	58	82
XHHW-2	1/0	8	19	25	31	48	69
	2/0	7	16	20	26	40	57
	3/0	6	13	17	22	33	47
	4/0	5	11	14	18	27	39
	250	4	9	11	15	22	32
	300	3	7	10	12	19	28
	350	3	6	9	11	17	24
	400	2	6	8	10	15	22
	500	1	5	6	8	12	18
	600	1	4	5	6	10	14
	700	1	3	4	6	9	12
	750	1	3	4	5	8	12
	800	1	3	4	5	8	11
	900	1	3	3	4	7	10
	1000	1	2	3	4	6	9
	1250	1	1	2	3	5	7
	1500	1	1	1	3	4	6
	1750	1	1	1	2	4	5
	2000	0	1	1	1	3	5

Note: This table is for concentric stranded conductors only. For compact stranded conductors, Table C12A should be used.

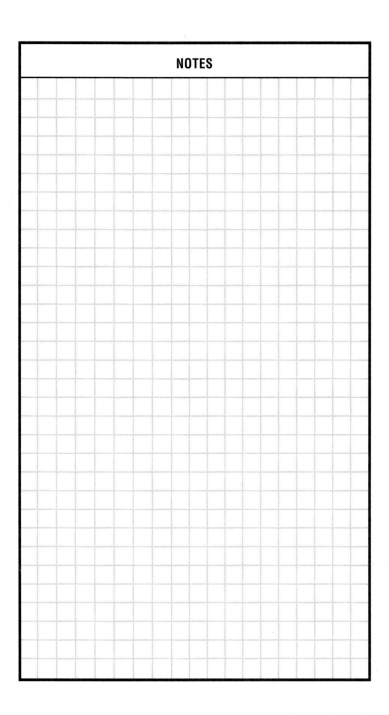

NOTES

NOTES

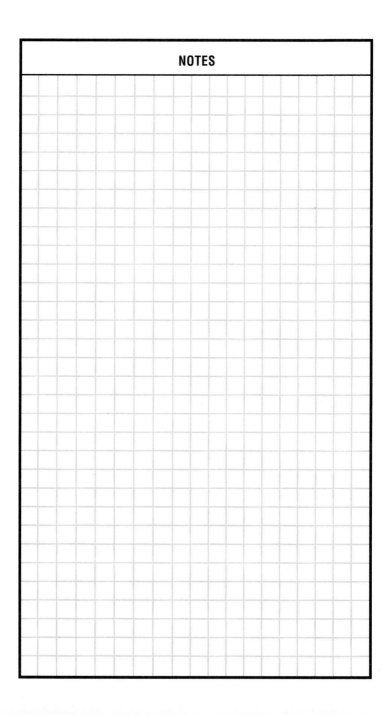

CHAPTER THREE
Service and Distribution

3.0 PRIMARY AND SECONDARY SERVICE AND SYSTEM CONFIGURATIONS

Introduction

To provide electrical service to a building or buildings, you must first determine what type of system is available from the utility company, or from a privately owned and operated system, such as might be found on a college or university campus, industrial or commercial complex, as the case may be. Once this is known, it is important to understand the characteristics of the system—not only voltage, capacity, and available fault current, but the operational, reliability, and relative cost characteristics inherent in the system by virtue of its configuration or arrangement. Knowing the characteristics associated with the system arrangement, the most appropriate service and distribution system for the application at hand can be determined.

Figures 3.1 through 3.10 feature the most frequently encountered system configurations and associated key characteristics attributable to their arrangement.

FIGURE 3.1   Radial circuit arrangements in commercial buildings.

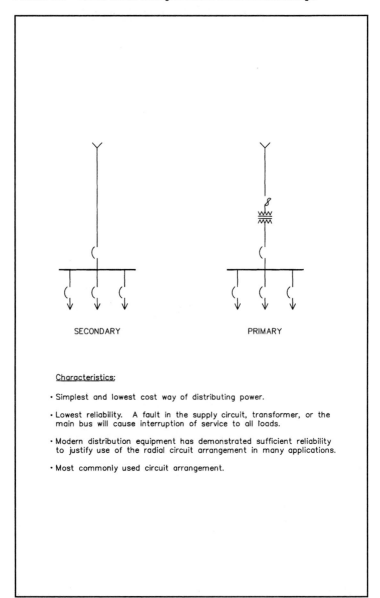

SECONDARY   PRIMARY

<u>Characteristics:</u>

• Simplest and lowest cost way of distributing power.

• Lowest reliability.   A fault in the supply circuit, transformer, or the main bus will cause interruption of service to all loads.

• Modern distribution equipment has demonstrated sufficient reliability to justify use of the radial circuit arrangement in many applications.

• Most commonly used circuit arrangement.

FIGURE 3.2 Radial circuit arrangement—common primary feeder to secondary unit substations.

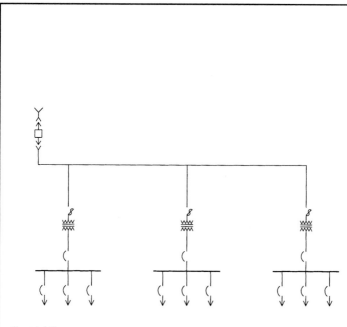

Characteristics:

• Multiple small rather than single large secondary substation.

• Used when demand, size of building, or both may be required to maintain adequate voltage at the utilization equipment.

• Smaller substations located close to center of load area.

• Provides better voltage conditions, lower system losses, less expensive installation cost than using relatively long, high–amperage, low–voltage feeder circuits.

• A primary feeder fault will cause the main protective device to operate and interrupt service to all loads. Service cannot be restored until the source of trouble has been eliminated.

• If a fault were in a transformer, service could be restored to all loads except those served by that transformer.

FIGURE 3.3   Radial circuit arrangement—individual primary feeders to secondary unit substations.

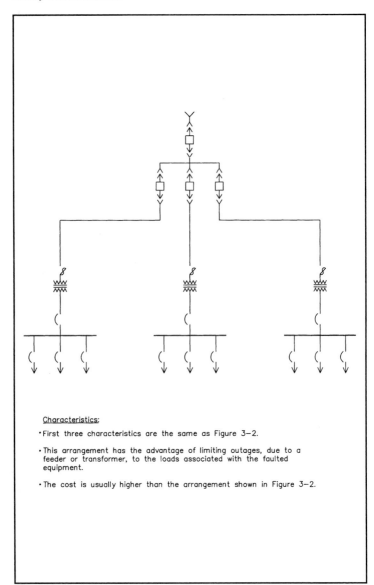

Characteristics:

• First three characteristics are the same as Figure 3–2.

• This arrangement has the advantage of limiting outages, due to a feeder or transformer, to the loads associated with the faulted equipment.

• The cost is usually higher than the arrangement shown in Figure 3–2.

FIGURE 3.4 Primary radial-selective arrangements.

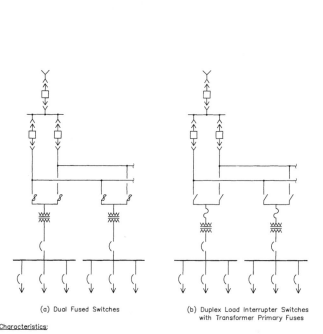

(a) Dual Fused Switches

(b) Duplex Load Interrupter Switches
with Transformer Primary Fuses

<u>Characteristics:</u>

• These circuit arrangements reduce both the extent and duration of an outage caused by a primary feeder fault.

• Operating feature — duplicate primary feeder circuits and load interrupter switches, permit connection to either primary feeder circuit.

• Each feeder must be capable of saving the entire load.

• Suitable safety interlocks usually required to prevent closing of both switches at the same time.

• Under normal operating conditions, appropriate switches are closed to balance loads between two primary feeder circuits.

• Primary–selective switches are usually manually operated, but can be automated for quicker restoration of service. Automated switching is more costly but may be justified in many applications.

• If a fault occurs in a secondary substation transformer, service can be restored to all loads except those served from the faulted transformer.

• The higher degree of service continuity afforded by the primary–selective arrangement is realized at a cost that is usually 10%–20% higher than the circuit arrangement of Figure 3–2.

FIGURE 3.5 Secondary-selective circuit arrangement (double-ended substation with single tie).

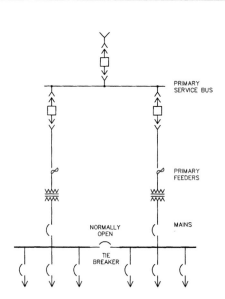

Characteristics:

• Under normal conditions, operates as two separate radial systems with the secondary bus—tie circuit breaker normally open.

• Loads should be divided equally between the two bus sections.

• If a fault occurs on a primary feeder or in a transformer, service is interrupted to all loads served from that half of the double—ended arrangement. Service can be restored to all secondary buses by opening the secondary main on the faulted side and closing the tie breaker.

• The main—tie —main breakers are normally interlocked to prevent paralleling the transformers and to prevent closing into a secondary bus fault. They can also be automated to transfer to standby operation and retransfer to normal operation.

• Cost of this arrangement will depend upon the spare capacity in the transformers and primary feeders. The minimum will be determined by the essential loads that need to be served under standby operating conditions. If service is to be provided for all loads under standby conditions, then the primary feeders and transformers must be capable of carrying the total load on both substation buses.

• This circuit arrangement is more expensive than either the radial or primary selective circuit configuration. This is primarily due to the redundant transformers.

FIGURE 3.6 Secondary-selective circuit arrangement (individual substations with interconnecting ties).

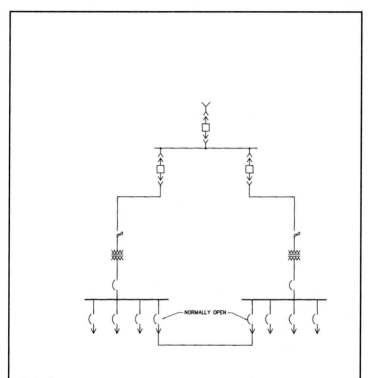

Characteristics:

• In this modification of the secondary—selective circuit arrangement shown if Figure 3—5, there is only one transformer in each secondary substation; but adjacent substations are interconnected in pairs by a normally open low—voltage tie circuit.

• When the primary feeder or transformer supplying one secondary substation bus is out of service, essential loads on that substation bus can be supplied over the tie circuit.

• Operating aspects of this system are somewhat complicated if the two substations are separated by distance.

• This would not be a desirable choice in a new building service design because a multiple key interlock system would be required to avoid tying the two substations together while they were both energized.

FIGURE 3.7 Primary- and secondary-selective circuit arrangement (double-ended substation with selective primary).

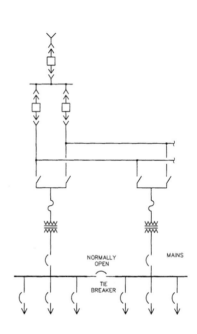

Characteristics:

• Used when highly reliable service is needed, such as hospital or data center loads.

• Has the combined benefits and characteristics of the arrangements shown in Figure 3–4 and 3–5.

• Small premium cost over configuration shown in Figure 3–5 for primary selector switches.

FIGURE 3.8   Looped primary circuit arrangement.

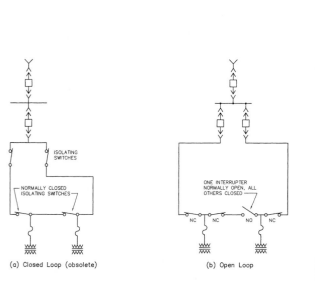

(a) Closed Loop (obsolete)   (b) Open Loop

Characteristics:

• Basically a two—circuit radial system with the ends connected together forming a continuous loop.

• Early versions of the closed loop in (a) above, although relatively inexpensive, fell into disfavor because of its apparent reliability advantages are offset by interruption of all service from a fault occurring anywhere in the loop, by the difficulty of locating primary faults, and by safety problems associated with the nonload break, or "dead break", isolating switches.

• Newer open—loop versions as shown in (b) above, designed for modern underground commercial and residential distribution systems, utilize fully rated air, oil, vacuum, and SF6 interrupter switches. Equipment available up to 34.5KV with interrupting ratings for both continuous load and fault currents to meet most system requirements. Certain equipment can close and latch on fault currents, equal to the equipment interrupting values, and still be operational without maintenance.

• Major advantages of the open—loop primary system over the simple radial system is the isolation of cable or transformer faults or both, while maintaining continuity of service to the remaining loads. With coordinated transformer fusing provided in the loop—tap position, transformer faults can be isolated without any interruption of primary service. Primary cable faults will temporarily drop service to half of the connected loads until the fault is located; then, by selective switching the unfaulted sections can be restored to service, leaving only the faulted section to be repaired.

• Disadvantages; increased costs to fully size cables, protective devices and interrupters to total capacity of the load, and the time delay necessary to locate the fault, isolate the section, and restore service. Safety considerations in maintaining a loop system are more complex than for a radial or a primary—selective system.

FIGURE 3.9 Distributed secondary network.

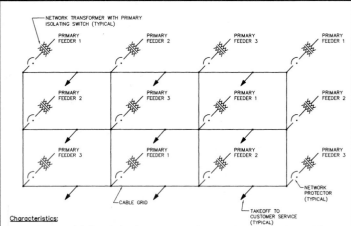

Characteristics:

•A secondary network is formed when two or more transformers having the same characteristics are supplied from separate feeders, and are connected to a common bus through network protectors.

•This arrangement is usually found in high load density urban areas where the highest level of service reliability is required. The cable grid shown typically represents a city block (for each square). Additional transformers are added as needed at locations where exceptionally high load customer service take-offs occur.

•Transformer and cable grid capacities are sized initially and added to as needed, to maintain voltage regulation, and load capacity.

•Transformer capacity and impedance characteristics are the same for equal load sharing. Likewise, the cable grid is sized for balanced load flow and to maintain voltage regulation.

•A typical grid voltage is 216Y/125—volts to provide preferred nominal 208Y/120—volt service and utilization. This helps to provide better voltage regulation.

•These systems are designed for 1st contingency operation, i.e., to provide full capacity with no interruption of service with the loss of one of the primary network feeders.

•Operational experience has shown that three primary feeders provides optimum reliability. Four or more primary network feeders provide virtually no additional reliability.

•216V secondary network cable grids are designed to operate so that faults at the grid are allowed to burn clear rather than incur a disruption of service. This is accomplished by providing cable limiters in each end of each conductor in a parallel cable grouping that forms the secondary grid.

•Network transformers have a higher impedance than conventional transformers (typically 7% vs 5.75% for a 500KVA unit) to help limit available fault current. Secondary networks typically have an available fault current in the order of 200,000 amperes RMS symmetrical.

FIGURE 3.10 Basic spot network.

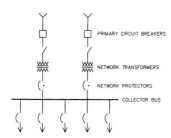

PRIMARY CIRCUIT BREAKERS

NETWORK TRANSFORMERS

NETWORK PROTECTORS

COLLECTOR BUS

Characteristics:

• The spot network is a localized distribution center consisting of two or more transformer/network protector units connected to a common bus called a "collector bus". A building may have one or more spot network services.

• Spot networks are employed to provide a reliable source of power to important electrical loads. Spare capacity is built in to allow for at least one contingency, i.e., loss of a transformer or primary network feeder will cause no interruption of service.

• Planning for service continuity should be extended beyond the consideration of using a utility primary feeder or transformer. The consequences of severe equipment damage including the resulting system downtime, should also be considered.

• The primary side of a spot network transformer usually contains an isolating/load interrupter switch, primary fuses, and a non—load break grounding switch located within the same enclosure. Although the grounding switch has a fault closing rating, it cannot be operated until the safety requirements of a key interlock scheme have been satisfied. The key interlocks prevent closing the grounding switch until all possible sources of supply to the feeder have been isolated.

• Conventional automatic network protectors are sophisticated devices. They are self—contained units consisting of an electrically operated circuit breaker, special network relays, control transformers, instrument transformers, and open—type fuse links. The protector will automatically close when the oncoming transformer voltage is greater than the collector bus voltage and will open when reverse current flows from the collector bus into the transformer. Reverse current flow can be the result of a fault beyond the line side of the protector, supplying load current back into the primary distribution system when the collector bus voltage is higher than the individual transformer voltage, or the opening of the transformer primary feeder breaker, which causes the collector bus to supply transformer magnetizing current via the transformer secondary winding.

• Most spot network applications for commercial buildings provide 480Y/277—volt utilization, thus requiring ground fault protection. Relay protection is the most common method of ground—fault protection. The fault current may be sensed by the ground return, residual, or zero—sequence method. Each of the methods have proved successful where appropriately applied; but they share a common limitation in that they cannot distinguish between in—zone and thru—zone ground faults unless incorporated in a complex protection scheme. One particular method of ground—fault detection that is not prone to unnecessary tripping is enclosure monitoring. This method offers the distinct advantage of not requiring coordination with other protective devices.

3.1 PRELIMINARY LOAD CALCULATIONS

Introduction

The electrical design professional should determine a building's electrical load characteristics early in the preliminary design stage of the building to select the proper power distribution system and equipment having adequate power capacity with proper voltage levels, and sufficient space and ventilation to maintain proper ambients. Once the power system is determined, it is often difficult to make major changes because of the coordination required with other disciplines. Architects and mechanical and structural engineers will be developing their designs simultaneously and making space and ventilation allocations. It is imperative, therefore, from the start that the electric systems be correctly based on realistic load data or best possible typical load estimates, or both because all final, finite load data are not available during the preliminary design stage of the project. When using estimated data, it should be remembered that the typical data applies only to the condition from which the data was taken, and most likely an adjustment to the particular application will be required.

Although many of the requirements of building equipment, such as ventilating, heating/cooling, lighting, and so forth, are furnished by other disciplines, the electrical design professional should also furnish to the other disciplines such data as space, accessibility, weight, and heat dissipation requirements for the electrical power distribution apparatus. This involves a continuing exchange of information that starts as preliminary data and is upgraded to be increasingly accurate as the design progresses. Documentation and coordination throughout the design process is imperative.

At the beginning of a project, the electrical design professional should review the utility's rate structure and the classes (system types) of service available. Information pertaining to demand, energy, and power factor should be developed to aid in evaluating, selecting, and specifying the most advantageous utility connection. As energy resources become more costly and scarce, items such as energy efficiency, power demand minimization, and energy conservation should be closely considered to reduce both energy consumption and utility cost.

System power (i.e., energy) losses should be considered as part of the total load in sizing service mains and service equipment. ANSI/NFPA 70-1996, NEC recommends that the total voltage drop from the electrical service to the load terminals of the farthest piece of equipment served should not exceed 5 percent of the system voltage and, thus, the energy loss, I^2R, will correspondingly be limited.

Listed hereafter are typical load groups and examples of classes of electrical equipment that should be considered when estimating initial and future loads.

- *Lighting:* Interior (general, task, exits, and stairwells), exterior (decorative, parking lot, security), normal, and emergency
- *Appliances:* Business and copying machines, receptacles for vending machines, and general use
- *Space conditioning:* Heating, cooling, cleaning, pumping, and air-handling units
- *Plumbing and sanitation:* Water pumps, hot water heaters, sump and sewage pumps, incinerators, and waste handling
- *Fire protection:* Fire detection, alarms, and pumps
- *Transportation:* Elevators, dumbwaiters, conveyors, escalators, and moving walkways
- *Data processing:* Desktop computers, central processing and peripheral equipment, and uninterruptible power supply (UPS) systems, including related cooling
- *Food preparation:* Cooling, cooking, special exhausts, dishwashing, disposing, and so forth
- *Special loads:* For equipment and facilities in mercantile buildings, restaurants, theaters, recreation and sports complexes, religious buildings, terminals and airports, health care facilities, laboratories, broadcasting stations, and so forth
- *Miscellaneous loads:* Security; central control systems; communications; audio-visual, snow-melting, recreational, or fitness equipment; incinerators, shredding devices, waste compactors, shop and maintenance equipment, and so forth

Load Estimates

There are several load estimates that should be made during the course of the project including:

1. Preliminary load estimate
2. Early design load estimate
3. NEC compliance load estimates that may be required
4. Energy compliance load estimates that may be appropriate
5. Final load estimates based on final design load information

The following tables are provided to assist the user in estimating preliminary loads for various building types. Considerable judgment should be used in the application of this data. Power densities are typically given in watts per square foot (W/ft^2) or volt-amps per square foot (VA/ft^2) and are used interchangeably because unity power factor is assumed for preliminary load calculations.

In the first of the tables that follow, criteria for controlling the energy consumption of lighting systems in, and connected with, building facilities have been prepared by the American Society of Heating, Refrigerating, and Air-Conditioning Engineers (ASHRAE) in concert with the Illuminating Engineering Society of North America (IESNA). They are identified in Section 6 of ASHRAE/IESNA 90.1-1989, Energy Efficient Design of New Buildings Except New Low-Rise Residential Buildings, which establishes an upper limit of power to be allowed for lighting systems plus guidelines for designing and managing those systems. A simplified method based on the above standard for determining the unit lighting power allowance for each building type is shown in Table 3.1.

The remaining tables provide power densities for various types of loads and building types. See Tables 3.2 through 3.10.

The foregoing tables give estimated connected loads for various types of buildings and spaces in buildings. To these the user must apply a demand factor to estimate the actual demand load. This requires experience and judgment. Applying a demand factor will help to design an economical power distribution system by designing to demand loads rather than connected loads. This will result in equipment that is appro-

TABLE 3.1   Prescriptive Unit Lighting Power Allowance (ULPA) (w/ft^2)—Gross Lighted Area of Total Building

Building Type or Space Activity	0 to 2000 ft^2	2001 to 10 000 ft^2	10 001 to 25 000 ft^2	25 001 to 50 000 ft^2	50 001 to 250 000 ft^2	>250 000 ft^2
Food Service						
Fast Food/Cafeteria	1.50	1.38	1.34	1.32	1.31	1.30
Leisure Dining/Bar	2.20	1.91	1.71	1.56	1.46	1.40
Offices	1.90	1.81	1.72	1.65	1.57	1.50
Retail*	3.30	3.08	2.83	2.50	2.28	2.10
Mall Concourse	1.60	1.58	1.52	1.46	1.43	1.40
Multiple-Store Service						
Service Establishment	2.70	2.37	2.08	1.92	1.80	1.70
Garages	0.30	0.28	0.24	0.22	0.21	0.20
Schools						
Preschool/Elementary	1.80	1.80	1.72	1.65	1.57	1.50
Jr. High/High School	1.90	1.90	1.88	1.83	1.76	1.70
Technical/Vocational	2.40	2.33	2.17	2.01	1.84	1.70
Warehouse/Storage	0.80	0.66	0.56	0.48	0.43	0.40

NOTE: * Includes general, merchandising, and display lighting.

This prescriptive table is intended primarily for core-and-shell (i.e., speculative) buildings or for use during the preliminary design phase (i.e., when the space uses are less than 80% defined). The values in this table are not intended to represent the needs of all buildings within the types listed.

TABLE 3.2 Typical Appliance/General-Purpose Receptacle Loads (Excluding Plug-In-Type A/C and Heating Equipment)

Type of Occupancy	Unit Load (VA/ft^2)		
	Low	High	Average
Auditoriums	0.1	0.3	0.2
Cafeterias	0.1	0.3	0.2
Churches	0.1	0.3	0.2
Drafting rooms	0.4	1.0	0.7
Gymnasiums	0.1	0.2	0.15
Hospitals	0.5	1.5	1.0
Hospitals, large	0.4	1.0	0.7
Machine shops	0.5	2.5	1.5
Office buildings	0.5	1.5	1.0
Schools, large	0.2	1.0	0.6
Schools, medium	0.25	1.2	0.7
Schools, small	0.3	1.5	0.9

Other Unit Loads:
  Specific appliances — ampere rating of
    appliance
  Supplying heavy-duty lampholders —
    5 A/outlet

TABLE 3.3 Typical Apartment Loads

Type	Load
Lighting and convenience outlets (except appliance)	3 VA/ft^2
Kitchen, dining appliance circuits	1.5 kVA each
Range	8 to 12 kW
Microwave oven	1.5 kW
Refrigerator	0.3 to 0.6 kW
Freezer	0.3 to 0.6 kW
Dishwasher	1.0 to 2.0 kW
Garbage disposal	0.33 to 0.5 hp
Clothes washer	0.33 to 0.5 hp
Clothes dryer	1.5 to 6.5 kW
Water heater	1.5 to 9.0 kW
Air conditioner (0.5 hp/room)	0.8 to 4.6 kW

TABLE 3.4 Typical Connected Electrical Load for Air Conditioning Only

Type of Building	Conditioned Area (VA/ft^2)
Bank	7
Department store	3 to 5
Hotel	6
Office building	6
Telephone equipment building	7 to 8
Small store (shoe, dress, etc.)	4 to 12
Restaurant (not including kitchen)	8

TABLE 3.5 Central Air Conditioning Watts per SF, BTUs per Hour per SF of Floor Area and SF per Ton of Air Conditioning

Type Building	Watts per S.F.	BTUH per S.F.	S.F. per Ton	Type Building	Watts per S.F.	BTUH per S.F.	S.F. per Ton	Type Building	Watts per S.F.	BTUH per S.F.	S.F. per Ton
Apartments, Individual	3	26	450	Dormitory, Rooms	4.5	40	300	Libraries	5.7	50	240
Corridors	2.5	22	550	Corridors	3.4	30	400	Low Rise Office, Ext.	4.3	38	320
Auditoriums & Theaters	3.3	40	300/18*	Dress Shops	4.9	43	280	Interior	3.8	33	360
Banks	5.7	50	240	Drug Stores	9	80	150	Medical Centers	3.2	28	425
Barber Shops	5.5	48	250	Factories	4.5	40	300	Motels	3.2	28	425
Bars & Taverns	15	133	90	High Rise Off.-Ext. Rms.	5.2	46	263	Office (small suite)	4.9	43	280
Beauty Parlors	7.6	66	180	Interior Rooms	4.2	37	325	Post Office, Int. Office	4.9	42	285
Bowling Alleys	7.8	68	175	Hospitals, Core	4.9	43	280	Central Area	5.3	46	260
Churches	3.3	36	330/20*	Perimeter	5.3	46	260	Residences	2.3	20	600
Cocktail Lounges	7.8	68	175	Hotels, Guest Rooms	5	44	275	Restaurants	6.8	60	200
Computer Rooms	16	141	85	Public Spaces	6.2	55	220	Schools & Colleges	5.3	46	260
Dental Offices	6	52	230	Corridors	3.4	30	400	Shoe Stores	6.2	55	220
Dept. Stores, Basement	4	34	350	Industrial Plants, Offices	4.3	38	320	Shop'g. Ctrs., Sup. Mkts.	4	34	350
Main Floor	4.5	40	300	General Offices	4	34	350	Retail Stores	5.5	48	250
Upper Floor	3.4	30	400	Plant Areas	4.5	40	300	Specialty Shops	6.8	60	200

*Persons per ton

12,000 BTUH = 1 ton of air conditioning

TABLE 3.6   All-Weather Comfort Standard Recommended Heat Loss Values

Degree Days	Design Heat Loss per Square Foot of Floor Area (Btu/h)	(watts)
Over 8000	40	11.7
7001 to 8000	38	11.3
6001 to 7000	35	10.3
5001 to 6000	32	9.4
3001 to 5000	30	8.8
Under 3001	28	8.2

TABLE 3.7   Typical Power Requirement (kW) for High-Rise Building Water Pressure–Boosting Systems

Building Type	Unit Quantity	Number of Stories 5	10	25	50
Apartments	10 apt./ floor	—	15	90	350
Hospitals	30 patients/ floor	10	45	250	—
Hotels/ Motels	40 rooms/ floor	7	35	175	450
Offices	10 000 ft^2/ floor	—	15	75	250

TABLE 3.8   Typical Power Requirement (kW) for Electric Hot Water–Heating System

Building Type	Unit Quantity	Load
Apartments/ Condominiums	20 apt/condo	30
Dormitories	100 residents	75
Elementary schools	100 students	6
High schools	100 students	12
Restaurant (full service)	100 servings/h	30
Restaurant (fast service)	100 servings/h	15
Nursing homes	100 residents	60
Hospitals	100 patient beds	200
Office buildings	10 000 ft^2	5

TABLE 3.9   Typical Power Requirement (kW) for Fire Pumps in Commercial Buildings (Light Hazard)

| Area/Floor (ft^2) | Number of Stories | | | |
	5	10	25	50
5000	40	65	150	250
10 000	60	100	200	400
25 000	75	150	275	550
50 000	120	200	400	800

*Based on zero pressure at floor 1.

priately sized rather than oversized to accommodate connected loads. Tables 3.11 and 3.12 give examples of demand loads.

Experience has shown that demand factors for buildings typically range between 50 and 80 percent of the connected load. For most building types, the demand factor at the service where the maximum diversity is experienced is usually 60 to 75 percent of the connected load. Specific portions of the system may have much higher demand factors, even approaching 100 percent.

The factors shown in Table 3.13 may be used in sizing the distribution system components shown for lighting demand and should result in a

TABLE 3.10   Typical Loads in Commercial Kitchens

	Number Served	Connected Load (kW)
Lunch counter (gas ranges, with 40 seats)		30
Cafeteria	800	150
Restaurant (gas cooking)		90
Restaurant (electric cooking)		180
Hospital (electric cooking)	1200	300
Diet kitchen (gas cooking)		200
Hotel (typical)		75
Hotel (modern, gas ranges, three kitchens)		150
Penitentiary (gas cooking)		175

TABLE 3.11 Comparison of Maximum Demand

Type of Store	Shopping Center A, New Jersey No Refrigeration*		Shopping Center B, New Jersey Refrigeration		Shopping Center C, New York Refrigeration	
	Gross Area (ft²)	(W/ft²)	Gross Area (ft²)	(W/ft²)	Gross Area (ft²)	(W/ft²)
Bank					4000	9.0
Book	3700	6.0	2500	6.7		
Candy	1600	6.9			2000	10.8
Department	343 500	4.7	222 000	7.3	226 900	8.0
	84 000	3.1	114 000	5.6		
Drug	7000	6.1	6000	7.7		
Men's wear	17 000	5.5	17 000	9.9	2000	10.8
	28 000	4.9	9100	8.8		
Paint					15 600	8.5
Pet					2000	12.1
Restaurant					4000	9.0
Shoe	11 000	6.3	7000	12.5	3300	15.4
	4000	8.0	4400	12.9	4400	9.0
Supermarket	32 000	5.7	25 000	8.6	37 600	11.5
Variety	31 000	4.6	24 000	6.8	37 400	7.1
	30 000	4.4			30 000	7.0
Women's wear	20 400	4.7	19 300	8.9	1360	13.0
	1000	5.8	4500	9.6	1000	11.7

*Loads include all lighting and power, but no power for air-conditioning refrigeration (chilled water), which is supplied from a central plant.

TABLE 3.12 Connected Load and Maximum Demand by Tenant Classification

Classification	Connected Load (W/ft²)	Maximum Demand (W/ft²)	Demand Factor
10 Women's wear	7.7	5.9	0.75
3 Men's wear	7.2	5.6	0.78
6 Shoe store	8.5	6.9	0.79
2 Department store	6.0	4.7	0.74
2 Variety store	10.5	4.5	0.45
2 Drug store	11.7	6.7	0.57
5 Household goods	5.4	3.9	0.76
10 Specialty shop	8.1	6.8	0.79
4 Bakery and candy	17.1	12.1	0.71
3 Food store (supermarkets)	9.9	5.9	0.60
5 Restaurant	15.9	7.1	0.45

NOTE: Connected load includes an allowance for spares.

TABLE 3.13 Factors Used in Sizing Distribution System Components

Distribution System Component	Lighting Demand Factor
Lighting panelboard buss and main overcurrent device	1.0
Lighting panelboard feeder and feeder overcurrent device	1.0
Distribution panelboard buss and main overcurrent device	
First 50 000 W or less	0.5
All over 50 000 W	0.4
Remaining components	0.4

conservative design. The factors should be applied to connected lighting load in the first step, and then to the product resulting from previous steps as the designer proceeds through the system.

The types of heating, ventilating, and air-conditioning systems chosen for a specific building will have the greatest single effect on electrical load. First, the choice of fuel will be critical. If natural gas, fuel oil, or coal is chosen, electrical loads will be lower than would be the case if electricity were chosen. Second, the choice of refrigeration cycle will have a considerable impact. If absorption chillers are chosen, electrical loads will be lower than those imposed by electric centrifugal or reciprocating chillers.

For initial estimates, before actual loads are known, the factors shown in Table 3.14 may be used to establish the major elements of the electrical system serving HVAC primary cooling systems.

In the writer's experience, a factor of 1.7 kVA/ton provides a good estimate for a primary cooling system made up of electric centrifugal chillers, chilled water pumps, condenser water pumps, and cooling tower fans.

TABLE 3.14 Factors Used to Establish Major Elements of the Electrical System Serving HVAC Systems

Item	Unit
Refrigeration Machines:	kVA/Ton of Chiller Capacity
Absorption	
Centrifugal	1.00
Reciprocating	
Auxiliary Pumps & Fans:	
Chilled Water Pumps	0.08
Condenser Water Pumps	
Absorption	
Centrifugal/Reciprocating	0.07
Cooling Tower Fans	
Absorption	
Centrifugal/Reciprocating	0.07
Boilers:	kVA/Boiler Horsepower
Natural Gas/Fuel Oil	0.07
Coal	
Boiler Auxiliary Pumps:	kVA/Boiler Horsepower
Deaerator	0.10
Auxiliary Equipment:	kVA/Bed
Clinical Vacuum Pumps	0.18
Clinical Air Compressors	0.10

To estimate loads for commercial kitchens, the choice of fuel in the kitchen is a major determinant. If natural gas is the primary fuel, electrical loads will be lower on a watts-per-square-foot basis than where electricity is the primary fuel. For estimating purposes, the following factors may be used as an alternative to those shown in Table 3.10. In calculating kitchen floor area include cooking and preparation, dishwashing, storage, walk-in refrigerators and freezers, food serving lines, tray assembly, and offices.

Primary Fuel	*Watts/Square Foot*
Natural gas	25
Electricity	125

A tabulation of actual service entrance demand per gross square foot is presented in Tables 3.15 and 3.16 for a group of health care facilities. Data used in preparation of these tables was obtained from the Veteran's Administration and Hospital Corporation of America. Refer to footnotes accompanying the tables for the criteria on which these tables are based.

The tables show the type of facility, the gross floor area and number of beds for each, the geographic location, and the major fuel type employed for HVAC systems in that facility. The derived factors may be used to estimate the anticipated demand for other facilities similar in size, location, and type of fuel. They also may be used to make initial estimates of service entrance capacity, switchgear size, and space required for service entrance equipment. It is important to recognize, however, that they will be useful principally in the schematic design

TABLE 3.15   Service Entrance Peak Demand (Veterans Administration)

Hospital	Floor Area Square Feet	Beds[*]	Degree Days[†] Cooling	Degree Days[†] Heating	Principal[‡] Fuel-HVAC	Watts Per Sq ft[§] Maximum	Watts Per Sq ft[§] Average
V.A. Hospital #1	821 000	922	234	3536	NG/FO	4.5	3.5
V.A. Hospital #2	334 000	500	863	5713	NG/FO	5.2	3.9
V.A. Hospital #3	645 995	670	3488	1488	NG/FO	3.8	2.8
V.A. Hospital #4	681 000	600	1016	654	NG/FO	6.1	4.0
V.A. Hospital #5	503 500	697	3495	841	NG/FO	7.2	5.5
V.A. Hospital #6	800 000	1050	600	7400	NG/FO	5.9	4.2

[*] Total beds shown. Beds actually occupied could affect values shown for watts per square foot.

[†] Degree Days: Normals, Base 65 °F, based on 1941-70 period. From *Local Climatological Data Series*, 1974, NOAA.

[‡] NG/FO = Natural Gas/Fuel Oil.   In all cases, electricity was the fuel used for refrigeration.

[§] Watts per square foot based on measured values at service entrance during metering periods ranging from 9 to 17 days, during cooling season in all instances, 1981.

TABLE 3.16 Service Entrance Peak Demand (Hospital Corporation of America)

Hospital and Location	Floor Area Square Feet	Beds*	Degree Days[†] Cooling	Heating	Principal[‡] Fuel–HVAC	Watts Per Sq ft[§] Maximum
#1 — East	273 000	458	1353	3939	NG/FO	6.8
#2 — Southeast	278 000	250	2294	2240	NG/FO	6.3
#3 — Central	123 000	157			NG/FO	7.5
#4 — Central	36 365	62	2029	3227‖	E	13.7
#5 — Central	318 000	300	1107	4306	NG/FO	4.6
#6 — Southeast	182 000	225	3786	299‖	NG/FO	5.3
#7 — East	283 523	320	1030	4307	NG/FO	6.8
#8 — Southwest	135 396	150	2250	2621‖	NG/FO	6.6
#9 — West	190 000	97	927	5983	NG/FO	2.8
#10 — Southeast	161 000	170	3226	733‖	NG/FO	6.3
#11 — Southeast	157 639	214	2078	2146	NG/FO	7.3
#12 — Southeast	162 187	222	2143	2378‖	NG/FO	4.3
#13 — East	109 617	146	1030	4307‖	NG/FO	5.7
#14 — East	76 000	153	1030	4307‖	E	8.8
#15 — Southeast	135 150	190	1995	2547‖	NG/FO	5.9
#16 — Southwest	75 769	131	2587	2382‖	NG/FO	7.4
#17 — Central	75 769	128	1636	3505‖	NG/FO	6.3
#18 — Northwest	129 000	150	714	5833‖	NG/FO	4.4
#19 — Central	54 938	108	1694	3696‖	E	13.3
#20 — West	144 000	160	2814	1752	NG/FO	4.5
#21 — Southeast	149 000	123	2078	2146‖	NG/FO	4.5
#22 — Central	89 000	128	2029	3227‖	E	8.4
#23 — Central	128 500	150	1197	4729‖	NG/FO	6.2
#24 — West	135 169	170	927	5983‖	NG/FO	4.7
#25 — Southeast	80 000	124	1722	2975‖	NG/FO	6.2
#26 — Southeast	83 117	126	3226	733‖	NG/FO	8.5
#27 — Central	51 000	97	1569	3478	E	8.8
#28 — Southeast	66 528	120	2929	902‖	E	9.7
#29 — East	112 000	140	1394	3514	NG/FO	4.3
#30 — Central	202 000	223	1636	3505	NG/FO	4.8
#31 — Southeast	56 000	51	3786	299‖	NG/FO	7.4
#32 — West	47 434	50	927	5983‖	NG/E	7.0
#33 — Central	23 835	32	1694	3696‖	E	10.8
#34 — Southeast	105 000	95	2706	1465‖	NG/FO	8.3
#35 — West	48 575	60	3042	108‖	NG/E	7.7
#36 — Southwest	133 000	185	2587	2382‖	NG/FO	6.3
#37 — Central	42 879	66	1694	3696‖	E	15.7

* Total beds shown. Beds actually occupied could affect values shown for watts per square foot.

[†] Degree Days: Normals, Base 65 °F, based on 1941–70 period. From *Local Climatological Data* Series, 1974, NOAA.

[‡] NG/FO = Natural Gas/Fuel Oil; E = Electricity. Principal fuel is defined as that used for heating. In all cases, electricity was the fuel used for refrigeration.

[§] Watts per square foot based on measured values by utility company meter at service entrance, 1977.

‖ Data shown for nearest recorded location.

(Each facility was self-contained, in that refrigeration and air conditioning equipment loads are included in power demands shown.)

phase. As the design proceeds through the preliminary and working drawing phases, these initial estimates should be modified by the actual conditions prevalent in the project.

3.2 SECONDARY VOLTAGE SELECTION

Introduction

Selection of the principal secondary utilization voltage is critical and should be made early in the preliminary design stage of a project. This is a critical decision because it has a significant impact on the cost of the distribution system, distribution equipment, and energy efficiency. The considerations are the same whether new service and distribution systems for a new building are to be considered or a renovation or addition to an existing building is considered. The options in the case of the latter, however, generally offer more limited choices.

Voltage Selection Considerations

The most prevalent secondary distribution voltage in commercial and institutional buildings today is 480Y/277 V, with a solidly grounded neutral. It is also a very common voltage in industrial plants and even in some high-rise, centrally air-conditioned and electrically heated residential buildings, because of the large loads.

The choice between 208Y/120-V and 480Y/277-V secondary distribution for commercial and institutional buildings depends on several factors. The most important of these are size and types of loads and the length of feeders. In general, large motor and fluorescent lighting loads, and long feeders, will tend to make the higher voltages, such as 480Y/ 277 V, more economical. Very large loads and long runs would indicate the use of medium-voltage distribution and load center unit substations close to the loads. Conversely, small loads, short runs, and a high percentage of incandescent lighting would favor lower utilization voltages such as 208Y/120 V.

The principal advantages of using higher secondary voltages in buildings are:

- Smaller conductors
- Lower voltage drop
- Fewer or smaller circuits
- Lower I^2R losses (thus, more energy efficient)
- Step-down transformers can be used for reregulation of voltage

Overall, the above advantages translate into a cost-effective, energy-efficient system design.

3.3   SHORT-CIRCUIT CALCULATIONS

Introduction

Several sections of the NEC relate to proper overcurrent protection. Safe and reliable application of overcurrent-protective devices based on these sections mandate that a short-circuit study and a selective-coordination study be conducted.

The protection for an electrical system should not only be safe under all service conditions but, to ensure continuity of service, it should be selectively coordinated as well. A coordinated system is one in which only the faulted circuit is isolated without disturbing any other part of the system. Overcurrent protection devices should also provide short-circuit as well as overload protection for system components, such as bus, wire, motor controllers, and so forth.

To obtain reliable, coordinated operation and assure that system components are protected from damage, it is necessary to first calculate the available fault current at various critical points in the electrical system.

Once the short-circuit levels are determined, the electrical design professional can specify proper interrupting rating requirements, selectively coordinate the system, and provide component protection.

General Comments on Short-Circuit Calculations

Short-circuit calculations should be done at all critical points in the electrical system, which would include the service entrance, panelboards, motor control centers, motor starters, transfer switches, and load centers.

Normally, short-circuit studies involve calculating a bolted three-phase fault condition. This can be characterized as all three phases "bolted" together to create a zero-impedance connection. This establishes a worst-case condition that results in maximum thermal and mechanical stress in the system. From this calculation, other types of fault conditions such as line-to-line and line-to ground can be obtained.

Sources of short-circuit current that are normally taken under consideration include utility generation, local generation, synchronous motors, and induction motors. Capacitor discharge currents can generally be neglected due to their short time duration.

Asymmetrical Components

Basically, the short-circuit current is determined by Ohm's law, except that the impedance is not constant because some reactance is included in the system. The effect of reactance in an AC system is to cause the initial current to be high and then decay toward steady-state (the Ohm's law) value. The fault current consists of an exponentially decreasing direct-current component superimposed upon a decaying alternating current. The rate of decay of both the DC and AC components depends upon the ratio of reactance to resistance (X/R) of the circuit. The greater this ratio, the longer the current remains higher than the steady-state value, which it will eventually reach.

The total fault current is not symmetrical with respect to the time axis because of the direct-current component; hence, it is called asymmetrical current. The DC component depends on the point on the voltage wave at which the fault is initiated (see Figure 3.11).

The AC component is not constant if rotating machines are connected to the system, because the impedance of this apparatus is not constant. The rapid variation of motor and generator impedance is due to these factors:

Subtransient reactance ($X_{d''}$): Determines fault current during the first cycle, and after about six cycles, this value increases to the transient reactance. It is used for the calculation of the momentary and interrupting duties of equipment and/or system.

Transient reactance ($X_{d'}$): Determines fault current after about six cycles, and in $\frac{1}{2}$ to 2 seconds this value increases to the value of the

FIGURE 3.11 Structure of asymmetrical current wave.

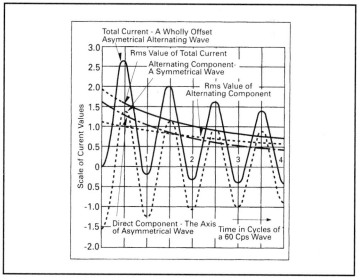

synchronous reactance. It is used in the setting of the phase over-current relays of generators.

Synchronous reactance (X_d): Determines fault current after steady-state condition is reached. It has no effect as far as short-circuit calculations are concerned, but it is useful in the determination of relay settings.

The calculation of asymmetrical currents is a laborious procedure because the degree of asymmetry is not the same on all three phases. It is common practice to calculate the root mean square (rms) symmetrical fault current, with the assumption being made that the DC component has decayed to zero, and then apply a multiplying factor to obtain the first half-cycle rms asymmetrical current, which is called the momentary current. For medium-voltage systems (defined by IEEE as greater than 1,000 V up to 69,000 V), the multiplying factor is established by NEMA and ANSI standards depending upon the operating speed of the breaker; for low-voltage systems (600 V and below), the multiplying factor is usually 1.17 (based on generally accepted use of an X/R ratio of 6.6, representing a source short-circuit power factor of 15 percent). These values take into account that medium-voltage breakers are rated on maximum asymmetry and low-voltage breakers are rated on average asymmetry.

To determine the motor contribution to the first half-cycle fault current when the system motor load is known, the following assumptions are generally made:

Induction motors: Use 4.0 times motor full-load current (impedance value of 25 percent).

Synchronous motors: Use 5.0 times motor full-load current (impedance value of 20 percent).

When the motor load is not known, the following assumptions are generally made:

208Y/120-V systems:
- Assume 50 percent lighting and 50 percent motor load.
- Assume motor feedback contribution of 2.0 times full-load current of transformer.

240-480-600-V three-phase, three-wire systems:
- Assume 100 percent motor load.
- Assume motors 25 percent synchronous and 75 percent induction.
- Assume motor feedback contribution of 4.0 times full-load current of transformer.

480Y/277-V systems in commercial buildings:
- Assume 50 percent induction motor load.
- Assume motor feedback contribution of 2.0 times full-load current of transformer or source.
- For industrial plants, make same assumptions as for three-phase, three-wire systems (above).

Medium-Voltage Motors:
- If known, use actual values. Otherwise, use the values indicated in the above for the same type of motor.

Procedures and Methods, Three-Phase Short-Circuit Calculations

Four basic methods are used to calculate short-circuit currents:

1. Ohmic method
2. Per-unit method
3. Computer software method
4. Point-to-point method

All four methods achieve essentially the same results with a reasonable degree of accuracy. The ohmic method is usually used for very simple systems. The per-unit and computer software methods are often used for more complex systems where there are many branches, buses, and critical points for fault calculations. The computer software method is by far the most popular method used today because of its speed and ability to run multiple system design condition scenarios. Computer software usually uses the per-unit method as the basis for computations.

For the purposes of this handbook, however, the point-to-point method offers a simple, effective, and quick way to determine available short-circuit levels in simple- to medium-complexity three-phase and single-phase electrical distribution systems with a reasonable degree of accuracy.

In any short-circuit calculation method, it must be understood that the calculations are performed without current-limiting devices in the system. Calculations are done as though these devices are replaced with copper bars, to determine the maximum available short-circuit current. This is necessary to project how the system and the current-limiting devices will perform.

Also, current-limiting devices do not operate in series to produce a "compounding" current-limiting effect. The downstream, or load-side, fuse/breaker will operate alone under a short-circuit condition if properly coordinated.

To start, first draw a one-line diagram showing all of the circuit components, parameters (including feeder lengths), and sources of fault current. Second, obtain the utility company–available short circuit in KVA, MVA, or SCA. With this information, the necessary calculations can be made to determine the fault current at any point in the electrical system.

The point-to-point method can best be illustrated by the following figures and table. Figure 3.12 shows the steps and equations needed in the point-to-point method. Figure 3.13 shows one-line diagrams of two systems (A and B) to be used as illustrative examples. Figures 3.14 and 3.15 show the calculations for these two examples. And, Table 3.17 provides the circuit constants needed in the equations for the point-to-point method.

How to Calculate Short-Circuit Currents at Ends of Conductors

Even the most exact methods for calculating fault energy (as in the point-to-point method) use some approximations and assumptions. Therefore, it is appropriate to select a method that is sufficiently accurate for the purpose, but not more burdensome than is justified. The following two methods make use of simplifications that are reasonable under most circumstances and will almost certainly yield answers that are on the safe side.

SHORT-CUT METHOD 1—ADDING Zs

This method uses the approximation of adding Zs instead of the accurate method of Rs and Xs (in complex form). Example:

- For a 480/277-V system with 30,000 amperes symmetrical available at the line side of a conductor run of 100 ft of 2–500 kcmil per phase and neutral, the approximate fault current at the load-side end of the conductors can be calculated as follows:

    277 V/30,000 A = 0.00923 Ω (source impedance).

- Conductor ohms for 500 kcmil conductor from Table 3.18 in magnetic conduit is 0.00546 Ω per 100 ft. For 100 ft and 2 conductors per phase we have:

    0.00546/2 = 0.00273 Ω (conductor impedance).

- Add source and conductor impedance or 0.00923 + 0.00273 = 0.01196 total ohms.

- Next, 277 V/0.01196 Ω = 23,160 A rms at load side of conductors.

For impedance values, refer to Tables 3.18, 3.19, and 3.20.

FIGURE 3.12   Point-to-point method, three-phase short-circuit calculations, basic calculation procedure and formulas.

The application of the point-to-point method permits the determination of available short-circuit currents with a reasonable degree of accuracy at various points for either 3Ø or 1Ø electrical distribution systems. This method can assume unlimited primary short-circuit current (infinite bus).

Basic Point-to-Point Calculation Procedure
Step 1. Determine the transformer full load amperes from either the nameplate or the following formulas:

3Ø Transformer $I_{f.l.} = \frac{KVA \times 1000}{E_{L-L} \times 1.732}$

1Ø Transformer $I_{f.l.} = \frac{KVA \times 1000}{E_{L-L}}$

Step 2. Find the transformer multiplier.

Multiplier $= \frac{100}{*\%Z_{trans}}$

* **Note.** Transformer impedance (Z) helps to determine what the short circuit current will be at the transformer secondary. Transformer impedance is determined as follows: The transformer secondary is short circuited. Voltage is applied to the primary which causes full load current to flow in the secondary. This applied voltage divided by the rated primary voltage is the impedance of the transformer.
Example: For a 480 volt rated primary, if 9.6 volts causes secondary full load current to flow through the shorted secondary, the transformer impedance is 9.6/480 = .02 = 2%Z.
In addition, UL listed transformer 25KVA and larger have a ± 10% impedance tolerance. Short circuit amperes can be affected by this tolerance.

Step 3. Determine the transformer let-thru short-circuit current**.

$I_{S.C.} = I_{f.l.} \times Multiplier$

** **Note.** Motor short-circuit contribution, if significant, may be added to the transformer secondary short-circuit current value as determined in Step 3. Proceed with this adjusted figure through Steps 4, 5 and 6. A practical estimate of motor short-circuit contribution is to multiply the total motor current in amperes by 4.

Step 4. Calculate the "f" factor.

3Ø Faults $f = \frac{1.732 \times L \times I}{C \times E_{L-L}}$

1Ø Line-to-Line (L-L) Faults on 1Ø Center Tapped Transformer $f = \frac{2 \times L \times I}{C \times E_{L-L}}$

1Ø Line-to-Neutral (L-N) Faults on 1Ø Center Tapped Transformer $f = \frac{2 \times L \times I}{C \times E_{L-N}}$†

Where:
L = length (feet) of circuit to the fault.
C = constant from Table 6, page 27. For parallel runs, multiply C values by the number of conductors per phase.
I = available short-circuit current in amperes at beginning of circuit.

† **Note.** The L-N fault current is higher than the L-L fault current at the secondary terminals of a single-phase center-tapped transformer. The short-circuit current available (I) for this case in Step 4 should be adjusted at the transformer terminals as follows.
At L-N center tapped transformer terminals.
I = 1.5 × L-L Short-Circuit Amperes at Transformer Terminals

At some distance from the terminals, depending upon wire size, the L-N fault current is lower than the L-L fault current. The 1.5 multiplier is an approximation and will theoretically vary from 1.33 to 1.67. These figures are based on change in turns ratio between primary and secondary, infinite source available, zero feet from terminals of transformer, and 1.2 x %X and 1.5 x %R for L-N vs. L-L resistance and reactance values. Begin L-N calculations at transformer secondary terminals, then proceed point-to-point.

Step 5. Calculate "M" (multiplier).

$M = \frac{1}{1 + f}$

Step 6. Calculate the available short-circuit symmetrical RMS current at the point of fault.

$I_{S.C. sym RMS} = I_{S.C.} \times M$

Calculation of Short-Circuit Currents at Second Transformer in System
Use the following procedure to calculate the level of fault current at the secondary of a second, downstream transformer in a system when the level of fault current at the transformer primary is known.

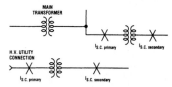

Procedure for Second Transformer in System

Step 1. Calculate the "f" factor ($I_{S.C. primary}$ known)

3Ø Transformer
($I_{S.C. primary}$ and $I_{S.C. secondary}$ are 3Ø fault values) $f = \frac{I_{S.C. primary} \times V_{primary} \times 1.73 (\%Z)}{100,000 \times KVA_{trans}}$

1Ø Transformer
($I_{S.C. primary}$ and $I_{S.C. secondary}$ are 1Ø fault values; $I_{S.C. secondary}$ is L-L) $f = \frac{I_{S.C. primary} \times V_{primary} \times (\%Z)}{100,000 \times KVA_{trans}}$

Step 2. Calculate "M" (multiplier).

$M = \frac{1}{1 + f}$

Step 3. Calculate the short-circuit current at the secondary of the transformer. (See Note under Step 3 of "Basic Point-to-Point Calculation Procedure".)

$I_{S.C. secondary} = \frac{V_{primary}}{V_{secondary}} \times M \times I_{S.C. primary}$

FIGURE 3.13 System A and system B circuit diagrams for sample calculations using point-to-point method.

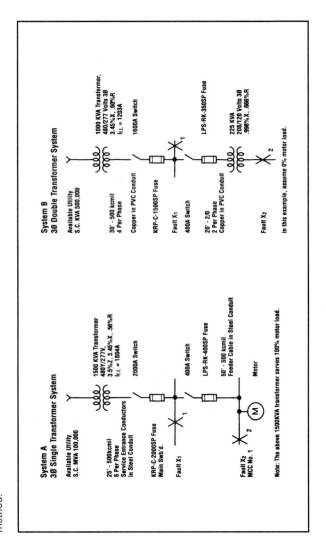

System A
3Ø Single Transformer System

Available Utility
S.C. MVA 100,000

1500 KVA Transformer
480Y/277V,
3.5%Z, 3.45%X, .56%R
I.f.l. = 1804A

2000A Switch

25' - 500kcmil
6 Per Phase
Service Entrance Conductors
in Steel Conduit

KRP-C-2000SP Fuse
Main Swb'd.

Fault X₁

400A Switch

LPS-RK-400SP Fuse

50' - 500 kcmil
Feeder Cable in Steel Conduit

Motor

Fault X₂
MCC No. 1

Note: The above 1500KVA transformer serves 100% motor load.

System B
3Ø Double Transformer System

Available Utility
S.C. KVA 500,000

1000 KVA Transformer,
480/277 Volts 3Ø
3.45%X, .60%R
I.f.l. = 1203A

1600A Switch

30' - 500 kcmil
4 Per Phase
Copper in PVC Conduit

KRP-C-1500SP Fuse

Fault X₁

400A Switch

LPS-RK-350SP Fuse

20' - 2/0
2 Per Phase
Copper in PVC Conduit

225 KVA
208/120 Volts 3Ø
.998%X, .666%R

Fault X₂

In this example, assume 0% motor load.

188

FIGURE 3.14 Point-to-point calculations for system A, to faults X_1 and X_2.

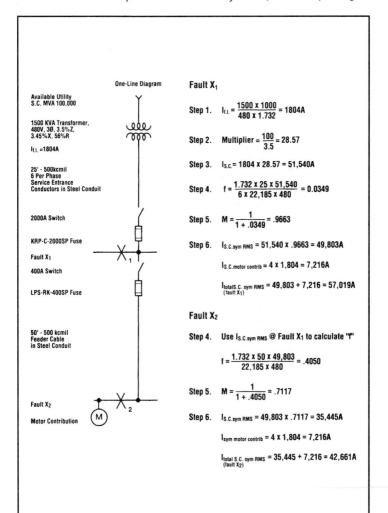

One-Line Diagram

Available Utility
S.C. MVA 100,000

1500 KVA Transformer,
480V, 3Ø, 3.5%Z,
3.45%X, 56%R

$I_{f.l.} = 1804A$

25' - 500kcmil
6 Per Phase
Service Entrance
Conductors in Steel Conduit

2000A Switch

KRP-C-2000SP Fuse

Fault X_1

400A Switch

LPS-RK-400SP Fuse

50' - 500 kcmil
Feeder Cable
in Steel Conduit

Fault X_2

Motor Contribution

Fault X_1

Step 1. $I_{f.l.} = \dfrac{1500 \times 1000}{480 \times 1.732} = 1804A$

Step 2. Multiplier $= \dfrac{100}{3.5} = 28.57$

Step 3. $I_{S.C.} = 1804 \times 28.57 = 51,540A$

Step 4. $f = \dfrac{1.732 \times 25 \times 51,540}{6 \times 22,185 \times 480} = 0.0349$

Step 5. $M = \dfrac{1}{1 + .0349} = .9663$

Step 6. $I_{S.C.sym \; RMS} = 51,540 \times .9663 = 49,803A$

$I_{S.C.motor \; contrib} = 4 \times 1,804 = 7,216A$

$I_{total S.C. \; sym \; RMS} = 49,803 + 7,216 = 57,019A$
(fault X_1)

Fault X_2

Step 4. Use $I_{S.C.sym \; RMS}$ @ Fault X_1 to calculate "f"

$f = \dfrac{1.732 \times 50 \times 49,803}{22,185 \times 480} = .4050$

Step 5. $M = \dfrac{1}{1 + .4050} = .7117$

Step 6. $I_{S.C.sym \; RMS} = 49,803 \times .7117 = 35,445A$

$I_{sym \; motor \; contrib} = 4 \times 1,804 = 7,216A$

$I_{total \; S.C. \; sym \; RMS} = 35,445 + 7,216 = 42,661A$
(fault X_2)

FIGURE 3.15 Point-to-point calculations for system B, to faults X_1 and X_2.

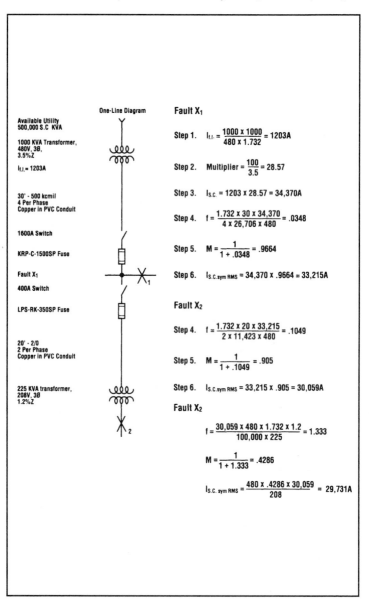

One-Line Diagram

Available Utility
500,000 S.C KVA

1000 KVA Transformer,
480V, 3Ø,
3.5%Z

$I_{f.l.} = 1203A$

30' - 500 kcmil
4 Per Phase
Copper in PVC Conduit

1600A Switch

KRP-C-1500SP Fuse

Fault X_1

400A Switch

LPS-RK-350SP Fuse

20' - 2/0
2 Per Phase
Copper in PVC Conduit

225 KVA transformer,
208V, 3Ø
1.2%Z

Fault X_1

Step 1. $I_{f.l.} = \dfrac{1000 \times 1000}{480 \times 1.732} = 1203A$

Step 2. Multiplier $= \dfrac{100}{3.5} = 28.57$

Step 3. $I_{s.c.} = 1203 \times 28.57 = 34,370A$

Step 4. $f = \dfrac{1.732 \times 30 \times 34,370}{4 \times 26,706 \times 480} = .0348$

Step 5. $M = \dfrac{1}{1 + .0348} = .9664$

Step 6. $I_{s.c.sym\,RMS} = 34,370 \times .9664 = 33,215A$

Fault X_2

Step 4. $f = \dfrac{1.732 \times 20 \times 33,215}{2 \times 11,423 \times 480} = .1049$

Step 5. $M = \dfrac{1}{1 + .1049} = .905$

Step 6. $I_{s.c.sym\,RMS} = 33,215 \times .905 = 30,059A$

Fault X_2

$f = \dfrac{30,059 \times 480 \times 1.732 \times 1.2}{100,000 \times 225} = 1.333$

$M = \dfrac{1}{1 + 1.333} = .4286$

$I_{s.c.\,sym\,RMS} = \dfrac{480 \times .4286 \times 30,059}{208} = 29,731A$

TABLE 3.17 "C" Values for Conductors and Busway

Copper AWG or kcmil	Three Single Conductors						Three-Conductor Cable					
	Conduit Steel			Nonmagnetic			Conduit Steel			Nonmagnetic		
	600V	5KV	15KV	600V	5KV	15KV	600V	5KV	15KV	600V	5KV	15KV
14	389	389	389	389	389	389	389	389	389	389	389	389
12	617	617	617	617	617	617	617	617	617	617	617	617
10	981	981	981	981	981	981	981	981	981	981	981	981
8	1557	1551	1557	1558	1555	1558	1559	1557	1559	1559	1558	1559
6	2425	2406	2389	2430	2417	2406	2431	2424	2414	2433	2428	2420
4	3806	3750	3695	3825	3789	3752	3830	3811	3778	3837	3823	3798
3	4760	4760	4760	4802	4802	4802	4760	4790	4760	4802	4802	4802
2	5906	5736	5574	6044	5926	5809	5989	5929	5827	6087	6022	5957
1	7292	7029	6758	7493	7306	7108	7454	7364	7188	7579	7507	7364
1/0	8924	8543	7973	9317	9033	8590	9209	9086	8707	9472	9372	9052
2/0	10755	10061	9389	11423	10877	10318	11244	11045	10500	11703	11528	11052
3/0	12843	11804	11021	13923	13048	12360	13656	13333	12613	14410	14118	13461
4/0	15082	13605	12542	16673	15351	14347	16391	15890	14813	17482	17019	16012
250	16483	14924	13643	18593	17120	15865	18310	17850	16465	19779	19352	18001
300	18176	16292	14768	20867	18975	17408	20617	20051	18318	22524	21938	20163
350	19703	17385	15678	22736	20526	18672	19557	21914	19821	22736	24126	21982
400	20565	18235	16365	24296	21786	19731	24253	23371	21042	26915	26044	23517
500	22185	19172	17492	26706	23277	21329	26980	25449	23125	30028	28712	25916
600	22965	20567	47962	28033	25203	22097	28752	27974	24896	32236	31258	27766
750	24136	21386	18888	28303	25430	22690	31050	30024	26932	32404	31338	28303
1000	25278	22539	19923	31490	28083	24887	33864	32688	29320	37197	35748	31959

TABLE 3.17 "C" Values for Conductors and Busway (*Continued*)

Aluminum

14	236	236	236	236	236	236	236	236	236	236	236	236
12	375	375	375	375	375	375	375	375	375	375	375	375
10	598	598	598	598	598	598	598	598	598	598	598	598
8	951	950	951	951	950	951	951	951	951	951	951	951
6	1480	1476	1472	1481	1478	1476	1481	1480	1478	1482	1481	1479
4	2345	2332	2319	2350	2341	2333	2351	2347	2339	2353	2349	2344
3	2948	2948	2948	2958	2958	2958	2948	2956	2948	2958	2958	2958
2	3713	3669	3626	3729	3701	3672	3733	3719	3693	3739	3724	3709
1	4645	4574	4497	4678	4631	4580	4686	4663	4617	4699	4681	4646
1/0	5777	5669	5493	5838	5766	5645	5852	5820	5717	5875	5851	5771
2/0	7186	6968	6733	7301	7152	6986	7327	7271	7109	7372	7328	7201
3/0	8826	8466	8163	9110	8851	8627	9077	8980	8750	9242	9164	8977
4/0	10740	10167	9700	11174	10749	10386	11184	11021	10642	11408	11277	10968
250	12122	11460	10848	12862	12343	11847	12796	12636	12115	13236	13105	12661
300	13909	13009	12192	14922	14182	13491	14916	14698	13973	15494	15299	14658
350	15484	14280	13288	16812	15857	14954	15413	16490	15540	16812	17351	16500
400	16670	15355	14188	18505	17321	16233	18461	18063	16921	19587	19243	18154
500	18755	16827	15657	21390	19503	18314	21394	20606	19314	22987	22381	20978
600	20093	18427	16484	23451	21718	19635	23633	23195	21348	25750	25243	23294
750	21766	19685	17686	23491	21769	19976	26431	25789	23750	25682	25141	23491
1000	23477	21235	19005	28778	26109	23482	29864	29049	26608	32938	31919	29135

TABLE 3.17   "C" Values for Conductors and Busway (*Continued*)

Ampacity	Busway					
	Plug-In	Feeder			High Impedance	
	Copper	Aluminum	Copper		Aluminum	Copper
225	28700	23000	18700		12000	—
400	38900	34700	23900		21300	—
600	41000	38300	36500		31300	—
800	46100	57500	49300		44100	—
1000	69400	89300	62900		56200	15600
1200	94300	97100	76900		69900	16100
1350	119000	104200	90100		84000	17500
1600	129900	120500	101000		90900	19200
2000	142900	135100	134200		125000	20400
2500	143800	156300	180500		166700	21700
3000	144900	175400	204100		188700	23800
4000	—	—	277800		256400	—

SHORT-CUT METHOD 2—CHART APPROXIMATE METHOD

The chart method is based on the following:

Motor Contribution Assumptions

120/208-V systems	50 percent motor load
	4 times motor FLA contribution
240- and 480-V systems	100 percent motor load
	4 times motor FLA contribution

Feeder Conductors
The conductor sizes most commonly used for feeders from molded case circuit breakers are shown. For conductor sizes not shown, the following table has been included for conversion to equivalent arrangements. In some cases, it may be necessary to interpolate for unusual feeder ratings. Table 3.21 is based on using copper conductor.

Short-Circuit Current Readout
The readout obtained from the charts is the rms symmetrical amperes available at the given distance from the transformer. The circuit breaker should have an interrupting capacity at least as large as this value.

HOW TO USE THE SHORT-CIRCUIT CHARTS

Step 1: Obtain the following data:
- System voltage
- Transformer kVA rating
- Transformer impedance
- Primary source fault energy available in KVA

TABLE 3.18 Average Characteristics of 600-Volt Conductors (Ohms per 100 ft)—Two or Three Single Conductors

Wire Size, AWG or kcmil	Copper Conductors						Aluminum Conductors					
	Magnetic Conduit			Nonmagnetic Conduit			Magnetic Conduit			Nonmagnetic Conduit		
	R	X	Z	R	X	Z	R	X	Z	R	X	Z
14	.3130	.00780	.3131	.3130	.00624	.3131	—	—	—	—	—	—
12	.1968	.00730	.1969	.1968	.00584	.1969	—	—	—	—	—	—
10	.1230	.00705	.1232	.1230	.00564	.1231	—	—	—	—	—	—
8	.0789	.00691	.0792	.0789	.00553	.0791	—	—	—	—	—	—
6	.0490	.00640	.0494	.0490	.00512	.0493	.0833	.00509	.0835	.0833	.00407	.0834
4	.0318	.00591	.0323	.0318	.00473	.0321	.0530	.00490	.0532	.0530	.00392	.0531
2	.0203	.00548	.0210	.0203	.00438	.0208	.0335	.00457	.0338	.0335	.00366	.0337
1	.0162	.00533	.0171	.0162	.00426	.0168	.0267	.00440	.0271	.0267	.00352	.0269
1/0	.0130	.00519	.01340	.0129	.00415	.01360	.0212	.00410	.0216	.0212	.00328	.0215
2/0	.0104	.00511	.01159	.0103	.00409	.01108	.0170	.00396	.0175	.0170	.00317	.0173
3/0	.00843	.00502	.00981	.00803	.00402	.00898	.01380	.00386	.0143	.01380	.00309	.01414
4/0	.00696	.00489	.00851	.00666	.00391	.00772	.01103	.00381	.0117	.01097	.00305	.01139
250	.00588	.00487	.00763	.00578	.00390	.00697	.00936	.00375	.01008	.00933	.00300	.00980
300	.00512	.00484	.00705	.00501	.00387	.00633	.00810	.00366	.00899	.00797	.00293	.00849
350	.00391	.00480	.00619	.00380	.00384	.00540	.00694	.00360	.00782	.00688	.00288	.00746
400	.00369	.00476	.00602	.00356	.00381	.00521	.00618	.00355	.00713	.00610	.00284	.00673
450	.00330	.00467	.00595	.00310	.00374	.00486	.00548	.00350	.00660	.00536	.00280	.00605
500	.00297	.00458	.00546	.00275	.00366	.00458	.00482	.00346	.00593	.00470	.00277	.00546
600	.00261	.00455	.00525	.00241	.00364	.00437	.00409	.00355	.00542	.00395	.00284	.00486
700	.00247	.00448	.00512	.00247	.00358	.00435	.00346	.00340	.00485	.00330	.00272	.00428
750	.00220	.00441	.00493	.00198	.00353	.00405	.00308	.00331	.00452	.00278	.00265	.00384
1000	—	—	—	—	—	—	.00250	.00330	.00414	.00230	.00264	.00350

TABLE 3.19 Average Characteristics of 600-Volt Conductors (Ohms per 100 ft)—Three Conductor Cables (and Interlocked Armored Cable)

Wire Size, AWG or kcmil	Copper Conductors						Aluminum Conductors					
	Magnetic Conduit			Nonmagnetic Conduit			Magnetic Conduit			Nonmagnetic Conduit		
	R	X	Z	R	X	Z	R	X	Z	R	X	Z
14	.3130	.00597	.3131	.3130	.00521	.3130	—	—	—	—	—	—
12	.1968	.00558	.1969	.1968	.00487	.1969	—	—	—	—	—	—
10	.1230	.00539	.1231	.1230	.00470	.1231	—	—	—	—	—	—
8	.0789	.00529	.0790	.0789	.00461	.0790	—	—	—	—	—	—
6	.0490	.00491	.0492	.0490	.00427	.0492	.0833	.00509	.0834	.0833	.00407	.0834
4	.0318	.00452	.0321	.0318	.00394	.0320	.0530	.00490	.0532	.0530	.00392	.0531
2	.0203	.00420	.0207	.0203	.00366	.0206	.0335	.00457	.0338	.0335	.00366	.0337
1	.0162	.00408	.0167	.0162	.00355	.0166	.0267	.00440	.0271	.0267	.00352	.0269
1/0	.0130	.00398	.0136	.0129	.00346	.0134	.0212	.00410	.0216	.0212	.00328	.0215
2/0	.0104	.00390	.0111	.0103	.00341	.0108	.0170	.00396	.0175	.0170	.00317	.0173
3/0	.00843	.00384	.00926	.00803	.00335	.00870	.01380	.00389	.0143	.01380	.00309	.01414
4/0	.00696	.00375	.00791	.00666	.00326	.00742	.01103	.00381	.0117	.01097	.00305	.01139
250	.00588	.00373	.00696	.00578	.00325	.00663	.00936	.00375	.01006	.00933	.00300	.00980
300	.00512	.00370	.00632	.00501	.00323	.00596	.00810	.00366	.00889	.00797	.00293	.00889
350	.00391	.00365	.00535	.00380	.00320	.00497	.00694	.00360	.00782	.00688	.00288	.00746
400	.00369	.00360	.00516	.00356	.00318	.00477	.00618	.00355	.00713	.00610	.00284	.00673
450	.00360	.00351	.00503	.00310	.00312	.00440	.00548	.00350	.00650	.00536	.00280	.00605
500	.00297	.00343	.00454	.00275	.00305	.00411	.00482	.00346	.00593	.00470	.00277	.00546
600	.00261	.00337	.00426	.00241	.00303	.00387	.00409	.00341	.00542	.00395	.00284	.00486
700	.00247	.00330	.00412	.00227	.00298	.00375	.00346	.00355	.00486	.00330	.00272	.00428
750	.00220	.00323	.00391	.00198	.00294	.00354	.00308	.00331	.00452	.00278	.00265	.00384
1000	—	—	—	—	—	—	.00250	.00330	.00414	.00230	.00264	.00350

① Resistance and reactance are phase-to-neutral values, based on 60 Hertz ac. 3-phase, 4-wire distribution, in ohms per 100 feet of circuit length (not total conductor lengths).

② Based upon conductivity of 100% for copper, 61% for aluminum.

③ Based on conductor temperatures of 75°C. Reactance values will have negligible variation with temperature. Resistance of both copper and aluminum conductors will be approximately 5% lower at 60°C or 5% higher at 90°C. Data shown in tables may be used without significant error between 60°C and 90°C.

④ For interlocked armored cable, use magnetic conduit data for steel armor and non-magnetic conduit data for aluminum armor.

⑤ $= \sqrt{X^2 + R}$

⑥ **For busway impedance data, see page 477.**

TABLE 3.20 LV Busway, R, X, and Z (Ohms per 100 ft)

Ampere Rating	Plug-in			Feeder		
	Resistance	Reactance	Impedance	Resistance	Reactance	Impedance
Aluminum						
225	.00737	.00323	.00805	.00737	.00323	.00805
400	.00371	.00280	.00465	.00371	.00280	.00465
600 ·	.00291	.00212	.00360	.00289	.00127	.00316
800	.00248	.00114	.00273	.00244	.000660	.00253
1000	.00188	.00100	.00213	.00197	.000552	.00205
1200	.00155	.000755	.00172	.00159	.000490	.00166
1350	.00130	.000600	.00143	.00134	.000385	.00139
1600	.00106	.000480	.00116	.00112	.000350	.00117
2000	.000841	.000449	.000953	.000864	.000310	.000918
2500	.000648	.000290	.000710	.000664	.000250	.000710
3000	.000521	.000183	.000552	.000558	.000197	.000592
4000	.000397	.000175	.000434	.000409	.000135	.000431
Copper						
225	.00425	.00323	.00534	.00425	.00323	.00534
400	.00291	.00301	.00419	.00291	.00301	.00419
600	.00212	.00234	.00316	.00202	.00170	.00264
800	.00169	.00212	.00271	.00188	.00149	.00240
1000	.00144	.00114	.00184	.00158	.000965	.00185
1200	.00112	.00100	.00150	.00120	.000552	.00132
1350	.00101	.000960	.00139	.00108	.000510	.00119
1600	.000898	.000716	.00115	.000920	.000480	.00104
2000	.000667	.000562	.000872	.000724	.000434	.000844
2500	.000494	.000449	.000668	.000520	.000305	.000603
3000	.000465	.000355	.000585	.000488	.000290	.000568
4000	.000336	.000242	.000414	.000378	.000203	.000429
5000	. . . . . .	. . . . . .	. . . . . .	.000264	.000139	.000298

TABLE 3.21 Conductor Conversion (Based on Using Copper Conductor)

If Your Conductor is:	Use Equivalent Arrangement
3 – No. 4/0 cables	2 – 500 MCM
4 – No. 2/0 cables	2 – 500 MCM
3 – 2000 MCM cables	4 – 750 MCM
5 – 400 MCM cables	4 – 750 MCM
6 – 300 MCM cables	4 – 750 MCM
800 Amp busway	2 – 500 MCM
1000 Amp busway	2 – 500 MCM
1600 Amp busway	4 – 750 MCM

This makes no sense. Nothing on these charts relates to the letters A thru F

Service and Distribution **197**

Step 2: Select the applicable chart from Figure 3.16 (Charts 1–13). The charts are grouped by secondary system voltage, which is listed with each transformer. Within each group, the chart for the lowest kVA transformer is shown first, followed in ascending order to the highest-rated transformer.

Step 3: Select the family of curves that is closest to the "available source kVA." The upper-value line family of curves is for a source of 500,000 kVA. The lower-value line family of curves is for a source of 50,000 kVA. You may interpolate between curves if necessary, but for values above 100,000 kVA, it is appropriate to use the 500,000 kVA curves.

Step 4: Select the specific curve for the conductor size being used. If your conductor size is something other than the sizes shown on the chart, refer to the conductor conversion Table 3.21.

Step 5: Enter the chart along the bottom horizontal scale with the distance (in feet) from the transformer to the fault point. Draw a vertical line up the chart to the point at which it intersects the

FIGURE 3.16 Charts 1 through 13 for calculating short-circuit currents using chart approximate method.

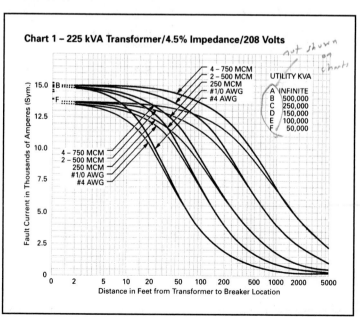

FIGURE 3.16   Charts 1 through 13 for calculating short-circuit currents using chart approximate method. (*Continued*)

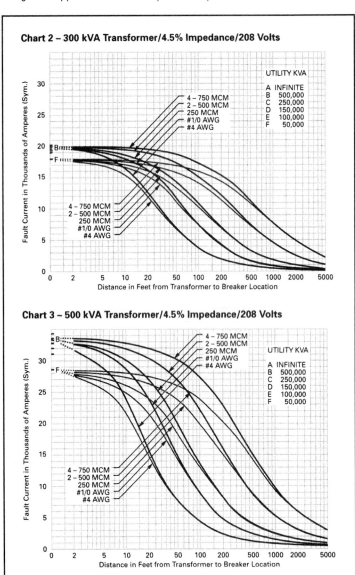

FIGURE 3.16   Charts 1 through 13 for calculating short-circuit currents using chart approximate method. (*Continued*)

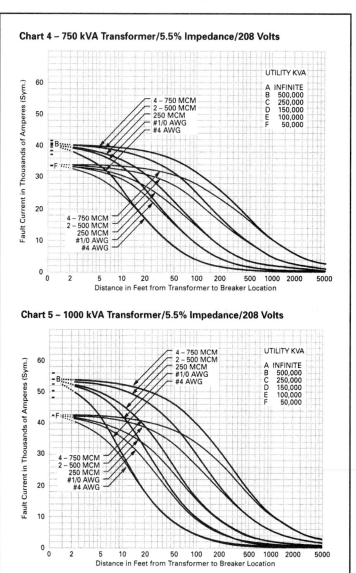

Chart 4 – 750 kVA Transformer/5.5% Impedance/208 Volts

Chart 5 – 1000 kVA Transformer/5.5% Impedance/208 Volts

FIGURE 3.16   Charts 1 through 13 for calculating short-circuit currents using chart approximate method. (*Continued*)

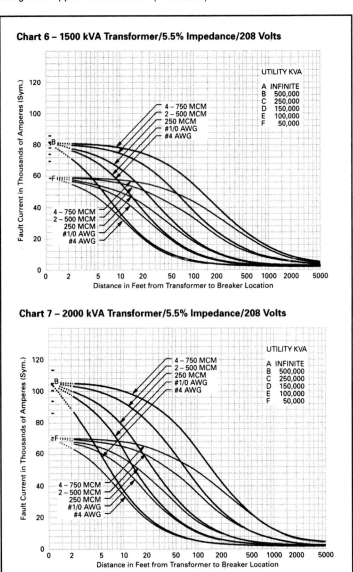

FIGURE 3.16   Charts 1 through 13 for calculating short-circuit currents using chart approximate method. (*Continued*)

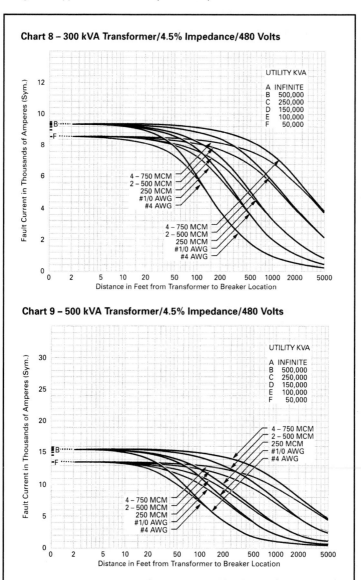

FIGURE 3.16 Charts 1 through 13 for calculating short-circuit currents using chart approximate method. (*Continued*)

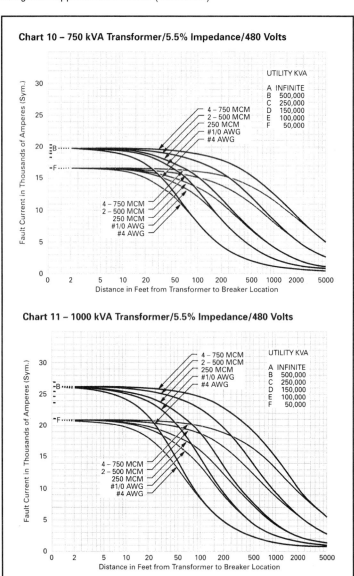

FIGURE 3.16   Charts 1 through 13 for calculating short-circuit currents using chart approximate method. (*Continued*)

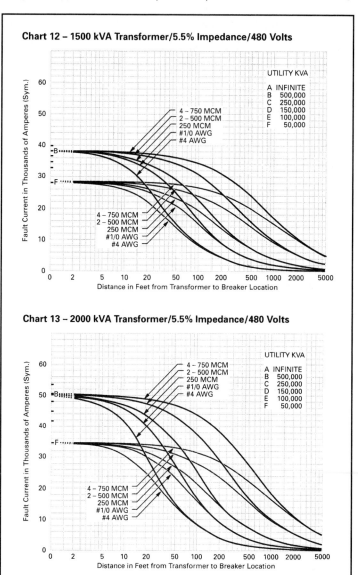

selected curve. Then draw a horizontal line to the left from this point to the scale along the left side of the chart.

Step 6: The value obtained from the left-hand vertical scale is the fault current (in thousands of amperes) available at the fault point.

Table 3.22 shows secondary short-circuit capacity of typical power transformers.

3.4 SELECTIVE COORDINATION OF OVERCURRENT-PROTECTIVE DEVICES

Introduction

It is not enough to select protective devices based solely on their ability to carry the system load current and interrupt the maximum fault current at their respective levels. A properly engineered system will allow *only* the protective device nearest the fault to open, leaving the remainder of the system undisturbed and preserving continuity of service.

We may then define selective coordination as *the act of isolating a faulted circuit from the remainder of the electrical system, thereby eliminating unnecessary power outages. The faulted circuit is isolated by the selective operation of only that overcurrent-protective device closest to the overcurrent condition.*

Popular Methods of Performing a Selective Coordination Study

Currently, two methods are most often used to perform a coordination study:

- Overlays of time-current curves, which use a light table and manufacturers' published data, then hand-plot on log-log paper.
- Computer programs, which use a PC and allow the designer to select time-current curves published by the manufacturers and transfer to a plotter or printer, following proper selections.

This text will apply to both methods.

Recommended Procedures

The following steps are recommended when conducting a selective coordination study.

1. *One-line diagram:* Obtain or develop the electrical system one-line diagram that identifies important system components, as given hereafter.

TABLE 3.22 Secondary Short-Circuit Capacity of Typical Power Transformers

Transformer Rating 3-Phase kVA and Impedance Percent	Maximum Short Circuit kVA Available From Primary System	208 Volts, 3-Phase Rated Load Continuous Current, Amps	Transformer Alone ②	50% Motor Load ②	Combined	240 Volts, 3-Phase Rated Load Continuous Current, Amps	Transformer Alone ②	100% Motor Load ②	Combined	480 Volts, 3-Phase Rated Load Continuous Current, Amps	Transformer Alone ②	100% Motor Load ①	Combined	600 Volts, 3-Phase Rated Load Continuous Current, Amps	Transformer Alone ②	100% Motor Load ①	Combined
300 5%	50000	834	14900	1700	16600	722	12900	2900	15800	361	6400	1400	7800	289	5200	1200	6400
	100000		15700		17400		13600		16500		6800		8200		5500		6700
	150000		16000		17700		13900		16800		6900		8300		5600		6800
	250000		16300		18000		14100		17000		7000		8400		5600		6800
	500000		16500		18200		14300		17200		7100		8500		5700		6900
	Unlimited		16700		18400		14400		17300		7200		8600		5800		7000
500 5%	50000	1388	21300	2800	25900	1203	20000	4800	24800	601	10000	2400	12400	481	8000	1900	9900
	100000		25200		28000		21900		26700		10900		13300		8700		10600
	150000		26000		28800		22500		27300		11300		13700		9000		10900
	250000		26700		29500		23100		27900		11600		14000		9300		11200
	500000		27200		30000		23600		28400		11800		14200		9400		11300
	Unlimited		27800		30600		24100		28900		12000		14400		9600		11500
750 5.75%	50000	2080	28700	4200	32900	1804	24900	7200	32100	902	12400	3600	16000	722	10000	2900	12900
	100000		32000		36200		27800		35000		13900		17500		11100		14000
	150000		33300		37500		28900		36100		14400		18000		11600		14500
	250000		34400		38600		29800		37000		14900		18500		11900		14800
	500000		35200		39400		30600		37800		15300		18900		12200		15100
	Unlimited		36200		40400		31400		38600		15700		19300		12600		15500
1000 5.75%	50000	2776	35900	5600	41500	2406	31000	9600	40600	1203	15500	4800	20300	962	12400	3900	16300
	100000		41200		46800		35600		45200		17800		22600		14300		18200
	150000		43300		48900		37500		47100		18700		23500		15000		18900
	250000		45200		50800		39100		48700		19600		24400		15600		19500
	500000		46700		52300		40400		50000		20200		25000		16200		20100
	Unlimited		48300		53900		41800		51400		20900		25700		16700		20600

TABLE 3.22 Secondary Short-Circuit Capacity of Typical Power Transformers (*Continued*)

Transformer Rating 3-Phase kVA and Impedance Percent	Maximum Short Circuit Available kVA From Primary System	208 Volts, 3-Phase Rated Load Cont. Current, Amps	Transformer Alone ①	50% Motor Load ②	Combined	240 Volts, 3-Phase Rated Load Cont. Current, Amps	Transformer Alone ①	100% Motor Load ②	Combined	480 Volts, 3-Phase Rated Load Cont. Current, Amps	Transformer Alone ②	100% Motor Load ①	Combined	600 Volts, 3-Phase Rated Load Cont. Current, Amps	Transformer Alone ②	100% Motor Load ①	Combined
1500 5.75%	50000	4164	47600	8300	55900	3609	41200	14400	55600	1804	20600	7200	27800	1444	16500	5800	22300
	100000		57500		65800		49800		64200		24900		32100		20000		25800
	150000		61800		70100		53500		67900		26700		33900		21400		27200
	250000		65600		73900		56800		71200		28400		35600		22700		28500
	500000		68800		77100		59600		74000		29800		37000		23900		29700
	Unlimited		72500		80800		62800		77200		31400		38600		25100		30900
2000 5.75%	50000									2406	24700	9600	34300	1924	19700	7800	27500
	100000										31000		40600		24800		32600
	150000										34000		43600		27200		35000
	250000										36700		46300		29400		37200
	500000										39100		48700		31300		39100
	Unlimited										41800		51400		33500		41300
2500 5.75%	50000									3008	28000	12000	40000	2405	22400	9600	32000
	100000										36500		48500		29200		38800
	150000										40500		52500		32400		42000
	250000										44600		56600		35600		45200
	500000										48100		60100		38500		48100
	Unlimited										52300		64300		41800		51400

① Short-circuit capacity values shown correspond to kVA and impedances shown in this table. For impedances other than these, short-circuit currents are inversely proportional to impedance.

② The motor's short-circuit current contributions are computed on the basis of motor characteristics that will give four times normal current. For 208 volts, 50% motor load is assumed while for other voltages 100% motor load is assumed. For other percentages, the motor short-circuit current will be in direct proportion.

a. *Transformers:* Obtain the following data for protection and coordination information of transformers:

- kVA rating
- Inrush points
- Primary and secondary connections
- Impedance
- Damage curves
- Primary and secondary voltages
- Liquid or dry type

b. *Conductors:* Check phase, neutral, and equipment grounding. The one-line diagram should include information such as:

- Conductor size
- Number of conductors per phase
- Material (copper or aluminum)
- Insulation
- Conduit (magnetic or nonmagnetic)

From this information, short-circuit withstand curves can be developed. This provides information on how overcurrent devices will protect conductors from overload *and* short-circuit damage.

c. *Motors:* The system one-line diagram should include motor information such as:

- Full-load currents
- Horsepower
- Voltage
- Type of starting characteristic (e.g., across the line)
- Type of overload relay (Class 10, 20, 30)

Overload protection of the motor and motor circuit can be determined from this data.

d. *Fuse characteristics:* Fuse types/classes should be identified on the one-line diagram.

e. *Circuit breaker characteristics:* Circuit breaker types should be identified on the one-line diagram.

f. *Relay characteristics:* Relay types should be identified on the one-line diagram.

2. *Short-circuit study:* Perform a short-circuit analysis, calculating maximum available short-circuit currents at critical points in the distribution system (such as transformers, main switchgear, panelboards, motor control centers, load centers, and large motors and generators). Refer to the previous section.

3. *Helpful hints:*

a. *Determine the ampere scale selection:* It is most convenient to place the time-current curves in the center of the log-log paper.

This is accomplished by multiplying or dividing the ampere scale by a factor of 10.

b. *Determine the reference (base) voltage:* The best reference voltage is the voltage level at which most of the devices being studied fall. On most low-voltage industrial and commercial studies, the reference voltage will be 208, 240, or 480 V. Devices at other voltage levels will be shifted by a multiplier based on the transformer turn ratio. The best reference voltage will require the least amount of manipulation. Most computer programs will automatically make these adjustments when the voltage levels of the devices are identified by the input data.

c. *Commencing the analysis:* The starting point can be determined by the designer. Typically, studies begin with the main circuit devices and work down through the feeders and branches. (Right to left on your log-log paper.)

d. *Multiple branches:* If many branches are taken off one feeder, and the branch loads are similar, the largest rated branch-circuit device should be checked for coordination with upstream devices. If the largest branch device will coordinate, and the branch devices are similar, they generally will coordinate as well. (The designer may wish to verify other areas of protection on those branches, conductors, and so forth.)

e. *Don't overcrowd the study:* Many computer-generated studies will allow a maximum of 10 device characteristics per page. It is good practice, however, to have a minimum of 3 devices in a coordination sequence, so that there is always one step of overlap.

f. *Existing systems:* The designer should be aware that when conducting a coordination study on an existing system, optimum coordination cannot always be achieved and compromise may be necessary. It is then necessary to exercise experience and judgment to achieve the best coordination possible to mitigate the effects of blackout conditions. The designer must set priorities within the constraints of the system under study.

g. *Conductor short-circuit protection:* In low-voltage (600 V or less) systems, it is generally safe to ignore possible damage to conductors from short circuits, because the philosophy is to isolate a fault as quickly as possible; thus, the I^2t energy damage curves don't have enough time to come into play (become a factor). In medium- and high-voltage systems, however, in which the philosophy is to have the overcurrent protection "hang in" as long as possible, the contrary is true; therefore, it can be a significant factor.

h. *One-line diagram:* A one-line diagram of the study should be drawn for future reference.

Example of Selective Coordination Study

INTRODUCTION

The following example will analyze in detail the system shown in Figure 3.17. It is understood that a short-circuit study has been completed, and all protective devices have adequate interrupting ratings. A selective coordination analysis is the next step.

FIGURE 3.17   Example system one-line diagram for selective coordination study.

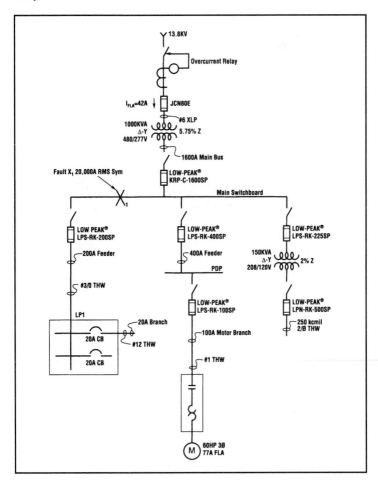

The simple radial system will involve three separate time-current studies, applicable to the three feeder/branches shown. The three time-current curves and their accompanying notes are self-explanatory (Figures 3.18 through 3.20).

FIGURE 3.18 Time-current curve No. 1 for system shown in Figure 3.17 with analysis notes and comments.

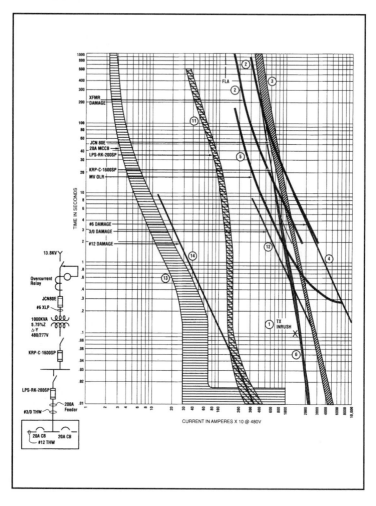

FIGURE 3.18 Time-current curve No. 1 for system shown in Figure 3.17 with analysis notes and comments. (*Continued*)

Notes:
1. TCC1 includes the primary fuse, secondary main fuse, 200 ampere feeder fuse, and 20 ampere branch circuit breaker from LP1.
2. Analysis will begin at the main devices and proceed down through the system.
3. Reference (base) voltage will be 480 volts, arbitrarily chosen since most of the devices are at this level.
4. Selective coordination between the feeder and branch circuit is not attainable for faults above 2500 amperes that occur on the 20 amp branch circuit, from LP1. Notice the overlap of the 200 ampere fuse and 20 ampere circuit breaker.
5. The required minimum ratio of 2:1 is easily met between the KRP-C-1600SP and the LPS-RK-200SP.

Device ID	Description	Comments
①	1000KVA XFMR Inrush Point	12 x FLA @ .1 Seconds
②	1000KVA XFMR Damage Curves	5.75%Z, liquid filled (Footnote 1) (Footnote 2)
③	JCN 80E	E-Rated Fuse
④	#6 Conductor Damage Curve	Copper, XLP Insulation
⑤	Medium Voltage Relay	Needed for XFMR Primary Overload Protection
⑥	KRP-C-1600SP	Class L Fuse
⑪	LPS-RK-200SP	Class RK1 Fuse
⑫	3/0 Conductor Damage Curve	Copper THW Insulation
⑬	20A CB	Thermal Magnetic Circuit Breaker
⑭	#12 Conductor Damage Curve	Copper THW Insulation

Footnote 1: Transformer damage curves indicate when it will be damaged, thermally and/or mechanically, under overcurrent conditions.

Transformer impedance, as well as primary and secondary connections, and type, all will determine their damage characteristics.

Footnote 2: A Δ-Y transformer connection requires a 15% shift, to the right, of the L-L thermal damage curve. This is due to a L-L secondary fault condition, which will cause 1.0 p.u. to flow through one primary phase, and .866 p.u. through the two faulted secondary phases. (These currents are p.u. of 3-phase fault current.)

FIGURE 3.19 Time-current curve No. 2 for system shown in Figure 3.17 with analysis notes and comments.

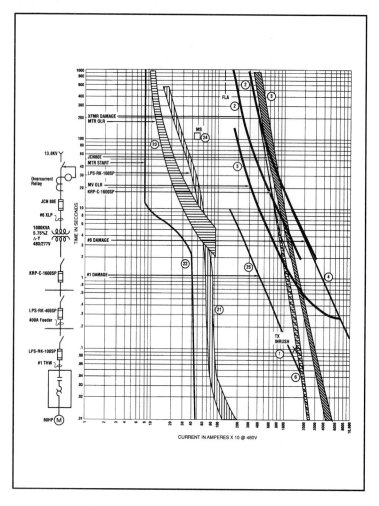

FIGURE 3.19   Time-current curve No. 2 for system shown in Figure 3.17 with analysis notes and comments. (*Continued*)

Notes:
1. TCC2 includes the primary fuse, secondary main fuse, 400 ampere feeder fuse, 100 ampere motor branch fuse, 77 ampere motor and overload relaying.
2. Analysis will begin at the main devices and proceed down through the system.
3. Reference (base) voltage will be 480 volts, arbitrarily chosen since most of the devices are at this level.

Device ID	Description	Comment
①	1000KVA XFMR Inrush Point	12 x FLA @ .1 seconds
②	1000KVA XFMR Damage Curves	5.75%Z, liquid filled (Footnote 1) (Footnote 2)
③	JCN 80E	E-Rated Fuse
④	#6 Conductor Damage Curve	Copper, XLP Insulation
⑤	Medium Voltage Relay	Needed for XFMR Primary Overload Protection
⑥	KRP-C-1600SP	Class L Fuse
㉑	LPS-RK-100SP	Class RK1 Fuse
㉒	Motor Starting Curve	Across the Line Start
㉓	Motor Overload Relay	Class 10
㉔	Motor Stall Point	Part of a Motor Damage Curve
㉕	#1 Conductor Damage Curve	Copper THW Insulation

Footnote 1: Transformer damage curves indicate when it will be damaged, thermally and/or mechanically, under overcurrent conditions.

Transformer impedance, as well as primary and secondary connections, and type, all will determine their damage characteristics.

Footnote 2: A Δ-Y transformer connection requires a 15% shift, to the right, of the L-L thermal damage curve. This is due to a L-L secondary fault condition, which will cause 1.0 p.u. to flow through one primary phase, and .866 p.u. through the two faulted secondary phases. (These currents are p.u. of 3-phase fault current.)

FIGURE 3.20 Time-current curve No. 3 for system shown in Figure 3.17 with analysis notes and comments.

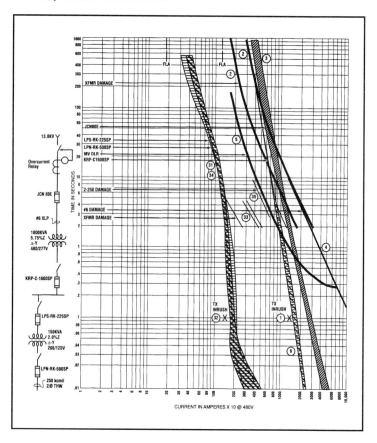

Short-Cut Ratio Method

INTRODUCTION

The selectivity ratio guide in Table 3.23 may be used for an easy check on fuse selectivity regardless of the short-circuit current levels involved. It may also be used for fixed thermal-magnetic trip circuit breakers (exercising good judgment) with a reasonable degree of accuracy. Where medium- and high-voltage primary fuses and relays are involved, the time-current characteristic curves should be plotted on standard log-log graph paper for proper study.

FIGURE 3.20 Time-current curve No. 3 for system shown in Figure 3.17
with analysis notes and comments. (*Continued*)

Notes:
1. TCC3 includes the primary fuse, secondary main fuse,
225 ampere feeder/transformer primary and secondary
fuses.
2. Analysis will begin at the main devices and proceed
down through the system.
3. Reference (base) voltage will be 480 volts, arbitrarily
chosen since most of the devices are at this level.
4. Relative to the 225 ampere feeder, coordination between
primary and secondary fuses is not attainable, noted by
overlap of curves.
5. Overload and short circuit protection for the 150 KVA
transformer is afforded by the LPS-RK-225SP fuse.

Device ID	Description	Comment
①	1000KVA XFMR Inrush Point	12 x FLA @ .1 seconds
②	1000KVA XFMR Damage Curves	5.75%Z, liquid filled (Footnote 1) (Footnote 2)
③	JCN 80E	E-Rated Fuse
④	#6 Conductor Damage Curve	Copper, XLP Insulation
⑤	Medium Voltage Relay	Needed for XFMR Primary Overload Protection
⑥	KRP-C-1600SP	Class L Fuse
㉛	LPS-RK-225SP	Class RK1 Fuse
㉜	150 KVA XFMR Inrush Point	12 x FLA @.1 Seconds
㉝	150 KVA XFMR Damage Curves	2.00% Dry Type (Footnote 3)
㉞	LPN-RK-500SP	Class RK1 Fuse
㉟	2-250kcmil Conductors Damage Curve	Copper THW Insulation

Footnote 1: Transformer damage curves indicate when it will be damaged,
thermally and/or mechanically, under overcurrent conditions.

Transformer impedance, as well as primary and secondary
connections, and type, all will determine their damage
characteristics.

Footnote 2: A Δ-Y transformer connection requires a 15% shift, to the right,
of the L-L thermal damage curve. This is due to a L-L
secondary fault condition, which will cause 1.0 p.u. to flow
through one primary phase, and .866 p.u. through the two
faulted secondary phases. (These currents are p.u. of 3-phase
fault current.)

Footnote 3: Damage curves for a small KVA (<500KVA) transformer,
illustrate thermal damage characteristics for Δ-Y connected.
From right to left, these reflect damage characteristics, for a
line-line fault, 3Ø fault, and L-G fault condition.

TABLE 3.23 Selectivity Ratio Guide

Circuit Current Rating Type	Trade Name & Class	Buss Symbol	601-6000A Time-Delay LOW-PEAK (L) KRP-CSP	601-4000A Time-Delay LIMITRON (L) KLU	0-600A Dual-Element Time-Delay LOW-PEAK (RK1) LPN-RKSP LPS-RKSP	LP-JSP (J)**	FUSETRON (RK5) FRN-R FRS-R	601-6000A Fast-Acting LIMITRON (L) KTU	0-600A Fast-Acting LIMITRON (RK1) KTN-R KTS-R	0-1200A T-TRON (T) JJN JJS	0-600A LIMITRON (J) JKS	0-60A Time-Delay SC (G) SC
601 to 6000A Time-Delay	LOW-PEAK (L)	KRP-CSP	2:1	2.5:1	2:1	2:1	4:1	2:1	2:1	2:1	2:1	N/A
601 to 4000A Time-Delay	LIMITRON (L)	KLU	2:1	2:1	2:1	2:1	4:1	2:1	2:1	2:1	2:1	N/A
0 to 600A Dual Element	LOW-PEAK (RK1)	LPN-RKSP LPS-RKSP	–	–	2:1	2:1	8:1	–	3:1	3:1	3:1	4:1
	(J)	LPJSP**	–	–	2:1	2:1	8:1	–	3:1	3:1	3:1	4:1
	FUSETRON (RK5)	FRN-R FRS-R	–	–	1.5:1	1.5:1	2:1	–	1.5:1	1.5:1	1.5:1	1.5:1
601 to 6000A	LIMITRON (L)	KTU	2:1	2.5:1	2:1	2:1	6:1	2:1	2:1	2:1	2:1	N/A
0 to 600A Fast-Acting	LIMITRON (RK1)	KTN-R KTS-R	–	–	3:1	3:1	8:1	–	3:1	3:1	3:1	4:1
0 to 1200A	T-TRON (T)	JJN JJS	–	–	3:1	3:1	8:1	–	3:1	3:1	3:1	4:1
0 to 600A	LIMITRON (J)	JKS	–	–	2:1	2:1	8:1	–	3:1	3:1	3:1	4:1
0 to 60A Time-Delay	SC (G)	SC	–	–	3:1	3:1	4:1	–	2:1	2:1	2:1	2:1

Note: At some values of fault current, specified ratios may be lowered to permit closer fuse sizing. Plot fuse curves or consult with Bussmann.

General Notes: Ratios given in this Table apply only to Buss fuses. When fuses are within the same case size, consult Bussmann.

** Consult Bussmann for latest LP-JSP ratios.

3.5 COMPONENT SHORT-CIRCUIT PROTECTION

Introduction

This section analyzes the protection of electrical system components from fault currents. It gives the specifier the necessary information regarding the withstand rating of electrical circuit components, such as wire, bus, motor starters, and so on. Proper protection of circuits will improve reliability and reduce the possibility of injury. Electrical systems can be destroyed if the overcurrent devices do not limit the short-circuit current to within the withstand rating of the system's components. Merely matching the ampere rating of a protective device will not assure component protection under short-circuit conditions.

The NEC covers component protection in several sections. The first section to note is NEC Section 110-10.

NEC SECTION 110-10: CIRCUIT IMPEDANCE AND OTHER CHARACTERISTICS

The overcurrent-protective devices, the total impedance, the component short-circuit withstand ratings, and other characteristics of the circuit to be protected shall be so selected and coordinated as to permit the circuit-protective devices used to clear a fault without the occurrence of extensive damage to the electrical components of the circuit. This fault shall be assumed to be either between two or more circuit conductors, or between any circuit conductor and the grounding conductor or enclosing metal raceway.

This requires that overcurrent-protective devices such as fuses and circuit breakers be selected in such a manner that the short-circuit withstand ratings of the system components will not be exceeded should a short circuit occur.

The *short-circuit withstand rating* is the maximum short-circuit current that a component can safely withstand. Failure to provide adequate protection may result in component destruction under short-circuit conditions.

CALCULATING SHORT-CIRCUIT CURRENTS

Before proceeding with a systems analysis of wire, cable, and other component protection requirements, it will be necessary to establish the short-circuit current levels available at various points in the electrical system. This can be accomplished by using the techniques given in Section 3.3 ("Short-Circuit Calculations"). After calculating the fault levels throughout the electrical system, the next step is to check the withstand ratings of wire and cable, bus, circuit breakers, transfer switches, motor starters, and so forth, not only under overload conditions, but also under short-circuit conditions.

NOTE The let-thru energy of the protective device must be equal to or less than the short-circuit withstand rating of the component being protected.

PROTECTING SYSTEM COMPONENTS—A PRACTICAL APPROACH

Most electrical equipment has a withstand rating that is defined in terms of a root mean square (rms) symmetrical short-circuit current, and in some cases, peak let-thru current. These values have been established through short-circuit testing of that equipment according to an accepted industry standard. Or, as is the case with conductors, the withstand rating is based on a mathematical calculation and is also expressed as an rms symmetrical short-circuit current.

The following provides the short-circuit withstand data of each system component. Please note that where industry standards are given (for example, NEMA), individual manufacturers of equipment often have withstand ratings that exceed industry standards.

- **A.** Wire and cable (Figures 3.21 through 3.26 and Table 3.24)
- **B.** Bus (busway, switchboards, motor control centers, and panelboards; Table 3.25)
- **C.** Low-voltage motor controllers (Table 3.26)
- **D.** Molded case circuit breakers (Table 3.27)
- **E.** Transformers (Table 3.28)
- **F.** Transfer switches (Table 3.29)
- **G.** HVAC equipment (Table 3.30)

Current Limitation
DEFINITION OF CURRENT LIMITATION

Today, most electrical distribution systems are capable of delivering very high short-circuit currents, some in excess of 200,000 A. If the components are not capable of handling these short-circuit currents, they could easily be damaged or destroyed. The current-limiting ability of today's modern fuses and current-limiting breakers (with current-limiting fuses) allows components with low short-circuit withstand ratings to be specified in spite of high available fault currents.

NEC Article 240-11 offers the following definition of a current-limiting device: "A current-limiting overcurrent-protective device is a device that, when interrupting currents in its current-limiting range, will reduce the current flowing in the faulted circuit to a magnitude substantially less than that obtainable in the same circuit if the device were replaced with a solid conductor having comparable impedance."

FIGURE 3.21   Short-circuit current withstand chart for copper cables with paper, rubber, or varnished cloth insulation.

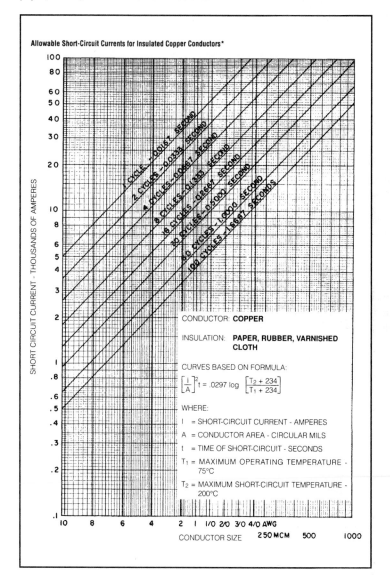

FIGURE 3.22 Short-circuit current withstand chart for copper cables with thermoplastic insulation.

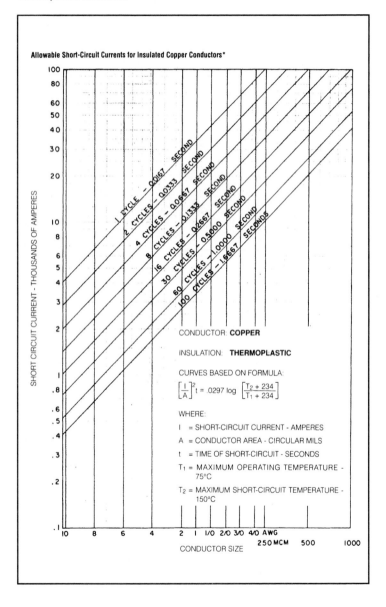

FIGURE 3.23   Short-circuit current withstand chart for copper cables with cross-linked polyethylene and ethylene propylene rubber insulation.

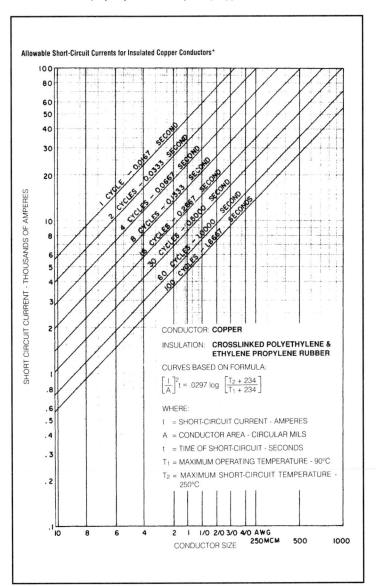

Allowable Short-Circuit Currents for Insulated Copper Conductors*

CONDUCTOR: **COPPER**

INSULATION: **CROSSLINKED POLYETHYLENE & ETHYLENE PROPYLENE RUBBER**

CURVES BASED ON FORMULA:

$$\left[\frac{I}{A}\right]^2 t = .0297 \log \left[\frac{T_2 + 234}{T_1 + 234}\right]$$

WHERE:

I = SHORT-CIRCUIT CURRENT - AMPERES

A = CONDUCTOR AREA - CIRCULAR MILS

t = TIME OF SHORT-CIRCUIT - SECONDS

T_1 = MAXIMUM OPERATING TEMPERATURE - 90°C

T_2 = MAXIMUM SHORT-CIRCUIT TEMPERATURE - 250°C

FIGURE 3.24   Short-circuit current withstand chart for aluminum cables with paper, rubber, or varnished cloth insulation.

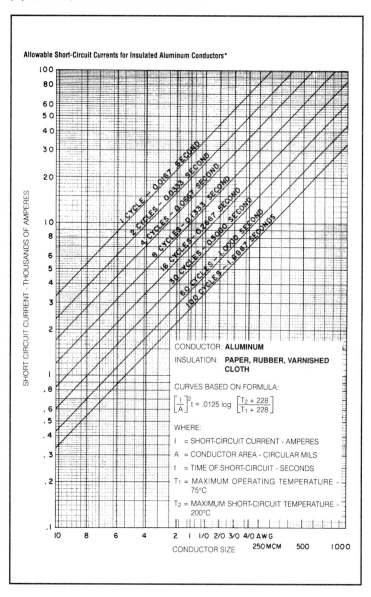

Allowable Short-Circuit Currents for Insulated Aluminum Conductors*

CONDUCTOR: **ALUMINUM**

INSULATION: **PAPER, RUBBER, VARNISHED CLOTH**

CURVES BASED ON FORMULA:

$$\left[\frac{I}{A}\right]^2 t = .0125 \log\left[\frac{T_2 + 228}{T_1 + 228}\right]$$

WHERE:

I   = SHORT-CIRCUIT CURRENT - AMPERES

A   = CONDUCTOR AREA - CIRCULAR MILS

t   = TIME OF SHORT-CIRCUIT - SECONDS

T_1 = MAXIMUM OPERATING TEMPERATURE - 75°C

T_2 = MAXIMUM SHORT-CIRCUIT TEMPERATURE - 200°C

FIGURE 3.25 Short-circuit current withstand chart for aluminum cables with thermoplastic insulation.

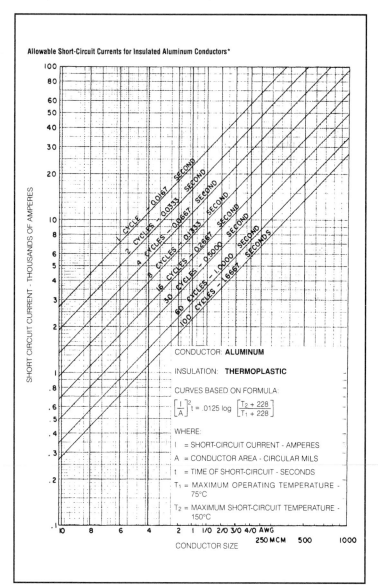

FIGURE 3.26 Short-circuit current withstand chart for aluminum cables with cross-linked polyethylene and ethylene propylene rubber insulation.

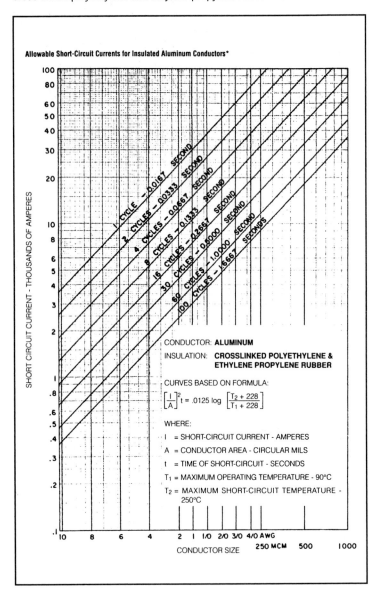

Allowable Short-Circuit Currents for Insulated Aluminum Conductors*

CONDUCTOR: **ALUMINUM**

INSULATION: **CROSSLINKED POLYETHYLENE & ETHYLENE PROPYLENE RUBBER**

CURVES BASED ON FORMULA:

$$\left[\frac{I}{A}\right]^2 t = .0125 \log \left[\frac{T_2 + 228}{T_1 + 228}\right]$$

WHERE:

I = SHORT-CIRCUIT CURRENT - AMPERES

A = CONDUCTOR AREA - CIRCULAR MILS

t = TIME OF SHORT-CIRCUIT - SECONDS

T_1 = MAXIMUM OPERATING TEMPERATURE - 90°C

T_2 = MAXIMUM SHORT-CIRCUIT TEMPERATURE - 250°C

TABLE 3.24 Comparison of Equipment Grounding Conductor Short-Circuit Withstand Ratings

Conductor Size	5 Sec. Rating (Amps)			I²t Rating x10⁶ (Ampere Squared Seconds)		
	ICEA P32-382 Insulation Damage 150°C	Soares 1 Amp/30 cm Validity 250°C	Onderdonk Melting Point 1,083°C	ICEA P32-382 Insulation Damage 150°C	Soares 1 Amp/30 cm Validity 250°C	Onderdonk Melting Point 1,083°C
14	97	137	253	.047	.094	.320
12	155	218	401	.120	.238	.804
10	246	346	638	.303	.599	2.03
8	391	550	1,015	.764	1.51	5.15
6	621	875	1,613	1.93	3.83	13.0
4	988	1,391	2,565	4.88	9.67	32.9
3	1,246	1,754	3,234	7.76	15.4	52.3
2	1,571	2,212	4,078	12.3	24.5	83.1
1	1,981	2,790	5,144	19.6	38.9	132.0
1/0	2,500	3,520	6,490	31.2	61.9	210.0
2/0	3,150	4,437	8,180	49.6	98.4	331.0
3/0	3,972	5,593	10,313	78.9	156.0	532.0
4/0	5,009	7,053	13,005	125.0	248.0	845.0
250	5,918	8,333	15,365	175.0	347.0	1,180.0
300	7,101	10,000	18,438	252.0	500.0	1,700.0
350	8,285	11,667	21,511	343.0	680.0	2,314.0
400	9,468	13,333	24,584	448.0	889.0	3,022.0
500	11,835	16,667	30,730	700.0	1,389.0	4,721.0
600	14,202	20,000	36,876	1,008.0	2,000.0	6,799.0
700	16,569	23,333	43,022	1,372.0	2,722.0	9,254.0
750	17,753	25,000	46,095	1,576.0	3,125.0	10,623.0
800	18,936	26,667	49,168	1,793.0	3,556.0	12,087.0
900	21,303	30,000	55,314	2,269.0	4,500.0	15,298.0
1,000	23,670	33,333	61,460	2,801.0	5,555.0	18,867.0

TABLE 3.25 NEMA (Standard Short-Circuit Ratings of Busway)

Continuous Current Rating of Busway (Amperes)	Short-Circuit Current Ratings (Symmetrical Amperes)	
	Plug-In Duct	Feeder Duct
100	10,000	–
225	14,000	–
400	22,000	–
600	22,000	42,000
800	22,000	42,000
1000	42,000	75,000
1200	42,000	75,000
1350	42,000	75,000
1600	65,000	100,000
2000	65,000	100,000
2500	65,000	150,000
3000	85,000	150,000
4000	85,000	200,000
5000	–	200,000

Table 3 pertains to feeder and plug-in busway. For switchboard and panelboard standard ratings refer to manufacturer.

U.L. Standard 891 details short-circuit durations for busway within switchboards for a minimum of three cycles, unless the main overcurrent device clears the short in less than three cycles.

*Reprinted with permission of NEMA, Pub. No. BU1-1988.

The concept of current limitation is pointed out in Figure 3.27, where the prospective available fault current is shown in conjunction with the limited current resulting when a current-limiting fuse clears. The area under the current curve indicates the amount of short-circuit energy being dissipated in the circuit. Because both magnetic forces and thermal energy are directly proportional to the square of the current, it is important to limit the short-circuit current to as small a value as possible. Magnetic forces vary as the square of the peak current, and thermal energy varies as the square of the rms current.

Thus, the current-limiting fuse in this example would limit the let-thru energy to a fraction of the value that is available from the system. In the first major loop of the fault current, standard non-current-limiting, electromechanical devices would let through approximately 100 times as much destructive energy as the fuse would let through.

TABLE 3.26   U.L. #508 Motor Controller Short-Circuit Test Ratings

Motor Controller HP Rating	Test Short Circuit Current Available
1HP or less and 300V or less	1,000A
50HP or less	5,000A
Greater than 50HP to 200HP	10,000A
201HP to 400HP	18,000A
401HP to 600HP	30,000A
601HP to 900HP	42,000A
901HP to 1600HP	85,000A

It should be noted that these are basic short-circuit requirements. Higher, combination ratings are attainable if tested to an applicable standard. However, damage is usually allowed.

ANALYSIS OF CURRENT-LIMITING FUSE LET-THRU CHARTS

The degree of current limitation of a given size and type of fuse depends, in general, upon the available short circuit that can be delivered by the electrical system. Current limitation of fuses is best described in the form of a let-thru chart, which, when applied from a practical point of view, is useful to determine the let-thru currents when a fuse opens.

Fuse let-thru charts are similar to the one shown in Figure 3.28 and are plotted from actual test data. The test circuit that establishes line A-B corresponds to a short-circuit power factor of 15 percent, which is associated with an X/R ratio of 6.6. The fuse curves represent the cutoff value of the prospective available short-circuit current under the given circuit conditions. Each type or class of fuse has its own family of let-thru curves.

The let-thru data has been generated by actual short-circuit tests of current-limiting fuses. It is important to understand how the curves are generated, and what circuit parameters affect the let-thru curve data. Typically, there are three circuit parameters that can affect fuse let-thru performance for a given available short-circuit current. These are:

1. Short-circuit power factor
2. Short-circuit closing angle
3. Applied voltage

Current-limiting fuse let-thru curves are generated under worst-case conditions, based on these three variable parameters. The benefit to the user is a conservative resultant let-thru current (both I_p and I_{rms}). Under actual field conditions, changing any one or a combination of these will result in lower let-thru currents. This provides for an additional degree of reliability when applying fuses for equipment protection.

TABLE 3.27 Molded-Case Circuit Breaker Interrupting Capacities

	SIEMENS					SQUARE D			
Frame Size	Maximum Voltage Rating	Breaker Type	Ampere Rating	UL Interruption Capacity Symm. RMS	Dimensions (inches)	Breaker Type	Ampere Rating	UL Interruption Capacity Symm. RMS (AC)	Dimensions (inches)
100A	Standard Interrupting 240V AC 250V DC	ED2	15-100	120V AC 10 kA-1 Pole 240V AC 10 kA-2,3 Pole 125V DC 10 kA-1 Pole 250V DC 5 kA-2 Pole	W=1 (1 Pole) W=2 (2 Pole) W=3 (3 Pole) H=6 11/32 D=4	FAL	15-100	120V AC 10 kA-1 Pole 240V AC 10 kA-2,3 Pole 125V DC 5 kA-1 Pole 250V DC 5 kA-2,3 Pole	W=1 1/2 (1Pole) W=3 (2 Pole) W=4 1/2 (3 Pole) H=6 D=3 5/32
125A	Standard Interrupting 480V AC 250V DC	ED4	15-125	120V AC 65 kA-1 Pole 240V AC 65 kA-2,3 Pole 277V AC 22 kA-1 Pole 480V AC 18 kA-2,3 Pole 125 V DC 5 kA-1 Pole 250V DC 30 kA-2 Pole	W=1 (1 Pole) W=2 (2 Pole) W=3 (3 Pole) H=6 11/32 D=4	FAL	15-100	120V AC 25 kA-1 Pole 240V AC 25 kA-2,3 Pole 277V AC 18 kA-1 Pole 480V AC 18 kA-2,3 Pole 125V DC 18 kA-1 Pole 250V DC 10 kA-2,3 Pole	W=1 1/2 (1Pole) W=3 (2 Pole) W=4 1/2 (3 Pole) H=6 D=3 5/32
	Standard Interrupting 600V AC 500V DC	ED6	15-125	240V AC 65 kA-2,3 Pole 480V AC 25 kA-2,3 Pole 600V AC 18 kA-2,3 Pole 250V DC 30 kA-2 Pole 500V DC 18 kA-3 Pole	W=2,3 H=6 11/32 D=4	FAL	15-100	240V AC 25 kA-2,3 Pole 480V AC 18 kA-2,3 Pole 600V AC 14 kA-2,3 Pole 250V DC 10 kA-2,3 Pole	W=3 (2 Pole) W=4 1/2 (3 Pole) H=6 D=3 5/32
	High Interrupting 480V AC 250V DC	HED4①	15-125	120V AC 100 kA-1 Pole 240V AC 100 kA-2,3 Pole 277V AC (15-30 Amperes) 65 kA-1 Pole 480V AC 42 kA-2,3 Pole 250V DC 30 kA-2 Pole	W=1,2,3 H=6 11/32 D=4	FCL	15-100	240V AC 100 kA-2,3 Pole 480V AC 65 kA-2,3 Pole	
	High Interrupting 600V AC 500V DC	HED6①	15-125	240V AC 100 kA-2,3 Pole 480V AC 30 kA-2,3 Pole 600V AC 18 kA-2,3 Pole 250V DC 30 kA-2 Pole 500V DC 25 kA-3 Pole	W=2,3 H=6 11/32 D=4	FHL FHL-DC	15-100	240V AC 65 kA-2,3 Pole 480V AC 25 kA-2,3 Pole 600V AC 18 kA-2,3 Pole 250V DC 10 kA-2 Pole 500V DC 20 kA-3 Pole	
	Current-Limiting 600V AC 500V DC	CED6	15-125	240V AC 200 kA-2,3 Pole 480V AC 200 kA-2,3 Pole 600V AC 30 kA-2,3 Pole 250V DC 30 kA-2 Pole 500V DC 50 kA-3 Pole	W=2,3 H=9 37/64 D=4	FIL	15-100	240V AC 200 kA-2,3 Pole 480V AC 200 kA-2,3 Pole 600V DC 100 kA-2,3 Pole	W=4 1/2 H=8 D=21/32
225A	240V AC 2 or 3 Pole Construction	QJ2	60-225	240V AC 10 kA-2,3 Pole	W=3 (2 Pole) W=4 1/2 (3 Pole) H=7 D=2 17/32	Q2L	100-225	240V AC 10 kA-2,3 Pole	W=3 (2 Pole) W=4 1/2 (3 Pole) H=6 7/16 D=3 59/64
		QJH2	60-225	240V AC 22 kA-2,3 Pole		Q2L-H	100-225	240V AC 22 kA-2,3 Pole	
		QJ2-H	60-225	240V AC 42 kA-2,3 Pole		Q2LH	100-225	240V AC 42 kA-2,3 Pole	
250A	Standard Interrupting 600V AC 500V DC	FXD6 (Fix.-Trip) FD6 (Interchangeable Trip)	70-250	240V AC 65 kA-2,3 Pole 480V AC 35 kA-2,3 Pole 600V AC 18 kA-2,3 Pole 250V DC 30 kA-2 Pole 500V DC 18 kA-3 Pole	W=4 1/2 H=9 1/2 D=4	KAL (Fix-Trip)	70-250	240V AC 42 kA-2,3 Pole 480V AC 25 kA-2,3 Pole 600V AC 22 kA-2,3 Pole 250V DC 10 kA-2 Pole	W=4 1/2 H=8 D=3 21/32
	High Interrupting 600V AC 500V DC	HFD6② (Interchangeable Trip) HFXD6 (Fix-Trip)	70-250	240V AC 100 kA-2,3 Pole 480V AC 65 kA-2,3 Pole 600V AC 25 kA-2,3 Pole 250V DC 30 kA-2 Pole 500V DC 25 kA-3 Pole		KHL (Fix-Trip) KHL-DC	70-250	240V AC 65 kA-2,3 Pole 480V AC 35 kA-2,3 Pole 600V AC 25 kA-2,3 Pole 250V DC 10 kA-2 Pole 500V DC 20 kA-3 Pole	
		HHFD6 (Interchangeable Trip) HHFXD6 (Fix-Trip)		240V AC 200 kA-3 Pole 480V AC 100 kA-3 Pole 600V AC 25 kA-3 Pole		Not Available			
	Current-Limiting 600V AC 500V DC	CFD6 (Fix-Trip)	70-250	240V AC 200 kA-2,3 Pole 480V AC 200 kA-2,3 Pole 600V DC 100 kA-2,3 Pole 250V DC 30 kA-2 Pole 500V DC 50 kA-3 Pole	W=4 1/2 H=14 1/8 D=4	KIL (Fix-Trip)	110-250	240V AC 200 kA-2,3 Pole 480V AC 200 kA-2,3 Pole 600V AC 100 kA-2,3 Pole	W=4 1/2 H=8 D=3 21/32
400A	Standard Interrupting 240V AC	JXD2 (Fix-Trip)	200-400	240V AC 65 kA-2,3 Pole 250V DC 30 kA-2 Pole	W=7 1/2 H=11 D=4	Q4L	250-400	240V AC 25 kA-2,3 Pole	W=6 H=11 D=4 1/16
	Standard Interrupting 600V AC 500V DC	JXD6 (Fix-Trip) JD6 (Interchangeable Trip)	200-400	240V AC 65 kA-2,3 Pole 480V AC 35 kA-2,3 Pole 600V AC 25 kA-2,3 Pole 250V DC 30 kA-2 Pole 500V DC 25 kA-3 Pole		LAL (Fix-Trip)	125-400	240V AC 42 kA-2,3 Pole 480V AC 30 kA-2,3 Pole 600V AC 22 kA-2,3 Pole 250V DC 10 kA-2 Pole	W=6 H=11 D=4 1/16
	High Interrupting 600V AC 500V DC	HJD6 (Interchangeable Trip) HJXD6 (Fix-Trip)	200-400	240V AC 100 kA-2,3 Pole 480V AC 65 kA-2,3 Pole 600V AC 35 kA-2,3 Pole 250V DC 30 kA-2 Pole 500V DC 35 kA-3 Pole		LHL (Fix-Trip) LHL-DC	125-400	240V AC 65 kA-2,3 Pole 480V AC 35 kA-2,3 Pole 600V AC 25 kA-2,3 Pole 250V DC 10 kA-2 Pole 500V DC 20 kA-3 Pole	
	High Interrupting① 600V AC	HHJXD6② (Fix-Trip) HHJD62② (Interchangeable Trip)	200-400	240V AC 200 kA-2,3 Pole 480V AC 100 kA-2,3 Pole 600V AC 50 kA-2,3 Pole		Not Available			
	Current-Limiting 600V AC 500V DC	CJD6 (Fix-Trip)	200-400	240V AC 200 kA-3 Pole 480V AC 150 kA-3 Pole 600V AC 100 kA-3 Pole 250V DC 30 kA-2 Pole 500V DC 50 kA-3 Pole	W=7 1/2 H=17 55/64 D=4	LIL	300-400	240V AC 200 kA-2,3 Pole 480V AC 200 kA-2,3 Pole 600V AC 100 kA-2,3 Pole	W=7 1/2 H=11 7/8 D=5 1/2
	Standard Interrupting 600V AC	Solid State SJD6	200-400	240V AC 65 kA-3 Pole 480V AC 35 kA-3 Pole 600V AC 25 kA-3 Pole	W=7 1/2 H=11 D=4	Not Available			
	High Interrupting 600V AC	Solid State SHJD6	200-400	240V AC 100 kA-3 Pole 480V AC 65 kA-3 Pole 600V AC 35 kA-3 Pole		Solid State LXL	300-400	240V AC 100 kA-3 Pole 480V AC 65 kA-3 Pole 600V AC 35 kA-3 Pole	W=7 1/2 H=11 7/8 D=5 1/2
	Current-Limiting 600V AC	Solid State SCJD6	200-400	240V AC 200 kA-3 Pole 480V AC 150 kA-3 Pole 600V AC 100 kA-3 Pole	W=7 1/2 H=17 55/64 D=4	Solid State LXIL	300-400	240V AC 200 kA,3 Pole 480V AC 200 kA-3 Pole 600V AC 100 kA-3 Pole	W=7 1/2 H=11 7/8 D=5 1/2

WESTINGHOUSE

Breaker Type	Ampere Rating	UL Interruption Capacity Symm. RMS (AC)	Dimensions (Inches)
EB	15-100	120V AC 10 kA-1 Pole 240V AC 10 kA-2,3 Pole 125V DC 5 kA-1 Pole 250V DC 5 kA-3 Pole	W=1 3/8 (1 Pole) W=2 3/4 (2 Pole) W=4 1/8 (3 Pole) H=6 D=3 3/8
EHD	15-100	240V AC 18 kA-2,3 Pole 277V AC 14 kA-1 Pole 480V AC 14 kA-2,3 Pole 125 V DC 10 kA-1 Pole 250V DC 10 kA-2,3 Pole	W=1 3/8 (1 Pole) W=2 3/4 (2 Pole) W=4 1/8 (3 Pole) H=6 D=3 3/8
FD	15-150	240V AC 65 kA-2,3 Pole 480V AC 25 kA-2,3 Pole 600V AC 18 kA-2,3 Pole 250V DC 10 kA-2,3 Pole	W=1 3/8 (1 Pole) W=2 3/4 (1 Pole) W=4 1/8 (3 Pole) H=6
HFD	15-150	240V AC 100 kA-2,3 Pole 277V AC 65 kA-1 Pole 480V AC 65 kA-2,3 Pole 125V DC 10 kA-1 Pole 250V DC 10 kA-2,3 Pole	
HFD	15-150	240V AC 100 kA-2,3 Pole 480V AC 65 kA-2,3 Pole 600V AC 25 kA-2,3 Pole 250V DC 10 kA-2,3 Pole	
FDC	15-150	240V AC 200 kA-2,3 Pole 480V AC 100 kA-2,3 Pole 600V AC 35 kA-2,3 Pole 250V DC 22 kA-2,3 Pole	
CA	125-225	240V AC 10 kA-2,3 Pole	W=2 3/4 (2 Pole) W=4 1/8 (3 Pole) H=6 1/2 D=2 11/16
CAH	125-225	240V AC 22 kA-2,3 Pole	
HCA	125-225	240V AC 42 kA-2,3 Pole	
JDB (Fix-Trip) JD (Interchangeable Trip)	70-250	240V AC 65 kA-2,3 Pole 480V AC 25 kA-2,3 Pole 600V AC 18 kA-2,3 Pole 250V DC 10 kA-2 Pole	W=4 1/8 H=10 D=4 1/16
HJD (Interchangeable Trip)	70-250	240V AC 100 kA-2,3 Pole 480V AC 65 kA-2,3 Pole 600V AC 25 kA-2,3 Pole 250V DC 22 kA-2 Pole 500V DC 20 kA-3 Pole	
Not Available			
JDC (Interchangeable Trip)	70-250	240V AC 200 kA-2,3 Pole 480V AC 100 kA-2,3 Pole 600V AC 35 kA-2,3 Pole 250V DC 22 kA-2 Pole	
DK (Fix-Trip)	250-400	240V AC 65 kA-2,3 Pole 250V DC 10 kA-2,3 Pole	W=5 1/2 H=10 1/8 D=4 1/16
KDB (Fix-Trip) KD (Interchangeable Trip)	100-400	240V AC 65 kA-2,3 Pole 480V AC 35 kA-2,3 Pole 600V AC 25 kA-2,3 Pole 250V DC 10 kA-2,3 Pole	W=5 1/2 H=10 1/8 D=4 1/16
HKD (Interchangeable Trip)	100-400	240V AC 100 kA-2,3 Pole 480V AC 65 kA-2,3 Pole 600V AC 35 kA-2,3 Pole 250V DC 22 kA-2,3 Pole 500V DC 35 kA-3 Pole	
Current Limiting KDC (Interchangeable Trip)	100-400	240V AC 200 kA-2,3 Pole 480V DC 100 kA-2,3 Pole 600V AC 50 kA-2,3 Pole 250V DC 22 kA-2,3 Pole	
Current Limiting LCL (Fix-Trip)	125-400	240V AC 200 kA-2,3 Pole 480V AC 200 kA-2,3 Pole 600V AC 100 kA-2,3 Pole	W=8 1/4 H=16 D=4 1/16
Solid State KD	125-400	240V AC 65 kA-3 Pole 480V AC 35 kA-3 Pole 600V DC 25 kA-3 Pole	W=5 1/2 H=10 1/8 D=4 1/16
Solid State HKD	125-400	240V AC 100 kA-3 Pole 480V AC 65 kA-3 Pole 600V AC 35 kA-3 Pole	
Solid State KDC	125-400	240V AC 200 kA,3 Pole 480V AC 100 kA-3 Pole 600V AC 50 kA-3 Pole	W=5 1/2 H=10 1/8 D=4 1/16

GENERAL ELECTRIC

Breaker Type	Ampere Rating	UL Interruption Capacity Symm. RMS (AC)	Dimensions (Inches)
TEB	15-100	120V AC 10 kA-1 Pole 240V AC 10 kA-2,3 Pole 125V DC 5 kA-1 Pole 250V DC 5 kA-2,3 Pole	W=1 3/8 (1 Pole) W=2 3/4 (2 Pole) W=4 1/8 (3 Pole) H=6 5/16 D=3 3/8
TED	15-100	240V AC 18 kA-2,3 Pole 277V AC 14 kA-1 Pole 480V AC 14 kA-2,3 Pole 250V DC 10 kA-2,3 Pole	W=1 3/8 (1 Pole) W=2 3/4 (2 Pole) W=4 1/8 (3 Pole) H=6 5/16 D=3 3/8
TED	15-100	240V AC 18 kA-2,3 Pole 480V AC 14 kA-2,3 Pole 600V AC 14 kA-2,3 Pole 250V DC 10 kA-2 Pole 500V DC 10 kA -3 Pole	
THED	15-150	240V AC (15-100 Amperes) 65 kA-2,3 Pole 240V AC (110-150 Amperes) 42 kA-2,3 Pole 277V AC (15-30 Amperes) 65 KA-1 Pole 480V AC 25 kA-2,3 Pole 250V DC 10 kA-2,3 Pole	
THED	15-150	240V AC (15-100 Amperes) 65 kA-2,3 Pole 240V AC (110-150 Amperes) 42 kA-2,3 Pole 480V AC 25 kA-2,3 Pole 600V AC 18 kA-2,3 Pole 250V DC 20 kA-2 Pole 500V DC 20 kA-3 Pole	
THLC1	15-150	240V AC 200 kA-3 Pole 480V AC (15-50 Amperes) 150 kA-3 Pole 480V AC (60-150 Amperes) 200 kA-3 Pole 600V AC 50 kA-3 Pole	
TQD		240V AC 10 kA-2,3 Pole	W=2 3/4 (2 Pole) W=4 1/8 (3 Pole) H=6 9/16 D=2 5/8
THQD		240V AC 22 kA-2,3 Pole	
Not Available			
TFJ (Fix-Trip)	70-250	240V AC 25 kA-2,3 Pole 480V AC 22 kA-2,3 Pole 600V AC 18 kA-3 Pole	W=4 1/8 H=10 1/8 D=3 3/16
THFK (Interchangeable Trip)	70-250	240V AC 65 kA-2,3 Pole 480V AC 25 kA-2,3 Pole 600V AC 18 kA-3 Pole 250 V DC 10 kA-2 Pole	
Current Limiting TFL (Fix-Trip)	70-250	240V AC 100 kA-3 Pole 480V AC 65 kA-3 Pole 600V AC 25 kA-3 Pole	
Not Available			
THLC2 (Fix-Trip)	125-225	240V AC 200 kA-3 Pole 480V AC 200 kA-3 Pole 600V DC 50 kA-3 Pole	W=5 33/32 H=11 7/16 D=4 7/8
TJD (Fix-Trip)	250-400	240V AC 22 kA-2,3 Pole 250V DC 10 kA-2 Pole	W=8 1/4 H=10 1/8 D=3 13/16
TJJ (Fix-Trip) TJK4 (Inchangeable Trip)	125-400	240V AC 42 kA-2,3 Pole 480V AC 30 kA-2,3 Pole 600V AC 22 kA-2,3 Pole 250V DC 10 kA-2 Pole 500V DC 20 kA-2 Pole	W=8 1/4 H=10 1/8 D=3 13/16
THJK4 (Interangeable Trip)	125-400	240V AC 65 kA-2,3 Pole 480V AC 35 kA-2,3 Pole 600V AC 25 kA-2,3 Pole 250V DC 10 kA-2 Pole	
Current Limiting TLB4 (Fix-Trip)	250-400	240V AC 100 kA-3 Pole 480V DC 65 kA-3 Pole 600V AC 25 kA-3 Pole	
THLC4 (Fix-Trip)	250-400	240V AC 200 kA-3 Pole 480V AC 200 kA-3 Pole 600V AC 50 kA-3 Pole	W=5 23/32 H=13 1/16 D=4 7/8
Solid State THJ4V	150-400	240V AC 65 kA-3 Pole 480V AC 35 kA-3 Pole 600V AC 25 kA-3 Pole	W=5 1/2 H=10 3/4 D=3 13/16
Solid State TJL4V	150-400	240V AC 100 kA-3 Pole 480V AC 65 kA-3 Pole 600V AC 30 kA-3 Pole	
Not Available			

CUTLER-HAMMER

Breaker Type	Ampere Rating	UL Interruption Capacity Symm. RMS (AC)	Dimensions (Inches)
FS	15-100	120V AC 65 kA-1 Pole (15-30 Amperes) 240V AC 10 kA-2,3 Pole 250V DC 120 kA-2 Pole	W=1 3/8 (1Pole) W=2 3/4 (2 Pole) W=4 1/8 (3 Pole) H=6 1/8 D=3 3/16
FS	15-150	2120V AC 65 kA-1 Pole 277V AC 22 kA-1 Pole (15-30 Amperes) 240V AC 22 kA-2,3 Pole 480V AC 14 kA-2,3 Pole 250V DC 10 kA-2 Pole	W=1 3/8 (1Pole) W=2 3/4 (2 Pole) W=4 1/8 (3 Pole) H=6 1/8 D=3 3/16
FS	15-150	240V AC 22 kA-3 Pole 480V AC 14 kA-3 Pole 600V AC 14 kA-3 Pole 250V DC 10 kA-3 Pole	
FH	15-150	240V AC 100 kA-3 Pole 480V AC 30 kA-3 Pole	
FH	15-150	240V AC 100 kA-3 Pole 480V AC 30 kA,3 Pole 600V AC 18 kA,3 Pole 250V DC 10 kA-3 Pole	
FL Not UL Current Limiting	15-100	240V AC 200 kA-2,3 Pole 480V AC 100 kA-2,3 Pole 600V AC 25 kA-2,3 Pole 250V DC 10 kA-2,3 Pole	W=4 1/8 H=9 5/16 D=3 3/16
CC	60-225	240V AC 10 KA-2,3 Pole	W=2 3/4 (2 Pole) W=4 1/8 (3 Pole) H=6 1/8 D=2 57/64
CCH	125-225	240V AC 25 KA-2.3 Pole	
CHH	60-225	240V AC 100 KA-2,3 Pole	
JS (Fix-Trip)	100-250	240V AC 25 kA-3 Pole 480V AC 22 kA-3 Pole 600V AC 14 kA-3 Pole 250V DC 10 kA-3 Pole	W=4 1/8 H=12 D=3 13/16
JH (Fix-Trip)	100-250	240V AC 100 kA-3 Pole 480V AC 30 kA-3 Pole 600V AC 18 kA-3 Pole 250V DC 10 kA-3 Pole	
Not Available			
(Fix-Trip) JL Not UL Current Limiting	100-250	240V AC 200 kA,3 Pole 480V AC 100 kA-3 Pole 600V AC 25 kA-3 Pole 250V DC10 kA-3 Pole	
KS (Fix-Trip)	250-400	240V AC 65 kA-2,3 Pole 250V DC 10 kA-2,3 Pole	W=5 1/2 H=10 1/8 100-300 H=12 15/32 400 D=3 13/16
KS (Fix-Trip)	100-400	240V AC 42 kA-3 Pole 480V AC 30 kA,3 Pole 600V AC 22 kA-3 Pole 250V DC 10 kA-3 Pole	
KH (Fix-Trip)	100-400	240V AC 65 kA-3 Pole 480V AC 35 kA,3 Pole 600V DC 25 kA,3 Pole 250V DC 10 kA-3 Pole	
Not Available			
Not Available			
Solid State KS	400	240V AC 42 kA-3 Pole 480V AC 30 kA-3 Pole 600V AC 22 kA-3 Pole	W=5 1/2 H=10 1/8 D=3 13/16
Solid State KH	400	240V AC 65 kA-3 Pole 480V AC 35 kA-3 Pole 600V AC 22 kA-3 Pole	
Not Available			

TABLE 3.27 Molded-Case Circuit Breaker Interrupting Capacities (*Continued*)

		SIEMENS				SQUARE D			
600A	Standard Interrupting 600V AC 500V DC	LXD6 (Fix Trip) LD6 (Interchangeable Trip)	250–600	240V AC 65 kA-2,3 Pole 480V AC 25 kA-2,3 Pole 600V AC 25 kA-2,3 Pole 250V DC 30 kA-2 Pole 500V DC 25 kA-3 Pole		Not Available			
	High Interrupting 600V AC 500V DC	HLXD6 (Fix Trip) HLD6 (Interchangeable Trip)	250–600	240V AC 100 kA-2,3 Pole 480V AC 65 kA-2,3 Pole 600V AC 35 kA-2,3 Pole 250V DC 30 kA-2 Pole 500V DC 35 kA-3 Pole	W=7¹/₂ H=11 D=4	LCL (Fix Trip)	300–600	240V AC 100 kA-2,3 Pole 480V AC 65 kA-2,3 Pole 600V AC 35 kA-3 Pole	W=7¹/₂ H=17⁵⁵/₆₄ D=4
	High Interrupting① 600 Ampere 600V AC	HHLXD6② (Fix Trip) HHLD6② (Interchangeable Trip)	250–600	240V AC 200 kA-2,3 Pole 480V AC 100 kA-2,3 Pole 600V AC 65 kA-2,3 Pole		Not Available			
	Current Limiting 600V AC 500V DC	CLD6 (Fix Trip)	450–600	240V AC 200 kA-3 Pole 480V AC 150 kA-3 Pole 600V AC 100 kA-3 Pole 250V DC 30 kA-2 Pole 500V DC 50 kA-3 Pole	W=7¹/₂ H=17⁵⁵/₆₄ D=4	LIL (Fix Trip)	450–600	240V AC 200 kA-3 Pole 480V AC 200 kA-3 Pole 600V AC 100 kA-3 Pole	W=7¹/₂ H=17⁵⁵/₆₄ D=4
	Standard Interrupting 600V AC	Solid State SLD6	300–600	240V AC 65 kA-3 Pole 480V AC 35 kA-3 Pole 600V AC 25 kA-3 Pole	W=7¹/₂ H=11 D=4	Not Available			
	High Interrupting 600V AC	Solid State SHLD6	300–600	240V AC 100 kA-3 Pole 480V AC 65 kA-3 Pole 600V AC 50 kA-3 Pole		Solid State LXL	400–600	240V AC 100 kA-3 Pole 480V AC 65 kA-3 Pole 600V AC 35 kA-3 Pole	W=7¹/₂ H=11¹/₈ D=5¹/₂
	Current Limiting 600V AC	Solid State SCLD6	300–600	240V AC 200 kA-3 Pole 480V AC 150 kA-3 Pole 600V AC 100 kA-3 Pole	W=7¹/₂ H=17⁵⁵/₆₄ D=4	Solid State LXL	400–600	240V AC 200 kA-3 Pole 480V AC 200 kA-3 Pole 600V AC 100 kA-3 Pole	
800A	Standard Interrupting 600V AC 500V DC	MXD6 (Fix Trip) MD6 (Interchangeable Trip)	500–800	240V AC 65 kA-2,3 Pole 480V AC 50 kA-2,3 Pole 600V AC 25 kA-2,3 Pole 250V DC 30 kA-2 Pole 500V DC 25 kA-3 Pole	W=9 H=16 D=6⁵/₁₆	MAL (Fix Trip)	300–1000	240V AC 42 kA-2,3 Pole 480V AC 30 kA-2,3 Pole 600V AC 22 kA-2,3 Pole 250V DC 14 kA-3 Pole	W=9 H=14
	High Interrupting 600V AC 500V DC	HMXD6 (Fix Trip) HMD6 (Interchangeable Trip)	500–800	240V AC 100 kA-2,3 Pole 480V AC 65 kA-2,3 Pole 600V AC 50 kA-2,3 Pole 250V DC 30 kA-2 Pole 500V DC 50 kA-3 Pole		MHL (Fix Trip) MHL-DC	300–1000	240V AC 65 kA-2,3 Pole 480V AC 65 kA-2,3 Pole 600V AC 65 kA-2,3 Pole 250V DC 14 kA-2 Pole 500V DC 20 kA-3 Pole	W=9 H=14 D=4¹¹/₃₂
	Current Limiting① 600V AC 500V DC	CMD6 (Fix Trip)	500–800	240V AC 200 kA-3 Pole 480V AC 100 kA-3 Pole 600V AC 65 kA-3 Pole 250V DC 30 kA-2 Pole 500V DC 50 kA-3 Pole	W=9 H=16 D=6⁵/₁₆	Not Available			
	Standard Interrupting 600V AC	Solid State SMD6	600–800	240V AC 65 kA-3 Pole 480V AC 50 kA-3 Pole 600V AC 25 kA-3 Pole		Solid State MXL	450–800	240V AC 65 kA-3 Pole 480V AC 65 kA-3 Pole 600V AC 50 kA-3 Pole	W=9 H=14³/₄ D=4¹⁵/₃₂
	High Interrupting 600V AC	Solid State SHMD6	600–800	240V AC 100 kA-3 Pole 480V AC 65 kA-3 Pole 600V AC 50 kA-3 Pole		Not Available			
	Current Limiting 600V AC	Solid State SCMD6	600–800	240V AC 200 kA-3 Pole 480V AC 100 kA-3 Pole 600V AC 65 kA-3 Pole		Not Available			
1200A	Standard Interrupting 600V AC 500V DC	NXD6 (Fix Trip) ND6 (Interchangeable Trip)	800–1200	240V AC 65 kA-2,3 Pole 480V AC 50 kA-2,3 Pole 600V AC 25 kA-2,3 Pole 250V DC 30 kA-3 Pole 500V DC 25 kA-3 Pole		NAL (Fix Trip)	600–1200	240V AC 100 kA-2,3 Pole 480V AC 50 kA-2,3 Pole 600V AC 25 kA-2,3 Pole	W=14⁶³/₆₄ H=12¹/₈ D=6¹³/₃₂
	High Interrupting 600V AC 500V DC	HNXD6 (Fix Trip) HND6 (Interchangeable Trip)	800–1200	240V AC 100 kA-2,3 Pole 480V AC 65 kA-2,3 Pole 600V AC 50 kA-2,3 Pole 250V DC 30 kA-2 Pole 500V DC 50 kA-3 Pole		NCL (Fix Trip)	600–1200	240V AC 125 kA-2,3 Pole 480V AC 100 kA-2,3 Pole 600V AC 65 kA-2,3 Pole	
	Current Limiting① 600V AC 500V DC	CND6 (Fix Trip)	900–1200	240V AC 200 kA-3 Pole 480V AC 100 kA-3 Pole 600V AC 65 kA-3 Pole 250V DC 30 kA-2 Pole 500V DC 50 kA-3 Pole	W=9 H=16 D=6⁵/₁₆	Not Available			
	Standard Interrupting 600V AC	Solid State SND6	800–1200	240V AC 65 kA-3 Pole 480V AC 50 kA-3 Pole 600V AC 25 kA-3 Pole		Not Available			
	High Interrupting 600V AC	Solid State SHND6	800–1200	240V AC 100 kA-3 Pole 480V AC 65 kA-3 Pole 600V AC 50 kA-3 Pole		Solid State NXL	600–1200	240V AC 125 kA-3 Pole 480V AC 100 kA-3 Pole 600V AC 65 kA-3 Pole	W=14⁶³/₆₄ H=12¹/₈ D=6¹³/₃₂
	Current Limiting 600V AC	Solid State SCND6	800–1200	240V AC 200 kA-3 Pole 480V AC 100 kA-3 Pole 600V AC 65 kA-3 Pole		Not Available			

WESTINGHOUSE	Amps	Ratings	W Dim	GENERAL ELECTRIC	Amps	Ratings	GE Dim	CUTLER-HAMMER	Amps	Ratings	CH Dim
LDB (Fix-Trip) / LD Interchangeable Trip	300-600	240V AC 65 kA-2,3 Pole / 480V AC 35 kA-2,3 Pole / 600V AC 25 kA-2,3 Pole / 250V DC 10 kA-2,3 Pole		TJK6 Interchangeable Trip	250-600	240V AC 42 kA-2,3 Pole / 480V AC 30 kA-2,3 Pole / 600V AC 22 kA-2,3 Pole / 250V DC 10 kA-2 Pole / 500V DC 20 kA-3 Pole	W=8¼ H=10⅛ D=3¹³/₁₆	LS—E (Fix-Trip) / LS—E Interchangeable Trip	250-600	240V AC 65 kA-3 Pole / 480V AC 35 kA-3 Pole / 600V AC 25 kA-3 Pole / 250V DC 10 kA-3 Pole	
HLD Interchangeable Trip	250-600	240V AC 100 kA-2,3 Pole / 480V AC 65 kA-2,3 Pole / 600V AC 35 kA-2,3 Pole / 250V DC 20 kA-2,3 Pole	W=8¼ H=10⅛ D=4¹/₁₆	THJK6 Interchangeable Trip	250-600	240V AC 65 kA-2,3 Pole / 480V AC 35 kA-2,3 Pole / 600V AC 25 kA-2,3 Pole / 250V DC 10 kA-2 Pole		LH—E Interchangeable Trip	250-600	240V AC 100 kA-3 Pole / 480V AC 65 kA-3 Pole / 600V AC 35 kA-3 Pole / 250V DC 10 kA-3 Pole	W=8¼ H=10⅛ D=3³/₁₆
Current Limiting LDC Interchangeable Trip	300-600	240V AC 200 kA-2,3 Pole / 480V AC 200 kA-2,3 Pole / 600V AC 50 kA-2,3 Pole / 250V DC 25 kA-2,3 Pole		Not Available				LL—E Not UL Current Limiting	250-600	240V AC 200 kA-3 Pole / 480V AC 150 kA-3 Pole / 600V AC 100 kA-3 Pole / 250V DC 50 kA-3 Pole	
Not Available				Not Available				Not Available			
Solid State LD	610	240V AC 65 kA-3 Pole / 480V AC 35 kA-3 Pole / 600V AC 25 kA-3 Pole		Solid State THJ4V	150-600	240V AC 65 kA-3 Pole / 480V AC 35 kA-3 Pole / 600V AC 25 kA-3 Pole	W=8¼ H=10⅛ D=3¹³/₁₆	Not Available			
Solid State HLD	600	240V AC 100 kA-3 Pole / 480V AC 65 kA-3 Pole / 600V AC 35 kA-3 Pole	W=8¼ H=10⅛ D=4¹/₁₆	Solid State TJL4V	150-600	240V AC 65 kA-3 Pole / 480V AC 65 kA-3 Pole / 600V AC 30 kA-3 Pole		Not Available			
Solid State LDC	600	240V AC 200 kA-3 Pole / 480V AC 100 kA-3 Pole / 600V AC 50 kA-3 Pole		Not Available				Not Available			
MA Interchangeable Trip	125-800	240V AC 42 kA-2,3 Pole / 480V AC 30 kA-2,3 Pole / 600V AC 22 kA-2,3 Pole / 250V DC 20 kA-2,3 Pole	W=8¼ H=16 D=4¹/₈	TKM8 Interchangeable Trip	300-800	240V AC 42 kA-2,3 Pole / 480V AC 30 kA-2,3 Pole / 600V AC 22 kA-2,3 Pole / 250V DC 10 kA-2 Pole / 500V DC 20 kA-3 Pole	W=8¼ H=15½ D=5½	MS (Fix-Trip)	350-800	240V AC 42 kA-2,3 Pole / 480V AC 30 kA-2,3 Pole / 600V AC 22 kA-2,3 Pole / 250V DC 10 kA-3 Pole (350-600 ONLY)	W=8¼ H=16 D=4¹/₈
HVA Interchangeable Trip	125-800	240V AC 65 kA-2,3 Pole / 480V AC 35 kA-2,3 Pole / 600V AC 25 kA-2,3 Pole / 250V DC 20 kA-2,3 Pole		THKM8 Interchangeable Trip	300-800	240V AC 65 kA-2,3 Pole / 480V AC 35 kA-2,3 Pole / 600V AC 25 kA-2,3 Pole / 250V DC 10 kA-2 Pole / 500V DC 10 kA-3 Pole		MH (Fix-Trip)	350-800	240V AC 65 kA-3 Pole / 480V AC 35 kA-3 Pole / 600V AC 25 kA-3 Pole / 250V DC 10 kA-3 Pole (350-600 ONLY)	
Not Available				Not Available				Not Available			
Solid State ND	600-800	240V AC 65 kA-3 Pole / 480V AC 50 kA-3 Pole / 600V AC 25 kA-3 Pole		Solid State TK4V	800	240V AC 42 kA-3 Pole / 480V AC 30 kA-3 Pole / 600V AC 22 kA-3 Pole	W=8¼ H=15½ D=3¹³/₁₆	Not Available			
Solid State HND	600-800	240V AC 100 kA-3 Pole / 480V AC 65 kA-3 Pole / 600V AC 35 kA-3 Pole	W=8¼ H=16 D=5¹/₂	Solid State TKL4V	800	240V AC 100 kA-3 Pole / 480V AC 65 kA-3 Pole / 600V AC 30 kA-3 Pole		Not Available			
Solid State NDC	600-800	240V AC 200 kA-3 Pole / 480V AC 100 kA-3 Pole / 600V AC 50 kA-3 Pole		Not Available				Not Available			
NB Interchangeable Trip	700-1200	240V AC 42 kA-2,3 Pole / 480V AC 30 kA-2,3 Pole / 600V AC 22 kA-2,3 Pole	W=8¼ H=16 D=5¹/₂	TKM12 Interchangeable Trip	600-1200	240V AC 42 kA-2,3 Pole / 480V AC 30 kA-2,3 Pole / 600V AC 22 kA-2,3 Pole	W=8¼ H=15½ D=3¹³/₁₆	NS (Fix-Trip)	700-1200	240V AC 42 kA-3 Pole / 480V AC 35 kA-3 Pole / 600V AC 23 kA-3 Pole	W=8¼ H=16 D=5¹/₂
HNB Interchangeable Trip	700-1200	240V AC 65 kA-2,3 Pole / 480V AC 35 kA-2,3 Pole / 600V AC 25 kA-2,3 Pole		THKM12 Interchangeable Trip	600-1200	240V AC 65 kA-2,3 Pole / 480V AC 35 kA-2,3 Pole / 600V AC 25 kA-2,3 Pole		NH (Fix-Trip)	700-1200	240V AC 65 kA-3 Pole / 480V AC 30 kA-3 Pole / 600V AC 22 kA-3 Pole	
Not Available				Not Available				Not Available			
Solid State ND	600-1200	240V AC 65 kA-3 Pole / 480V AC 50 kA-3 Pole / 600V AC 25 kA-3 Pole		Solid State TKRV	800-1200	240V AC 42 kA-3 Pole / 480V AC 30 kA-3 Pole / 600V AC 25 kA-3 Pole	W=8¼ H=15½ D=5¹/₂	Not Available			
Solid State HND	600-1200	240V AC 100 kA-3 Pole / 480V AC 65 kA-3 Pole / 600V AC 35 kA-3 Pole	W=8¼ H=16 D=5¹/₂	Solid State TKL4V	800-1200	240V AC 100 kA-3 Pole / 480V AC 65 kA-3 Pole / 600V AC 30 kA-3 Pole		Not Available			
Solid State NDC	600-1200	240V AC 200 kA-3 Pole / 480V AC 100 kA-3 Pole / 600V AC 50 kA-3 Pole		Not Available				Not Available			

TABLE 3.27 Molded-Case Circuit Breaker Interrupting Capacities (*Continued*)

		SIEMENS				SQUARE D			
1600A	Standard Interrupting 600V AC 500V DC	PXD6 (Fix Trip) PD6 (Interchangeable Trip)	1200-1600	240V AC 65 kA-3 Pole 480V AC 50 kA-3 Pole 600V AC 25 kA-3 Pole 250V DC 30 kA-2 Pole 500V DC 25 kA-3 Pole		Not Available			
	High Interrupting 600V AC 500V DC	HPXD6 (Fix Trip) HPD6 (Interchangeable Trip)		240V AC 100 kA-3 Pole 480V AC 65 kA-3 Pole 600V AC 50 kA-3 Pole 250V DC 30 kA-2 Pole 500V DC 50 kA-3 Pole		Not Available			
	Current Limiting 600V AC 500V DC	CPD6 (Fix Trip)	1200-1600	240V AC 200 kA-3 Pole 480V AC 100 kA-3 Pole 600V AC 65 kA-3 Pole 250V DC 30 kA-2 Pole 500V DC 50 kA-3 Pole	W=9 H=16 D=6⅝	Not Available			
	Standard Interrupting 600V AC	Solid State SPD6	1400-1600	240V AC 65 kA-3 Pole 480V AC 50 kA-3 Pole 600V AC 25 kA-3 Pole		Not Available			
	High Interrupting 600V AC	Solid State SHPD6	1400-1600	240V AC 100 kA-3 Pole 480V AC 65 kA-3 Pole 600V AC 65 kA-3 Pole		Solid State PXF	1400-1600	240V AC 125 kA-3 Pole 480V AC 100 kA-3 Pole 600V AC 65 kA-3 Pole	W=23¹/₁₆ H=26⅞ D=13²⁷/₆₄
2000A	Standard Interrupting 600V AC 500V DC	RXD6 (Fix Trip) RD6 (Interchangeable Trip)	1600-2000	240V AC 65 kA-3 Pole 480V AC 50 kA-3 Pole 600V AC 25 kA-3 Pole 250V DC 30 kA-2 Pole 500V DC 25 kA-3 Pole	W=8 H=16 D=6⅝	PAF (Fix Trip) PAF-DC	600-2000	240V AC 65 kA-2,3 Pole 480V AC 50 kA-2,3 Pole 600V AC 42 kA-2,3 Pole 500V DC 25 kA-3 Pole	W=13¾ H=20¹/₁₆ D=7⅞
	High Interrupting 600V AC 500V DC	HRXD6 (Fix Trip) HRD6 (Interchangeable Trip)	1600-2000	240V AC 100 kA-3 Pole 480V AC 65 kA-3 Pole 600V AC 50 kA-3 Pole 250V DC 30 kA-2 Pole 500V DC 50 kA-3 Pole		PHF (Fix Trip)	600-2000	240V AC 125 kA-2,3 Pole 480V AC 100 kA-3 Pole 600V AC 65 kA-2,3 Pole	

① Meets UL criteria for current limiting @ 240 VAC
② Meets UL criteria for current limiting @ 240 and 480 VAC
③ Current limiting @ 240 and 480 VAC

TABLE 3.28 NEC Article 450-3(a)(1): Maximum Rating or Setting for Overcurrent Device—Transformers over 600 Volts

Maximum Rating or Setting for Overcurrent Device					
	Primary		**Secondary**		
	Over 600 Volts		**Over 600 Volts**		**600 Volts or Below**
Transformer Rated Impedance	**Circuit Breaker Setting**	**Fuse Rating**	**Circuit Breaker Setting**	**Fuse Rating**	**Circuit Breaker Setting or Fuse Rating**
Not more than 6%	600%	300%	300%	250%	125%
More than 6% and not more than 10%	400%	300%	250%	225%	125%

WESTINGHOUSE				GENERAL ELECTRIC				CUTLER-HAMMER
Not Available				Not Available				Not Available
Not Available				Not Available				Not Available
Not Available				Not Available				Not Available
Solid State RD	800-1600	240V AC 125 kA-3 Pole 480V AC 65 kA-3 Pole 600V AC 50 kA-3 Pole	W=15½ H=16 D=9¾	Solid State TRLA	600-1600	240V AC 100 kA-3 Pole 480V AC 65 kA-3 Pole 600V AC 50 kA-3 Pole	W=13½ H=17½ D=8⅝	Not Available
Solid State RDC	800-1600	240V AC 200 kA-3 Pole 480V AC 100 kA-3 Pole 600V AC 65 kA-3 Pole		Solid State TRPA	600-1200	240V AC 125 kA-3 Pole 480V AC 100 kA-3 Pole 600V AC 65 kA-3 Pole		Not Available
Not Available				Not Available				Not Available
Not Available				Not Available				Not Available

TABLE 3.29   U.L. 1008 Minimum Withstand Test Requirement

Automatic Transfer Switch Rating	U.L. Minimum Current Amps	U.L. Test Current Power Factor
100 Amps or less	5,000	40% to 50%
101-400 Amps	10,000	40% to 50%
401 Amps and greater	20 times rating but not less than 10,000 Amps	40% to 50% for current of 10,000 Amps. OR 25% to 30% for currents of 20,000 Amps or less. OR 20% or less for current greater than 20,000 Amps.

TABLE 3.30   Short-Circuit Test Currents—Table 55.1 of U.L. Standard 1995

Product Ratings, A				Circuit Capacity, A
Single-Phase				
110-120V	200-208V	220-240V	254-277V	
9.8 or less	5.4 or less	4.9 or less	–	200
9.9-16.0	5.5-8.8	5.0-8.0	6.65 or less	1000
16.1-34.0	8.9-18.6	8.1-17.0	–	2000
34.1-80.0	18.7-44.0	17.1-40.0	–	3500
Over 80.0	Over 44.0	Over 40.0	Over 6.65	5000
3-Phase				Circuit Capacity, A
200-208V	220-240V	440-480V	550-600V	
2.12 or less	2.0 or less	–	–	200
2.13-3.7	2.1-3.5	1.8 or less	1.4 or less	1000
3.8-9.5	3.6-9.0	–	–	2000
9.6-23.3	9.1-22.0	–	–	3500
Over 23.3	Over 22.0	Over 1.8	Over 1.4	5000

[*]Table 55.1 of U.L. Standard 1995.

FIGURE 3.27   Current-limiting effect of fuses.

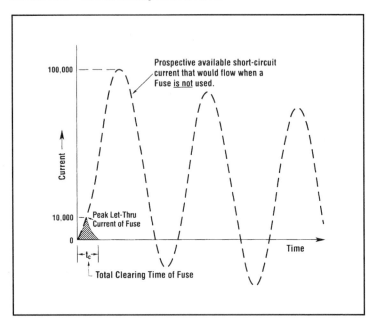

FIGURE 3.28 Analysis of a current-limiting fuse.

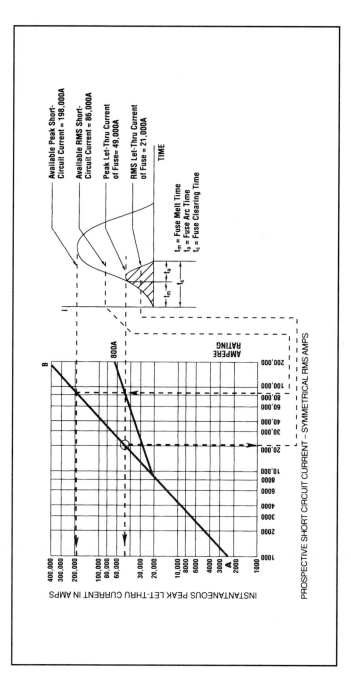

LET-THRU DATA PERTINENT TO EQUIPMENT WITHSTAND

Prior to using the Fuse Let-Thru Charts, it must be determined what let-thru data is pertinent to equipment withstand ratings.

Equipment withstand ratings can be describe as: How much fault current can the equipment handle, and for how long? Based on standards currently available, the most important data that can be obtained from the Fuse Let-Thru Charts and their physical effects are the following:

- *Peak let-thru current:* mechanical forces
- *Apparent prospective rms symmetrical let-thru current:* heating effect

Figure 3.29 is a typical example showing the short-circuit current available to an 800-A circuit, an 800-A Bussmann Low-Peak current-limiting time-delay fuse, and the let-thru data of interest.

HOW TO USE THE LET-THRU CHARTS

Using the example given in Figure 3.29, one can determine the pertinent let-thru data for the Bussmann KRP-C800SP ampere Low-Peak fuse. The Let-Thru Chart pertaining to the 800-A Low-Peak fuse is illustrated in Figure 3.30.

Determine the Peak Let-Thru Current

Step 1: Enter the chart on the Prospective Short-Circuit current scale at 86,000 A and proceed vertically until the 800-A fuse curve is intersected.

Step 2: Follow horizontally until the Instantaneous Peak Let-Thru Current scale is intersected.

Step 3: Read the Peak Let-Thru Current as 49,000 A. (If a fuse had not been used, the peak current would have been 198,000 A.)

Determine the Apparent Prospective rms
Symmetrical Let-Thru Current

Step 1: Enter the chart on the Prospective Short-Circuit Current scale at 86,000 A and proceed vertically until the 800-A fuse curve is intersected.

Step 2: Follow horizontally until line A-B is intersected.

Step 3: Proceed vertically down to the Prospective Short-Circuit Current.

Step 4: Read the Apparent Prospective RMS Symmetrical Let-Thru Current as 21,000 A. (The RMS Symmetrical Let-Thru Current would be 86,000 A if there were no fuse in the circuit.)

FIGURE 3.29 800-A Low-Peak® current-limiting time-delay fuse and associated let-thru data.

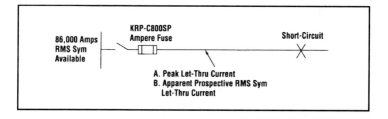

FIGURE 3.30 Current-limitation curves-Bussmann Low-Peak® time-delay fuse KRP-C800SP.

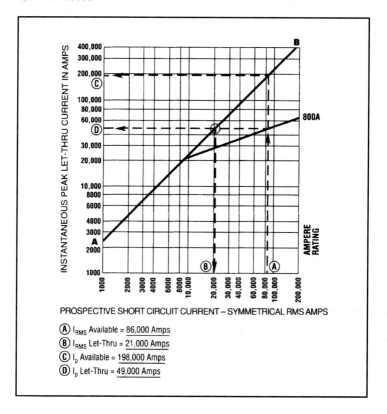

Refer to different fuse manufacturers' current limitation characteristics for applications of different fuse types and sizes under various circuit conditions.

3.6 TRANSFORMER ELECTRICAL CHARACTERISTICS

Introduction

Transformers are a critical part of electrical distribution systems because they are most often used to change voltage levels. This affects voltage, current (both load and fault current levels), and system capacity. They can also be used to isolate, suppress harmonics, derive neutrals through a zig-zag grounding arrangement, and reregulate voltage. Their electrical characteristics are as follows (see Tables 3.31–3.34 and Figure 3.31).

Auto Zig-Zag Grounding Transformers

Three single-phase transformers can be connected in an autotransformer arrangement for developing a neutral from a three-phase, three-wire supply (phase-shifting). For proper overcurrent protection, refer to NEC Article 450-5. Figure 3.32 shows the one line and wiring diagrams for this arrangement.

Table 3.35 shows the nameplate kVA for each transformer, number of transformers required, three-phase kVA rating, and maximum continuous amp load per phase (@ 277 V) for a primary input of 480 V, three-phase, three-wire, to a secondary output of 480Y/277 V, three-phase, four-wire.

Buck-Boost/Autotransformers

INTRODUCTION

Buck-boost transformers are small, single-phase transformers designed to reduce (buck) or raise (boost) line voltage from 5 to 20 percent. The most common example is boosting 208 V to 230 V, usually to operate a 230-V motor, such as an air-conditioner compressor, from a 208-V supply line.

Buck-boosts are a standard type of single-phase distribution transformer, with primary voltages of 120, 240, or 480 V and secondaries typically of 12, 16, 24, 32, or 48 V. They are available in sizes ranging from 50 VA to 10 kVA.

Buck-boost transformers are insulating-type transformers. When their primary and secondary lead wires are connected together electrically in a recommended bucking or boosting connection, however, they are in all respects an autotransformer.

TABLE 3.31 Transformer Full-Load Current, Three-Phase, Self-Cooled Ratings

kVA	Voltage, Line-to-Line												
	208	240	480	600	2,400	4,160	7,200	12,000	12,470	13,200	13,800	22,900	34,400
30	83.3	72.2	36.1	28.9	7.22	4.16	2.41	1.44	1.39	1.31	1.26	0.75	0.50
45	125	108	54.1	43.3	10.8	6.25	3.61	2.17	2.08	1.97	1.88	1.13	0.76
75	208	180	90.2	72.2	18.0	10.4	6.01	3.61	3.47	3.28	3.14	1.89	1.26
112½	312	271	135	108	27.1	15.6	9.02	5.41	5.21	4.92	4.71	2.84	1.89
150	416	361	180	144	36.1	20.8	12.0	7.22	6.94	6.56	6.28	3.78	2.52
225	625	541	271	217	54.1	31.2	18.0	10.8	10.4	9.84	9.41	5.67	3.78
300	833	722	361	289	72.2	41.6	24.1	14.4	13.9	13.1	12.6	7.56	5.04
500	1,388	1,203	601	481	120	69.4	40.1	24.1	23.1	21.9	20.9	12.6	8.39
750	2,082	1,804	902	722	180	104	60.1	36.1	34.7	32.8	31.4	18.9	12.6
1,000	2,776	2,406	1,203	962	241	139	80.2	48.1	46.3	43.7	41.8	25.2	16.8
1,500	4,164	3,608	1,804	1,443	361	208	120	72.2	69.4	65.6	62.8	37.8	25.2
2,000	....	4,811	2,406	1,925	481	278	160	96.2	92.6	87.5	83.7	50.4	33.6
2,500	....	....	3,007	2,406	601	347	200	120	116	109	105	63.0	42.0
3,000	....	....	3,609	2,887	722	416	241	144	139	131	126	75.6	50.4
3,750	....	....	4,511	3,608	902	520	301	180	174	164	157	94.5	62.9
5,000	....	....	....	4,811	1,203	694	401	241	231	219	209	126	83.9
7,500	....	....	....	....	1,804	1,041	601	361	347	328	314	189	126
10,000	....	....	....	....	2,406	1,388	802	481	463	437	418	252	168

TABLE 3.32 Typical Impedances, Three-Phase Transformers

kVA	Liquid-Filled	
	Network	**Padmount**
37.5	....	....
45	....	....
50	....	....
75	....	3.4
112.5	....	3.2
150	....	2.4
225	....	3.3
300	5.00	3.4
500	5.00	4.6
750	5.00	5.75
1000	5.00	5.75
1500	7.00	5.75
2000	7.00	5.75
2500	7.00	5.75
3000	....	6.50
3750	....	6.50
5000	....	6.50

① Values are typical. For guaranteed values, refer to transformer manufacturer.

APPLICATIONS

Electrical and electronic equipment is designed to operate on standard supply voltage. When the supply voltage is constantly too low or too high (usually more than ±5 percent), the equipment fails to operate at maximum efficiency. A buck-and-boost transformer is a simple and economical means of correcting such an off-standard voltage.

Buck-boost transformers are commonly used for boosting 208 V to 230 or 240 V and vice versa for commercial and industrial air-conditioning systems, boosting 110 V to 120 V and 240 V to 277 V for lighting systems, and voltage correction for heating systems and induction motors of all types.

Buck-boost transformers can also be used to power low-voltage circuits for control, lighting, and other applications requiring 12, 16, 24, 32, or 48 V. The unit is connected as an insulating transformer and the nameplate kVA rating is the transformer's capacity.

OPERATION AND CONSTRUCTION

Buck-boost transformers have four windings to make them versatile. Their two primary and two secondary windings can be connected eight different ways to provide a multitude of voltage and kVA outputs. They cannot be used to stabilize voltage, however, because the output voltage

TABLE 3.33 Approximate Transformer Loss and Impedance Data

15 kV Class Oil Liquid-Filled Transformers

65°C Rise						
kVA	No Load Watts Loss	Full Load Watts Loss	%Z	%R	%X	X/R
112.5	550	2470	5.00	1.71	4.70	2.75
150	545	3360	5.00	1.88	4.63	2.47
225	650	4800	5.00	1.84	4.65	2.52
300	950	5000	5.00	1.35	4.81	3.57
500	1200	8700	5.00	1.50	4.77	3.18
750	1600	12160	5.75	1.41	5.57	3.96
1000	1800	15100	5.75	1.33	5.59	4.21
1500	3000	19800	5.75	1.12	5.64	5.04
2000	4000	22600	5.75	0.93	5.67	6.10
2500	4500	26000	5.75	0.86	5.69	6.61

15 kV Class Primary – Dry-Type Transformers Class H

150°C Rise						
kVA	No Load Watts Loss	Full Load Watts Loss	%Z	%R	%X	X/R
300	1600	10200	4.50	2.87	3.47	1.21
500	1900	15200	5.75	2.66	5.10	1.92
750	2700	21200	5.75	2.47	5.19	2.11
1000	3400	25000	5.75	2.16	5.33	2.47
1500	4500	32600	5.75	1.87	5.44	2.90
2000	5700	44200	5.75	1.93	5.42	2.81
2500	7300	50800	5.75	1.74	5.48	3.15
80°C Rise						
300	1800	7600	4.50	1.93	4.06	2.10
500	2300	9500	5.75	1.44	5.57	3.87
750	3400	13000	5.75	1.28	5.61	4.38
1000	4200	13500	5.75	0.93	5.67	6.10
1500	5900	19000	5.75	0.87	5.68	6.51
2000	6900	20000	5.75	0.66	5.71	8.72
2500	7200	21200	5.75	0.56	5.72	10.22

is a function of the input voltage; i.e., if the input voltage varies, the output voltage will also vary by the same percentage.

LOAD DATA

The fact that a buck-boost transformer can operate a kVA load many times larger than the kVA rating on its nameplate may seem paradoxical, and consequently, sometimes causes confusion in sizing.

TABLE 3.33 Approximate Transformer Loss and Impedance Data (*Continued*)

600-Volt Primary Class Dry-Type Transformers

150°C Rise						
kVA	No Load Watts Loss	Full Load Watts Loss	%Z	%R	%X	X/R
3	33	231	7.93	6.60	4.40	0.67
6	58	255	3.70	3.28	1.71	0.52
9	77	252	3.42	1.94	2.81	1.45
15	150	875	5.20	4.83	1.92	0.40
30	200	1600	5.60	4.67	3.10	0.66
45	300	1900	4.50	3.56	2.76	0.78
75	400	3000	4.90	3.47	3.46	1.00
112.5	500	4900	5.90	3.91	4.42	1.13
150	600	6700	6.20	4.07	4.68	1.15
225	700	8600	6.40	3.51	5.35	1.52
300	800	10200	7.10	3.13	6.37	2.03
500	1700	9000	5.50	1.46	5.30	3.63
750	2200	11700	6.30	1.27	6.17	4.87
1000	2800	13600	6.50	1.08	6.41	5.93

600-Volt Primary Class Dry-Type Transformers

115°C Rise						
kVA	No Load Watts Loss	Full Load Watts Loss	%Z	%R	%X	X/R
15	150	700	5.20	3.67	3.69	1.01
30	200	1500	4.60	4.33	1.54	0.36
45	300	1700	3.70	3.11	2.00	0.64
75	400	2300	4.60	2.53	3.84	1.52
112.5	500	3100	6.50	2.31	6.08	2.63
150	600	5900	6.20	3.53	5.09	1.44
225	700	6000	7.20	2.36	6.80	2.89
300	800	6600	6.30	1.93	6.00	3.10
500	1700	6800	5.50	1.02	5.40	5.30
750	1500	9000	4.10	1.00	3.98	3.98

600-Volt Primary Class Dry-Type Transformers

80°C Rise						
kVA	No Load Watts Loss	Full Load Watts Loss	%Z	%R	%X	X/R
15	200	500	2.30	2.00	1.14	0.57
30	300	975	2.90	2.25	1.83	0.81
45	300	1100	2.90	1.78	2.29	1.29
75	400	1950	3.70	2.07	3.07	1.49
112.5	600	3400	4.30	2.49	3.51	1.41
150	700	3250	4.10	1.70	3.73	2.19
225	800	4000	5.30	1.42	5.11	3.59
300	1300	4300	3.30	1.00	3.14	3.14
500	2200	5300	4.50	0.62	4.46	7.19

TABLE 3.34 Transformer Primary (480-Volt, Three-Phase, Delta) and Secondary (208Y/120-Volt, Three-Phase, Four-Wire) Overcurrent Protection, Conductors and Grounding

Primary CB/Fuse	Recommended Primary Wire and Ground (CB/Fuse)	Primary FLA	Transformer KVA	Secondary FLA	Recommended Secondary Wire and Ground (CB/Fuse)	Secondary CB/Fuse
15/10	#12 and #12	9.0	7.5	20.8	#10 and #10/#16	25/25
25/20	#12 and #10 or #12	18.1	15	41.7	#8 and #10	60/50
40/35	#10 and #10	30.1	25	69.4	#4 and #8	90/80
50/45	#6 and #10	36.1	30	83.3	#3 and #6/#8	110/100
70/60	#4 and #8 or #10	54.2	45	125	#1 and #8	150/125
125/110	#1 and #6	90.3	75	208.3	#4/0 and #4	250/225
175/150	#2/0 and #6	135.4	112.5	312.5	350 and #3	400/350
225/200	#4/0 and #4 or #6	180.5	150	416.6	500 and #2	500/450
350/300	350 and #3 or #4	270.8	225	625	2 sets [350 and #1/0]	800/700
450/400	500 and #2 or #3	361.0	300	833.3	3 sets [500 and #2/0]	1000/1000
800/700	2 sets [350 and #1/0]	601.7	500	1388.8	5 sets [500 and #4/0]	1600/1600
1200/1000	3 sets [500 and #3/0 or #2/0]	902.5	750	2083.3	7 sets [500 and 350]	2500/2500
1600/1200	4 sets [350 and #4/0 or #3/0]	1203.4	1000	2777.8	8 sets [500 and 400]	3000/3000
2000/1600	5 sets [500 and 250/#4/0]	1504.2	1250	3472.2	11 sets [500 and 500]	4000/4000
2500/2000	6 sets [500 and 350/250]	1805.1	1500	4166.7	14 sets [500 and 700]	5000/5000

Notes:
1. C/B sizes based on maximum permissible per 450-36 and are for 80% rated thermal magnetic C/Bs. Primary/Secondary coordination is taken into consideration. Fuse sizes are for dual element time delay RK-5 fuses.
2. Use fuse sizes for 100% rated C/Bs.
3. Lower C/B values are commonly used to align with other equipment ratings. In particular, 30 KVA: primary = 45A, secondary = 100A with 80% rated C/Bs; 75 KVA: primary = 100A, secondary = 225A with 80% rated C/Bs; 112.5 KVA: primary = 150A, secondary = 350A with 80% rated C/Bs; 150 KVA: primary = 200A, secondary = 400A with 80% rated C/Bs; 500 KVA: primary = 400A, secondary = 800A with 80% rated C/Bs.
4. Equipment grounding conductor size is based on Table 250-95 and assumes separately derived ground at the transformer secondary.
5. Recommended wire size is based on Table 310-1b (75°C types THWN, THHN and XHHW), 240-3b and 240-3c. Wire sizes may need to be increased due to voltage drop.
6. Maximum conduit fill depends on the wire insulation type and the type of conduit. These values have been calculated for conductors of the same size and are given in tabular form in Appendix C of NFPA 70.

EMT - Table 2.21
IMC - Table 2.24
RMC - Table 2.28
Schedule 80 rigid PVC - Table 2.29
Schedule 40 rigid PVC - Table 2.30

FIGURE 3.31 Electrical connection diagrams.

CONNECTION DIAGRAMS
For Transformers

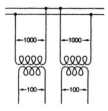

Single-phase transformers on
a single phase system.

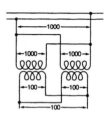

Single-phase transformers,
secondaries in parallel.

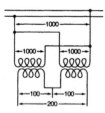

Single-phase transformers
secondaries in series.

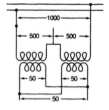

Single-phase transformers
primaries in series, secondaries
in parallel.

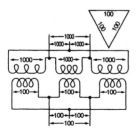

Three single-phase transformers
connected delta-delta to
a three-phase system.

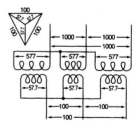

Three single-phase transformers
connected star-star to
a three-phase system.

FIGURE 3.31   Electrical connection diagrams. (*Continued*)

CONNECTION DIAGRAMS
For Transformers

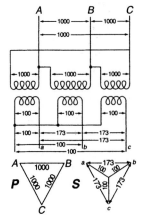

Three single-phase transformers
connected delta-star to
a three-phase system.

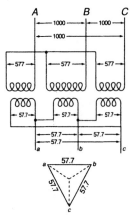

Three single-phase transformers
connected star-delta to
a three phase system.

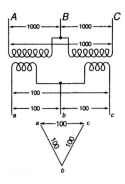

Two single-phase transformers
connected open-delta to a three-phase
system.

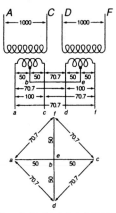

Two single-phase transformers
connected star to a four-wire
two-phase system.

FIGURE 3.31 Electrical connection diagrams. (*Continued*)

CONNECTION DIAGRAMS
For Transformers

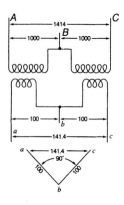

Two single-phase transformers
connected to a three-wire
two-phase system.

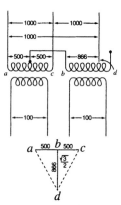

Two single-phase transformers
connected T to a three-phase
two-phase system. Scott Connection.

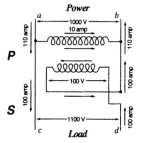

Single phase transformer
used as a booster.

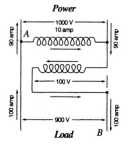

Single phase transformer
connected to lower the E.M.F.

FIGURE 3.32 Auto zig-zag grounding transformers for deriving a neutral—schematic and wiring diagram.

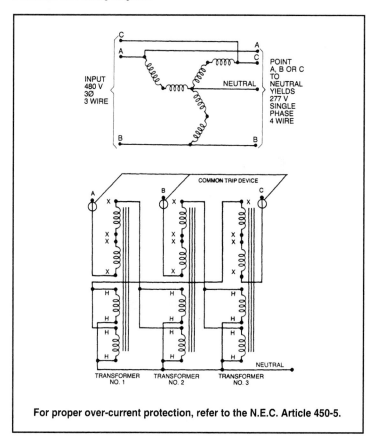

For proper over-current protection, refer to the N.E.C. Article 450-5.

To cite an example, a buck-boost transformer has a nameplate rating of 1 kVA, but when it's connected as an autotransformer boosting 208 V to 230 V, its kVA capacity increases to 9.58 kVA. The key to understanding the operation of buck-boost transformers lies in the fact that the secondary windings are the only parts of the transformer that do the work of transforming voltage and current. In the example given, only 22 V are being transformed (boosted): 208 V + 22 V = 230 V. This 22-V transformation is carried out by the secondary windings, which are designed to operate at a maximum current of 41.67 A (determined by wire size of windings).

TABLE 3.35 Auto Zig-Zag Grounding Transformer Ratings

3Ø, 3 WIRE   ***50/60 Hz**   **3Ø, 4 WIRE**

Use 3 Pieces of Type No.	Available In	Nameplate KVA For Each Trfmr.	No. of Trfmr. Required	Three Phase KVA	Max. Continuous Amp. Load Per Phase (277 Volts)
T-2-53010-S	No Taps Only	1.0	3	10.8	12.50
T-2-53011-S	No Taps Only	1.5	3	15.6	18.75
T-2-53012-S	No Taps Only	2.0	3	20.7	25.00
T-2-53013-4S	Taps & No Taps	3.0	3	31.2	37.50
T-2-53014-4S	Taps & No Taps	5.0	3	51.9	62.50
T-2-53515-3S	With Taps Only	7.5	3	78.0	93.50
T-2-53516-3S	With Taps Only	10.0	3	103.8	125.00
T-2-53517-3S	With Taps Only	15.0	3	156.0	187.50
T-2-53518-3S	With Taps Only	25.0	3	259.5	312.00
T-1-53019-3S	With Taps Only	37.5	3	390.0	468.00
T-1-53020-3S	With Taps Only	50.0	3	519.0	625.00
T-1-53021-3S	With Taps Only	75.0	3	780.0	935.00
T-1-53022-3S	With Taps Only	100.0	3	1038.0	1250.00
T-1-53023-3S	With Taps Only	167.0	3	1734.0	2085.00

Connection diagram (using 3 pieces of 1 phase 60 hertz transformers connected zig-zag auto) for developing a neutral (4th wire) from a 3 phase 3 wire supply
*Applicable for the above connection only

Maximum secondary amps = nameplate kVA × 1000/secondary volts
Maximum secondary amps = 1.0 kVA × 1000/24 V = 41.67 A

Because the transformer has been autoconnected in such a fashion that the 22-V secondary voltage is added to the 208-V primary voltage, it produces a 230-V output.

The autotransformer kVA is calculated thus:

$$kVA = \text{output volts} \times \text{secondary amps}/1000$$
$$kVA = 230\text{ V} \times 41.67\text{ A}/1000 = 9.58\text{ kVA}$$

THREE-PHASE

To this point, we have only discussed single-phase applications. Buck-boost transformers can be used on three-phase systems. Two or three units are used to buck or boost three-phase voltage. The number of units to be used in a three-phase installation depends on the number of wires in the supply line. If the three-phase supply is four-wire Y, use three buck-boost transformers. If the three-phase supply is three-wire Y (neutral not available), use two buck-boost transformers.

A three-phase wye buck-boost transformer connection should be used only on a four-wire source of supply. A delta-to-wye connection does not provide adequate current capacity to accommodate unbalanced currents flowing in the neutral wire of the four-wire circuit.

A closed delta buck-boost autotransformer connection requires more transformer kilovolt-amperes than a wye or open delta connection, and phase shifting occurs on the output. Consequently, the closed delta connection is more expensive and electrically inferior to other three-phase connections.

The do's and don'ts of three-phase connections are summarized in Table 3.36.

TABLE 3.36   Buck-Boost Transformer Three-Phase Connection Summary

INPUT (SUPPLY SYSTEM)	DESIRED OUTPUT CONNECTION	
DELTA 3 wire	WYE 3 or 4 wire	DO NOT USE
OPEN DELTA 3 wire	WYE 3 or 4 wire	DO NOT USE
WYE 3 or 4 wire	CLOSED DELTA 3 wire	DO NOT USE
WYE 4 wire	WYE 3 or 4 wire	OK
WYE 3 or 4 wire	OPEN DELTA 3 wire	OK
CLOSED DELTA 3 wire	OPEN DELTA 3 wire	OK

SOUND LEVELS, LIFE EXPECTANCY, AND COST

The sound levels and life expectancy of buck-boost transformers are the same as any other insulating transformer. However, an autoconnected buck-boost transformer will be quieter than an insulating transformer capable of handling the same load. The insulating unit would have to be physically larger than the buck-boost transformer, and smaller transformers are quieter than larger ones. Using a similar rationale, for the most common buck-boost applications, the dollar savings are generally in the order of 75 percent compared with the use of an insulating-type distribution transformer for the same application.

DIAGRAMS

Figure 3.33 shows typical connection diagrams for single-phase buck-boost transformers used for low-voltage power supply applications.

Figures 3.34 and 3.35 show typical connection diagrams for single-phase and three-phase, respectively, buck-boost transformers connected in an autotransformer arrangement.

FIGURE 3.33 Wiring diagrams for low-voltage single-phase buck-boost transformers.

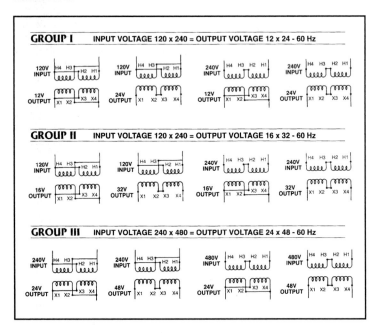

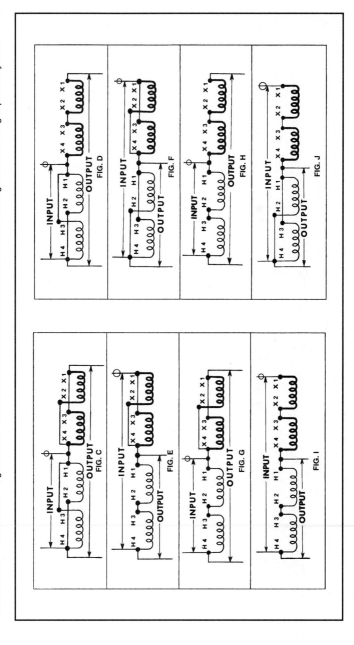

FIGURE 3.34 Connection diagrams for buck-boost transformers in autotransformer arrangement for single-phase system.

FIGURE 3.35 Connection diagrams for buck-boost transformers in autotransformer arrangement for three-phase system.

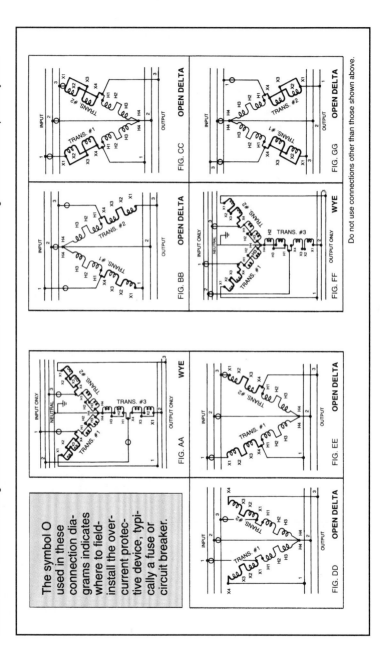

3.7 TRANSFORMER THERMAL AND SOUND CHARACTERISTICS

In addition to transformer electrical characteristics, their thermal and sound level characteristics are very important. The thermal characteristics are determined by industry standards (UL/ANSI 1561-1987) and are generally only of concern to the electrical design professional. Sound levels, on the other hand, are of concern to everyone, especially the architect and occupants of the building. Electrical design professionals must be sensitive and aware of the sound levels of electrical equipment and their impact on the occupants of the building and exercise appropriate measures to mitigate their effects. These could include remotely locating the equipment, sound attenuation techniques, and/or structural isolation. To assist you in evaluating these considerations, Figure 3.36 shows the thermal characteristics of dry-type distribution transformers, and Tables 3.37 and 3.38, respectively, show the maximum average sound levels of dry-type and liquid-filled transformers and typical ambient sound levels.

k-Rated Transformers

Transformers used for supplying the nonsinusoidal high harmonic (>5 percent) content loads that are increasingly prevalent must be designed and listed for these loads. ANSI C57.110-1986, "Recommended Practice for Establishing Transformer Capability When Supplying Non-Sinusoidal Load Currents," provides a method for calculating the heating effect in a transformer when high harmonic currents are present. This method generates a number called the *k*-factor, which is a multiplier that

FIGURE 3.36 Transformer insulation system temperature ratings.

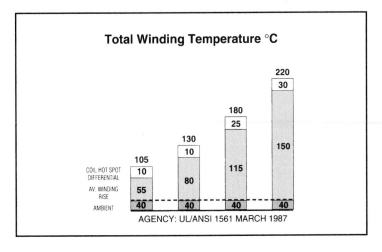

TABLE 3.37 Typical Building Sound Levels

Radio, Recording and TV Studios	25-30 db
Theatres and Music Rooms	30-35
Hospitals, Auditoriums and Churches	35-40
Classrooms and Lecture Rooms	35-40
Apartments and Hotels	35-45
Private Offices and Conference Rooms	40-45
Stores	45-55
Residence (Radio, TV Off) and Small Offices	53
Medium Office (3 to 10 Desks)	58
Residence (Radio, TV On)	60
Large Store (5 or More Clerks)	61
Factory Office	61
Large Office	64
Average Factory	70
Average Street	80

related eddy current losses in the transformer core due to harmonics to increased transformer heating. Transformer manufacturers use this information to design transformer core/coil and insulation systems that are more tolerant of the higher internal heating load than a standard design. Simply put, a k-rated transformer can tolerate approximately k times more internal heat than a similar, standard-design transformer (for example, a k-4 transformer can handle approximately four times the internal heating load of a similar ANSI standard nonharmonic rated transformer with no life expectancy reduction).

TABLE 3.38 Maximum Average Sound Levels for Transformers

kVA	Dry-Type		Liquid-Filled	
	Self-Cooled Rating (AA)	Forced-Air Cooling (FA)	Self-Cooled Rating (OA)	Forced-Air Cooling (FA)
0-50	50	..	..	..
51-150	55	..	..	..
151-300	58	67	55	67
301-500	60	67	56	67
501-700	62	67	57	67
701-1000	64	67	58	67
1001-1500	65	68	60	67
1501-2000	66	69	61	67
2001-2500	68	71	62	67
2501-3000	70	71	63	67
3001-4000	71	73	64	67
4001-5000	72	74	65	67
5001-6000	73	75	66	68
6001-7500	..	76	67	69
7501-10000	..	76	68	70

The *k*-rating of a transformer addresses only increased internal heating. It does not address mitigation of the harmonic content of the transformer load.

3.8 MOTOR FEEDERS AND STARTERS

Introduction

Motors comprise a significant portion of a building's electrical system loads. They are needed to power fans and pumps for basic mechanical building infrastructure, such as heating, ventilation, air-conditioning, plumbing, fire protection, elevators, and escalators. They are also needed to power equipment endemic to the occupancy, such as commercial kitchen equipment in an institutional facility, CT and MRI scanners in a hospital, and process equipment such as conveyors and machinery in an industrial plant or stone quarry. Consequently, designing motor-circuit feeders is very much in the mainstream of the electrical design professional's daily work. To save time in this process, the following information is provided.

Sizing Motor-Circuit Feeders and Their Overcurrent Protection

I. *For AC single-phase motors, polyphase motors other than wound-rotor (synchronous and induction other than Code E):*[1,2]

  1. Feeder wire size is 125 percent of motor full-load (FL) current *minimum.*
  2. Feeder breaker (thermal-magnetic fixed-trip type) is 250 percent of FL current *maximum.*
  3. Feeder breaker (instantaneous magnetic-only type) is 800 percent of FL current *maximum.*
  4. Feeder fuse (dual-element time-delay type) is 175 percent of FL current *maximum.*
  5. Feeder fuse (NEC non-time-delay type) is 300 percent of FL current *maximum.*

II. *For wound-rotor motors:*

  1. Feeder wire size is 125 percent of motor FL current *minimum.*
  2. Feeder breaker (thermal-magnetic fixed-trip type) is 150 percent of FL current *maximum.*
  3. Feeder breaker (instantaneous magnetic-only type) is 800 percent of FL current *maximum.*

[1] Synchronous motors of the low-torque, low-speed type (usually 450 rpm or lower), such as those used to drive reciprocating compressors, pumps, and so forth, that start unloaded, do not require a fuse rating or circuit breaker setting in excess of 200 percent of full-load current.

[2] For Code Letter E induction motors, everything is the same as above except if an instantaneous magnetic-only-type circuit breaker is used, it shall have a *maximum* setting of 1100 percent.

   4. Feeder fuse (dual-element time-delay type) is 150 percent of FL current *maximum.*
   5. Feeder fuse (NEC non-time-delay type) is 150 percent of FL current *maximum.*

III. *For hermetic motors (special case):* Hermetic motors are actually a combination consisting of a compressor and motor, both of which are enclosed in the same housing, with no external shaft or shaft seals, the motor operating in the refrigerant; thus, their characteristics are different than standard induction motors. Calculating their feeder size and overcurrent protection is based on their nameplate branch-circuit selection current (BCSC) or their rated-load current (RLC), whichever is greater. The BCSC is always equal to or greater than the RLC. Hence, the following:

   1. Feeder wire size is 125 percent of BCSC/RLC *maximum.*
   2. Feeder breaker (thermal-magnetic fixed-trip type) is between 175 and 225 percent of BCSC/RLC *maximum.*
   3. Feeder breaker (instantaneous magnetic-only type) is 800 percent of BCSC/RLC *maximum.*
   4. Feeder fuse (dual-element time-delay type) is between 175 and 225 percent of BCSC/RLC *maximum.*
   5. Feeder fuse (NEC non-time-delay type) is NOT RECOMMENDED—DO NOT USE.

IV. *Direct-current (constant-voltage) motors:*

   1. Feeder wire size is 125 percent of motor FL current *maximum.*
   2. Feeder breaker (thermal-magnetic fixed-trip type) is 150 percent of FL current *maximum.*
   3. Feeder breaker (instantaneous magnetic-only type) is 250 percent of FL current *maximum.*
   4. Feeder fuse (dual-element time-delay type) is 150 percent of FL current *maximum.*
   5. Feeder fuse (NEC non-time-delay type) is 150 percent of FL current *maximum.*

V. *For multiple motors on one feeder:* First, size the feeder and overcurrent protection for the largest motor and add the full-load current of the remaining motors to size the overall feeder and overcurrent protection.

VI. *Application tips:*

   1. Refer to NEC Articles 430 and 440 for further details on sizing motor feeders and overcurrent protection.
   2. For elevator motors, always try to get the full-load current, because the nameplate horsepower on many machines is about 10 to 25 percent below the actual rating.
   3. For packaged-type evaporative condensers with many small fans nominally rated 1 hp (for example), be sure to get the full-load current, because these are really equivalent to about

2 hp (for example) each, and feeders sized on nominal horse-power ratings will be inadequate. Remember to size the feeder and overcurrent protection as a multiple-motor load. Also refer to NEC Article 440.

4. Note that *maximum* and *minimum* have precise meanings: feeder sizes shall be not less than the calculated minimum within 3 or 4 percent (e.g., 30 A-rated No. 10 wire is okay for a 31-A load), and breaker sizes shall be not more than the maximum indicated. In general, for larger motor sizes, the overcurrent protection needed decreases considerably from the maximum limit.

5. In sizing nonfused disconnects for motors, use the horsepower rating table in the manufacturer's catalog or realize that in general, a nonfused disconnect switch should be rated the same as a switch fused with a dual-element time-delay fuse.

6. When sizing feeders for tape drives in mainframe data centers, it is usually necessary to oversize both the overcurrent protection and the feeder to accommodate the long acceleration time characteristic of this equipment.

7. Today's highly energy-efficient motors are characterized by low losses and high inrush currents, thus requiring overcurrent protection sized at or near the maximum limit prescribed by the NEC when these motors are used.

8. For NEC Locked-Rotor Indicating Code Letters, refer to Table 3.39 [NEC Table 430-7(b)].

TABLE 3.39   NEC Table 430-7(b): Locked-Rotor Indicating Code Letters

Code Letter	Kilovolt-Amperes per Horsepower with Locked Rotor		
A	0	—	3.14
B	3.15	—	3.54
C	3.55	—	3.99
D	4.0	—	4.49
E	4.5	—	4.99
F	5.0	—	5.59
G	5.6	—	6.29
H	6.3	—	7.09
J	7.1	—	7.99
K	8.0	—	8.99
L	9.0	—	9.99
M	10.0	—	11.19
N	11.2	—	12.49
P	12.5	—	13.99
R	14.0	—	15.99
S	16.0	—	17.99
T	18.0	—	19.99
U	20.0	—	22.39
V	22.4	—	and up

Motor Circuit Data Sheets

The following motor circuit data sheets provide recommended design standards for branch-circuit protection and wiring of squirrel cage induction motors of the sizes and voltages most frequently encountered in commercial, institutional, and industrial facilities. Experience has shown that most facilities of this type use copper wire, and use No. 12 AWG wire and $\frac{3}{4}$-in conduit as minimum sizes for power distribution. These standards are reflected in the tables that follow. Refer also to the notes to Tables 3.40–3.44 for assumptions and other criteria used.

Motor Starter Characteristics (for Squirrel Cage Motors)

There are fundamentally two types of motor starters: full-voltage (both reversing and nonreversing) and reduced-voltage. In the information that follows, their characteristics and selection criteria are briefly summarized.

FULL-VOLTAGE STARTERS

A squirrel cage motor draws high starting current (inrush) and produces high starting torque when started at full voltage. Although these values differ for different motor designs, for a typical NEMA design B motor, the inrush will be approximately 600 percent of the motor full-load amperage (FLA) rating and the starting torque will be approximately 150 percent of full-load torque at full voltage. High-current inrush and starting torque can cause problems in the electrical and mechanical systems and may even cause damage to the utilization equipment or materials being processed.

REDUCED-VOLTAGE STARTERS

When a motor is started at reduced voltage, the current at the motor terminals is reduced in direct proportion to the voltage reduction, whereas the torque is reduced by the square of the voltage reduction. If the "typical" NEMA B motor is started at 70 percent of line voltage, the starting current would be 70 percent of the full-voltage value (i.e., $0.70 \times 600\% = 420\%$ FLA). The torque would then be $(0.70)^2$ or 49 percent of the normal starting torque (i.e., $0.49 \times 150\% = 74\%$ full-load torque). Therefore, reduced-voltage starting provides an effective means of reducing both inrush current and starting torque.

If the motor has a high inertia or if the motor rating is marginal for the applied load, reducing the starting torque may prevent the motor from reaching full speed before the thermal overloads trip. Applications that require high starting torque should be reviewed carefully to determine if reduced-voltage starting is suitable. As a rule, motors with a

TABLE 3.40   480-Volt System (460-Volt Motors) Three-Phase Motor Circuit Feeders

HP	FLA	Safety Switch	Fuse	C/B	Wire	Ground (fuse)	Ground (C/B)	NEMA Starter Size
½	1.1	30	1.4	15	3 #12	#12	#12	00
3/4	1.6	30	2	15	3 #12	#12	#12	00
1	2.1	30	2.5	15	3 #12	#12	#12	00
1 ½	3.0	30	3.5	15	3 #12	#12	#12	00
2	3.4	30	4	15	3 #12	#12	#12	00
3	4.8	30	5.6	15	3 #12	#12	#12	0
5	7.6	30	9	20	3 #12	#12	#12	0
7 ½	11	30	15	30	3 #10	#12	#10	1
10	14	30	17.5	30	3 #10	#12	#10	1
15	21	30	25	50	3 #10	#10	#10	2
20	27	60	35	60	3 #10	#10	#10	2
25	34	60	40	70	3 #8	#10	#8	2
30	40	60	50	90	3 #8	#10	#8	3
40	52	100	60	125	3 #6	#10	#6	3
50	65	100	75	150	3 #4	#8	#6	3
60	77	100	90	175	3 #3	#8	#6	4
75	96	200	125	200	3 #1	#6	#6	4
100	124	200	150	250	3 #2/0	#6	#4	4
125	156	200	200	300	3 #3/0	#6	#4	5
150	180	400	225	350	3 #4/0	#4	#3	5
200	240	400	300	400	3-350	#4	#3	5
250	302	400	350	600	3-500	#3	#1	6
300	361	600	450	800	2 sets [3 #4/0]	#2	#1/0	6
350	414	600	500	800	2 sets [3-300]	#2	#1/0	--
400	477	800	600	1000	2 sets [3-350]	#1	#2/0	--
450	515	800	600	1000	2 sets [3-400]	#1	#2/0	--

Notes for Tables 3-40 through 3-44:

1. This table assumes that motor overload protection per 430 Part C is provided separately as part of the motor starter.
2. Horse powers are for atypical induction type squirrel cage motor and are taken from Table 430-150. Sizing characteristics are based on a FVNR, Design B, Code letter G motor. Multi speed, low power factor, high torque, high locked rotor or design E motors may have different FLA values and sizing requirements.
3. FLA (full load amperes) from Table 430-150
4. Fuses and circuit breakers are sized for back up overload and short circuit protection per 430 Part D.
5. Fuses are dual element time delay RK-1 type sized at a minimum 115% FLA based on service factors less than 1.15 or temperature greater than 40 degrees C.
6. Circuit breakers are general purpose thermal magnetic (inverse time) type sized at approximately 200% FLA.
7. Part 430-52 and Table 430-152 establish maximum sizes.
8. Wire sizes are based on 75 degrees C copper with THWN, THHN or XHHW insulation. Size may need to be increased because of voltage drop.
9. Maximum conduit fill depends on the wire insulation type and the type of conduit. These values have been calculated for conductors of the same size and are given in tabular form in Appendix C of NFPA 70 [EMT - Table 2.21; IMC - Table 2.24; RMC - Table 2.28; Schedule 40 rigid PVC - Table 2.29; Schedule 80 rigid PVC - Table 2.30]

horsepower rating in excess of 15 percent of the kilovolt-ampere rating of the transformer feeding it should use a reduced-voltage start.

There are several types of electromechanical as well as solid-state reduced-voltage starters that provide different starting characteristics. The following tables from Square D Company are a good representation of industry standard characteristics. Table 3.45(a) shows the starting characteristics for Square D's class 8600 series of reduced-voltage

TABLE 3.41 208-Volt System (200-Volt Motors) Three-Phase Motor Circuit Feeders

HP	FLA	Safety Switch	Fuse	C/B	Wire	Ground (fuse)	Ground (C/B)	NEMA Starter Size
½	2.5	30	3	15	3 #12	#12	#12	00
3/4	3.7	30	4.5	15	3 #12	#12	#12	00
1	4.8	30	5.6	15	3 #12	#12	#12	00
1 ½	6.9	30	8	20	3 #12	#12	#12	00
2	7.8	30	9	20	3 #12	#12	#12	0
3	11.0	30	15	25	3 #10	#12	#10	0
5	17.5	30	25	40	3 #10	#10	#10	1
7 ½	25.3	60	30	50	3 #10	#10	#10	1
10	32.2	60	40	70	3 #8	#10	#8	2
15	48.3	100	60	100	3 #6	#10	#8	3
20	62.1	100	75	150	3 #4	#8	#6	3
25	78.2	100	90	175	3 #3	#8	#6	3
30	92	200	110	200	3 #2	#6	#6	4
40	120	200	150	250	3 #1/0	#6	#4	4
50	150	200	175	300	3 #3/0	#6	#4	5
60	177	400	225	400	3 #4/0	#4	#3	5
75	221	400	300	400	3 #300	#4	#3	5
100	285	400	350	500	3 #500	#3	#2	6
125	359	600	450	600	2 sets [3 #4/0]	#2	#1	6
150	414	600	500	600	2 sets [3-300]	#2	#1	6

See Notes, Table 3.40

starters compared with full-voltage starting, along with the advantages and disadvantages of each type. Table 3.45(b) provides an aid in the selection of the starter best suited for a particular application and desired starting characteristic.

3.9 STANDARD VOLTAGES AND VOLTAGE DROP

Introduction

An understanding of system voltage nomenclature and preferred voltage ratings of distribution apparatus and utilization equipment is essential to ensure the proper design and operation of a power distribution system. The dynamic characteristics of the system should be recognized and the proper principles of voltage regulation applied so that satisfactory voltages will be supplied to utilization equipment under all normal conditions of operation.

TABLE 3.42 115-Volt Single-Phase Motor Circuit Feeders

HP	FLA	Safety Switch	Fuse	C/B	Wire	Ground (fuse)	Ground (C/B)	NEMA Starter Size
1/6	4.4	30	5.6/10	15	2 #12	#12	#12	00
1/4	5.8	30	7	15	2 #12	#12	#12	00
1/3	7.2	30	9	20	2 #12	#12	#12	0
1/2	9.8	30	12	20	2 #12	#12	#12	0
3/4	13.8	30	17.5	30	2 #10	#12	#10	0
1	16	30	20	40	2 #10	#12	#10	0
1-1/2	20	30	25	50	2 #12	#10	#10	1
2	24	30	30	60	2 #10	#10	#10	1

See Notes, Table 3.40

261

TABLE 3.43   200-Volt Single-Phase Motor Circuit Feeders

HP	FLA	Safety Switch	Fuse	C/B	Wire	Ground (fuse)	Ground (C/B)	NEMA Starter Size
1/6	2.5	30	3	15	2 #12	#12	#12	00
1/4	3.3	30	4	15	2 #12	#12	#12	00
1/3	4.1	30	5	15	2 #12	#12	#12	00
1/2	5.6	30	7	15	2 #12	#12	#12	00
3/4	7.0	30	10	20	2 #12	#12	#12	00
1	9.2	30	12	20	2 #12	#12	#12	00
1-1/2	11.5	30	15	30	2 #10	#12	#10	0
2	13.8	30	17.5	30	2 #10	#12	#10	0
3	19.6	30	25	40	2 #10	#10	#10	1
5	32.2	60	40	70	2 #8	#10	#8	2
7-1/2	46	100	60	100	2 #6	#10	#8	2
10	57.5	100	70	125	2 #4	#8	#6	3

See Notes, Table 3.40

TABLE 3.44 230-Volt Single-Phase Motor Circuit Feeders

HP	FLA	Safety Switch	Fuse	C/B	Wire	Ground (fuse)	Ground (C/B)	NEMA Starter Size
1/6	2.2	30	2.8	15	2 #12	#12	#12	00
1/4	2.9	30	3.5	15	2 #12	#12	#12	00
1/0	3.6	30	4.5	15	2 #12	#12	#12	00
1/2	4.9	30	6	15	2 #12	#12	#12	00
3/4	6.9	30	8	15	2 #12	#12	#12	00
1	8	30	10	20	2 #12	#12	#12	00
1-1/2	10	30	12	20	2 #12	#12	#12	0
2	12	30	15	30	2 #10	#12	#10	0
3	17	30	20	40	2 #10	#12	#10	1
5	28	60	35	60	2 #10	#10	#10	2
7-1/2	40	60	50	80	2 #8	#10	#8	2
10	50	100	60	100	2 #6	#10	#8	3

See Notes, Table 3.40

263

TABLE 3.45(a)   Reduced-Voltage Starter Characteristics

Characteristic	Full Voltage	Autotransformer Class 8606	Wye-Delta Class 8630	Part Winding Class 8640	Primary Resistance Class 8647	Solid State ATS23
Voltage at Motor	100%	50% / 65% / 80% (tap setting)	100%	100%	70%	Ramped Up
Line Current (% Full Load Current)	600%	150% / 250% / 380%	200%	390%	420%	200% to 500% (potentiometer adjustment)
Starting Torque (% Rated Torque)	150%	40% / 60% / 100%	50%	70%	75%	10% to 105% (function of i & V)
Start Time (Factory Setting)		6 - 7 sec	10 sec / 15 sec (open / closed transition)	1 - 1.5 sec	4 - 5 sec	10 sec (adjustable 5 to 30 sec)
Advantages	- Simple - Economical - High Starting Torque	- High torque/amp - High inertial loads - Flexibility	- High inertial loads - Long acceleration loads - Good torque/amp	- Simple - Small size	- Smooth acceleration Motor voltage increases with speed	- Greatest flexibility - Smooth ramp - Solid state O/L - Diagnostics
Disadvantages	- Abrupt starts - Large current inrush	- Large size	- Low torque - No flexibility	- Not suitable for: High inertial loads Frequent starting	- Low current limitation - Heat dissipation - Short start time	- SCR heat dissipation - Ambient limitations - Sensitive to power quality
Motor	Standard	Standard	Special	Special	Standard	Standard

TABLE 3.45(b) Reduced-Voltage Starter Selection Table

Application	Need		Comments
	Smooth Acceleration	Minimum Line Current	
High Inertial Loading	1. Solid State 2. Autotransformer 3. Primary Resistor 4. Wye Delta 5. Part Winding	1. Autotransformer 2. Solid State 3. Wye-Delta 4. Part Winding 5. Primary Resistor	
Long Acceleration Time	1. Solid State 2. Wye-Delta 3. Autotransformer 4. Primary Resistor	1. Solid State 2. Wye-Delta 3. Autotransformer 4. Primary Resistor	* For acceleration times greater than 5 sec primary resistor requires non-std resistors * Part winding not suitable for acceleration time greater than 2 seconds
Frequent Starting	1. Solid State 2. Wye-Delta 3. Primary Resistor 4. Autotransformer	1. Solid State 2. Wye-Delta 3. Primary Resistor 4. Autotransformer	* Part winding is unsuitable for frequent starts
Flexibility in Selecting Starter Characteristics	1. Solid State 2. Autotransformer 3. Primary Resistor 4. Part Winding	1. Solid State 2. Autotransformer 3. Primary Resistor 4. Part Winding	* For primary resistor, resistor change required to change starting characteristics * Starting characteristics cannot be changed for Wye-Delta starters

System Voltage Classes

- *Low voltage:* A class of nominal system voltages 1,000 V or less
- *Medium voltage:* A class of nominal system voltages greater than 1,000 V but less than 100,000 V
- *High voltage:* A class of nominal system voltages equal to or greater than 100,000 V and equal to or less than 230,000 V

Standard Nominal System Voltages in the United States

These voltages and their associated tolerance limits are listed in ANSI C84.1-1989 for voltages from 120 to 230,000 V, and ANSI C92.2-1987, *Power Systems—Alternating Current Electrical Systems and Equipment Operating at Voltages Above 230 kV Nominal-Preferred Voltage Ratings.* The nominal system voltages and their associated tolerance limits and notes in the two standards have been combined in Table 3.46 to provide a single table, listing all the nominal system voltages and their associated tolerance limits for the United States. Preferred nominal system voltages and voltage ranges are shown in boldface type, whereas other systems in substantial use that are recognized as standard voltages are shown in medium type. Other voltages may be encountered on older systems, but they are not recognized as standard voltages. The transformer connections from which these voltages are derived are shown in Figure 3.37.

Application of Voltage Classes

1. Low-voltage-class voltages are used to supply utilization equipment.
2. Medium-voltage-class voltages are used as primary distribution voltages to supply distribution transformers that step the medium voltage down to a low voltage to supply utilization equipment. Medium voltages of 13,800 V and below are also used to supply utilization equipment, such as large motors.
3. High-voltage-class voltages are used to transmit large amounts of electric power over transmission lines that interconnect transmission substations.

Voltage Systems Outside of the United States

Voltage systems in other countries (including Canada) generally differ from those in the United States. Also, the frequency in many countries is 50 Hz instead of 60 Hz, which affects the operation of some equipment, such as motors, which will run approximately 17 percent slower. Plugs and receptacles are generally different, which helps to prevent utilization equipment from the United States from being connected to the wrong voltage.

FIGURE 3.37   Principal transformer connections to supply the system voltages of Table 3.46.

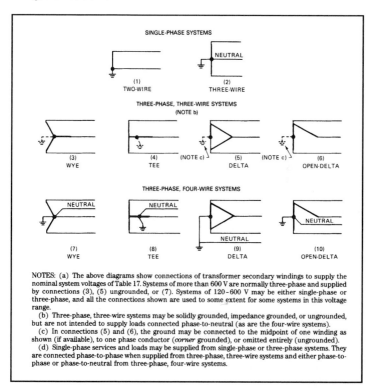

NOTES: (a) The above diagrams show connections of transformer secondary windings to supply the nominal system voltages of Table 17. Systems of more than 600 V are normally three-phase and supplied by connections (3), (5) ungrounded, or (7). Systems of 120–600 V may be either single-phase or three-phase, and all the connections shown are used to some extent for some systems in this voltage range.

(b) Three-phase, three-wire systems may be solidly grounded, impedance grounded, or ungrounded, but are not intended to supply loads connected phase-to-neutral (as are the four-wire systems).

(c) In connections (5) and (6), the ground may be connected to the midpoint of one winding as shown (if available), to one phase conductor (*corner* grounded), or omitted entirely (ungrounded).

(d) Single-phase services and loads may be supplied from single-phase or three-phase systems. They are connected phase-to-phase when supplied from three-phase, three-wire systems and either phase-to-phase or phase-to-neutral from three-phase, four-wire systems.

In general, equipment rated for use in the United States cannot be used outside of the United States, and vice versa. If electrical equipment made for use in the United States must be used outside the United States, and vice versa, information on the voltage, frequency, and type of plug required should be obtained. If the difference is only in the voltage, transformers are generally available to convert the supply voltage to the equipment voltage.

System Voltage Tolerance Limits

Table 3.46 lists two voltage ranges to provide a practical application of voltage tolerance limits to distribution systems.

Electric supply systems are to be designed and operated so that most service voltages fall within the Range A limits. User systems are to be

TABLE 3.46 Standard Nominal System Voltages and Voltage Ranges

VOLTAGE CLASS (Note 1)	NOMINAL SYSTEM VOLTAGE (Note e) Two-wire	Three-wire	Four-wire	Nominal Utilization Voltage (Note j) Two-wire	Three-wire / Four-wire	VOLTAGE RANGE A (Note b) Maximum Utilization and Service Voltage (Note c)	Service Voltage	Utilization Voltage (Minimum)	VOLTAGE RANGE B (Note b) Maximum Utilization and Service Voltage	Service Voltage (Minimum)	Utilization Voltage (Minimum)
Low Voltage (Note i)	120	120/240		115	115/230						
						Single-Phase Systems					
	120					126	114	110	127	110	106
		120/240				126/252	114/228	110/220	127/254	110/220	106/212
						Three-Phase Systems					
			208Y/120 (Note d)	200		218Y/126	197Y/114	191Y/110 (Note 2)	220Y/127	191Y/110 (Note 2)	184Y/106 (Note 2)
			240/120	230/115		252/126	228/114	220/110	254/127	220/110	212/106
		240		230		252	228	220	254	220	212
	480		480Y/277	460		504Y/291	456Y/263	440Y/254	508Y/293	440Y/254	424Y/245
		480		460		504	456	440	508	440	424
	600 (Note e)			575		630 (Note e)	570	550	635 (Note e)	550	530
Medium Voltage	2400					2520	2340	2160	2540	2280	2080
			4160Y/2400			4370/2520	4050/2340	3740/2160	4400Y/2540	3950Y/2280	3600/2080
	4160					4370	4050	3740	4400	3950	3600
	4800					5040	4680	4320	5080	4560	4160
	6900					7240	6730	6210	7260	6560	5940
			8320Y/4800			8730Y/5040	8110Y/4680	(Note f)	8800Y/5080	7900Y/4560	(Note f)
			12000Y/6930			12600/7270	11700/6750		12700/7330	11400/6580	
			12470Y/7200			13090Y/7560	12160Y/7020		13200Y/7620	11850Y/6840	
			13200Y/7620			13860Y/8000	12870Y/7430		13870Y/8070	12504Y/7240	
			13800Y/7970			14490Y/8370	13460Y/7770		14520Y/8380	13110Y/7570	
	13800					14490	13460	12420	14520	13110	11880
			20780Y/12000			21820/12600	20260Y/11700	(Note f)	22000Y/12700	19740Y/11400	(Note f)
			22860Y/13200			24000Y/13860	22290Y/12870		24200Y/13970	21720Y/12540	
	23000					24150	22430		24340	21850	
			24940Y/14400			26190Y/15120	24320Y/14040		26400Y/15240	23990Y/13680	
			34500Y/19920			36230Y/20920	33640Y/19420		36510Y/21060	32780Y/18930	
	34500					36230	33640		36510	32780	
High Voltage						**Maximum Voltage**					
	46000					48300					
	69000					72500					
	115000					121000					
	138000					145000					
	161000					169000					
	230000 (Note h)					242000					
Extra-High Voltage	345000					362000					
	500000					550000					
	765000					800000					
Ultra-High Voltage	1100000					1200000					

NOTES: (1) Minimum utilization voltages for 120–600 volt circuits not supplying lighting loads are as follows

Nominal System Voltage	Range A	Range B
120	110	106
120/240	110/220	106/212
(Note 2) 208Y/120	187Y/108	180Y/104
240/120	216/108	208/104
240	216	208
480Y/277	432Y/249	416Y/240
480	432	416
600	540	520

(2) Many 220 volt motors were applied on existing 208 volt systems on the assumption that the utilization voltage would not be less than 187 volts. Caution should be exercised in applying the Range B minimum voltages of Table 17 and Note (1) to existing 208 volt systems supplying such motors

(a) Three-phase, three-wire systems are systems in which only the three-phase conductors are carried out from the source for connection of loads. The source may be derived from any type of three-phase transformer connection, grounded or ungrounded. Three-phase, four-wire systems are systems in which a grounded neutral conductor is also carried out from the source for connection of loads. Four-wire systems in this table are designated by the phase-to-phase voltage, followed by the letter Y (except for the 240/120 V delta system), a slant line, and the, and the phase-to-neutral voltage. Single-phase services and loads may be supplied from either single-phase or three-phase systems. The principal transformer connections that are used to supply single-phase and three-phase systems are illustrated in Fig 3.

(b) The voltage ranges in this table are illustrated in ANSI C84.1-1989, Appendix B [2].

(c) For 120–600 V nominal systems, voltages in this column are maximum service voltages. Maximum utilization voltages would not be expected to exceed 125 V for the nominal system voltage of 120, nor appropriate multiples thereof for other nominal system voltages through 600 V.

(d) A modification of this three-phase, four-wire system is available as a 120/208Y-volt service for single-phase, three-wire, open-wye applications.

(e) Certain kinds of control and protective equipment presently available have a maximum voltage limit of 600 V; the manufacturer or power supplier or both should be consulted to assure proper application.

(f) Utilization equipment does not generally operate directly at these voltages. For equipment supplied through transformers, refer to limits for nominal system voltage of transformer output.

(g) For these systems, Range A and Range B limits are not shown because, where they are used as service voltages, the operating voltage level on the user's system is normally adjusted by means of voltage regulation to suit their requirements.

(h) Standard voltages are reprinted from ANSI C92.2-1987 [3] for convenience only.

(i) Nominal utilization voltages are for low-voltage motors and control. See ANSI C84.1-1989, Appendix C [2] for other equipment nominal utilization voltages (or equipment nameplate voltage ratings).

designed and operated so that, when the service voltages are within Range A, the utilization voltages are within Range A. Utilization equipment is to be designed and rated to give fully satisfactory performance within Range A limits for utilization voltages.

Range B is provided to allow limited excursions of voltage outside the Range A limits that necessarily result from practical design and operating conditions. The supplying utility is expected to take action within a reasonable time to restore service voltages to Range A limits. The user is expected to take action within a reasonable time to restore utilization voltages to Range A limits. Insofar as practical, utilization equipment may be expected to give acceptable performance outside Range A but within Range B. When voltages occur outside the limits of Range B, prompt corrective action should be taken.

The voltage tolerance limits in ANSI C84.1-1989 are based on ANSI/NEMA MG1-1978, *Motors and Generators,* which establishes the voltage tolerance limits of the standard low-voltage induction motor at ± 10 percent of nameplate voltage ratings of 230 and 460 V. Because motors represent the major component of utilization equipment, they were given primary consideration in the establishment of this voltage standard.

The best way to show the voltages in a distribution system is by using a 120-V base. This cancels the transformation ratios between systems, so that the actual voltages vary solely on the basis of voltage drops in the system. Any voltage may be converted to a 120-V base by dividing the actual voltage by the ratio of transformation to the 120-V base. For example, the ratio of transformation for the 480-V system is 480/120, or 4, so 460 V in a 480-V system would be 460/4, or 115 V.

The tolerance limits of the 460-V motor as they relate to the 120-V base become 115 V + 10 percent or 126.5 V, and 115 V − 10 percent, or 103.5 V. The problem is to decide how this tolerance range of 23 V should be divided between the primary distribution system, the distribution transformer, and the secondary distribution system that make up the regulated distribution system. The solution adopted by the American National Standards Committee C84 is shown in Table 3.47.

Voltage Profile Limits for a Regulated Distribution System

Figure 3.38 shows the voltage profile of a regulated power distribution system using the limits of Range A in Table 3.46. This table assumes a standard nominal distribution voltage of 13,200 V, Range A in Table 3.46, for the example profile shown.

System Voltage Nomenclature

The nominal system voltages in Table 3.46 are designated in the same way as the designation on the nameplate of the transformer for the winding or windings supplying the system.

TABLE 3.47 Standard Voltage Profile for a Regulated Power Distribution System, 120-Volt Base

	Range A	Range B
Maximum allowable voltage	126(125*)	127
Voltage-drop allowance for the primary distribution feeder	9	13
Minimum primary service voltage	117	114
Voltage-drop allowance for the distribution transformer	3	4
Minimum low-voltage service voltage	114	110
Voltage-drop allowance for the building wiring	6(4†)	6(4†)
Minimum utilization voltage	108(110†)	104(106†)

*For utilization voltages of 120–600 V.
†For building wiring circuits supplying lighting equipment.

1. Single-phase systems
   - *120 V:* Indicates a single-phase, two-wire system in which the nominal voltage between the two wires is 120 V.
   - *120/240 V:* Indicates a single-phase, three-wire system in which the nominal voltage between the two-phase conductors is 240 V, and from each phase conductor to the neutral is 120 V.
2. Three-phase systems
   - *240/120 V:* Indicates a three-phase, four-wire system supplied from a delta-connected transformer. The midtap of one winding is connected to a neutral. The three phase conductors provide a

FIGURE 3.38 Voltage profile of the limits of range A, ANSI C84.1-1989.

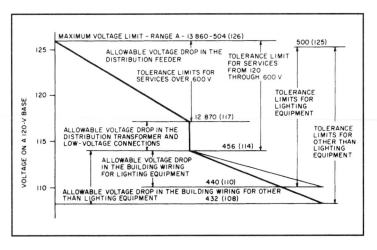

nominal 240-V three-phase, three-wire system, and the neutral and two adjacent phase conductors provide a nominal 120/240-V single-phase, three-wire system.

- *Single number:* Indicates a three-phase, three-wire system in which the number designates the nominal voltage between phases.
- *Two numbers separated by Y/:* Indicates a three-phase, four-wire system from a wye-connected transformer in which the first number indicates the nominal phase-to-phase voltage and the second the nominal phase-to-neutral voltage.

NOTES

1. All single-phase systems and all three-phase, four-wire systems are suitable for the connection of phase-to-neutral load.
2. See Chapter 4 for methods of system grounding.
3. See Figure 3.37 for transformer connections.

Voltage Ratings for Utilization Equipment

According to the IEEE, utilization equipment is defined as "electrical equipment that converts electric power into some other form of energy, such as light, heat, or mechanical motion." Every item of utilization equipment should have a nameplate listing, which includes, among other things, the rated voltage for which the equipment is designed. With one major exception, most electrical utilization equipment carries a nameplate rating that is the same as the voltage system on which it is to be used; that is, equipment to be used on 120-V systems is rated 120 V, and so on. The major exception is motors and equipment containing motors. See Table 3.48 for the proper selection of the motor nameplate voltage that is compatible with the specific available nominal system voltage. Motors are also about the only utilization equipment used on systems over 600 V.

Effect of Voltage Variation on Utilization Equipment

Whenever the voltage at the terminals of utilization equipment varies from its nameplate rating, the performance of the equipment and its life expectancy change. The effect may be minor or serious, depending on the characteristics of the equipment and the amount of voltage deviation from the nameplate rating. NEMA standards provide tolerance limits within which performance will be acceptable. In precise operations, however, closer voltage control may be required. In general, a change in the applied voltage causes a proportional change in the current. Because the effect on the load equipment is proportional to the

TABLE 3.48   Voltage Ratings of Standard Motors

Nominal System Voltage	Nameplate Voltage
Single-phase motors	
120	115
240	230
Three-phase motors	
208	200
240	230
480	460
600	575
2400	2300
4160	4000
4800	4600
6900	6600
13 800	13 200

voltage and current, and because the current is proportional to the voltage, the total effect is approximately proportional to the square of the voltage.

However, the change is only approximately proportional and not exact, because the change in the current affects the operation of the equipment, so the current will continue to change until a new equilibrium position is established. For example, when the load is a resistance heater, the increase in current will increase the temperature of the heater, which will increase its resistance, which will in turn reduce the current. This effect will continue until a new equilibrium current and temperature are established. In the case of an induction motor, a reduction in the voltage will cause a reduction in the current flowing to the motor, causing the motor to slow down. This reduces the impedance of the motor, causing an increase in the current until a new equilibrium position is established between the current and the motor speed.

EXAMPLES OF EFFECTS OF VOLTAGE VARIATION

The variations in characteristics of induction motors as a function of voltage are given in Table 3.49.

The light output and life of incandescent filament lamps are critically affected by the impressed voltage. The variation of life and light output with voltage is given in Table 3.50. The variation figures for 125- and 130-V lamps are also included, because these ratings are useful in locations where long life is more important than light output.

TABLE 3.49 General Effect of Voltage Variations on Induction Motor Characteristics

Characteristic	Function of Voltage	Voltage Variation	
		90% Voltage	110% Voltage
Starting and maximum running torque	(Voltage)^2	Decrease 19%	Increase 21%
Synchronous speed	Constant	No change	No change
Percent slip	1/(Voltage)^2	Increase 23%	Decrease 17%
Full-load speed	Synchronous speed-slip	Decrease 1.5%	Increase 1%
Efficiency			
Full load	—	Decrease 2%	Increase 0.5 to 1%
¾ load	—	Practically no change	Practically no change
½ load	—	Increase 1 to 2%	Decrease 1 to 2%
Power factor			
Full load	—	Increase 1%	Decrease 3%
¾ load	—	Increase 2 to 3%	Decrease 4%
½ load	—	Increase 4 to 5%	Decrease 5 to 6%
Full-load current	Voltage	Increase 11%	Decrease 7%
Starting current	—	Decrease 10 to 12%	Increase 10 to 12%
Temperature rise, full load	(Voltage)^2	Increase 6 to 7 °C	Decrease 1 to 2 °C
Maximum overload capacity	—	Decrease 19%	Increase 21%
Magnetic noise - no load in particular	—	Decrease slightly	Increase slightly

TABLE 3.50 Effect of Voltage Variations on Incandescent Lamps

Applied Voltage (volts)	Lamp Rating					
	120 V		125 V		130 V	
	Percent Life	Percent Light	Percent Life	Percent Light	Percent Life	Percent Light
105	575	64	880	55	—	—
110	310	74	525	65	880	57
115	175	87	295	76	500	66
120	100	100	170	88	280	76
125	58	118	100	100	165	88
130	34	132	59	113	100	100

Fluorescent lamps, unlike incandescent lamps, operate satisfactorily over a range of ±10 percent of the ballast nameplate voltage rating. Light output varies approximately in direct proportion to the applied voltage. Thus, a 1 percent increase in applied voltage will increase the light output by 1 percent, and, conversely, a decrease of 1 percent in the applied voltage will reduce the light output by 1 percent. The life of fluorescent lamps is affected less by voltage variation than the life of incandescent lamps.

The voltage-sensitive component of the fluorescent fixture is the ballast, which is a small reactor, or transformer, that supplies the starting and operating voltages to the lamp and limits the lamp current to design values. These ballasts may overheat when subjected to above-normal voltage and operating temperature, and ballasts with integral thermal protection may be required.

Mercury lamps that use the conventional unregulated ballast will have a 30 percent decrease in the light output for a 10 percent decrease in terminal voltage. When a constant wattage ballast is used, the decrease in light output for a 10 percent decrease in terminal voltage will be about 2 percent.

Mercury lamps require between 4 and 8 min to vaporize the mercury in the lamp and reach full brilliance. At about 20 percent undervoltage, the mercury arc will be extinguished and the lamp cannot be restarted until the mercury condenses, which takes between 4 and 8 min, unless the lamps have special cooling controls. The lamp life is related inversely to the number of starts; so that, if low-voltage conditions require repeated starting, lamp life will be affected adversely. Excessively high voltage raises the arc temperature, which could damage the glass enclosure when the temperature approaches the glass-softening point.

Sodium and metal-halide lamps have similar characteristics to mercury lamps; however, the starting and operating voltages may be somewhat different. See the manufacturers' catalogs for detailed information.

In resistance heating devices, the energy input and, therefore, the heat output of resistance heaters varies approximately as the square of the impressed voltage. Thus, a 10 percent drop in voltage will cause a drop of approximately 19 percent in heat output. This, however, holds true only for an operating range over which the resistance remains approximately constant.

The foregoing gives some idea of how critical proper voltage is, and thus the need for voltage drop calculations.

Voltage Drop Calculations

Electrical design professionals designing building wiring systems should have a working knowledge of voltage drop calculations, not only to meet NEC, Articles 210-19(a) and 215-2, requirements (recommended, not mandatory), but also to ensure that the voltage applied to utilization equipment is maintained within proper limits. Due to the vector relationships of the circuit parameters, a working knowledge of trigonometry is needed, especially for making exact calculations. Fortunately, most voltage drop calculations are based on assumed limiting conditions, and approximate formulas are adequate. Within the context of this book, voltage drop tables and charts are sufficiently accurate to determine the approximate voltage drop for most problems, thus formulas will not be needed.

VOLTAGE DROP TABLES

These tables (Tables 3.51 through 3.72), reading directly in volts, give values for the voltage drop found in aluminum and copper cables under various circumstances.

1. In magnetic conduit—AC
   a. 70 percent power factor
   b. 80 percent power factor
   c. 90 percent power factor
   d. 95 percent power factor
   e. 100 percent power factor

2. In nonmagnetic conduit—AC
   a. 70 percent power factor
   b. 80 percent power factor
   c. 90 percent power factor
   d. 95 percent power factor
   e. 100 percent power factor

3. In direct-current circuits

All voltage drops are calculated at 60 Hz and 60°C. This temperature represents a typical conductor temperature encountered in service. No error of practical significance is involved in using the table for any conductor temperature of 75°C or less.

Space limitations make it necessary to prepare the following pages with the "Ampere Feet" column in abbreviated form. For example, reference to the proper table will show that the voltage drop encountered in a 253,000-ampere-foot circuit using 1,000-kcmil aluminum cable would be (for 80 percent power factor, magnetic conduit) 17.6 + 4.4 + 0.3, or 22.3 V. These voltage drops are the individual drops given by the table for 200,000 ampere feet, 50,000 ampere feet, and 3,000 ampere feet, respectively, for a total of 253,000 ampere feet. *Note that the length of run refers to the length of the physical circuit (i.e., circuit feet, not the footage of conductor).*

Factors are given at the bottom of each table to make the tables usable in any of the common AC circuits.

TABLE 3.51 Volts Drop for AL Conductor—Direct Current

Volts Drop — Wire Size (AWG or MCM). Row label = Ampere Feet.

Ampere Feet	12*	10*	8*	6	4	2	1	1/0	2/0	3/0	4/0	250	300	350	400	500	600	700	750	800	900	1000
500,000	—	—	—	—	483.0	303.0	240.0	191.0	151.0	120.0	95.9	80.5	67.2	57.5	50.4	40.2	33.6	28.8	26.9	25.2	22.3	20.2
400,000	—	—	—	—	386.0	241.0	192.0	153.0	121.0	96.0	76.8	64.4	53.7	46.0	40.4	32.1	26.9	23.0	21.5	20.2	17.8	16.1
300,000	—	—	—	460.0	290.0	182.0	144.0	115.0	90.6	72.0	57.6	48.3	40.3	34.5	30.3	24.1	20.2	17.3	16.1	15.1	13.4	12.1
200,000	—	—	478.0	307.0	193.0	121.0	96.0	76.4	60.4	48.0	38.4	32.2	26.9	23.0	20.2	16.1	13.4	11.5	10.8	10.1	8.9	8.1
100,000	—	380.0	239.0	153.0	96.6	60.6	48.0	38.2	30.2	24.0	19.2	16.1	13.4	11.5	10.1	8.0	6.7	5.8	5.4	5.0	4.5	4.0
90,000	—	342.0	215.0	138.0	87.0	54.6	43.2	34.4	27.2	21.6	17.3	14.5	12.1	10.3	9.1	7.2	6.1	5.2	4.8	4.5	4.0	3.6
80,000	483.0	304.0	191.0	123.0	77.0	48.5	38.4	30.5	24.2	19.2	15.4	12.9	10.8	9.2	8.1	6.4	5.4	4.6	4.3	4.0	3.6	3.2
70,000	423.0	266.0	167.0	107.0	67.6	42.4	33.6	26.7	21.1	16.8	13.4	11.3	9.4	8.1	7.1	5.6	4.7	4.0	3.8	3.5	3.1	2.8
60,000	362.0	228.0	144.0	92.0	58.0	36.4	28.8	22.9	18.1	14.4	11.5	9.7	8.1	6.9	6.1	4.8	4.0	3.5	3.2	3.0	2.7	2.4
50,000	302.0	190.0	120.0	76.7	48.3	30.3	24.0	19.1	15.1	12.0	9.6	8.1	6.7	5.8	5.0	4.0	3.4	2.9	2.7	2.5	2.2	2.0
40,000	242.0	152.0	95.6	61.4	38.6	24.1	19.2	15.3	12.1	9.6	7.7	6.4	5.4	4.6	4.0	3.2	2.7	2.3	2.2	2.0	1.8	1.6
30,000	181.0	114.0	71.7	46.0	29.0	18.2	14.4	11.5	9.1	7.2	5.8	4.8	4.0	3.5	3.0	2.4	2.0	1.7	1.6	1.5	1.3	1.2
20,000	121.0	76.0	47.8	30.7	19.3	12.1	9.6	7.6	6.0	4.8	3.8	3.2	2.7	2.3	2.0	1.6	1.3	1.2	1.1	1.0	0.9	0.8
10,000	60.4	38.0	23.9	15.3	9.7	6.1	4.8	3.8	3.0	2.4	1.9	1.6	1.3	1.2	1.0	0.8	0.7	0.6	0.5	0.5	0.5	0.4
9,000	54.4	34.2	21.5	13.8	8.7	5.5	4.3	3.4	2.7	2.2	1.7	1.5	1.2	1.0	0.9	0.7	0.6	0.5	0.5	0.5	0.4	0.4
8,000	48.3	30.4	19.1	12.3	7.7	4.9	3.8	3.1	2.4	1.9	1.5	1.3	1.1	0.9	0.8	0.6	0.5	0.5	0.4	0.4	0.4	0.3
7,000	42.3	26.6	16.7	10.7	6.8	4.2	3.4	2.7	2.1	1.7	1.3	1.1	0.9	0.8	0.7	0.6	0.5	0.4	0.4	0.4	0.3	0.3
6,000	36.2	22.8	14.4	9.2	5.8	3.6	2.9	2.3	1.8	1.4	1.2	1.0	0.8	0.7	0.6	0.5	0.4	0.3	0.3	0.3	0.3	0.2
5,000	30.2	19.0	12.0	7.7	4.8	3.0	2.4	1.9	1.5	1.2	1.0	0.8	0.7	0.6	0.5	0.4	0.3	0.3	0.3	0.3	0.2	0.2
4,000	24.2	15.2	9.6	6.1	3.9	2.4	1.9	1.5	1.2	1.0	0.8	0.6	0.5	0.5	0.4	0.3	0.3	0.2	0.2	0.2	0.2	0.1
3,000	18.1	11.4	7.2	4.6	2.9	1.8	1.4	1.1	0.9	0.7	0.6	0.5	0.4	0.3	0.3	0.2	0.2	0.2	0.2	0.2	0.1	0.1
2,000	12.1	7.6	4.8	3.1	1.9	1.2	1.0	0.8	0.6	0.5	0.4	0.3	0.3	0.2	0.2	0.2	0.1	0.1	0.1	0.1	0.1	—
1,000	6.0	3.8	2.4	1.5	1.0	0.6	0.5	0.4	0.3	0.2	0.2	0.2	0.1	0.1	0.1	0.1	0.1	0.1	0.1	0.1	0.1	—
900	5.4	3.4	2.2	1.4	0.9	0.6	0.4	0.3	0.3	0.2	0.2	0.2	0.1	0.1	0.1	0.1	0.1	0.1	0.1	0.1	—	—
800	4.8	3.0	1.9	1.2	0.8	0.5	0.4	0.3	0.2	0.2	0.2	0.1	0.1	0.1	0.1	0.1	0.1	0.1	—	—	—	—
700	4.2	2.7	1.7	1.1	0.7	0.4	0.3	0.3	0.2	0.2	0.1	0.1	0.1	0.1	0.1	0.1	—	—	—	—	—	—
600	3.6	2.3	1.4	0.9	0.6	0.4	0.3	0.2	0.2	0.1	0.1	0.1	0.1	0.1	0.1	—	—	—	—	—	—	—
500	3.0	1.9	1.2	0.8	0.5	0.3	0.2	0.2	0.2	0.1	0.1	0.1	0.1	0.1	0.1	—	—	—	—	—	—	—
400	2.4	1.5	1.0	0.6	0.4	0.2	0.2	0.2	0.1	0.1	0.1	0.1	0.1	—	—	—	—	—	—	—	—	—
300	1.8	1.1	0.7	0.5	0.3	0.2	0.1	0.1	0.1	0.1	0.1	—	—	—	—	—	—	—	—	—	—	—
200	1.2	0.8	0.5	0.3	0.2	0.1	0.1	0.1	—	—	—	—	—	—	—	—	—	—	—	—	—	—
100	0.6	0.4	0.2	0.2	0.1	0.1	0.1	—	—	—	—	—	—	—	—	—	—	—	—	—	—	—

* Solid Conductors. Other conductors are stranded.

Note 1—The footage employed in the tabulated ampere feet refers to the length of run of the circuit rather than to the footage of individual conductor.

Note 2—The above table is figured at 60°C since this is an estimate of the average temperature which may be anticipated in service. The table may be used without significant error for conductor temperatures up to and including 75°C.

TABLE 3.52 Volts Drop for AL Conductor in Magnetic Conduit—70 Percent PF

Volts Drop — WIRE SIZE AWG or MCM

Ampere Feet	1000	900	800	750	700	600	500	400	350	300	250	4/0	3/0	2/0	1/0	1	2	4	6	8*	10*	12*
500,000	46.4	48.0	50.2	51.5	53.0	56.5	60.8	68.5	73.6	80.9	96.9	101.0	119.0	141.0	170.0	205.0	249.0	373.0	—	—	—	—
400,000	37.1	38.4	40.2	41.2	42.4	45.2	48.6	54.8	58.9	64.7	77.5	80.8	95.2	113.0	136.0	164.0	199.0	298.0	463.0	—	—	—
300,000	27.8	28.8	30.1	30.9	31.8	33.9	36.5	41.1	44.2	48.5	58.1	60.6	71.4	84.6	102.0	123.0	149.0	224.0	347.0	—	—	—
200,000	18.6	19.2	20.1	20.6	21.2	22.6	24.4	27.4	29.5	32.4	38.8	40.4	47.6	56.4	68.0	82.0	99.6	149.0	232.0	352.0	—	—
100,000	9.3	9.6	10.0	10.3	10.6	11.3	12.2	13.7	14.7	16.2	19.4	20.2	23.8	28.2	34.0	41.0	49.8	74.7	116.0	176.0	275.0	431.0
90,000	8.4	8.6	9.0	9.3	9.5	10.2	10.9	12.3	13.2	14.6	17.4	18.2	21.4	25.4	30.6	36.9	44.8	67.2	104.0	159.0	247.0	389.0
80,000	7.4	7.7	8.0	8.2	8.5	9.0	9.7	11.0	11.8	12.9	15.5	16.2	19.0	22.6	27.2	32.8	39.8	59.7	92.6	141.0	210.0	346.0
70,000	6.5	6.7	7.0	7.2	7.4	7.9	8.5	9.6	10.3	11.3	13.6	14.1	16.7	19.7	23.8	28.7	34.9	52.2	81.0	123.0	192.0	302.0
60,000	5.6	5.8	6.0	6.2	6.4	6.8	7.3	8.2	8.8	9.7	11.6	12.1	14.3	16.9	20.4	24.6	29.9	44.8	69.4	106.0	165.0	259.0
50,000	4.6	4.8	5.0	5.2	5.3	5.7	6.1	6.9	7.4	8.1	9.7	10.1	11.9	14.1	17.0	20.5	24.9	37.3	57.9	88.1	137.0	216.0
40,000	3.7	3.8	4.0	4.1	4.2	4.5	4.9	5.5	5.9	6.5	7.8	8.1	9.5	11.3	13.6	16.4	19.9	29.8	46.3	70.5	110.0	173.0
30,000	2.8	2.9	3.0	3.1	3.2	3.4	3.6	4.1	4.4	4.9	5.8	6.1	7.1	8.5	10.2	12.3	14.9	22.4	34.7	52.8	82.6	129.0
20,000	1.9	1.9	2.0	2.1	2.1	2.3	2.4	2.7	2.9	3.2	3.9	4.0	4.8	5.6	6.8	8.2	10.0	14.9	23.2	35.2	55.0	86.4
10,000	0.9	1.0	1.0	1.0	1.1	1.1	1.2	1.4	1.5	1.6	1.9	2.0	2.4	2.8	3.4	4.1	5.0	7.5	11.6	17.6	27.5	43.2
9,000	0.8	0.9	0.9	0.9	1.0	1.0	1.1	1.2	1.3	1.5	1.7	1.8	2.1	2.5	3.1	3.7	4.5	6.7	10.4	15.9	24.7	38.9
8,000	0.7	0.8	0.8	0.8	0.8	0.9	1.0	1.1	1.2	1.3	1.6	1.6	1.9	2.3	2.7	3.3	4.0	6.0	9.3	14.1	21.0	34.6
7,000	0.6	0.7	0.7	0.7	0.7	0.8	0.9	1.0	1.0	1.1	1.4	1.4	1.7	2.0	2.4	2.9	3.5	5.2	8.1	12.3	19.2	30.2
6,000	0.6	0.6	0.6	0.6	0.6	0.7	0.7	0.8	0.9	1.0	1.2	1.2	1.4	1.7	2.0	2.5	3.0	4.5	6.9	10.6	16.5	25.9
5,000	0.5	0.5	0.5	0.5	0.5	0.6	0.6	0.7	0.7	0.8	1.0	1.0	1.2	1.4	1.7	2.1	2.5	3.7	5.8	8.8	13.7	21.6
4,000	0.4	0.4	0.4	0.4	0.4	0.5	0.5	0.5	0.6	0.6	0.8	0.8	1.0	1.1	1.4	1.6	2.0	3.0	4.6	7.0	11.0	17.3
3,000	0.3	0.3	0.3	0.3	0.3	0.3	0.4	0.4	0.4	0.5	0.6	0.6	0.7	0.8	1.0	1.2	1.5	2.2	3.5	5.3	8.5	12.9
2,000	0.2	0.2	0.2	0.2	0.2	0.2	0.2	0.3	0.3	0.3	0.4	0.4	0.5	0.6	0.7	0.8	1.0	1.5	2.3	3.5	5.5	8.6
1,000	0.1	0.1	0.1	0.1	0.1	0.1	0.1	0.1	0.1	0.2	0.2	0.2	0.2	0.3	0.3	0.4	0.5	0.7	1.2	1.8	2.8	4.3
900	0.1	0.1	0.1	0.1	0.1	0.1	0.1	0.1	0.1	0.1	0.2	0.2	0.2	0.3	0.3	0.4	0.4	0.7	1.0	1.6	2.5	3.9
800	0.1	0.1	0.1	0.1	0.1	0.1	0.1	0.1	0.1	0.1	0.2	0.2	0.2	0.2	0.3	0.3	0.4	0.6	0.9	1.4	2.1	3.5
700	0.1	0.1	0.1	0.1	0.1	0.1	0.1	0.1	0.1	0.1	0.1	0.1	0.2	0.2	0.2	0.3	0.3	0.5	0.8	1.2	1.9	3.0
600	0.1	0.1	0.1	0.1	0.1	0.1	0.1	0.1	0.1	0.1	0.1	0.1	0.1	0.2	0.2	0.2	0.3	0.4	0.7	1.1	1.7	2.6
500	—	—	0.1	0.1	0.1	0.1	0.1	0.1	0.1	0.1	0.1	0.1	0.1	0.1	0.2	0.2	0.2	0.4	0.6	0.9	1.4	2.2
400	—	—	—	—	—	—	—	0.1	0.1	0.1	0.1	0.1	0.1	0.1	0.1	0.2	0.2	0.3	0.5	0.7	1.1	1.7
300	—	—	—	—	—	—	—	—	—	—	0.1	0.1	0.1	0.1	0.1	0.1	0.1	0.2	0.3	0.5	0.8	1.3
200	—	—	—	—	—	—	—	—	—	—	—	—	—	0.1	0.1	0.1	0.1	0.1	0.2	0.4	0.5	0.9
100	—	—	—	—	—	—	—	—	—	—	—	—	—	—	—	—	0.1	0.1	0.1	0.2	0.3	0.4

* Solid Conductors. Other conductors are stranded.

Note 1—The above table gives voltage drops encountered in a single phase two-wire system. The voltage drops in other systems may be obtained through multiplication by appropriate factors listed below:

System for Which Voltage Drop Is Desired	Multiplying Factors for Modification of Values in Table
Single Phase—3 Wire—Line to Line	1.00
Single Phase—3 Wire—Line to Neutral	0.50
Three Phase—3 Wire—Line to Line	0.866
Three Phase—4 Wire—Line to Line	0.866
Three Phase—4 Wire—Line to Neutral	0.50

Note 2—Allowable voltage drops for systems other than single phase, two wire cannot be used directly in the above table. Such drops should be modified through multiplication by the appropriate factor listed below. The voltage thus modified may then be used to obtain the proper wire size directly from the table.

System for Which Allowable Voltage Drop Is Known	Multiplying Factor for Modification of Known Value to Permit Direct Use of Table
Single Phase—3 Wire—Line to Line	1.00
Single Phase—3 Wire—Line to Neutral	2.00
Three Phase—3 Wire—Line to Line	1.155
Three Phase—4 Wire—Line to Line	1.155
Three Phase—4 Wire—Line to Neutral	2.00

Note 3—The footage employed in the tabulated ampere feet refers to the length of run of the circuit rather than to the footage of individual conductor.

Note 4—The above table is figured at 60°C since this is an estimate of the average temperature which may be anticipated in service. The table may be used without significant error for conductor temperatures up to and including 75°C.

TABLE 3.53 Volts Drop for AL Conductor in Magnetic Conduit—80 Percent PF

Volts Drop

WIRE SIZE AWG or MCM / Ampere Feet	1000	900	800	750	700	600	500	400	350	300	250	4/0	3/0	2/0	1/0	1	2	4	6	8*	10*	12*
500,000	44.0	46.0	48.0	49.0	51.0	55.0	60.0	69.0	74.0	82.0	94.0	105.0	125.0	150.0	183.0	223.0	273.0	419.0	—	—	—	—
400,000	35.2	36.8	38.4	39.2	40.8	44.0	48.0	55.2	59.2	65.6	75.2	84.0	100.0	120.0	146.0	178.0	218.0	335.0	—	—	—	—
300,000	26.4	27.6	28.8	29.4	30.6	33.0	36.0	41.4	44.4	49.2	56.4	63.0	75.0	90.0	110.0	134.0	164.0	251.0	389.0	—	—	—
200,000	17.6	18.4	19.2	19.6	20.4	22.0	24.0	27.6	29.6	32.8	37.6	42.0	50.0	60.0	73.2	89.2	109.0	168.0	259.0	398.0	—	—
100,000	8.8	9.2	9.6	9.8	10.2	11.0	12.0	13.8	14.8	16.4	18.8	21.0	25.0	30.0	36.6	44.6	54.6	83.8	130.0	199.0	311.0	492.0
90,000	7.9	8.3	8.6	8.8	9.2	9.9	10.8	12.4	13.3	14.8	16.9	18.9	22.5	27.0	32.9	40.1	49.1	75.4	117.0	179.0	280.0	443.0
80,000	7.0	7.4	7.7	7.8	8.2	8.8	9.6	11.0	11.8	13.1	15.0	16.8	20.0	24.0	29.3	35.7	43.7	67.0	104.0	159.0	249.0	394.0
70,000	6.2	6.4	6.7	6.9	7.1	7.7	8.4	9.7	10.4	11.5	13.2	14.7	17.5	21.0	25.6	31.2	38.2	58.7	90.9	139.0	218.0	345.0
60,000	5.3	5.5	5.8	5.9	6.1	6.6	7.2	8.3	8.9	9.8	11.3	12.6	15.0	18.0	22.0	26.8	32.8	50.3	77.9	119.0	187.0	295.0
50,000	4.4	4.6	4.8	4.9	5.1	5.5	6.0	6.9	7.4	8.2	9.4	10.5	12.5	15.0	18.3	22.3	27.3	41.9	64.9	99.4	156.0	246.0
40,000	3.5	3.7	3.8	3.9	4.1	4.4	4.8	5.5	5.9	6.6	7.5	8.4	10.0	12.0	14.6	17.8	21.8	33.5	51.9	79.6	124.0	197.0
30,000	2.6	2.8	2.9	2.9	3.1	3.3	3.6	4.1	4.4	4.9	5.6	6.3	7.5	9.0	11.0	13.4	16.4	25.1	38.9	59.7	93.3	148.0
20,000	1.8	1.8	1.9	2.0	2.0	2.2	2.4	2.8	3.0	3.3	3.8	4.2	5.0	6.0	7.3	8.9	10.9	16.8	25.9	39.8	62.2	98.4
10,000	0.9	0.9	1.0	1.0	1.0	1.1	1.2	1.4	1.5	1.6	1.9	2.1	2.5	3.0	3.7	4.5	5.5	8.4	13.0	19.9	31.1	49.2
9,000	0.8	0.8	0.9	0.9	0.9	1.0	1.1	1.2	1.3	1.5	1.7	1.9	2.3	2.7	3.3	4.0	4.9	7.5	11.7	17.9	28.0	44.3
8,000	0.7	0.7	0.8	0.8	0.8	0.9	1.0	1.1	1.2	1.3	1.5	1.7	2.0	2.4	2.9	3.6	4.4	6.7	10.4	15.9	24.9	39.4
7,000	0.6	0.6	0.7	0.7	0.7	0.8	0.8	1.0	1.0	1.1	1.3	1.5	1.8	2.1	2.6	3.1	3.8	5.9	9.1	13.9	21.8	34.5
6,000	0.5	0.6	0.6	0.6	0.6	0.7	0.7	0.8	0.9	1.0	1.1	1.3	1.5	1.8	2.2	2.7	3.3	5.0	7.8	11.9	18.7	29.5
5,000	0.4	0.5	0.5	0.5	0.5	0.6	0.6	0.7	0.7	0.8	0.9	1.1	1.3	1.5	1.8	2.2	2.7	4.2	6.5	9.9	15.6	24.6
4,000	0.4	0.4	0.4	0.4	0.4	0.4	0.5	0.6	0.6	0.7	0.8	0.8	1.0	1.2	1.5	1.8	2.2	3.4	5.2	8.0	12.4	19.7
3,000	0.3	0.3	0.3	0.3	0.3	0.3	0.4	0.4	0.4	0.5	0.6	0.6	0.8	0.9	1.1	1.3	1.6	2.5	3.9	5.9	9.3	14.8
2,000	0.2	0.2	0.2	0.2	0.2	0.2	0.2	0.3	0.3	0.3	0.4	0.4	0.5	0.6	0.7	0.9	1.1	1.7	2.6	4.0	6.2	9.8
1,000	0.1	0.1	0.1	0.1	0.1	0.1	0.1	0.1	0.1	0.2	0.2	0.2	0.3	0.3	0.4	0.4	0.5	0.8	1.3	2.0	3.1	4.9
900	0.1	0.1	0.1	0.1	0.1	0.1	0.1	0.1	0.1	0.1	0.2	0.2	0.2	0.3	0.3	0.4	0.5	0.8	1.2	1.8	2.8	4.4
800	0.1	0.1	0.1	0.1	0.1	0.1	0.1	0.1	0.1	0.1	0.2	0.2	0.2	0.2	0.3	0.4	0.4	0.7	1.0	1.6	2.5	3.9
700	0.1	0.1	0.1	0.1	0.1	0.1	0.1	0.1	0.1	0.1	0.1	0.1	0.2	0.2	0.3	0.3	0.4	0.6	0.9	1.4	2.2	3.4
600	0.1	0.1	0.1	0.1	0.1	0.1	0.1	0.1	0.1	0.1	0.1	0.1	0.2	0.2	0.2	0.3	0.3	0.5	0.8	1.2	1.9	3.0
500					0.1	0.1	0.1	0.1	0.1	0.1	0.1	0.1	0.1	0.2	0.2	0.2	0.3	0.4	0.7	1.0	1.6	2.5
400								0.1	0.1	0.1	0.1	0.1	0.1	0.1	0.1	0.2	0.2	0.3	0.5	0.8	1.2	2.0
300											0.1	0.1	0.1	0.1	0.1	0.1	0.2	0.3	0.4	0.6	0.9	1.5
200													0.1	0.1	0.1	0.1	0.1	0.2	0.3	0.4	0.6	1.0
100																	0.1	0.1	0.1	0.2	0.3	0.5

* Solid Conductors. Other conductors are stranded.

Note 1—The above table gives voltage drops encountered in a single phase two-wire system. The voltage drops in other systems may be obtained through multiplication by appropriate factors listed below:

System for Which Voltage Drop is Desired	Multiplying Factors for Modification of Values in Table
Single Phase—3 Wire—Line to Line	1.00
Single Phase—3 Wire—Line to Neutral	0.50
Three Phase—3 Wire—Line to Line	0.866
Three Phase—4 Wire—Line to Line	0.866
Three Phase—4 Wire—Line to Neutral	0.50

Note 2—Allowable voltage drops for systems other than single phase, two wire cannot be used directly in the above table. Such drops should be modified through multiplication by the appropriate factor listed below. The voltage thus modified may then be used to obtain the proper wire size directly from the table.

System for Which Allowable Voltage Drop is Known	Multiplying Factor for Modification of Known Value to Permit Direct Use of Table
Single Phase—3 Wire—Line to Line	1.00
Single Phase—3 Wire—Line to Neutral	2.00
Three Phase—3 Wire—Line to Line	1.155
Three Phase—4 Wire—Line to Line	1.155
Three Phase—4 Wire—Line to Neutral	2.00

Note 3—The footage employed in the tabulated ampere feet refers to the length of run of the circuit rather than to the footage of individual conductor.

Note 4—The above table is figured at 60°C since this is an estimate of the average temperature which may be anticipated in service. The table may be used without significant error for conductor temperatures up to and including 75°C,

TABLE 3.54 Volts Drop for AL Conductor in Magnetic Conduit—90 Percent PF

WIRE SIZE AWG or MCM	1000	900	800	750	700	600	500	400	350	300	250	4/0	3/0	2/0	1/0	1	2	4	6	8*	10*	12*
Ampere Feet	Volts Drop																					
500,000	39.6	41.5	44.0	45.6	47.4	51.0	57.3	66.5	73.0	81.7	93.9	107.0	130.0	157.0	194.0	239.0	295.0	459.0	—	—	—	—
400,000	31.7	33.2	35.2	36.4	39.9	40.8	45.8	53.2	58.4	65.4	75.1	85.6	104.0	125.0	155.0	191.0	236.0	337.0	—	—	—	—
300,000	23.8	24.9	26.4	27.4	28.4	30.6	34.4	39.9	43.8	49.0	56.3	64.2	78.0	94.2	116.0	143.0	177.0	275.0	430.0	—	—	—
200,000	15.9	16.6	17.6	18.2	18.9	20.4	22.9	26.6	29.2	32.7	37.5	42.8	52.0	62.8	77.6	95.6	118.0	184.0	287.0	440.0	—	—
100,000	7.9	8.3	8.8	9.1	9.5	10.2	11.5	13.3	14.6	16.3	18.8	21.4	26.0	31.4	38.8	47.8	59.0	91.8	143.0	220.0	247.0	—
90,000	7.1	7.5	7.9	8.2	8.5	9.2	10.3	12.0	13.1	14.7	16.9	19.3	23.4	28.2	35.0	43.0	53.1	82.7	128.0	199.0	313.0	495.0
80,000	6.3	6.6	7.0	7.3	7.5	8.2	9.2	10.6	11.7	13.1	15.0	17.1	20.8	25.1	31.0	38.2	47.2	73.5	115.0	177.0	278.0	440.0
70,000	5.5	5.8	6.2	6.4	6.6	7.1	8.0	9.3	10.2	11.4	13.1	15.0	18.2	22.0	27.2	33.5	41.8	64.3	100.0	155.0	243.0	385.0
60,000	4.8	5.0	5.3	5.5	5.7	6.1	6.9	8.0	8.8	9.8	11.3	12.8	15.6	18.8	23.3	28.7	35.4	55.1	86.0	133.0	208.0	330.0
50,000	4.0	4.2	4.4	4.6	4.7	5.1	5.7	6.7	7.3	8.2	9.4	10.7	13.0	15.7	19.4	23.9	29.5	45.9	71.6	110.0	174.0	275.0
40,000	3.2	3.3	3.5	3.6	4.0	4.1	4.6	5.3	5.8	6.5	7.5	8.6	10.4	12.5	15.5	19.1	23.6	36.7	57.3	88.0	139.0	220.0
30,000	2.4	2.5	2.6	2.7	2.8	3.1	3.4	4.0	4.4	4.9	5.6	6.4	7.8	9.4	11.6	14.3	17.7	27.5	43.0	66.0	104.0	165.0
20,000	1.6	1.7	1.8	1.8	1.9	2.0	2.3	2.7	2.9	3.3	3.8	4.3	5.2	6.3	7.8	9.6	11.8	18.4	28.7	44.0	69.4	110.0
10,000	0.8	0.8	0.9	0.9	1.0	1.0	1.2	1.3	1.5	1.6	1.9	2.1	2.6	3.1	3.9	4.8	5.9	9.2	14.3	22.0	34.7	55.0
9,000	0.7	0.8	0.8	0.8	0.9	0.9	1.0	1.2	1.3	1.5	1.7	1.9	2.3	2.8	3.5	4.3	5.3	8.3	12.8	19.9	31.3	49.5
8,000	0.6	0.7	0.7	0.7	0.8	0.8	0.9	1.1	1.2	1.3	1.5	1.7	2.1	2.5	3.1	3.8	4.7	7.4	11.5	17.7	27.8	44.0
7,000	0.6	0.6	0.6	0.6	0.7	0.7	0.8	0.9	1.0	1.1	1.3	1.5	1.8	2.2	2.7	3.4	4.2	6.4	10.0	15.5	24.3	38.5
6,000	0.5	0.5	0.5	0.6	0.6	0.6	0.7	0.8	0.9	1.0	1.1	1.3	1.6	1.9	2.3	2.9	3.5	5.5	8.6	13.3	20.8	33.0
5,000	0.4	0.4	0.4	0.5	0.5	0.5	0.6	0.7	0.7	0.8	0.9	1.1	1.3	1.6	1.9	2.4	3.0	4.5	7.2	11.0	17.4	27.5
4,000	0.3	0.3	0.4	0.4	0.4	0.4	0.5	0.5	0.6	0.7	0.8	0.9	1.0	1.3	1.6	1.9	2.4	3.7	5.7	8.8	13.9	22.0
3,000	0.2	0.2	0.3	0.3	0.3	0.3	0.3	0.4	0.4	0.5	0.6	0.6	0.8	0.9	1.2	1.4	1.8	2.8	4.3	6.6	10.4	16.5
2,000	0.2	0.2	0.2	0.2	0.2	0.2	0.2	0.3	0.3	0.3	0.4	0.4	0.5	0.6	0.8	1.0	1.2	1.8	2.9	4.4	6.9	11.0
1,000	0.1	0.1	0.1	0.1	0.1	0.1	0.1	0.1	0.2	0.2	0.2	0.2	0.3	0.3	0.4	0.5	0.6	0.9	1.4	2.2	3.5	5.5
900	0.1	0.1	0.1	0.1	0.1	0.1	0.1	0.1	0.1	0.2	0.2	0.2	0.2	0.3	0.4	0.4	0.5	0.8	1.3	2.0	3.1	5.0
800	0.1	0.1	0.1	0.1	0.1	0.1	0.1	0.1	0.1	0.1	0.2	0.2	0.2	0.3	0.3	0.4	0.5	0.7	1.2	1.8	2.8	4.4
700	0.1	0.1	0.1	0.1	0.1	0.1	0.1	0.1	0.1	0.1	0.1	0.2	0.2	0.2	0.3	0.3	0.4	0.6	1.0	1.6	2.4	3.9
600	0.1	0.1	0.1	0.1	0.1	0.1	0.1	0.1	0.1	0.1	0.1	0.1	0.2	0.2	0.2	0.3	0.4	0.6	0.9	1.3	2.1	3.3
500	—	—	—	0.1	0.1	0.1	0.1	0.1	0.1	0.1	0.1	0.1	0.1	0.2	0.2	0.2	0.3	0.5	0.7	1.1	1.7	2.8
400	—	—	—	—	—	—	0.1	0.1	0.1	0.1	0.1	0.1	0.1	0.1	0.2	0.2	0.2	0.4	0.6	0.9	1.4	2.2
300	—	—	—	—	—	—	—	—	—	0.1	0.1	0.1	0.1	0.1	0.1	0.1	0.2	0.3	0.4	0.7	1.0	1.7
200	—	—	—	—	—	—	—	—	—	—	—	—	0.1	0.1	0.1	0.1	0.1	0.2	0.3	0.4	0.7	1.1
100	—	—	—	—	—	—	—	—	—	—	—	—	—	—	—	0.1	0.1	0.1	0.1	0.2	0.4	0.6

* Solid Conductors. Other conductors are stranded.

Note 1—The above table gives voltage drops encountered in a single phase two-wire system. The voltage drops in other systems may be obtained through multiplication by appropriate factors listed below:

System for Which Voltage Drop is Desired	Multiplying Factors for Modification of Values in Table
Single Phase—3 Wire—Line to Line	1.00
Single Phase—3 Wire—Line to Neutral	0.50
Three Phase—3 Wire—Line to Line	0.866
Three Phase—4 Wire—Line to Line	0.866
Three Phase—4 Wire—Line to Neutral	0.50

Note 2—Allowable voltage drops for systems other than single phase, two wire cannot be used directly in the above table. Such drops should be modified through multiplication by the appropriate factor listed below. The voltage thus modified may then be used to obtain the proper wire size directly from the table.

System for Which Allowable Voltage Drop is Known	Multiplying Factor for Modification of Known Value to Permit Direct Use of Table
Single Phase—3 Wire—Line to Line	1.00
Single Phase—3 Wire—Line to Neutral	2.00
Three Phase—3 Wire—Line to Line	1.155
Three Phase—4 Wire—Line to Line	1.155
Three Phase—4 Wire—Line to Neutral	2.00

Note 3—The footage employed in the tabulated ampere feet refers to the length of run of the circuit rather than to the footage of individual conductor.

Note 4—The above table is figured at 60°C since this is an estimate of the average temperature which may be anticipated in service. The table may be used without significant error for conductor temperatures up to and including 75°C.

TABLE 3.55 Volts Drop for AL Conductor in Magnetic Conduit—95 Percent PF

Values in table are **Volts Drop**. Wire sizes (AWG or MCM) run across the top; Ampere Feet run down the side.

Ampere Feet	1000	900	800	750	700	600	500	400	350	300	250	4/0	3/0	2/0	1/0	1	2	4	6	8*	10*	12*
500,000	35.7	37.7	40.3	41.9	43.8	48.0	54.0	63.6	70.4	79.4	92.2	106.0	130.0	158.0	197.0	244.0	304.0	475.0	—	—	—	—
400,000	28.6	30.2	32.2	33.5	35.0	38.4	43.2	50.9	56.3	63.5	73.8	84.8	104.0	126.4	157.6	195.2	243.2	380.0	—	—	—	—
300,000	21.4	22.6	24.2	25.1	26.3	28.8	32.4	38.2	42.2	47.6	55.3	63.6	78.0	94.8	118.2	146.4	182.4	285.0	447.0	—	—	—
200,000	14.3	15.1	16.1	16.8	17.5	19.2	21.6	25.4	28.2	31.8	36.9	42.4	52.0	63.2	78.8	97.6	121.6	190.0	298.0	462.0	—	—
100,000	7.1	7.5	8.1	8.4	8.8	9.6	10.8	12.7	14.1	15.9	18.4	21.2	26.0	31.6	39.4	48.8	60.8	95.0	149.0	231.0	365.0	—
90,000	6.4	6.8	7.3	7.5	7.9	8.6	9.7	11.4	12.7	14.3	16.6	19.1	23.4	28.4	35.4	43.9	54.7	85.5	134.0	208.0	328.0	—
80,000	5.7	6.0	6.4	6.7	7.0	7.7	8.6	10.2	11.3	12.7	14.8	17.0	20.8	25.3	31.5	39.0	48.6	76.0	119.0	185.0	292.0	463.0
70,000	5.0	5.3	5.6	5.9	6.1	6.7	7.6	8.9	9.9	11.1	12.9	14.8	18.2	22.1	27.6	34.2	42.6	66.5	104.0	162.0	255.0	404.0
60,000	4.3	4.5	4.8	5.0	5.3	5.8	6.5	7.6	8.4	9.5	11.1	12.7	15.6	19.0	23.6	29.3	36.5	57.0	89.4	139.0	219.0	347.0
50,000	3.6	3.8	4.0	4.2	4.4	4.8	5.4	6.4	7.0	7.9	9.2	10.6	13.0	15.8	19.7	24.4	30.4	47.5	74.5	115.5	182.5	289.0
40,000	2.9	3.0	3.2	3.4	3.5	3.8	4.3	5.1	5.6	6.4	7.4	8.5	10.4	12.6	15.8	19.5	24.3	38.0	59.6	92.4	146.0	231.0
30,000	2.1	2.3	2.4	2.5	2.6	2.9	3.2	3.8	4.2	4.8	5.5	6.4	7.8	9.5	11.8	14.6	18.2	28.5	44.7	69.3	109.5	173.0
20,000	1.4	1.5	1.6	1.7	1.8	1.9	2.2	2.5	2.8	3.2	3.7	4.2	5.2	6.3	7.9	9.8	12.2	19.0	29.8	46.2	73.0	116.0
10,000	0.7	0.8	0.8	0.8	0.9	1.0	1.1	1.3	1.4	1.6	1.8	2.1	2.6	3.2	3.9	4.9	6.1	9.5	14.9	23.1	36.5	57.8
9,000	0.6	0.7	0.7	0.8	0.8	0.9	1.0	1.1	1.3	1.4	1.7	1.9	2.3	2.8	3.5	4.4	5.5	8.6	13.4	20.8	32.8	52.0
8,000	0.6	0.6	0.6	0.7	0.7	0.8	0.9	1.0	1.1	1.3	1.5	1.7	2.1	2.5	3.2	3.9	4.9	7.6	11.9	18.5	29.2	46.3
7,000	0.5	0.5	0.6	0.6	0.6	0.7	0.8	0.9	1.0	1.1	1.3	1.5	1.8	2.2	2.8	3.4	4.3	6.7	10.4	16.2	25.5	40.4
6,000	0.4	0.5	0.5	0.5	0.5	0.6	0.6	0.8	0.8	1.0	1.1	1.3	1.6	1.9	2.4	2.9	3.6	5.7	8.9	13.9	21.9	34.7
5,000	0.4	0.4	0.4	0.4	0.4	0.5	0.5	0.6	0.7	0.8	0.9	1.1	1.3	1.6	2.0	2.4	3.0	4.8	7.5	11.6	18.2	28.9
4,000	0.3	0.3	0.3	0.3	0.4	0.4	0.4	0.5	0.6	0.6	0.7	0.8	1.0	1.3	1.6	2.0	2.4	3.8	6.0	9.2	14.6	23.1
3,000	0.2	0.2	0.2	0.3	0.3	0.3	0.3	0.4	0.4	0.5	0.6	0.6	0.8	0.9	1.2	1.5	1.8	2.9	4.5	6.9	10.9	17.3
2,000	0.1	0.2	0.2	0.2	0.2	0.2	0.2	0.3	0.3	0.3	0.4	0.4	0.5	0.6	0.8	1.0	1.2	1.9	3.0	4.6	7.3	11.6
1,000	0.1	0.1	0.1	0.1	0.1	0.1	0.1	0.1	0.1	0.2	0.2	0.2	0.3	0.3	0.4	0.5	0.6	1.0	1.5	2.3	3.7	5.8
900	0.1	0.1	0.1	0.1	0.1	0.1	0.1	0.1	0.1	0.1	0.2	0.2	0.2	0.3	0.4	0.4	0.5	0.9	1.3	2.1	3.3	5.2
800	0.1	0.1	0.1	0.1	0.1	0.1	0.1	0.1	0.1	0.1	0.1	0.2	0.2	0.3	0.3	0.4	0.5	0.8	1.2	1.8	2.9	4.6
700	0.1	0.1	0.1	0.1	0.1	0.1	0.1	0.1	0.1	0.1	0.1	0.1	0.2	0.2	0.3	0.3	0.4	0.7	1.0	1.6	2.6	4.0
600	—	—	—	0.1	0.1	0.1	0.1	0.1	0.1	0.1	0.1	0.1	0.2	0.2	0.2	0.3	0.4	0.6	0.9	1.4	2.2	3.5
500	—	—	—	—	—	—	0.1	0.1	0.1	0.1	0.1	0.1	0.1	0.2	0.2	0.2	0.3	0.5	0.7	1.2	1.8	2.9
400	—	—	—	—	—	—	—	0.1	0.1	0.1	0.1	0.1	0.1	0.1	0.2	0.2	0.2	0.4	0.6	0.9	1.5	2.3
300	—	—	—	—	—	—	—	—	—	—	0.1	0.1	0.1	0.1	0.1	0.1	0.2	0.3	0.4	0.7	1.1	1.7
200	—	—	—	—	—	—	—	—	—	—	—	—	0.1	0.1	0.1	0.1	0.1	0.2	0.3	0.5	0.7	1.2
100	—	—	—	—	—	—	—	—	—	—	—	—	—	—	—	—	0.1	0.1	0.1	0.2	0.4	0.6

* Solid Conductors. Other conductors are stranded.

Note 1—The above table gives voltage drops encountered in a single phase two-wire system. The voltage drops in other systems may be obtained through multiplication by appropriate factors listed below:

System for Which Voltage Drop Is Desired	Multiplying Factors for Modification of Values in Table
Single Phase—3 Wire—Line to Line	1.00
Single Phase—3 Wire—Line to Neutral	0.50
Three Phase—3 Wire—Line to Line	0.866
Three Phase—4 Wire—Line to Line	0.866
Three Phase—4 Wire—Line to Neutral	0.50

Note 2—Allowable voltage drops for systems other than single phase, two wire cannot be used directly in the above table. Such drops should be modified through multiplication by the appropriate factor listed below. The voltage thus modified may then be used to obtain the proper wire size directly from the table.

System for Which Allowable Voltage Drop Is Known	Multiplying Factor for Modification of Known Value to Permit Direct Use of Table
Single Phase—3 Wire—Line to Line	1.00
Single Phase—3 Wire—Line to Neutral	2.00
Three Phase—3 Wire—Line to Line	1.155
Three Phase—4 Wire—Line to Line	1.155
Three Phase—4 Wire—Line to Neutral	2.00

Note 3—The footage employed in the tabulated ampere feet refers to the length of run of the circuit rather than to the footage of individual conductor.

Note 4—The above table is figured at 60°C since this is an estimate of the average temperature which may be anticipated in service. The table may be used without significant error for conductor temperatures up to and including 75°C.

TABLE 3.56 Volts Drop for AL Conductor in Magnetic Conduit—100 Percent PF

WIRE SIZE AWG or MCM	1000	900	800	750	700	600	500	400	350	300	250	4/0	3/0	2/0	1/0	1	2	4	6	8*	10*	12*
Ampere Feet	Volts Drop																					
500,000	24.0	25.9	28.5	30.1	32.0	36.2	42.6	52.4	59.2	68.5	81.8	96.8	121.0	151.0	191.0	240.0	303.0	483.0	—	—	—	—
400,000	19.2	20.7	22.8	24.1	25.6	29.0	34.1	41.9	47.4	54.8	65.4	77.4	96.8	121.0	153.0	192.0	241.0	386.0	—	—	—	—
300,000	14.4	15.5	17.1	18.1	19.2	21.7	25.5	31.4	35.5	41.1	49.1	58.1	72.6	90.6	115.0	144.0	182.0	290.0	460.0	—	—	—
200,000	9.6	10.3	11.4	12.0	12.8	14.5	17.0	21.0	27.7	27.4	32.7	38.7	48.4	60.4	76.4	96.0	121.0	193.0	307.0	478.0	—	—
100,000	4.8	5.2	5.7	6.0	6.4	7.2	8.5	10.5	11.8	13.7	16.4	19.4	24.2	30.2	38.2	48.0	60.6	96.6	153.0	239.0	380.0	—
90,000	4.3	4.7	5.1	5.4	5.8	6.5	7.7	9.4	10.7	12.3	14.7	17.4	21.8	27.2	34.4	43.2	54.6	87.0	138.0	215.0	342.0	—
80,000	3.8	4.1	4.6	4.8	5.1	5.8	6.8	8.4	9.5	11.0	13.1	15.5	19.4	24.2	30.5	38.4	48.5	77.3	123.0	191.0	304.0	483.0
70,000	3.4	3.6	4.0	4.2	4.5	5.1	6.0	7.3	8.3	9.6	11.5	13.5	16.9	21.1	26.7	33.6	42.4	67.6	107.0	167.0	266.0	423.0
60,000	2.9	3.1	3.4	3.6	3.8	4.3	5.1	6.3	7.1	8.2	9.8	11.6	14.5	18.1	22.9	28.8	36.4	58.0	92.0	144.0	228.0	362.0
50,000	2.4	2.6	2.9	3.0	3.2	3.6	4.3	5.2	5.9	6.9	8.2	9.7	12.1	15.1	19.1	24.0	30.3	48.3	76.7	120.0	190.0	302.0
40,000	1.9	2.1	2.3	2.4	2.6	2.9	3.4	4.2	4.7	5.5	6.5	7.7	9.7	12.1	15.3	19.2	24.1	38.6	61.4	95.6	152.0	242.0
30,000	1.4	1.6	1.7	1.8	1.9	2.2	2.6	3.1	3.6	4.1	4.9	5.8	7.3	9.1	11.5	14.4	18.2	29.0	46.0	71.7	114.0	181.0
20,000	1.0	1.0	1.1	1.2	1.3	1.5	1.7	2.1	2.8	2.7	3.3	3.9	4.8	6.0	7.6	9.6	12.1	19.3	30.7	47.8	76.0	121.0
10,000	0.5	0.5	0.6	0.6	0.6	0.7	0.9	1.1	1.2	1.4	1.6	1.9	2.4	3.0	3.8	4.8	6.1	9.7	15.3	23.9	38.0	60.4
9,000	0.4	0.5	0.5	0.5	0.6	0.7	0.8	0.9	1.1	1.2	1.5	1.7	2.2	2.7	3.4	4.3	5.5	8.7	13.8	21.5	34.2	54.4
8,000	0.4	0.4	0.5	0.5	0.5	0.6	0.7	0.8	1.0	1.1	1.3	1.6	1.9	2.4	3.1	3.8	4.9	7.7	12.3	19.1	30.4	48.3
7,000	0.3	0.4	0.4	0.4	0.5	0.5	0.6	0.7	0.8	1.0	1.2	1.4	1.7	2.1	2.7	3.4	4.2	6.8	10.7	16.7	26.6	42.3
6,000	0.3	0.3	0.3	0.4	0.4	0.4	0.5	0.6	0.7	0.8	1.0	1.2	1.5	1.8	2.3	2.9	3.6	5.8	9.2	14.4	22.8	36.2
5,000	0.2	0.3	0.3	0.3	0.3	0.4	0.4	0.5	0.6	0.7	0.8	1.0	1.2	1.5	1.9	2.4	3.0	4.8	7.7	12.0	19.0	30.2
4,000	0.2	0.2	0.2	0.2	0.3	0.3	0.3	0.4	0.5	0.6	0.7	0.8	1.0	1.2	1.5	1.9	2.4	3.9	6.1	9.6	15.2	24.2
3,000	0.1	0.2	0.2	0.2	0.2	0.2	0.3	0.3	0.4	0.4	0.5	0.6	0.7	0.9	1.2	1.4	1.8	2.9	4.6	7.2	11.4	18.1
2,000	0.1	0.1	0.1	0.1	0.1	0.2	0.2	0.2	0.3	0.3	0.3	0.4	0.5	0.6	0.8	1.0	1.2	1.9	3.1	4.8	7.6	12.1
1,000	0.1	0.1	0.1	0.1	0.1	0.1	0.1	0.1	0.1	0.1	0.2	0.2	0.2	0.3	0.4	0.5	0.6	1.0	1.5	2.4	3.8	6.0
900	—	0.1	0.1	0.1	0.1	0.1	0.1	0.1	0.1	0.1	0.2	0.2	0.2	0.3	0.3	0.4	0.6	0.9	1.4	2.2	3.4	5.4
800	—	—	0.1	0.1	0.1	0.1	0.1	0.1	0.1	0.1	0.1	0.2	0.2	0.2	0.3	0.4	0.5	0.8	1.2	1.9	3.0	4.8
700	—	—	—	—	0.1	0.1	0.1	0.1	0.1	0.1	0.1	0.1	0.2	0.2	0.3	0.3	0.4	0.7	1.1	1.7	2.7	4.2
600	—	—	—	—	—	—	0.1	0.1	0.1	0.1	0.1	0.1	0.1	0.2	0.2	0.3	0.4	0.6	0.9	1.4	2.3	3.6
500	—	—	—	—	—	—	—	0.1	0.1	0.1	0.1	0.1	0.1	0.2	0.2	0.2	0.3	0.5	0.8	1.2	1.9	3.0
400	—	—	—	—	—	—	—	—	0.1	0.1	0.1	0.1	0.1	0.1	0.2	0.2	0.2	0.4	0.6	1.0	1.5	2.4
300	—	—	—	—	—	—	—	—	—	—	0.1	0.1	0.1	0.1	0.1	0.1	0.2	0.3	0.5	0.7	1.1	1.8
200	—	—	—	—	—	—	—	—	—	—	—	—	0.1	0.1	0.1	0.1	0.1	0.2	0.3	0.5	0.8	1.2
100	—	—	—	—	—	—	—	—	—	—	—	—	—	—	—	0.1	0.1	0.1	0.2	0.2	0.4	0.6

* Solid Conductors. Other conductors are stranded.

Note 1—The above table gives voltage drops encountered in a single phase two-wire system. The voltage drops in other systems may be obtained through multiplication by appropriate factors listed below:

System for Which Voltage Drop is Desired	Multiplying Factors for Modification of Values in Table
Single Phase—3 Wire—Line to Line	1.00
Single Phase—3 Wire—Line to Neutral	0.50
Three Phase—3 Wire—Line to Line	0.866
Three Phase—4 Wire—Line to Line	0.866
Three Phase—4 Wire—Line to Neutral	0.50

Note 2—Allowable voltage drops for systems other than single phase, two wire cannot be used directly in the above table. Such drops should be modified through multiplication by the appropriate factor listed below. The voltage thus modified may then be used to obtain the proper wire size directly from the table.

System for Which Allowable Voltage Drop is Known	Multiplying Factor for Modification of Known Value to Permit Direct Use of Table
Single Phase—3 Wire—Line to Line	1.00
Single Phase—3 Wire—Line to Neutral	2.00
Three Phase—3 Wire—Line to Line	1.155
Three Phase—4 Wire—Line to Line	1.155
Three Phase—4 Wire—Line to Neutral	2.00

Note 3—The footage employed in the tabulated ampere feet refers to the length of run of the circuit rather than to the footage of individual conductor.

Note 4—The above table is figured at 60°C since this is an estimate of the average temperature which may be anticipated in service. The table may be used without significant error for conductor temperatures up to and including 75°C.

TABLE 3.57 Volts Drop for AL Conductor in Nonmagnetic Conduit—70 Percent PF

WIRE SIZE AWG or MCM (Ampere Feet)	12*	10*	8*	6	4	2	1	1/0	2/0	3/0	4/0	250	300	350	400	500	600	700	760	800	900	1000
											Volts Drop											
500,000	—	—	—	—	370.0	242.0	198.0	153.0	134.0	101.0	83.3	83.0	73.3	66.2	60.8	53.0	47.9	44.9	43.4	42.1	39.3	38.1
400,000	—	—	—	456.0	296.0	194.0	158.0	122.0	107.0	80.8	66.6	66.4	58.7	52.9	48.6	42.4	38.3	35.9	34.7	33.6	31.9	30.6
300,000	—	—	—	342.0	222.0	146.0	118.0	91.7	80.4	60.6	49.9	44.0	44.0	39.6	36.5	31.8	28.7	26.9	26.0	25.3	23.9	22.8
200,000	—	—	348.0	228.0	148.0	98.0	79.2	61.2	53.6	40.4	33.2	29.3	29.3	26.4	24.4	21.2	19.2	17.9	17.4	16.8	16.0	15.4
100,000	430.0	273.0	174.0	114.0	74.0	48.4	39.6	32.6	26.8	20.2	16.6	16.6	14.7	13.2	12.2	10.6	9.6	9.0	8.7	8.4	8.0	7.6
90,000	387.0	246.0	157.0	103.0	66.6	43.6	35.6	29.1	24.1	18.2	16.8	14.9	13.2	11.9	11.0	9.5	8.6	8.1	7.8	7.6	7.2	6.8
80,000	344.0	219.0	139.0	91.2	59.2	38.7	31.7	26.1	21.4	16.2	14.9	13.3	11.7	10.6	9.7	8.5	7.7	7.2	7.0	6.7	6.6	6.1
70,000	301.0	191.0	122.0	79.8	51.8	33.9	27.7	22.8	18.8	14.1	13.1	11.6	10.3	9.3	8.5	7.4	6.7	6.3	6.1	5.9	5.6	5.3
60,000	258.0	164.0	105.0	68.4	44.4	29.1	23.8	19.6	16.1	12.1	11.2	10.0	8.8	8.0	7.3	6.4	5.8	5.4	5.2	5.1	4.8	4.6
50,000	215.0	137.0	87.2	57.0	37.0	24.2	19.8	16.3	13.4	10.1	9.3	8.3	7.3	6.6	6.1	5.3	4.8	4.5	4.3	4.2	4.0	3.8
40,000	172.0	109.0	69.6	45.6	29.2	19.4	15.8	13.0	10.7	8.1	7.5	6.6	5.9	5.3	4.9	4.2	3.8	3.6	3.5	3.4	3.2	3.1
30,000	129.0	81.9	52.6	34.2	22.2	14.5	11.9	9.8	8.0	6.1	5.6	5.0	4.4	4.0	3.7	3.2	2.9	2.7	2.6	2.5	2.4	2.3
20,000	86.0	54.6	34.8	22.8	14.8	9.7	7.9	6.5	5.4	4.0	3.3	3.3	2.9	2.6	2.4	2.1	1.9	1.8	1.7	1.7	1.6	1.5
10,000	43.0	27.3	17.4	11.4	7.4	4.8	4.0	3.3	2.7	2.0	1.9	1.7	1.5	1.3	1.2	1.1	1.0	0.9	0.9	0.8	0.8	0.8
9,000	38.7	24.6	15.7	10.3	6.7	4.4	3.6	2.9	2.4	1.8	1.7	1.5	1.3	1.2	1.1	1.0	0.9	0.8	0.8	0.8	0.7	0.7
8,000	34.4	21.9	13.9	9.1	5.9	3.9	3.2	2.3	2.1	1.4	1.5	1.3	1.2	1.1	1.0	0.8	0.8	0.7	0.7	0.6	0.6	0.6
7,000	30.1	19.1	12.2	8.0	5.2	3.4	2.8	2.3	1.8	1.2	1.1	1.2	1.0	0.9	0.9	0.7	0.7	0.6	0.6	0.6	0.5	0.5
6,000	25.8	16.4	10.5	6.8	4.4	2.9	2.4	2.0	—	—	—	1.0	0.9	0.8	0.7	0.6	0.6	0.5	0.5	0.5	0.4	0.5
5,000	21.5	13.7	8.7	5.7	3.7	2.4	2.0	1.8	1.3	1.0	0.9	0.8	0.7	0.7	0.6	0.5	0.5	0.5	0.4	0.4	0.4	0.4
4,000	17.2	10.9	7.0	4.6	3.0	1.9	1.6	1.2	1.1	0.8	0.8	0.7	0.6	0.5	0.5	0.4	0.4	0.4	0.3	0.3	0.3	0.3
3,000	12.9	8.2	5.3	3.4	2.2	1.5	1.2	1.0	0.8	0.6	0.4	0.5	0.4	0.4	0.4	0.3	0.3	0.2	0.3	0.3	0.2	0.2
2,000	8.6	5.5	3.5	2.3	1.5	1.0	0.8	0.7	0.5	0.4	0.2	0.3	0.3	0.3	0.2	0.2	0.2	0.2	0.2	0.1	0.1	0.2
1,000	4.3	2.7	1.7	1.1	0.7	0.5	0.4	0.3	—	0.2	—	0.2	0.1	0.1	0.1	0.1	0.1	0.1	0.1	0.1	0.1	0.1
900	3.9	2.5	1.6	1.0	0.7	0.4	0.4	0.3	0.2	0.2	0.2	0.2	0.1	0.1	0.1	0.1	0.1	0.1	—	—	—	—
800	3.4	2.2	1.4	0.9	0.6	0.4	0.3	0.3	0.2	0.2	0.2	0.1	0.1	0.1	0.1	0.1	0.1	0.1	—	—	—	—
700	3.0	1.9	1.2	0.8	0.5	0.3	0.3	0.2	0.2	0.1	0.1	0.1	0.1	0.1	0.1	0.1	0.1	0.1	—	—	—	—
600	2.6	1.6	1.1	0.7	0.4	0.3	0.2	0.2	0.1	—	—	0.1	0.1	0.1	0.1	0.1	0.1	0.1	—	—	—	—
500	2.2	1.4	0.9	0.6	0.4	0.2	0.2	0.2	0.1	0.1	0.1	0.1	0.1	0.1	0.1	0.1	0.1	0.1	—	—	—	—
400	1.7	1.1	0.7	0.5	0.3	0.2	0.2	0.1	0.1	0.1	0.1	0.1	0.1	0.1	0.1	—	—	—	—	—	—	—
300	1.3	0.8	0.5	0.3	0.2	0.1	0.1	0.1	0.1	—	0.1	—	—	—	—	—	—	—	—	—	—	—
200	0.9	0.6	0.4	0.2	0.1	0.1	0.1	0.1	—	—	—	—	—	—	—	—	—	—	—	—	—	—
100	0.4	0.3	0.2	0.1	0.1	0.1	—	—	—	—	—	—	—	—	—	—	—	—	—	—	—	—

* Solid Conductors. Other conductors are stranded.

Note 1—The above table gives voltage drops encountered in a single phase two-wire system. The voltage drops in other systems may be obtained through multiplication by appropriate factors listed below:

Multiplication Factors for Modification of Values in Table

System for Which Voltage Drop Is Desired	
Single Phase—2 Wire—Line to Line	1.00
Single Phase—3 Wire—Line to Neutral	0.50
Three Phase—3 Wire—Line to Line	0.866
Three Phase—4 Wire—Line to Line	0.866
Three Phase—4 Wire—Line to Neutral	0.50

Note 2—Allowable voltage drops for systems other than single phase, two wire cannot be used directly in the above table. Such drops should be modified through multiplication by the appropriate factor listed below. The voltage thus modified may then be used to obtain the proper wire size directly from the table.

System for Which Allowable Voltage Drop Is Known — **Multiplying Factor for Modification of Known Value to Permit Direct Use of Table**

Single Phase—3 Wire—Line to Line	1.00
Single Phase—3 Wire—Line to Neutral	2.00
Three Phase—3 Wire—Line to Line	1.165
Three Phase—4 Wire—Line to Line	1.165
Three Phase—4 Wire—Line to Neutral	2.00

Note 3—The footage employed in the tabulated ampere feet refers to the length of run of the circuit rather than to the footage of individual conductor.

Note 4—The above table is figured at 60°C since this is an estimate of the average temperature which may be anticipated in service. The table may be used without significant error for conductor temperatures up to and including 75°C.

TABLE 3.58 Volts Drop for AL Conductor in Nonmagnetic Conduit—80 Percent PF

Volts Drop

WIRE SIZE AWG or MCM — Ampere Feet	1000	900	800	750	700	600	500	400	350	300	250	4/0	3/0	2/0	1/0	1	2	4	6	8*	10*	12*
500,000	36.5	38.3	40.8	42.2	43.9	48.0	53.1	61.9	67.7	76.8	86.8	98.7	119.0	145.0	177.0	217.0	267.0	413.1	—	—	—	—
400,000	29.3	30.6	32.6	33.8	35.3	38.4	42.8	49.5	54.1	60.7	69.4	78.9	95.2	116.0	142.0	174.0	214.0	330.0	—	—	—	—
300,000	21.9	23.0	24.5	25.3	26.3	28.8	31.9	37.1	40.6	46.1	52.1	59.2	71.4	87.0	106.0	130.0	160.0	248.0	385.1	—	—	—
200,000	14.6	15.3	16.3	16.9	17.6	19.2	21.2	24.8	27.1	30.7	34.7	39.5	47.6	58.0	70.8	86.8	106.8	165.0	257.0	394.0	—	—
100,000	7.3	7.7	8.2	8.4	8.8	9.6	10.6	12.4	13.5	15.4	17.4	19.7	23.8	29.0	35.4	43.4	53.4	82.6	128.5	197.0	310.0	490.0
90,000	6.6	6.9	7.3	7.6	7.9	8.6	9.6	11.1	12.2	13.8	15.6	17.8	21.4	26.1	31.9	39.1	48.1	74.3	116.0	177.0	279.0	441.0
80,000	5.8	6.1	6.5	6.8	7.0	7.7	8.5	9.9	10.8	12.3	13.9	15.8	19.0	23.2	28.3	34.7	42.7	66.1	103.0	158.0	248.0	392.0
70,000	5.1	5.4	5.7	5.9	6.1	6.7	7.4	8.7	9.5	10.8	12.2	13.8	16.7	20.3	24.8	30.4	37.4	57.8	89.9	138.0	217.0	343.0
60,000	4.4	4.6	4.9	5.1	5.3	5.8	6.4	7.4	8.1	9.2	10.4	11.8	14.3	17.4	21.2	26.0	32.0	49.6	77.1	118.0	186.0	294.0
50,000	3.7	3.8	4.1	4.2	4.4	4.8	5.3	6.2	6.8	7.7	8.7	9.9	11.9	14.5	17.7	21.7	26.7	41.3	64.2	98.6	155.0	245.0
40,000	2.9	3.1	3.3	3.4	3.5	3.8	4.2	5.0	5.4	6.1	6.9	7.9	9.5	11.6	14.2	17.4	21.4	33.0	51.4	78.8	124.0	196.0
30,000	2.2	2.3	2.4	2.5	2.6	2.9	3.2	3.7	4.1	4.6	5.2	5.9	7.1	8.7	10.6	13.0	16.0	24.8	38.5	59.1	93.0	147.0
20,000	1.5	1.5	1.6	1.7	1.8	1.9	2.1	2.5	2.7	3.1	3.5	3.9	4.8	5.8	7.1	8.7	10.7	16.5	25.7	39.4	62.0	98.0
10,000	0.7	0.8	0.8	0.8	0.9	1.0	1.1	1.2	1.4	1.5	1.7	2.0	2.4	2.9	3.5	4.3	5.3	8.3	12.8	19.7	31.0	49.0
9,000	0.7	0.7	0.7	0.8	0.8	0.9	1.0	1.1	1.2	1.4	1.6	1.8	2.1	2.6	3.2	3.9	4.8	7.4	11.6	17.7	27.9	44.1
8,000	0.6	0.6	0.7	0.7	0.7	0.8	0.8	1.0	1.1	1.2	1.4	1.6	1.9	2.3	2.8	3.5	4.3	6.6	10.3	15.8	24.8	39.2
7,000	0.5	0.5	0.6	0.6	0.6	0.7	0.7	0.9	0.9	1.1	1.2	1.4	1.7	2.0	2.5	3.0	3.7	5.8	9.0	13.8	21.7	34.3
6,000	0.4	0.5	0.5	0.5	0.5	0.6	0.6	0.7	0.8	0.9	1.0	1.2	1.4	1.7	2.1	2.6	3.2	5.0	7.7	11.8	18.6	29.4
5,000	0.4	0.4	0.4	0.4	0.4	0.5	0.5	0.6	0.7	0.8	0.9	1.0	1.2	1.5	1.8	2.2	2.7	4.1	6.4	9.9	15.5	24.5
4,000	0.3	0.3	0.3	0.3	0.4	0.4	0.4	0.5	0.5	0.6	0.7	0.8	1.0	1.2	1.4	1.7	2.1	3.3	5.1	7.9	12.4	19.6
3,000	0.2	0.2	0.2	0.3	0.3	0.3	0.3	0.4	0.4	0.5	0.5	0.6	0.7	0.9	1.1	1.3	1.6	2.5	3.9	5.9	9.3	14.7
2,000	0.1	0.2	0.2	0.2	0.2	0.2	0.2	0.2	0.3	0.3	0.3	0.4	0.5	0.6	0.7	0.9	1.1	1.7	2.6	3.9	6.2	9.8
1,000	0.1	0.1	0.1	0.1	0.1	0.1	0.1	0.1	0.1	0.2	0.2	0.2	0.2	0.3	0.4	0.4	0.5	0.8	1.3	2.0	3.1	4.9
900	0.1	0.1	0.1	0.1	0.1	0.1	0.1	0.1	0.1	0.1	0.2	0.2	0.2	0.3	0.3	0.4	0.5	0.7	1.2	1.8	2.8	4.4
800	0.1	0.1	0.1	0.1	0.1	0.1	0.1	0.1	0.1	0.1	0.1	0.2	0.2	0.2	0.3	0.3	0.4	0.7	1.0	1.6	2.5	3.9
700	0.1	0.1	0.1	0.1	0.1	0.1	0.1	0.1	0.1	0.1	0.1	0.1	0.2	0.2	0.2	0.3	0.4	0.6	0.9	1.4	2.2	3.4
600	—	—	—	0.1	0.1	0.1	0.1	0.1	0.1	0.1	0.1	0.1	0.1	0.2	0.2	0.3	0.3	0.5	0.8	1.2	1.9	2.9
500	—	—	—	—	—	—	0.1	0.1	0.1	0.1	0.1	0.1	0.1	0.1	0.2	0.2	0.3	0.4	0.6	1.0	1.6	2.5
400	—	—	—	—	—	—	—	0.1	0.1	0.1	0.1	0.1	0.1	0.1	0.1	0.2	0.2	0.3	0.5	0.8	1.2	2.0
300	—	—	—	—	—	—	—	—	—	—	0.1	0.1	0.1	0.1	0.1	0.1	0.2	0.2	0.4	0.6	0.9	1.5
200	—	—	—	—	—	—	—	—	—	—	—	—	—	0.1	0.1	0.1	0.1	0.2	0.3	0.4	0.6	1.0
100	—	—	—	—	—	—	—	—	—	—	—	—	—	—	—	—	0.1	0.1	0.1	0.2	0.3	0.5

* Solid Conductors. Other conductors are stranded.

Note 1—The above table gives voltage drops encountered in a single phase two-wire system. The voltage drops in other systems may be obtained through multiplication by appropriate factors listed below:

Multiplying Factors for Modification of Values in Table

System for Which Voltage Drop is Desired	
Single Phase—3 Wire—Line to Line	1.00
Single Phase—3 Wire—Line to Neutral	0.50
Three Phase—3 Wire—Line to Line	0.866
Three Phase—4 Wire—Line to Line	0.866
Three Phase—4 Wire—Line to Neutral	0.50

Note 2—Allowable voltage drops for systems other than single phase, two wire cannot be used directly in the above table. Such drops should be modified through multiplication by the appropriate factor listed below. The voltage thus modified may then be used to obtain the proper wire size directly from the table.

Multiplying Factor for Modification of Known Value to Permit Direct Use of Table

System for Which Allowable Voltage Drop is Known	
Single Phase—3 Wire—Line to Line	1.00
Single Phase—3 Wire—Line to Neutral	2.00
Three Phase—3 Wire—Line to Line	1.155
Three Phase—4 Wire—Line to Line	1.155
Three Phase—4 Wire—Line to Neutral	2.00

Note 3—The footage employed in the tabulated ampere feet refers to the length of run of the circuit rather than to the footage of individual conductor.

Note 4—The above table is figured at 60°C since this is an estimate of the average temperature which may be anticipated in service. The table may be used without significant error for conductor temperatures up to and including 75°C.

TABLE 3.59 Volts Drop for AL Conductor in Nonmagnetic Conduit—90 Percent PF

WIRE SIZE AWG or MCM — Ampere Feet	12*	10*	8*	6	4	2	1	1/0	2/0	3/0	4/0	250	300	350	400	500	600	700	750	800	900	1000
										Volts Drop												
600,000	—	—	—	—	—	290.0	234.0	189.0	153.0	125.0	102.0	88.7	76.5	67.5	61.1	51.5	45.7	41.2	39.4	37.8	35.1	33.0
400,000	—	—	—	—	363.0	232.0	187.0	151.0	122.0	100.0	81.6	71.0	61.2	54.0	48.8	41.2	36.5	32.9	31.5	30.2	28.1	26.4
300,000	—	—	—	427.0	272.0	174.0	140.0	113.0	91.8	75.0	61.2	53.3	45.9	40.8	36.6	30.9	27.4	24.6	23.6	22.8	21.0	19.8
200,000	—	—	438.0	284.0	182.0	116.0	93.6	75.6	61.2	50.0	40.8	35.6	30.6	27.2	24.4	20.6	18.2	16.4	15.8	15.2	14.0	13.2
100,000	—	346.0	219.0	142.0	90.8	58.0	46.8	37.8	30.6	25.0	20.4	17.7	15.3	13.6	12.2	10.3	9.1	8.2	7.9	7.6	7.0	6.6
90,000	484.0	312.0	197.0	128.0	81.7	52.2	42.1	34.0	27.5	22.5	18.4	16.0	13.8	12.2	11.0	9.3	8.2	7.4	7.1	6.8	6.3	5.9
80,000	438.0	277.0	175.0	114.0	72.6	46.4	37.4	30.2	24.5	20.0	16.3	14.2	12.2	10.8	9.8	8.2	7.3	6.6	6.3	6.0	5.5	5.2
70,000	384.0	242.0	153.0	99.6	63.5	40.6	32.8	26.4	21.4	17.5	14.3	12.4	10.7	9.5	8.6	7.2	6.4	5.8	5.5	5.3	4.9	4.6
60,000	329.0	208.0	132.0	85.4	54.4	34.8	28.1	22.5	18.4	15.0	12.2	10.6	9.2	8.1	7.3	6.1	5.5	4.9	4.7	4.5	4.2	3.9
50,000	274.0	173.0	109.0	71.1	45.4	29.0	23.4	18.9	15.3	12.5	10.2	8.9	7.7	6.8	6.1	5.2	4.6	4.1	3.9	3.8	3.6	3.3
40,000	219.0	138.0	87.6	56.8	36.3	23.2	18.7	15.1	12.2	10.0	8.2	7.1	6.1	5.4	4.9	4.1	3.7	3.3	3.2	3.0	2.9	2.6
30,000	165.0	104.0	65.7	42.7	27.2	17.4	14.1	11.3	9.2	7.5	6.1	5.3	4.6	4.1	3.7	3.1	2.7	2.5	2.4	2.3	2.1	2.0
20,000	110.0	69.2	43.8	28.4	18.2	11.6	9.4	7.6	6.1	5.0	4.1	3.6	3.1	2.7	2.4	2.1	1.9	1.7	1.6	1.5	1.4	1.3
10,000	54.9	34.6	21.9	14.2	9.1	5.8	4.7	3.8	3.1	2.5	2.0	1.8	1.5	1.4	1.2	1.0	0.9	0.8	0.8	0.8	0.7	0.7
9,000	49.4	31.2	19.7	12.8	8.2	5.2	4.2	3.4	2.8	2.3	1.8	1.6	1.4	1.2	1.1	0.9	0.8	0.7	0.7	0.7	0.6	0.6
8,000	43.8	27.7	17.5	11.4	7.3	4.6	3.7	3.0	2.5	2.0	1.7	1.4	1.2	1.1	1.0	0.8	0.7	0.7	0.6	0.6	0.6	0.5
7,000	38.4	24.2	15.3	10.4	6.4	4.1	3.3	2.7	2.1	1.8	1.4	1.2	1.1	1.0	0.9	0.7	0.6	0.6	0.5	0.5	0.5	0.5
6,000	32.9	20.8	13.2	8.5	5.4	3.5	2.8	2.3	1.8	1.5	1.2	1.1	0.9	0.8	0.7	0.6	0.6	0.5	0.5	0.5	0.4	0.4
5,000	27.4	17.3	10.9	7.1	4.5	2.9	2.3	1.9	1.5	1.3	1.0	0.9	0.8	0.7	0.6	0.5	0.5	0.4	0.4	0.4	0.3	0.3
4,000	21.9	13.8	8.8	5.7	3.6	2.3	1.9	1.5	1.2	1.0	0.8	0.7	0.6	0.5	0.5	0.4	0.4	0.3	0.3	0.3	0.3	0.2
3,000	16.5	10.4	6.6	4.3	2.7	1.7	1.4	1.1	0.9	0.8	0.6	0.5	0.5	0.4	0.4	0.3	0.2	0.2	0.2	0.2	0.2	0.2
2,000	11.0	6.9	4.4	2.8	1.8	1.2	0.9	0.7	0.6	0.5	0.4	0.4	0.3	0.3	0.2	0.2	0.2	0.2	0.1	0.2	0.1	0.1
1,000	5.5	3.5	2.2	1.4	0.9	0.6	0.5	0.4	0.3	0.3	0.2	0.2	0.2	0.1	0.1	0.1	0.1	0.1	—	0.1	0.1	0.1
900	4.9	3.1	2.0	1.3	0.8	0.5	0.4	0.3	0.3	0.2	0.2	0.2	0.1	0.1	0.1	0.1	—	—	—	—	—	—
800	4.4	2.8	1.8	1.1	0.7	0.5	0.3	0.3	0.3	0.2	0.2	0.2	0.1	0.1	0.1	—	—	—	—	—	—	—
700	3.8	2.4	1.5	1.0	0.6	0.4	0.3	0.3	0.2	0.2	0.2	0.1	0.1	0.1	0.1	—	—	—	—	—	—	—
600	3.3	2.1	1.3	0.9	0.5	0.4	0.3	0.2	0.2	0.2	0.1	0.1	0.1	0.1	0.1	—	—	—	—	—	—	—
500	2.7	1.7	1.1	0.7	0.5	0.3	0.2	0.2	0.2	0.1	0.1	0.1	0.1	0.1	0.1	—	—	—	—	—	—	—
400	2.2	1.4	0.9	0.6	0.4	0.2	0.2	0.1	0.1	0.1	0.1	0.1	0.1	—	—	—	—	—	—	—	—	—
300	1.6	1.0	0.7	0.4	0.3	0.2	0.1	0.1	0.1	0.1	0.1	—	—	—	—	—	—	—	—	—	—	—
200	1.1	0.7	0.4	0.3	0.2	0.1	0.1	0.1	0.1	—	—	—	—	—	—	—	—	—	—	—	—	—
100	0.6	0.4	0.2	0.1	0.1	0.1	0.1	—	—	—	—	—	—	—	—	—	—	—	—	—	—	—

* Solid Conductors. Other conductors are stranded.

Note 1— The above table gives voltage drops encountered in a single phase two-wire system. The voltage drops in other systems may be obtained through multiplication by appropriate factors listed below:

System for Which Voltage Drop is Desired	Multiplying Factors for Modification of Values in Table
Single Phase—3 Wire—Line to Line	1.00
Single Phase—3 Wire—Line to Neutral	0.50
Three Phase—3 Wire—Line to Line	0.866
Three Phase—4 Wire—Line to Line	0.866
Three Phase—4 Wire—Line to Neutral	0.50

Note 2— Allowable voltage drops for systems other than single phase, two wire cannot be used directly in the above table. Such drops should be modified through multiplication by the appropriate factor listed below. The voltage thus modified may then be used to obtain the proper wire size directly from the table.

System for Which Allowable Voltage Drop is Known	Multiplying Factor for Modification of Known Value to Permit Direct Use of Table
Single Phase—3 Wire—Line to Line	1.00
Single Phase—3 Wire—Line to Neutral	2.00
Three Phase—3 Wire—Line to Line	1.155
Three Phase—4 Wire—Line to Line	1.155
Three Phase—4 Wire—Line to Neutral	2.00

Note 3— The footage employed in the tabulated ampere feet refers to the length of run of the circuit rather than to the footage of individual conductor.

Note 4— The above table is figured at 60°C since this is an estimate of the average temperature which may be anticipated in service. The table may be used without significant error for conductor temperatures up to and including 75°C.

TABLE 3.60 Volts Drop for AL Conductor in Nonmagnetic Conduit—95 Percent PF

WIRE SIZE AWG or MCM — Volts Drop

Ampere Feet	1000	900	800	750	700	600	500	400	350	300	250	4/0	3/0	2/0	1/0	1	2	4	6	8*	10*	12*
500,000	30.0	32.2	34.9	36.6	38.6	43.0	49.2	59.3	65.9	75.3	88.1	103.0	126.0	156.0	194.0	241.0	301.0	473.0	—	—	—	—
400,000	24.0	25.8	27.9	29.3	30.9	34.4	39.4	47.4	52.7	60.2	70.5	82.4	100.8	125.0	155.0	193.0	241.0	378.0	—	—	—	—
300,000	18.0	19.3	21.0	22.0	23.2	25.8	29.5	35.6	39.5	45.2	52.9	61.8	75.6	93.6	116.0	145.0	181.0	284.0	447.0	—	—	—
200,000	12.0	12.9	14.0	14.6	15.4	17.2	19.7	23.7	26.4	30.1	35.2	41.2	50.4	62.4	77.6	96.4	120.0	189.0	298.0	460.0	—	—
100,000	6.0	6.4	7.0	7.3	7.7	8.6	9.8	11.9	13.2	15.1	17.6	20.6	25.2	31.2	38.8	48.2	60.2	94.6	149.0	230.0	364.0	—
90,000	5.4	5.8	6.3	6.6	6.9	7.7	8.9	10.7	11.9	13.6	15.9	18.5	22.7	28.1	34.9	43.4	54.2	85.1	134.0	207.0	329.0	—
80,000	4.8	5.2	5.6	5.9	6.2	6.9	7.9	9.5	10.5	12.0	14.1	16.5	20.2	25.0	31.0	38.6	48.2	75.6	119.0	184.0	291.0	462.0
70,000	4.2	4.5	4.9	5.1	5.4	6.0	6.9	8.3	9.2	10.5	12.3	14.4	17.6	21.8	27.2	33.7	42.1	66.2	104.0	161.0	255.0	404.0
60,000	3.6	3.9	4.2	4.4	4.6	5.2	5.9	7.1	7.9	9.0	10.6	12.4	15.1	18.7	23.3	28.9	36.1	56.8	89.3	138.0	218.0	346.0
50,000	3.0	3.2	3.5	3.7	3.9	4.3	4.9	5.9	6.6	7.5	8.8	10.3	12.6	15.6	19.4	24.1	30.1	47.3	74.4	115.0	182.0	289.0
40,000	2.4	2.6	2.8	2.9	3.1	3.4	3.9	4.7	5.3	6.0	7.0	8.2	10.1	12.5	15.5	19.3	24.1	37.8	59.5	92.0	146.0	231.0
30,000	1.8	1.9	2.1	2.2	2.3	2.6	3.0	3.6	4.0	4.5	5.3	6.2	7.6	9.4	11.6	14.5	18.1	28.4	44.7	69.0	109.0	173.0
20,000	1.2	1.3	1.4	1.5	1.5	1.7	2.0	2.4	2.6	3.0	3.5	4.1	5.0	6.2	7.8	9.6	12.0	18.9	29.8	46.0	72.8	115.0
10,000	0.6	0.6	0.7	0.7	0.8	0.9	1.0	1.2	1.3	1.5	1.8	2.1	2.5	3.1	3.9	4.8	6.0	9.5	14.9	23.0	36.4	57.7
9,000	0.5	0.6	0.6	0.7	0.7	0.8	0.9	1.1	1.2	1.4	1.6	1.9	2.3	2.8	3.5	4.3	5.4	8.5	13.4	20.7	32.8	52.0
8,000	0.5	0.5	0.6	0.6	0.6	0.7	0.8	1.0	1.1	1.2	1.4	1.6	2.0	2.5	3.1	3.9	4.8	7.6	11.9	18.4	29.1	46.2
7,000	0.4	0.5	0.5	0.5	0.5	0.6	0.7	0.8	0.9	1.1	1.2	1.4	1.8	2.2	2.8	3.4	4.2	6.6	10.4	16.1	25.5	40.4
6,000	0.4	0.4	0.4	0.4	0.5	0.5	0.6	0.7	0.8	0.9	1.1	1.2	1.5	1.9	2.3	2.9	3.6	5.7	8.9	13.8	21.8	34.6
5,000	0.3	0.3	0.4	0.4	0.4	0.4	0.5	0.6	0.7	0.8	0.9	1.0	1.3	1.6	1.9	2.4	3.0	4.7	7.4	11.5	18.2	28.9
4,000	0.2	0.2	0.3	0.3	0.3	0.3	0.4	0.5	0.5	0.6	0.7	0.8	1.0	1.2	1.5	1.9	2.4	3.8	6.0	9.2	14.6	23.1
3,000	0.2	0.2	0.2	0.2	0.2	0.3	0.3	0.4	0.4	0.5	0.5	0.6	0.8	0.9	1.2	1.5	1.8	2.8	4.5	6.9	10.9	17.3
2,000	0.1	0.1	0.1	0.1	0.2	0.2	0.2	0.2	0.3	0.3	0.4	0.4	0.5	0.6	0.8	1.0	1.2	1.9	3.0	4.6	7.3	11.5
1,000	0.1	0.1	0.1	0.1	0.1	0.1	0.1	0.1	0.1	0.2	0.2	0.2	0.3	0.3	0.4	0.5	0.6	0.9	1.5	2.3	3.6	5.8
900	0.1	0.1	0.1	0.1	0.1	0.1	0.1	0.1	0.1	0.1	0.2	0.2	0.2	0.3	0.3	0.4	0.5	0.9	1.3	2.1	3.3	5.2
800	—	0.1	0.1	0.1	0.1	0.1	0.1	0.1	0.1	0.1	0.1	0.2	0.2	0.2	0.3	0.4	0.5	0.8	1.2	1.8	2.9	4.6
700	—	—	—	0.1	0.1	0.1	0.1	0.1	0.1	0.1	0.1	0.1	0.2	0.2	0.3	0.3	0.4	0.7	1.0	1.6	2.6	4.0
600	—	—	—	—	—	0.1	0.1	0.1	0.1	0.1	0.1	0.1	0.2	0.2	0.2	0.3	0.4	0.6	0.9	1.4	2.2	3.5
500	—	—	—	—	—	—	—	0.1	0.1	0.1	0.1	0.1	0.1	0.2	0.2	0.2	0.3	0.5	0.7	1.2	1.8	2.9
400	—	—	—	—	—	—	—	—	0.1	0.1	0.1	0.1	0.1	0.1	0.2	0.2	0.2	0.4	0.6	0.9	1.5	2.3
300	—	—	—	—	—	—	—	—	—	—	0.1	0.1	0.1	0.1	0.1	0.1	0.2	0.3	0.4	0.7	1.1	1.7
200	—	—	—	—	—	—	—	—	—	—	—	—	0.1	0.1	0.1	0.1	0.1	0.2	0.3	0.5	0.7	1.2
100	—	—	—	—	—	—	—	—	—	—	—	—	—	—	—	—	0.1	0.1	0.1	0.2	0.4	0.6

* Solid Conductors. Other conductors are stranded.

Note 1—The above table gives voltage drops encountered in a single phase two-wire system. The voltage drops in other systems may be obtained through multiplication by appropriate factors listed below:

System for Which Voltage Drop is Desired	Multiplying Factors for Modification of Values in Table
Single Phase—3 Wire—Line to Line	1.00
Single Phase—3 Wire—Line to Neutral	0.50
Three Phase—3 Wire—Line to Line	0.866
Three Phase—4 Wire—Line to Line	0.866
Three Phase—4 Wire—Line to Neutral	0.50

Note 2—Allowable voltage drops for systems other than single phase, two wire cannot be used directly in the above table. Such drops should be modified through multiplication by the appropriate factor listed below. The voltage thus modified may then be used to obtain the proper wire size directly from the table.

System for Which Allowable Voltage Drop is Known	Multiplying Factor for Modification of Known Value to Permit Direct Use of Table
Single Phase—3 Wire—Line to Line	1.00
Single Phase—3 Wire—Line to Neutral	2.00
Three Phase—3 Wire—Line to Line	1.155
Three Phase—4 Wire—Line to Line	1.155
Three Phase—4 Wire—Line to Neutral	2.00

Note 3—The footage employed in the tabulated ampere feet refers to the length of run of the circuit rather than to the footage of individual conductor.

Note 4—The above table is figured at 60°C since this is an estimate of the average temperature which may be anticipated in service. The table may be used without significant error for conductor temperatures up to and including 75°C.

TABLE 3.61 Volts Drop for AL Conductor in Nonmagnetic Conduit—100 Percent PF

WIRE SIZE AWG or MCM	1000	900	800	750	700	600	500	400	350	300	250	4/0	3/0	2/0	1/0	1	2	4	6	8*	10*	12*
Ampere Feet	Volts Drop																					
500,000	20.7	22.8	25.6	27.3	29.2	33.9	40.4	50.7	57.5	67.2	80.5	95.9	120.0	151.0	191.0	240.0	303.0	483.0	—	—	—	—
400,000	16.5	18.2	20.5	21.8	23.4	27.1	32.3	40.6	46.0	53.7	64.4	76.8	96.0	121.0	153.0	192.0	241.0	386.0	—	—	—	—
300,000	12.4	13.7	15.3	16.4	17.5	20.3	24.3	30.4	34.5	40.6	48.3	57.6	72.0	90.6	115.0	144.0	182.0	290.0	460.0	—	—	—
200,000	8.2	9.1	10.2	10.9	11.7	13.5	16.2	20.3	23.0	26.9	32.2	38.4	48.0	60.4	76.4	96.0	121.0	193.0	307.0	478.0	—	—
100,000	4.1	4.6	5.1	5.5	5.8	6.8	8.1	10.1	11.5	13.4	16.1	19.2	24.0	30.2	38.2	48.0	60.6	96.6	153.0	239.0	380.0	—
90,000	3.7	4.1	4.6	4.9	5.3	6.1	7.3	9.1	10.3	12.1	14.5	17.3	21.6	27.2	34.4	43.2	54.6	87.0	138.0	215.0	342.0	—
80,000	3.3	3.7	4.1	4.4	4.7	5.4	6.5	8.1	9.2	10.8	12.9	15.3	19.2	24.2	30.5	38.4	48.5	77.3	123.0	191.0	304.0	483.0
70,000	2.9	3.2	3.6	3.8	4.1	4.8	5.7	7.1	8.1	9.4	11.3	13.4	16.8	21.1	27.6	33.6	42.4	67.6	107.0	167.0	266.0	423.0
60,000	2.5	2.7	3.1	3.3	3.5	4.1	4.9	6.1	6.9	8.1	9.7	11.5	14.4	18.1	22.9	28.8	36.4	58.0	92.0	144.0	228.0	362.0
50,000	2.1	2.3	2.6	2.7	2.9	3.4	4.0	5.1	5.8	6.7	8.1	9.6	12.0	15.1	19.1	24.0	30.3	48.3	76.7	120.0	190.0	302.0
40,000	1.7	1.8	2.1	2.2	2.3	2.7	3.2	4.1	4.6	5.4	6.4	7.7	9.6	12.1	15.3	19.2	24.1	38.6	61.4	95.6	152.0	242.0
30,000	1.2	1.4	1.5	1.6	1.8	2.0	2.4	3.0	3.5	4.1	4.8	5.8	7.2	9.1	11.5	14.4	18.2	29.0	46.0	71.7	114.0	181.0
20,000	0.8	0.9	1.0	1.1	1.2	1.4	1.6	2.0	2.3	2.7	3.2	3.8	4.8	6.0	7.6	9.6	12.1	19.3	30.7	47.8	76.0	121.0
10,000	0.4	0.5	0.5	0.6	0.6	0.7	0.8	1.0	1.2	1.3	1.6	1.9	2.4	3.0	3.8	4.8	6.1	9.7	15.3	23.9	38.0	60.4
9,000	0.4	0.4	0.5	0.5	0.5	0.6	0.7	0.9	1.0	1.2	1.5	1.7	2.2	2.7	3.4	4.3	5.5	8.7	13.8	21.5	34.2	54.4
8,000	0.3	0.4	0.4	0.4	0.5	0.5	0.7	0.8	0.9	1.1	1.3	1.5	1.9	2.4	3.1	3.8	4.9	7.7	12.3	19.1	30.4	48.3
7,000	0.3	0.3	0.4	0.4	0.4	0.5	0.6	0.7	0.8	0.9	1.1	1.3	1.7	2.1	2.8	3.4	4.2	6.8	10.7	16.7	26.6	42.3
6,000	0.3	0.3	0.3	0.3	0.4	0.4	0.5	0.6	0.7	0.8	1.0	1.2	1.4	1.8	2.3	2.9	3.6	5.8	9.2	14.4	22.8	36.2
5,000	0.2	0.2	0.3	0.3	0.3	0.3	0.4	0.5	0.6	0.7	0.8	1.0	1.2	1.5	1.9	2.4	3.0	4.8	7.7	12.0	19.0	30.2
4,000	0.2	0.2	0.2	0.2	0.2	0.3	0.3	0.4	0.5	0.5	0.6	0.7	1.0	1.2	1.5	1.9	2.4	3.9	6.1	9.6	15.2	24.2
3,000	0.1	0.1	0.2	0.2	0.2	0.2	0.2	0.3	0.4	0.4	0.5	0.6	0.7	0.9	1.2	1.4	1.8	2.9	4.6	7.2	11.4	18.1
2,000	0.1	0.1	0.1	0.1	0.1	0.1	0.2	0.2	0.2	0.3	0.3	0.4	0.5	0.6	0.8	1.0	1.2	1.9	3.1	4.8	7.6	12.1
1,000	—	0.1	0.1	0.1	0.1	0.1	0.1	0.1	0.1	0.1	0.2	0.2	0.2	0.3	0.4	0.5	0.6	1.0	1.5	2.4	3.8	6.0
900	—	—	0.1	0.1	0.1	0.1	0.1	0.1	0.1	0.1	0.2	0.2	0.2	0.3	0.3	0.4	0.6	0.9	1.4	2.2	3.4	5.4
800	—	—	—	—	0.1	0.1	0.1	0.1	0.1	0.1	0.1	0.2	0.2	0.2	0.3	0.4	0.5	0.8	1.2	1.9	3.0	4.8
700	—	—	—	—	—	0.1	0.1	0.1	0.1	0.1	0.1	0.1	0.2	0.2	0.3	0.3	0.4	0.7	1.1	1.7	2.7	4.2
600	—	—	—	—	—	—	0.1	0.1	0.1	0.1	0.1	0.1	0.1	0.2	0.2	0.3	0.4	0.6	0.9	1.4	2.3	3.6
500	—	—	—	—	—	—	—	0.1	0.1	0.1	0.1	0.1	0.1	0.2	0.2	0.2	0.3	0.5	0.8	1.2	1.9	3.0
400	—	—	—	—	—	—	—	—	0.1	0.1	0.1	0.1	0.1	0.1	0.2	0.2	0.2	0.3	0.6	1.0	1.5	2.4
300	—	—	—	—	—	—	—	—	—	—	0.1	0.1	0.1	0.1	0.1	0.1	0.2	0.3	0.5	0.7	1.1	1.8
200	—	—	—	—	—	—	—	—	—	—	—	—	0.1	0.1	0.1	0.1	0.1	0.2	0.3	0.5	0.8	1.2
100	—	—	—	—	—	—	—	—	—	—	—	—	—	—	—	0.1	0.1	0.1	0.2	0.2	0.4	0.6

* Solid Conductors. Other conductors are stranded.

Note 1—The above table gives voltage drops encountered in a single phase two-wire system. The voltage drops in other systems may be obtained through multiplication by appropriate factors listed below:

System for Which Voltage Drop is Desired	Multiplying Factors for Modification of Values in Table
Single Phase—3 Wire—Line to Line	1.00
Single Phase—3 Wire—Line to Neutral	0.50
Three Phase—3 Wire—Line to Line	0.866
Three Phase—4 Wire—Line to Line	0.866
Three Phase—4 Wire—Line to Neutral	0.50

Note 2—Allowable voltage drops for systems other than single phase, two wire cannot be used directly in the above table. Such drops should be modified through multiplication by the appropriate factor listed below. The voltage thus modified may then be used to obtain the proper wire size directly from the table.

System for Which Allowable Voltage Drop is Known	Multiplying Factor for Modification of Known Value to Permit Direct Use of Table
Single Phase—3 Wire—Line to Line	1.00
Single Phase—3 Wire—Line to Neutral	2.00
Three Phase—3 Wire—Line to Line	1.155
Three Phase—4 Wire—Line to Line	1.155
Three Phase—4 Wire—Line to Neutral	2.00

Note 3—The footage employed in the tabulated ampere feet refers to the length of run of the circuit rather than to the footage of individual conductor.

Note 4—The above table is figured at 60°C since this is an estimate of the average temperature which may be anticipated in service. The table may be used without significant error for conductor temperatures up to and including 75°C.

TABLE 3.62 Volts Drop for CU Conductor—Direct Current

WIRE SIZE AWG or MCM (column headings) · Volts Drop (body) · Ampere Feet (row labels)

Ampere Feet	1000	900	800	750	700	600	500	400	350	300	250	4/0	3/0	2/0	1/0	1	2	4	6	8*	10*	12*	14*
500,000	13.0	14.0	16.0	17.0	18.0	21.0	26.0	32.0	36.0	42.0	50.0	59.0	76.0	96.0	121.0	152.0	192.0	305.0	483.0	—	—	—	—
400,000	10.4	11.2	12.8	13.6	14.4	16.8	20.8	25.6	28.8	33.6	40.0	47.2	60.8	76.8	96.8	122.0	153.0	244.0	386.0	—	—	—	—
300,000	7.8	8.4	9.6	10.2	10.8	12.6	15.6	19.2	21.6	25.2	30.0	35.4	45.6	57.6	72.6	91.2	115.0	183.0	290.0	450.0	—	—	—
200,000	5.2	5.6	6.4	6.8	7.2	8.4	10.4	12.8	14.4	16.8	20.0	23.6	30.4	38.4	48.4	60.8	76.8	122.0	193.0	300.0	480.0	—	—
100,000	2.6	2.8	3.2	3.4	3.6	4.2	5.2	6.4	7.2	8.4	10.0	11.8	15.2	19.2	24.2	30.4	38.4	61.2	96.6	150.0	240.0	384.0	—
90,000	2.3	2.5	2.9	3.1	3.2	3.8	4.7	5.8	6.5	7.6	9.0	10.6	13.7	17.3	21.8	27.4	34.6	55.0	87.0	135.0	216.0	345.0	547.0
80,000	2.1	2.2	2.6	2.7	2.9	3.4	4.2	5.1	5.8	6.7	8.0	9.4	12.2	15.4	19.4	24.3	30.7	49.0	77.3	120.0	192.0	307.0	486.0
70,000	1.8	2.0	2.2	2.4	2.5	2.9	3.6	4.5	5.0	5.9	7.0	8.3	10.6	13.4	16.9	21.3	26.9	42.8	67.6	105.0	168.0	269.0	426.0
60,000	1.6	1.7	1.9	2.0	2.2	2.5	3.1	3.8	4.3	5.0	6.0	7.1	9.1	11.5	14.5	18.2	23.0	36.7	58.0	90.0	144.0	230.0	365.0
50,000	1.3	1.4	1.6	1.7	1.8	2.1	2.6	3.2	3.6	4.2	5.0	5.9	7.6	9.6	12.1	15.2	19.2	30.6	48.3	75.0	120.0	192.0	304.0
40,000	1.0	1.1	1.3	1.4	1.4	1.7	2.1	2.6	2.9	3.4	4.0	4.7	6.1	7.7	9.7	12.2	15.4	24.4	38.6	60.0	96.0	154.0	243.0
30,000	0.8	0.8	1.0	1.0	1.1	1.3	1.6	1.9	2.2	2.5	3.0	3.5	4.6	5.8	7.3	9.1	11.5	18.3	29.0	45.0	72.0	115.0	182.0
20,000	0.5	0.6	0.6	0.7	0.7	0.8	1.0	1.3	1.4	1.7	2.0	2.4	3.0	3.8	4.8	6.1	7.7	12.2	19.3	30.0	48.0	76.8	122.0
10,000	0.3	0.3	0.3	0.3	0.4	0.4	0.5	0.6	0.7	0.8	1.0	1.2	1.5	1.9	2.4	3.0	3.8	6.1	9.7	15.0	24.0	38.4	60.8
9,000	0.2	0.3	0.3	0.3	0.3	0.4	0.5	0.6	0.6	0.8	0.9	1.1	1.4	1.7	2.2	2.7	3.5	5.5	8.7	13.5	21.6	34.6	54.7
8,000	0.2	0.2	0.3	0.3	0.3	0.3	0.4	0.5	0.6	0.7	0.8	0.9	1.2	1.5	1.9	2.4	3.1	4.9	7.7	12.0	19.2	30.7	48.7
7,000	0.2	0.2	0.2	0.2	0.3	0.3	0.4	0.4	0.5	0.6	0.7	0.8	1.1	1.3	1.7	2.1	2.7	4.3	6.8	10.5	16.8	26.9	42.6
6,000	0.2	0.2	0.2	0.2	0.2	0.3	0.3	0.4	0.4	0.5	0.6	0.7	0.9	1.2	1.5	1.8	2.3	3.7	5.8	9.0	14.4	23.0	36.5
5,000	0.1	0.1	0.2	0.2	0.2	0.2	0.3	0.3	0.4	0.4	0.5	0.6	0.8	1.0	1.2	1.5	1.9	3.1	4.8	7.5	12.0	19.2	30.4
4,000	0.1	0.1	0.1	0.1	0.1	0.2	0.2	0.3	0.3	0.3	0.4	0.5	0.6	0.8	1.0	1.2	1.5	2.4	3.9	6.0	9.6	15.4	24.3
3,000	0.1	0.1	0.1	0.1	0.1	0.1	0.2	0.2	0.2	0.3	0.3	0.4	0.5	0.6	0.7	0.9	1.2	1.8	2.9	4.5	7.2	11.5	18.2
2,000	0.1	0.1	0.1	0.1	0.1	0.1	0.1	0.1	0.1	0.2	0.2	0.2	0.3	0.4	0.5	0.6	0.8	1.2	1.9	3.0	4.8	7.7	12.2
1,000	—	—	—	—	—	—	0.1	0.1	0.1	0.1	0.1	0.1	0.2	0.2	0.2	0.3	0.4	0.6	1.0	1.5	2.4	3.8	6.1
900	—	—	—	—	—	—	—	0.1	0.1	0.1	0.1	0.1	0.1	0.2	0.2	0.3	0.3	0.6	0.9	1.4	2.2	3.5	5.5
800	—	—	—	—	—	—	—	0.1	0.1	0.1	0.1	0.1	0.1	0.2	0.2	0.2	0.3	0.5	0.8	1.2	1.9	3.1	4.9
700	—	—	—	—	—	—	—	—	0.1	0.1	0.1	0.1	0.1	0.1	0.2	0.2	0.3	0.4	0.7	1.1	1.7	2.7	4.3
600	—	—	—	—	—	—	—	—	—	0.1	0.1	0.1	0.1	0.1	0.1	0.2	0.2	0.4	0.6	0.9	1.4	2.3	3.7
500	—	—	—	—	—	—	—	—	—	—	0.1	0.1	0.1	0.1	0.1	0.2	0.2	0.3	0.5	0.8	1.2	1.9	3.0
400	—	—	—	—	—	—	—	—	—	—	—	—	0.1	0.1	0.1	0.1	0.2	0.2	0.4	0.6	1.0	1.5	2.4
300	—	—	—	—	—	—	—	—	—	—	—	—	—	0.1	0.1	0.1	0.1	0.2	0.3	0.5	0.7	1.2	1.8
200	—	—	—	—	—	—	—	—	—	—	—	—	—	—	—	0.1	0.1	0.1	0.2	0.3	0.5	0.8	1.2
100	—	—	—	—	—	—	—	—	—	—	—	—	—	—	—	—	—	0.1	0.1	0.2	0.2	0.4	0.6

* Solid Conductors. Other conductors are stranded.

Note 1—The footage employed in the tabulated ampere feet refers to the length of run of the circuit rather than to the footage of individual conductor.

Note 2—The above table is figured at 60°C since this is an estimate of the average temperature which may be anticipated in service. The table may be used without significant error for conductor temperatures up to and including 75°C.

TABLE 3.63 Volts Drop for CU Conductor in Magnetic Conduit—70 Percent PF

WIRE SIZE AWG or MCM / Ampere Feet	14*	12*	10*	8*	6	4	2	1	1/0	2/0	3/0	4/0	250	300	350	400	500	600	700	750	800	900	1000
600,000	—	—	—	466.0	380.0	254.0	173.0	144.0	122.0	104.0	89.0	77.0	71.0	64.0	60.0	56.0	51.0	48.0	46.0	45.0	44.0	42.0	41.0
400,000	—	—	—	311.0	253.0	169.0	115.0	96.0	81.0	69.0	59.0	51.0	47.0	43.0	40.0	37.0	34.0	32.0	31.0	30.0	29.0	28.0	27.0
300,000	—	—	—	233.0	190.0	127.0	86.0	72.0	61.0	52.0	44.0	38.0	35.5	32.0	30.0	28.0	25.5	24.0	23.0	22.5	22.0	21.0	20.5
200,000	—	—	354.0	155.0	127.0	85.0	58.0	48.0	41.0	35.0	30.0	26.0	23.7	21.3	20.0	18.7	17.0	16.0	15.3	15.0	14.7	14.0	13.7
100,000	436.0	278.0	177.0	78.0	63.3	42.3	28.8	24.0	20.3	17.3	14.8	12.8	11.8	10.7	10.0	9.3	8.5	8.0	7.7	7.5	7.3	7.0	6.8
90,000	392.0	250.0	159.0	69.9	57.0	38.1	25.9	21.6	18.3	15.6	13.3	11.5	10.6	9.6	9.0	8.4	7.7	7.2	6.9	6.8	6.6	6.3	6.1
80,000	349.0	222.0	142.0	62.2	50.6	33.8	23.0	19.2	16.2	13.8	11.8	10.2	9.5	8.5	8.0	7.5	6.8	6.4	6.1	6.0	5.9	5.6	5.5
70,000	305.0	194.0	124.0	54.4	44.3	29.6	20.2	16.8	14.2	12.1	10.4	9.0	8.3	7.5	7.0	6.5	6.0	5.6	5.4	5.3	5.1	4.9	4.8
60,000	262.0	167.0	106.0	46.6	38.0	25.4	17.3	14.4	12.2	10.4	8.9	7.7	7.1	6.4	6.0	5.6	5.1	4.8	4.6	4.5	4.4	4.2	4.1
50,000	218.0	139.0	88.5	38.9	31.7	21.2	14.4	12.0	10.2	8.7	7.4	6.4	5.9	5.3	5.0	4.7	4.3	4.0	3.8	3.8	3.7	3.5	3.4
40,000	174.0	111.0	70.8	31.1	25.3	16.9	11.5	9.6	8.1	6.9	5.9	5.1	4.7	4.3	4.0	3.7	3.4	3.2	3.1	3.0	2.9	2.8	2.7
30,000	131.0	83.0	53.1	23.3	19.0	12.7	8.6	7.2	6.1	5.2	4.4	3.8	3.5	3.2	3.0	2.8	2.6	2.4	2.3	2.3	2.2	2.1	2.0
20,000	87.2	55.6	35.4	15.5	12.7	8.5	5.8	4.8	4.1	3.5	3.0	2.6	2.4	2.1	2.0	1.9	1.7	1.6	1.5	1.5	1.5	1.4	1.4
10,000	43.6	27.8	17.7	7.8	6.3	4.2	2.9	2.4	2.0	1.7	1.5	1.3	1.2	1.1	1.0	0.9	0.9	0.8	0.8	0.8	0.7	0.7	0.7
9,000	39.2	25.0	15.9	7.0	5.7	3.8	2.6	2.2	1.8	1.6	1.3	1.2	1.1	1.0	0.9	0.8	0.8	0.7	0.7	0.7	0.7	0.6	0.6
8,000	34.9	22.2	14.2	6.2	5.1	3.4	2.3	1.9	1.6	1.4	1.2	1.0	1.0	0.9	0.8	0.7	0.7	0.6	0.6	0.6	0.6	0.6	0.5
7,000	30.5	19.4	12.4	5.4	4.4	3.0	2.0	1.7	1.4	1.2	1.0	0.9	0.8	0.7	0.7	0.7	0.6	0.6	0.5	0.5	0.5	0.5	0.5
6,000	26.2	16.7	10.6	4.7	3.8	2.5	1.7	1.4	1.2	1.0	0.9	0.8	0.7	0.6	0.6	0.6	0.5	0.5	0.5	0.5	0.4	0.4	0.4
5,000	21.8	13.9	8.9	3.9	3.2	2.1	1.4	1.2	1.0	0.9	0.7	0.6	0.6	0.5	0.5	0.5	0.4	0.4	0.4	0.4	0.4	0.4	0.3
4,000	17.4	11.1	7.1	3.1	2.5	1.7	1.2	1.0	0.8	0.7	0.6	0.5	0.5	0.4	0.4	0.4	0.3	0.3	0.3	0.3	0.3	0.3	0.3
3,000	13.1	8.3	5.3	2.3	1.9	1.3	0.9	0.7	0.6	0.5	0.4	0.4	0.4	0.3	0.3	0.3	0.3	0.2	0.2	0.2	0.2	0.2	0.2
2,000	8.7	5.6	3.5	1.6	1.3	0.8	0.6	0.5	0.4	0.3	0.3	0.3	0.2	0.2	0.2	0.2	0.2	0.2	0.2	0.2	0.1	0.1	0.1
1,000	4.4	2.8	1.8	0.8	0.6	0.4	0.3	0.2	0.2	0.2	0.1	0.1	0.1	0.1	0.1	0.1	0.1	0.1	0.1	0.1	0.1	0.1	0.1
900	3.9	2.5	1.6	0.7	0.6	0.4	0.3	0.2	0.2	0.2	0.1	0.1	0.1	0.1	0.1	0.1	0.1	0.1	0.1	0.1	0.1	0.1	0.1
800	3.5	2.2	1.4	0.6	0.5	0.3	0.2	0.2	0.2	0.1	0.1	0.1	0.1	0.1	0.1	0.1	0.1	0.1	0.1	0.1	0.1	0.1	0.1
700	3.1	1.9	1.2	0.5	0.4	0.3	0.2	0.2	0.1	0.1	0.1	0.1	0.1	0.1	0.1	0.1	0.1	0.1	0.1	0.1	0.1	0.1	0.1
600	2.6	1.7	1.1	0.5	0.4	0.3	0.2	0.1	0.1	0.1	0.1	0.1	0.1	0.1	0.1	0.1	—	—	—	—	—	—	—
500	2.2	1.4	0.9	0.4	0.3	0.2	0.1	0.1	0.1	0.1	0.1	0.1	0.1	0.1	0.1	—	—	—	—	—	—	—	—
400	1.7	1.1	0.7	0.3	0.3	0.2	0.1	0.1	0.1	0.1	0.1	0.1	—	—	—	—	—	—	—	—	—	—	—
300	1.3	0.8	0.5	0.2	0.2	0.1	0.1	0.1	0.1	0.1	—	—	—	—	—	—	—	—	—	—	—	—	—
200	0.9	0.6	0.4	0.2	0.1	0.1	0.1	—	—	—	—	—	—	—	—	—	—	—	—	—	—	—	—
100	0.4	0.3	0.2	0.1	0.1	—	—	—	—	—	—	—	—	—	—	—	—	—	—	—	—	—	—

* Solid Conductors. Other conductors are stranded.

Note 1—The above table gives voltage drops encountered in a single phase two-wire system. The voltage drops in other systems may be obtained through multiplication by appropriate factors listed below.

System for Which Voltage Drop is Desired	Multiplying Factors for Modification of Values in Table
Single Phase—3 Wire—Line to Line	1.00
Single Phase—3 Wire—Line to Neutral	0.50
Three Phase—3 Wire—Line to Line	0.866
Three Phase—4 Wire—Line to Line	0.866
Three Phase—4 Wire—Line to Neutral	0.50

Note 2—Allowable voltage drops for systems other than single phase, two-wire cannot be used directly in the above table. Such drops should be modified through multiplication by the appropriate factor listed below. The voltage thus modified may then be used to obtain the proper wire size directly from the table.

System for Which Allowable Voltage Drop is Known	Multiplying Factor for Modification of Known Value to Permit Direct Use of Table
Single Phase—3 Wire—Line to Line	1.00
Single Phase—3 Wire—Line to Neutral	2.00
Three Phase—3 Wire—Line to Line	1.155
Three Phase—4 Wire—Line to Line	1.155
Three Phase—4 Wire—Line to Neutral	2.00

Note 3—The footage employed in the tabulated ampere feet refers to the length of run of the circuit rather than to the footage of individual conductor.

Note 4—The above table is figured at 60°C since this is an estimate of the average temperature which may be anticipated in service. The table may be used without significant error for conductor temperatures up to and including 75°C.

TABLE 3.64 Volts Drop for CU Conductor in Magnetic Conduit—80 Percent PF

Columns 1000 through 14* below are the **Volts Drop** values. Rows are labeled by **Ampere Feet** in the WIRE SIZE AWG or MCM column.

WIRE SIZE AWG or MCM (Ampere Feet)	1000	900	800	750	700	600	500	400	350	300	250	4/0	3/0	2/0	1/0	1	2	4	6	8*	10*	12*	14*
500,000	38.0	39.0	41.0	42.0	43.0	46.0	49.0	55.0	58.0	64.0	71.0	78.0	92.0	108.0	130.0	153.0	186.0	278.0	421.0	—	—	—	—
400,000	30.4	31.2	32.8	33.6	34.4	36.8	39.2	44.0	46.4	51.2	56.8	62.4	73.6	86.4	104.0	122.0	148.0	222.0	336.0	—	—	—	—
300,000	22.8	23.4	24.6	25.2	25.8	27.6	29.4	33.0	34.8	38.4	42.6	46.8	55.2	64.8	78.0	91.8	111.0	166.0	252.0	381.0	—	—	—
200,000	15.2	15.6	16.4	16.8	17.2	18.4	19.6	22.0	23.2	25.6	28.4	31.2	36.8	43.2	52.0	61.2	74.0	111.0	168.0	254.0	398.0	—	—
100,000	7.6	7.8	8.2	8.4	8.6	9.2	9.8	11.0	11.6	12.8	14.2	15.6	18.4	21.6	26.0	30.6	37.2	55.6	84.2	127.0	199.0	314.0	494.0
90,000	6.9	7.0	7.4	7.6	7.7	8.3	8.8	9.9	10.4	11.5	12.8	14.0	16.6	19.4	23.4	27.5	33.5	50.0	75.8	115.8	179.0	283.0	445.0
80,000	6.1	6.2	6.6	6.7	6.9	7.4	7.8	8.8	9.3	10.2	11.4	12.5	14.7	17.3	20.8	24.5	29.8	44.5	67.4	102.0	160.0	252.0	395.0
70,000	5.3	5.5	5.7	5.9	6.0	6.4	6.9	7.7	8.1	8.9	9.9	10.9	12.9	15.1	18.2	21.4	26.0	38.9	58.9	89.3	140.0	220.0	346.0
60,000	4.6	4.7	4.9	5.0	5.2	5.5	5.9	6.6	6.9	7.7	8.5	9.4	11.0	12.9	15.6	18.4	22.3	33.4	50.5	76.5	120.0	188.0	296.0
50,000	3.8	3.9	4.1	4.2	4.3	4.6	4.9	5.5	5.8	6.4	7.1	7.8	9.2	10.8	13.0	15.3	18.6	27.8	42.1	63.7	99.7	157.0	247.0
40,000	3.0	3.1	3.3	3.3	3.4	3.6	3.9	4.4	4.6	5.1	5.7	6.2	7.4	8.6	10.4	12.2	14.8	22.2	33.6	50.9	79.6	126.0	198.0
30,000	2.3	2.3	2.5	2.5	2.6	2.8	2.9	3.3	3.5	3.8	4.3	4.7	5.5	6.5	7.8	9.2	11.1	16.7	25.2	38.1	59.7	94.2	149.0
20,000	1.5	1.5	1.6	1.7	1.7	1.8	2.0	2.2	2.3	2.6	2.8	3.1	3.7	4.3	5.2	6.1	7.4	11.1	16.8	25.4	39.8	62.8	98.8
10,000	0.8	0.8	0.8	0.8	0.9	0.9	1.0	1.1	1.2	1.3	1.4	1.6	1.8	2.2	2.6	3.1	3.7	5.6	8.4	12.7	19.9	31.4	49.4
9,000	0.7	0.7	0.7	0.8	0.8	0.8	0.9	1.0	1.0	1.2	1.3	1.4	1.7	1.9	2.3	2.8	3.4	5.0	7.6	11.5	17.9	28.3	44.5
8,000	0.6	0.6	0.7	0.7	0.7	0.7	0.8	0.9	0.9	1.0	1.1	1.2	1.5	1.7	2.1	2.5	3.0	4.5	6.7	10.2	16.0	25.2	39.5
7,000	0.5	0.5	0.6	0.6	0.6	0.6	0.7	0.8	0.8	0.9	1.0	1.1	1.3	1.5	1.8	2.1	2.6	3.9	5.9	8.9	14.0	22.0	34.6
6,000	0.5	0.5	0.5	0.5	0.5	0.6	0.6	0.7	0.7	0.8	0.9	0.9	1.1	1.3	1.6	1.8	2.2	3.3	5.1	7.7	12.0	18.8	29.6
5,000	0.4	0.4	0.4	0.4	0.4	0.5	0.5	0.6	0.6	0.6	0.7	0.8	0.9	1.1	1.3	1.5	1.9	2.8	4.2	6.4	10.0	15.7	24.7
4,000	0.3	0.3	0.3	0.3	0.3	0.4	0.4	0.4	0.5	0.5	0.6	0.6	0.7	0.9	1.0	1.2	1.5	2.2	3.4	5.1	8.0	12.6	19.8
3,000	0.2	0.2	0.2	0.2	0.3	0.3	0.3	0.3	0.3	0.4	0.4	0.5	0.6	0.6	0.8	0.9	1.1	1.7	2.5	3.8	6.0	9.4	14.8
2,000	0.2	0.2	0.2	0.2	0.2	0.2	0.2	0.2	0.2	0.3	0.3	0.3	0.4	0.4	0.5	0.6	0.7	1.1	1.7	2.5	4.0	6.3	9.9
1,000	0.1	0.1	0.1	0.1	0.1	0.1	0.1	0.1	0.1	0.1	0.1	0.2	0.2	0.2	0.3	0.3	0.4	0.6	0.8	1.3	2.0	3.1	4.9
900	0.1	0.1	0.1	0.1	0.1	0.1	0.1	0.1	0.1	0.1	0.1	0.1	0.2	0.2	0.2	0.3	0.3	0.5	0.8	1.2	1.8	2.8	4.5
800	0.1	0.1	0.1	0.1	0.1	0.1	0.1	0.1	0.1	0.1	0.1	0.1	0.1	0.2	0.2	0.2	0.3	0.4	0.7	1.0	1.6	2.5	3.9
700	0.1	0.1	0.1	0.1	0.1	0.1	0.1	0.1	0.1	0.1	0.1	0.1	0.1	0.1	0.2	0.2	0.3	0.4	0.6	0.9	1.4	2.2	3.5
600	—	—	—	0.1	0.1	0.1	0.1	0.1	0.1	0.1	0.1	0.1	0.1	0.1	0.2	0.2	0.2	0.3	0.5	0.8	1.2	1.9	2.9
500	—	—	—	—	—	—	—	0.1	0.1	0.1	0.1	0.1	0.1	0.1	0.1	0.2	0.2	0.3	0.4	0.6	1.0	1.6	2.5
400	—	—	—	—	—	—	—	—	—	0.1	0.1	0.1	0.1	0.1	0.1	0.1	0.2	0.2	0.3	0.5	0.8	1.3	2.0
300	—	—	—	—	—	—	—	—	—	—	—	—	0.1	0.1	0.1	0.1	0.1	0.2	0.3	0.4	0.6	0.9	1.5
200	—	—	—	—	—	—	—	—	—	—	—	—	—	—	0.1	0.1	0.1	0.1	0.2	0.3	0.4	0.6	1.0
100	—	—	—	—	—	—	—	—	—	—	—	—	—	—	—	—	—	0.1	0.1	0.1	0.2	0.3	0.5

* Solid Conductors. Other conductors are stranded.

Note 1—The above table gives voltage drops encountered in a single phase two-wire system. The voltage drops in other systems may be obtained through multiplication by appropriate factors listed below:

Multiplying Factors for Modification of Values in Table

System for Which Voltage Drop is Desired	
Single Phase—3 Wire—Line to Line	1.00
Single Phase—3 Wire—Line to Neutral	0.50
Three Phase—3 Wire—Line to Line	0.866
Three Phase—4 Wire—Line to Line	0.866
Three Phase—4 Wire—Line to Neutral	0.50

Note 2—Allowable voltage drops for systems other than single phase, two-wire cannot be used directly in the above table. Such drops should be modified through multiplication by the appropriate factor listed below. The voltage thus modified may then be used to obtain the proper wire size directly from the table.

System for Which Allowable Voltage Drop is Known	Multiplying Factor for Modification of Known Value to Permit Direct Use of Table
Single Phase—3 Wire—Line to Line	1.00
Single Phase—3 Wire—Line to Neutral	2.00
Three Phase—3 Wire—Line to Line	1.155
Three Phase—4 Wire—Line to Line	1.155
Three Phase—4 Wire—Line to Neutral	2.00

Note 3—The footage employed in the tabulated ampere feet refers to the length of run of the circuit rather than to the footage of individual conductor.

Note 4—The above table is figured at 60°C since this is an estimate of the average temperature which may be anticipated in service. The table may be used without significant error for conductor temperatures up to and including 75°C.

TABLE 3.65 Volts Drop for CU Conductor in Magnetic Conduit—90 Percent PF

Ampere Feet / WIRE SIZE AWG or MCM	14*	12*	10*	8*	6	4	2	1	1/0	2/0	3/0	4/0	250	300	350	400	500	600	700	750	800	900	1000
											Volts Drop												
500,000	—	—	—	—	461.0	300.0	198.0	161.0	133.0	110.0	91.0	76.0	68.0	61.0	55.0	51.0	45.0	41.0	38.0	37.0	36.0	34.0	33.0
400,000	—	—	—	—	369.0	240.0	158.0	129.0	106.0	88.0	72.8	60.8	54.4	48.8	44.0	40.8	36.0	32.8	30.4	29.6	28.8	26.2	26.4
300,000	—	—	—	—	277.0	180.0	119.0	96.6	80.0	66.0	54.6	45.6	40.8	36.6	33.0	30.6	27.0	24.6	22.8	22.2	21.6	20.4	19.8
200,000	—	—	442.0	420.0	184.0	120.0	79.2	64.4	53.2	44.0	36.4	30.4	27.2	24.4	22.0	20.4	18.0	16.4	15.2	14.8	14.4	13.6	13.2
100,000	—	351.0	221.0	140.0	92.2	60.0	39.6	32.2	26.6	22.0	18.2	15.2	13.6	12.2	11.0	10.2	9.0	8.2	7.6	7.4	7.2	6.8	6.6
90,000	498.0	316.0	199.0	125.0	82.9	54.0	35.8	28.9	23.8	19.8	16.3	13.6	12.3	10.9	9.9	9.2	8.1	7.4	6.8	6.8	6.4	6.2	5.9
80,000	443.0	277.0	177.0	112.0	73.7	48.0	31.8	25.5	21.1	17.6	14.5	12.1	10.9	9.7	8.8	8.2	7.2	6.6	6.1	6.0	5.7	5.5	5.2
70,000	388.0	246.0	155.0	98.4	64.5	42.0	27.8	22.6	18.6	15.4	12.7	10.6	9.5	8.5	7.7	7.1	6.3	5.7	5.3	5.2	5.0	4.8	4.6
60,000	332.0	210.0	133.0	84.3	55.3	36.0	23.8	19.3	16.0	13.2	10.9	9.1	8.1	7.3	6.6	6.1	5.4	4.9	4.6	4.4	4.4	4.1	4.0
50,000	277.0	176.0	111.0	70.2	46.1	30.0	19.8	16.5	13.3	11.0	9.1	7.6	6.8	6.1	5.5	5.1	4.5	4.1	3.8	3.7	3.6	3.4	3.3
40,000	222.0	140.0	88.4	56.0	36.9	24.0	15.8	12.9	10.6	8.8	7.3	6.1	5.4	4.9	4.4	4.1	3.6	3.3	3.0	3.0	2.9	2.6	2.6
30,000	166.0	105.0	66.3	42.0	27.7	18.0	11.9	9.7	7.9	6.6	5.5	4.6	4.1	3.7	3.3	3.0	2.7	2.5	2.3	2.2	2.2	2.0	1.9
20,000	111.0	70.2	44.2	28.0	18.4	12.0	7.9	6.4	5.3	4.4	3.6	3.0	2.7	2.4	2.2	2.0	1.8	1.6	1.5	1.5	1.4	1.4	1.3
10,000	55.4	35.1	22.1	14.0	9.2	6.0	3.9	3.2	2.7	2.2	1.8	1.5	1.4	1.2	1.1	1.0	0.9	0.8	0.7	0.7	0.7	0.7	0.7
9,000	49.8	31.6	19.9	12.6	8.3	5.4	3.6	2.9	2.4	1.9	1.6	1.4	1.2	1.0	1.0	0.9	0.8	0.7	0.7	0.7	0.6	0.6	0.6
8,000	44.3	28.1	17.7	11.2	7.4	4.8	3.2	2.6	2.1	1.8	1.5	1.2	1.1	1.0	0.9	0.8	0.7	0.6	0.6	0.6	0.6	0.5	0.5
7,000	38.8	24.6	15.5	9.8	6.5	4.2	2.8	2.3	1.9	1.5	1.3	1.1	0.9	0.9	0.8	0.7	0.6	0.6	0.5	0.5	0.5	0.4	0.4
6,000	33.2	21.0	13.3	8.4	5.5	3.6	2.4	1.9	1.6	1.3	1.1	0.9	0.8	0.7	0.7	0.6	0.5	0.5	0.5	0.5	0.4	0.4	0.4
5,000	27.7	17.6	11.1	7.0	4.6	3.0	1.9	1.6	1.3	1.1	0.9	0.8	0.7	0.6	0.6	0.5	0.5	0.4	0.4	0.4	0.4	0.3	0.3
4,000	22.2	14.0	8.8	5.6	3.7	2.4	1.6	1.3	1.1	0.9	0.7	0.6	0.5	0.5	0.4	0.4	0.4	0.3	0.3	0.3	0.3	0.3	0.3
3,000	16.6	10.5	6.6	4.2	2.8	1.8	1.2	1.0	0.8	0.7	0.5	0.5	0.4	0.4	0.3	0.3	0.2	0.2	0.2	0.2	0.2	0.2	0.2
2,000	11.1	7.0	4.4	2.8	1.8	1.2	0.8	0.6	0.5	0.4	0.4	0.3	0.3	0.2	0.2	0.2	0.2	0.2	0.1	0.1	0.1	0.1	0.1
1,000	5.5	3.5	2.2	1.4	0.9	0.6	0.4	0.3	0.3	0.2	0.2	0.2	0.1	0.1	0.1	0.1	0.1	0.1	0.1				
900	5.0	3.2	2.0	1.3	0.8	0.5	0.4	0.3	0.3	0.2	0.1	0.1	0.1	0.1	0.1	0.1	0.1	0.1					
800	4.4	2.8	1.8	1.1	0.7	0.4	0.3	0.2	0.2	0.2	0.1	0.1	0.1	0.1	0.1	0.1	0.1	0.1					
700	3.9	2.5	1.6	1.0	0.6	0.4	0.3	0.2	0.2	0.1	0.1	0.1	0.1	0.1	0.1	0.1	0.1						
600	3.3	2.1	1.3	0.8	0.6	0.4	0.2	0.2	0.2	0.1	0.1	0.1	0.1	0.1	0.1	0.1	0.1						
500	2.8	1.8	1.1	0.7	0.5	0.3	0.2	0.2	0.1	0.1	0.1	0.1	0.1	0.1	0.1	0.1							
400	2.2	1.4	0.9	0.6	0.4	0.2	0.2	0.1	0.1	0.1		0.1	0.1	0.1	0.1								
300	1.7	1.1	0.7	0.4	0.3	0.2	0.1	0.1	0.1														
200	1.1	0.7	0.4	0.3	0.2	0.1	0.1	0.1															
100	0.6	0.4	0.2	0.1	0.1	0.1	0.1																

* Solid Conductors. Other conductors are stranded.

Note 1—The above table gives voltage drops encountered in a single phase phase two-wire system. The voltage drops in other systems may be obtained through multiplication by appropriate factors listed below:

Multiplying Factors for Modification of Values in Table

System for Which Voltage Drop is Desired	
Single Phase—3 Wire—Line to Line	1.00
Single Phase—3 Wire—Line to Neutral	0.50
Three Phase—3 Wire—Line to Line	0.866
Three Phase—4 Wire—Line to Line	0.866
Three Phase—4 Wire—Line to Neutral	0.50

Note 2—Allowable voltage drops for systems other than single phase, two-wire cannot be used directly in the above table. Such voltage drops should be modified through multiplication by the appropriate factor listed below. The voltage thus modified may then be used to obtain the proper wire size directly from the table.

System for Which Allowable Voltage Drop is Known	Multiplying Factor for Modification of Known Value to Permit Direct Use of Table
Single Phase—3 Wire—Line to Line	1.00
Single Phase—3 Wire—Line to Neutral	2.00
Three Phase—3 Wire—Line to Line	1.155
Three Phase—4 Wire—Line to Line	1.155
Three Phase—4 Wire—Line to Neutral	2.00

Note 3—The footage employed in the tabulated ampere feet refers to the length of run of the circuit rather than to the footage of individual conductor.

Note 4—The above table is figured at 60°C since this is an estimate of the average temperature which may be anticipated in service. The table may be used without significant error for conductor temperatures up to and including 75°C.

TABLE 3.66 Volts Drop for CU Conductor in Magnetic Conduit—95 Percent PF

WIRE SIZE AWG or MCM	1000	900	800	750	700	600	500	400	350	300	250	4/0	3/0	2/0	1/0	1	2	4	6	8*	10*	12*	14*
Ampere Feet	Volts Drop																						
500,000	29.0	30.0	32.0	33.0	34.0	37.0	41.0	47.0	51.0	58.0	65.0	74.0	89.0	109.0	133.0	161.0	200.0	308.0	476.0	—	—	—	—
400,000	23.2	24.0	25.6	26.4	26.2	29.6	32.8	37.6	40.8	46.4	52.0	59.2	71.2	87.2	106.0	129.0	160.0	245.0	380.0	—	—	—	—
300,000	17.4	18.0	19.2	19.8	20.4	22.2	24.6	28.2	30.6	34.8	39.0	44.4	63.4	65.4	79.8	96.6	120.0	184.0	285.0	438.0	—	—	—
200,000	11.6	12.0	12.8	13.2	13.6	14.8	16.4	18.8	20.4	23.2	26.0	29.6	35.6	43.6	53.2	64.4	80.0	123.0	190.0	292.0	464.0	—	—
100,000	5.8	6.0	6.4	6.6	6.8	7.4	8.2	9.4	10.2	11.6	13.0	14.8	17.8	21.8	26.6	32.2	40.0	61.6	95.2	146.0	232.0	369.0	—
90,000	5.3	5.4	5.8	5.9	6.2	6.8	7.3	8.6	9.1	10.5	11.7	13.4	16.1	19.7	23.8	28.9	36.0	55.6	85.4	132.0	209.0	332.0	—
80,000	4.7	4.8	5.2	5.2	5.5	6.0	6.5	7.6	8.1	9.3	10.4	11.9	14.3	17.5	21.2	25.7	32.0	49.4	76.0	117.0	185.0	295.0	466.0
70,000	4.1	4.2	4.5	4.6	4.8	5.2	5.7	6.6	7.1	8.1	9.1	10.4	12.5	15.3	18.6	22.5	28.0	43.2	66.6	103.0	162.0	258.0	408.0
60,000	3.5	3.6	3.8	4.0	4.1	4.4	4.9	5.6	6.1	6.9	7.8	8.9	10.7	13.1	16.0	19.3	24.0	37.0	57.2	88.0	139.0	221.0	349.0
50,000	2.9	3.0	3.2	3.3	3.4	3.7	4.1	4.7	5.1	5.8	6.5	7.4	8.9	10.9	13.3	16.1	20.0	30.8	47.6	73.2	116.0	184.0	291.0
40,000	2.3	2.4	2.5	2.6	2.6	2.9	3.3	3.8	4.1	4.6	5.2	5.9	7.1	8.7	10.6	12.9	16.0	24.5	38.0	58.4	92.8	148.0	233.0
30,000	1.7	1.8	1.9	2.0	2.0	2.2	2.5	2.8	3.1	3.5	3.9	4.4	5.3	6.5	7.9	9.7	12.0	18.4	28.5	43.8	69.6	111.0	175.0
20,000	1.2	1.2	1.3	1.3	1.4	1.5	1.6	1.9	2.0	2.3	2.6	2.9	3.6	4.4	5.3	6.4	8.0	12.3	19.0	29.2	46.4	73.8	116.0
10,000	0.6	0.6	0.6	0.7	0.7	0.7	0.8	0.9	1.0	1.2	1.3	1.5	1.8	2.2	2.7	3.2	4.0	6.2	9.5	14.6	23.2	36.9	58.2
9,000	0.5	0.5	0.6	0.6	0.6	0.7	0.7	0.9	0.9	1.1	1.2	1.3	1.6	1.9	2.4	2.9	3.6	5.6	8.5	13.2	20.9	33.2	52.4
8,000	0.5	0.5	0.5	0.5	0.6	0.6	0.7	0.8	0.8	0.9	1.0	1.2	1.4	1.8	2.1	2.6	3.2	4.9	7.6	11.7	18.5	29.5	46.6
7,000	0.4	0.4	0.5	0.5	0.5	0.5	0.6	0.7	0.7	0.8	0.9	1.0	1.3	1.5	1.9	2.3	2.8	4.3	6.7	10.3	16.2	25.8	40.8
6,000	0.4	0.4	0.4	0.4	0.4	0.4	0.5	0.6	0.6	0.7	0.8	0.9	1.1	1.3	1.6	1.9	2.4	3.7	5.7	8.8	13.9	22.1	34.9
5,000	0.3	0.3	0.3	0.3	0.3	0.4	0.4	0.5	0.5	0.6	0.7	0.7	0.9	1.1	1.3	1.6	2.0	3.1	4.8	7.3	11.6	18.4	29.1
4,000	0.2	0.2	0.3	0.3	0.3	0.3	0.3	0.4	0.4	0.5	0.5	0.6	0.7	0.9	1.1	1.3	1.6	2.5	3.8	5.8	9.3	14.8	23.1
3,000	0.1	0.2	0.2	0.2	0.2	0.2	0.3	0.3	0.3	0.4	0.4	0.4	0.5	0.7	0.8	1.0	1.2	1.8	2.9	4.4	7.0	11.1	17.5
2,000	0.1	0.1	0.1	0.1	0.1	0.2	0.2	0.2	0.2	0.2	0.3	0.3	0.4	0.4	0.5	0.6	0.8	1.2	1.9	2.9	4.6	7.4	11.6
1,000	0.1	0.1	0.1	0.1	0.1	0.1	0.1	0.1	0.1	0.1	0.1	0.2	0.2	0.2	0.3	0.3	0.4	0.6	1.0	1.5	2.3	3.7	5.8
900	0.1	0.1	0.1	0.1	0.1	0.1	0.1	0.1	0.1	0.1	0.1	0.1	0.2	0.2	0.2	0.3	0.4	0.6	0.9	1.3	2.1	3.3	5.2
800	0.1	0.1	0.1	0.1	0.1	0.1	0.1	0.1	0.1	0.1	0.1	0.1	0.1	0.2	0.2	0.3	0.3	0.5	0.8	1.2	1.9	2.9	4.7
700	—	—	—	—	—	0.1	0.1	0.1	0.1	0.1	0.1	0.1	0.1	0.2	0.2	0.2	0.3	0.4	0.7	1.0	1.6	2.6	4.1
600	—	—	—	—	—	—	—	0.1	0.1	0.1	0.1	0.1	0.1	0.1	0.2	0.2	0.2	0.4	0.6	0.9	1.4	2.2	3.5
500	—	—	—	—	—	—	—	—	0.1	0.1	0.1	0.1	0.1	0.1	0.1	0.2	0.2	0.3	0.5	0.7	1.2	1.8	2.9
400	—	—	—	—	—	—	—	—	—	0.1	0.1	0.1	0.1	0.1	0.1	0.1	0.2	0.3	0.4	0.6	0.9	1.5	2.3
300	—	—	—	—	—	—	—	—	—	—	—	—	0.1	0.1	0.1	0.1	0.1	0.2	0.3	0.4	0.7	1.1	1.8
200	—	—	—	—	—	—	—	—	—	—	—	—	—	—	0.1	0.1	0.1	0.1	0.2	0.3	0.5	0.7	1.2
100	—	—	—	—	—	—	—	—	—	—	—	—	—	—	—	—	—	0.1	0.1	0.2	0.2	0.4	0.6

* Solid Conductors. Other conductors are stranded.

Note 1—The above table gives voltage drops encountered in a single phase two-wire system. The voltage drops in other systems may be obtained through multiplication by appropriate factors listed below:

System for Which Voltage Drop Is Desired	Multiplying Factors for Modification of Values in Table
Single Phase—3 Wire—Line to Line	1.00
Single Phase—3 Wire—Line to Neutral	0.50
Three Phase—3 Wire—Line to Line	0.866
Three Phase—4 Wire—Line to Line	0.866
Three Phase—4 Wire—Line to Neutral	0.50

Note 2—Allowable voltage drops for systems other than single phase, two-wire cannot be used directly in the above table. Such drops should be modified through multiplication by the appropriate factor listed below. The voltage thus modified may then be used to obtain the proper wire size directly from the table.

System for Which Allowable Voltage Drop Is Known	Multiplying Factor for Modification of Known Value to Permit Direct Use of Table
Single Phase—3 Wire—Line to Line	1.00
Single Phase—3 Wire—Line to Neutral	2.00
Three Phase—3 Wire—Line to Line	1.155
Three Phase—4 Wire—Line to Line	1.155
Three Phase—4 Wire—Line to Neutral	2.00

Note 3—The footage employed in the tabulated ampere feet refers to the length of run of the circuit rather than to the footage of individual conductor.

Note 4—The above table is figured at 60°C since this is an estimate of the average temperature which may be anticipated in service. The table may be used without significant error for conductor temperatures up to and including 75°C.

TABLE 3.67 Volts Drop for CU Conductor in Magnetic Conduit—100 Percent PF

Values below are volts drop for the indicated wire size and ampere-feet.

WIRE SIZE AWG or MCM — Ampere Feet	1000	900	800	750	700	600	500	400	350	300	250	4/0	3/0	2/0	1/0	1	2	4	6	8*	10*	12*	14*
500,000	16.4	17.9	19.6	20.3	21.5	24.4	28.8	34.8	39.0	45.0	53.4	62.4	78.6	98.5	123.0	153.0	194.0	306.0	483.0				
400,000	13.1	14.3	15.7	16.2	17.2	19.5	23.0	27.8	31.2	36.0	42.7	49.9	62.9	78.8	98.4	122.4	155.2	244.0	386.0				
300,000	9.8	10.7	11.8	12.2	12.9	14.6	17.3	20.9	23.4	27.0	32.0	37.4	47.2	59.1	73.8	91.8	116.4	184.0	290.0	450.0			
200,000	6.6	7.2	7.8	8.1	8.6	9.8	11.5	13.9	15.6	18.0	21.4	25.0	31.4	39.4	49.2	61.2	77.6	122.0	193.0	300.0	480.0		
100,000	3.3	3.6	3.9	4.1	4.3	4.9	5.8	7.0	7.8	9.0	10.7	12.5	15.7	19.7	24.6	30.6	38.8	61.2	96.6	150.0	240.0	384.0	
90,000	2.9	3.2	3.5	3.7	3.9	4.4	5.2	6.3	7.0	8.1	9.6	11.2	14.1	17.7	22.1	27.5	34.9	55.1	87.0	135.0	216.0	345.0	
80,000	2.6	2.9	3.1	3.2	3.4	3.9	4.6	5.6	6.2	7.2	8.5	10.0	12.6	15.8	19.7	24.5	31.0	49.0	77.3	120.0	192.0	307.0	487.0
70,000	2.3	2.5	2.7	2.8	3.0	3.4	4.0	4.9	5.5	6.3	7.5	8.7	11.0	13.8	17.2	21.4	27.2	42.8	67.6	105.0	168.0	269.0	425.0
60,000	2.0	2.1	2.3	2.4	2.6	2.9	3.5	4.2	4.7	5.4	6.4	7.5	9.4	11.8	14.8	18.4	23.3	36.7	58.0	90.0	144.0	230.0	366.0
50,000	1.6	1.8	2.0	2.0	2.2	2.4	2.9	3.5	3.9	4.5	5.3	6.2	7.9	9.9	12.3	15.3	19.4	30.6	48.3	75.0	120.0	192.0	304.0
40,000	1.3	1.4	1.6	1.6	1.7	2.0	2.3	2.8	3.1	3.6	4.3	5.0	6.3	7.9	9.8	12.2	15.5	24.5	38.6	60.0	96.0	154.0	243.0
30,000	1.0	1.1	1.2	1.2	1.3	1.5	1.7	2.1	2.3	2.7	3.2	3.7	4.7	5.9	7.4	9.2	11.6	18.4	29.0	45.0	72.0	115.0	182.0
20,000	0.7	0.7	0.8	0.8	0.9	1.0	1.2	1.4	1.6	1.8	2.1	2.5	3.1	3.9	4.9	6.1	7.8	12.2	19.3	30.0	48.0	76.8	122.0
10,000	0.3	0.4	0.4	0.4	0.4	0.5	0.6	0.7	0.8	0.9	1.1	1.2	1.6	2.0	2.5	3.1	3.9	6.1	9.7	15.0	24.0	38.4	60.8
9,000	0.3	0.3	0.4	0.4	0.4	0.4	0.5	0.6	0.7	0.8	1.0	1.1	1.4	1.8	2.2	2.8	3.5	5.5	8.7	13.5	21.6	34.5	54.7
8,000	0.3	0.3	0.3	0.3	0.3	0.4	0.5	0.6	0.6	0.7	0.9	1.0	1.3	1.6	2.0	2.4	3.1	4.9	7.7	12.0	19.2	30.7	48.7
7,000	0.2	0.3	0.3	0.3	0.3	0.3	0.4	0.5	0.5	0.6	0.7	0.9	1.1	1.4	1.7	2.1	2.7	4.3	6.8	10.5	16.8	26.9	42.6
6,000	0.2	0.2	0.2	0.2	0.3	0.3	0.3	0.4	0.5	0.5	0.6	0.7	0.9	1.2	1.5	1.8	2.3	3.7	5.8	9.0	14.4	23.0	36.5
5,000	0.2	0.2	0.2	0.2	0.2	0.2	0.3	0.3	0.4	0.5	0.5	0.6	0.8	1.0	1.2	1.5	1.9	3.1	4.8	7.5	12.0	19.2	30.4
4,000	0.1	0.1	0.2	0.2	0.2	0.2	0.2	0.3	0.3	0.4	0.4	0.5	0.6	0.8	1.0	1.2	1.6	2.4	3.9	6.0	9.6	15.4	24.3
3,000	0.1	0.1	0.1	0.1	0.1	0.1	0.2	0.2	0.2	0.3	0.3	0.4	0.5	0.6	0.7	0.9	1.2	1.8	2.9	4.5	7.2	11.5	18.2
2,000	0.1	0.1	0.1	0.1	0.1	0.1	0.1	0.1	0.2	0.2	0.2	0.2	0.3	0.4	0.5	0.6	0.8	1.2	1.9	3.0	4.8	7.7	12.2
1,000							0.1	0.1	0.1	0.1	0.1	0.1	0.2	0.2	0.2	0.3	0.4	0.6	1.0	1.5	2.4	3.8	6.1
900							0.1	0.1	0.1	0.1	0.1	0.1	0.1	0.2	0.2	0.3	0.3	0.6	0.9	1.4	2.2	3.5	5.5
800								0.1	0.1	0.1	0.1	0.1	0.1	0.2	0.2	0.2	0.3	0.5	0.8	1.2	1.9	3.1	4.9
700								0.1	0.1	0.1	0.1	0.1	0.1	0.1	0.2	0.2	0.3	0.4	0.7	1.1	1.7	2.7	4.3
600									0.1	0.1	0.1	0.1	0.1	0.1	0.1	0.2	0.2	0.4	0.6	0.9	1.4	2.3	3.7
500										0.1	0.1	0.1	0.1	0.1	0.1	0.2	0.2	0.3	0.5	0.8	1.2	1.9	3.0
400												0.1	0.1	0.1	0.1	0.1	0.2	0.2	0.4	0.6	1.0	1.5	2.4
300														0.1	0.1	0.1	0.1	0.2	0.3	0.5	0.7	1.2	1.8
200																0.1	0.1	0.1	0.2	0.3	0.5	0.8	1.2
100																		0.1	0.1	0.2	0.2	0.4	0.6

* Solid Conductors. Other conductors are stranded.

Note 1—The above table gives voltage drops encountered in a single phase two-wire system. The voltage drops in other systems may be obtained through multiplication by appropriate factors listed below:

System for Which Voltage Drop Is Desired	Multiplying Factors for Modification of Values in Table
Single Phase—3 Wire—Line to Line	1.00
Single Phase—3 Wire—Line to Neutral	0.50
Three Phase—3 Wire—Line to Line	0.866
Three Phase—4 Wire—Line to Line	0.866
Three Phase—4 Wire—Line to Neutral	0.50

Note 2—Allowable voltage drops for systems other than single phase, two-wire cannot be used directly in the above table. Such drops should be modified through multiplication by the appropriate factor listed below. The voltage thus modified may then be used to obtain the proper wire size directly from the table.

System for Which Allowable Voltage Drop Is Known	Multiplying Factor for Modification of Known Value to Permit Direct Use of Table
Single Phase—3 Wire—Line to Line	1.00
Single Phase—3 Wire—Line to Neutral	2.00
Three Phase—3 Wire—Line to Line	1.155
Three Phase—4 Wire—Line to Line	1.155
Three Phase—4 Wire—Line to Neutral	2.00

Note 3—The footage employed in the tabulated ampere feet refers to the length of run of the circuit rather than to the footage of individual conductor.

Note 4—The above table is figured at 60°C since this is an estimate of the average temperature which may be anticipated in service. The table may be used without significant error for conductor temperatures up to and including 75°C.

TABLE 3.68 Volts Drop for CU Conductor in Nonmagnetic Conduit—70 Percent PF

WIRE SIZE AWG or MCM — Volts Drop

Ampere Feet	1000	900	800	750	700	600	500	400	350	300	250	4/0	3/0	2/0	1/0	1	2	4	6	8*	10*	12*	14*
500,000	33.0	34.0	36.0	37.0	38.0	40.0	43.0	48.0	51.0	56.0	62.0	68.0	81.0	95.0	114.0	136.0	164.0	246.0	372.0	—	—	—	—
400,000	26.4	27.2	28.8	29.6	30.4	32.0	34.4	38.4	40.8	44.8	49.6	54.4	64.8	76.0	91.2	109.0	131.0	196.0	297.0	448.0	—	—	—
300,000	19.8	20.4	21.6	22.2	22.8	24.0	25.8	28.8	30.6	33.6	37.2	40.8	48.6	57.0	68.4	81.6	98.4	147.0	223.0	336.0	—	—	—
200,000	13.2	13.6	14.4	14.8	15.2	16.0	17.2	19.2	20.4	22.4	24.8	27.2	32.4	38.0	45.6	54.4	65.6	98.4	148.0	224.0	350.0	—	—
100,000	6.6	6.8	7.2	7.4	7.6	8.0	8.6	9.6	10.2	11.2	12.4	13.6	16.2	19.0	22.8	27.2	32.8	49.2	74.4	112.0	175.0	276.0	434.0
90,000	5.9	6.1	6.5	6.7	6.8	7.2	7.7	8.6	9.2	10.1	11.2	12.2	14.6	17.1	20.5	24.5	29.5	44.2	67.1	101.0	158.0	248.0	390.0
80,000	5.3	5.4	5.8	5.9	6.1	6.4	6.9	7.7	8.2	9.0	9.9	10.9	13.0	15.2	18.3	21.8	26.2	39.3	59.6	89.6	140.0	221.0	347.0
70,000	4.6	4.8	5.0	5.2	5.3	5.6	6.0	6.7	7.1	7.9	8.7	9.5	11.3	13.3	16.0	19.0	22.9	34.4	52.1	78.4	123.0	193.0	304.0
60,000	4.0	4.1	4.3	4.4	4.6	4.8	5.2	5.8	6.1	6.7	7.4	8.2	9.7	11.4	13.7	16.3	19.7	29.6	44.6	67.3	105.0	166.0	260.0
50,000	3.3	3.4	3.6	3.7	3.8	4.0	4.3	4.8	5.1	5.6	6.2	6.8	8.1	9.5	11.4	13.6	16.4	24.6	37.2	56.0	87.6	138.0	217.0
40,000	2.6	2.7	2.9	3.0	3.0	3.2	3.4	3.8	4.1	4.5	5.0	5.4	6.5	7.6	9.1	10.9	13.1	19.7	29.8	44.8	70.0	110.0	174.0
30,000	2.0	2.0	2.2	2.2	2.3	2.4	2.6	2.9	3.1	3.4	3.7	4.1	4.9	5.7	6.8	8.2	9.8	14.8	22.3	33.6	52.5	82.8	130.0
20,000	1.3	1.4	1.4	1.5	1.5	1.6	1.7	1.9	2.0	2.2	2.5	2.7	3.2	3.8	4.6	5.4	6.6	9.8	14.9	22.4	35.0	55.2	86.8
10,000	0.7	0.7	0.7	0.7	0.8	0.8	0.9	1.0	1.0	1.1	1.2	1.4	1.6	1.9	2.3	2.7	3.3	4.9	7.4	11.2	17.5	27.6	43.4
9,000	0.6	0.6	0.6	0.7	0.7	0.7	0.8	0.9	0.9	1.0	1.1	1.2	1.5	1.7	2.1	2.5	2.9	4.4	6.7	10.1	15.8	24.8	39.0
8,000	0.5	0.5	0.6	0.6	0.6	0.6	0.7	0.8	0.8	0.9	1.0	1.1	1.3	1.5	1.8	2.2	2.6	3.9	6.0	9.0	14.0	22.1	34.7
7,000	0.5	0.5	0.5	0.5	0.5	0.6	0.6	0.7	0.7	0.8	0.9	1.0	1.1	1.3	1.6	1.9	2.3	3.4	5.2	7.8	12.3	19.3	30.4
6,000	0.4	0.4	0.4	0.4	0.5	0.5	0.5	0.6	0.6	0.7	0.7	0.8	1.0	1.1	1.4	1.6	2.0	3.0	4.5	6.7	10.5	16.6	26.0
5,000	0.3	0.3	0.4	0.4	0.4	0.4	0.4	0.5	0.5	0.6	0.6	0.7	0.8	1.0	1.1	1.4	1.6	2.5	3.7	5.6	8.8	13.8	21.7
4,000	0.3	0.3	0.3	0.3	0.3	0.3	0.3	0.4	0.4	0.4	0.5	0.5	0.6	0.8	0.9	1.1	1.3	2.0	3.0	4.5	7.0	11.0	17.4
3,000	0.2	0.2	0.2	0.2	0.2	0.2	0.3	0.3	0.3	0.3	0.4	0.4	0.5	0.6	0.7	0.8	1.0	1.5	2.2	3.4	5.3	8.3	13.0
2,000	0.1	0.1	0.1	0.1	0.2	0.2	0.2	0.2	0.2	0.2	0.2	0.3	0.3	0.4	0.5	0.5	0.7	1.0	1.5	2.2	3.5	5.5	8.7
1,000	0.1	0.1	0.1	0.1	0.1	0.1	0.1	0.1	0.1	0.1	0.1	0.1	0.2	0.2	0.2	0.3	0.3	0.5	0.7	1.1	1.8	2.8	4.3
900	0.1	0.1	0.1	0.1	0.1	0.1	0.1	0.1	0.1	0.1	0.1	0.1	0.1	0.2	0.2	0.2	0.3	0.4	0.7	1.0	1.6	2.5	3.9
800	0.1	0.1	0.1	0.1	0.1	0.1	0.1	0.1	0.1	0.1	0.1	0.1	0.1	0.2	0.2	0.2	0.3	0.4	0.6	0.9	1.4	2.2	3.5
700	—	—	0.1	0.1	0.1	0.1	0.1	0.1	0.1	0.1	0.1	0.1	0.1	0.1	0.2	0.2	0.2	0.3	0.5	0.8	1.2	1.9	3.0
600	—	—	—	—	—	—	0.1	0.1	0.1	0.1	0.1	0.1	0.1	0.1	0.1	0.2	0.2	0.3	0.4	0.7	1.1	1.7	2.6
500	—	—	—	—	—	—	—	—	0.1	0.1	0.1	0.1	0.1	0.1	0.1	0.1	0.2	0.2	0.4	0.6	0.9	1.4	2.2
400	—	—	—	—	—	—	—	—	—	—	—	0.1	0.1	0.1	0.1	0.1	0.1	0.2	0.3	0.4	0.7	1.1	1.7
300	—	—	—	—	—	—	—	—	—	—	—	—	—	0.1	0.1	0.1	0.1	0.1	0.2	0.3	0.5	0.8	1.3
200	—	—	—	—	—	—	—	—	—	—	—	—	—	—	—	0.1	0.1	0.1	0.1	0.2	0.4	0.6	0.9
100	—	—	—	—	—	—	—	—	—	—	—	—	—	—	—	—	—	—	0.1	0.1	0.2	0.3	0.4

* Solid Conductors. Other conductors are stranded.

Note 1—The above table gives voltage drops encountered in a single phase two-wire system. The voltage drops in other systems may be obtained through multiplication by appropriate factors listed below:

System for Which Voltage Drop is Desired	Multiplying Factors for Modification of Values in Table
Single Phase—3 Wire—Line to Line	1.00
Single Phase—3 Wire—Line to Neutral	0.50
Three Phase—3 Wire—Line to Line	0.866
Three Phase—4 Wire—Line to Line	0.866
Three Phase—4 Wire—Line to Neutral	0.50

Note 2—Allowable voltage drops for systems other than single phase, two wire cannot be used directly in the above table. Such drops should be modified through multiplication by the appropriate factor listed below. The voltage thus obtained may then be used to obtain the proper wire size directly from the table.

System for Which Allowable Voltage Drop is Known	Multiplying Factor for Modification of Known Value to Permit Direct Use of Table
Single Phase—2 Wire—Line to Line	1.00
Single Phase—3 Wire—Line to Neutral	2.00
Three Phase—3 Wire—Line to Line	1.155
Three Phase—4 Wire—Line to Line	1.155
Three Phase—4 Wire—Line to Neutral	2.00

Note 3—The footage employed in the tabulated ampere feet refers to the length of run of the circuit rather than to the footage of individual conductor.

Note 4—The above table is figured at 80°C since this is an estimate of the average temperature which may be anticipated in service. The table may be used without significant error for conductor temperatures up to and including 75°C.

TABLE 3.69 Volts Drop for CU Conductor in Nonmagnetic Conduit—80 Percent PF

WIRE SIZE AWG or MCM — Ampere Feet	14*	12*	10*	8*	6	4	2	1	1/0	2/0	3/0	4/0	250	300	350	400	500	600	700	750	800	900	1000
												Volts Drop											
500,000	—	—	—	—	414.0	272.0	179.0	147.0	121.0	100.0	83.0	70.0	63.0	56.0	51.0	47.0	42.0	38.0	36.0	34.0	33.0	32.0	31.0
400,000	—	—	—	—	331.0	217.6	143.2	117.6	96.8	80.0	66.4	56.0	50.4	44.8	40.8	37.6	33.6	30.4	28.8	27.2	26.4	25.6	24.8
300,000	—	—	—	378.0	248.0	163.2	107.4	88.2	72.6	60.0	49.8	42.0	37.8	33.6	30.6	28.2	25.2	22.8	21.6	20.4	19.8	19.2	18.6
200,000	—	—	396.0	252.0	166.0	108.8	71.6	58.8	48.4	40.0	33.2	28.0	25.2	22.4	20.4	18.8	16.8	15.2	14.4	13.6	13.2	12.8	12.4
100,000	492.0	313.0	198.0	125.0	82.8	54.4	35.8	29.4	24.2	20.0	16.6	14.0	12.6	11.2	10.2	9.4	8.4	7.6	7.2	6.8	6.6	6.4	6.2
90,000	443.0	282.0	178.0	113.0	74.5	49.0	32.2	26.5	21.8	18.0	14.9	12.6	11.3	10.1	9.2	8.5	7.6	6.8	6.5	6.1	5.9	5.8	5.6
80,000	394.0	250.0	158.0	101.0	66.2	43.5	28.6	23.5	19.4	16.0	13.3	11.2	10.1	9.0	8.2	7.5	6.7	6.1	5.8	5.4	5.3	5.1	5.0
70,000	345.0	219.0	139.0	88.2	58.0	38.1	25.1	20.6	16.9	14.0	11.6	9.8	8.8	7.8	7.1	6.6	5.9	5.3	5.0	4.8	4.6	4.5	4.3
60,000	295.0	188.0	119.0	75.6	49.7	32.6	21.5	17.6	14.5	12.0	10.0	8.4	7.6	6.7	6.1	5.6	5.0	4.6	4.3	4.1	4.0	3.8	3.7
50,000	246.0	157.0	99.0	62.9	41.4	27.2	17.9	14.7	12.1	10.0	8.3	7.0	6.3	5.6	5.1	4.7	4.2	3.8	3.6	3.4	3.3	3.2	3.1
40,000	197.0	125.0	79.2	50.4	33.1	21.8	14.3	11.8	9.7	8.0	6.6	5.6	5.0	4.5	4.1	3.8	3.4	3.0	2.9	2.7	2.6	2.6	2.5
30,000	148.0	93.9	59.4	37.8	24.8	16.3	10.7	8.8	7.3	6.0	5.0	4.2	3.8	3.4	3.1	2.8	2.5	2.3	2.2	2.0	2.0	1.9	1.9
20,000	98.4	62.6	39.6	25.2	16.6	10.9	7.2	5.9	4.8	4.0	3.3	2.8	2.5	2.2	2.0	1.9	1.7	1.5	1.4	1.4	1.3	1.3	1.2
10,000	49.2	31.3	19.8	12.6	8.3	5.4	3.6	2.9	2.4	2.0	1.7	1.4	1.3	1.1	1.0	0.9	0.8	0.8	0.7	0.7	0.7	0.6	0.6
9,000	44.3	28.2	17.8	11.3	7.5	4.9	3.2	2.6	2.2	1.8	1.5	1.3	1.1	1.0	0.9	0.8	0.8	0.7	0.6	0.6	0.6	0.6	0.6
8,000	39.4	25.0	15.8	10.1	6.6	4.4	2.9	2.4	1.9	1.6	1.3	1.1	1.0	0.9	0.8	0.8	0.7	0.6	0.6	0.5	0.5	0.5	0.5
7,000	34.5	21.9	13.9	8.8	5.8	3.8	2.5	2.1	1.7	1.4	1.2	1.0	0.9	0.8	0.7	0.7	0.6	0.5	0.5	0.5	0.5	0.4	0.4
6,000	29.5	18.8	11.9	7.6	5.0	3.3	2.1	1.8	1.5	1.2	1.0	0.8	0.8	0.7	0.6	0.6	0.5	0.5	0.4	0.4	0.4	0.4	0.4
5,000	24.6	15.7	9.9	6.3	4.1	2.7	1.8	1.5	1.2	1.0	0.8	0.7	0.6	0.6	0.5	0.5	0.4	0.4	0.4	0.3	0.3	0.3	0.3
4,000	19.7	12.5	7.9	5.0	3.3	2.2	1.4	1.2	1.0	0.8	0.7	0.6	0.5	0.4	0.4	0.4	0.3	0.3	0.3	0.3	0.3	0.3	0.2
3,000	14.8	9.4	5.9	3.8	2.5	1.6	1.1	0.9	0.7	0.6	0.5	0.4	0.4	0.3	0.3	0.3	0.3	0.2	0.2	0.2	0.2	0.2	0.2
2,000	9.8	6.3	4.0	2.5	1.7	1.1	0.7	0.6	0.5	0.4	0.3	0.3	0.3	0.2	0.2	0.2	0.2	0.2	0.1	0.1	0.1	0.1	0.1
1,000	4.9	3.1	2.0	1.3	0.8	0.5	0.4	0.3	0.2	0.2	0.2	0.1	0.1	0.1	0.1	0.1	0.1	0.1	0.1	0.1	0.1	0.1	0.1
900	4.4	2.8	1.8	1.1	0.7	0.5	0.3	0.3	0.2	0.2	0.1	0.1	0.1	0.1	0.1	0.1	0.1	0.1	0.1	0.1	0.1	0.1	0.1
800	3.9	2.5	1.6	1.0	0.7	0.4	0.3	0.2	0.2	0.2	0.1	0.1	0.1	0.1	0.1	0.1	0.1	0.1	0.1	0.1	0.1	0.1	0.1
700	3.4	2.2	1.4	0.9	0.6	0.4	0.3	0.2	0.2	0.1	0.1	0.1	0.1	0.1	0.1	0.1	0.1	0.1	0.1	—	—	—	—
600	2.9	1.9	1.2	0.8	0.5	0.3	0.2	0.2	0.1	0.1	0.1	0.1	0.1	0.1	0.1	0.1	0.1	—	—	—	—	—	—
500	2.5	1.6	1.0	0.6	0.4	0.3	0.2	0.1	0.1	0.1	0.1	0.1	0.1	0.1	0.1	0.1	0.1	—	—	—	—	—	—
400	2.0	1.3	0.8	0.5	0.3	0.2	0.1	0.1	0.1	0.1	0.1	0.1	0.1	0.1	0.1	—	—	—	—	—	—	—	—
300	1.5	0.9	0.6	0.4	0.2	0.2	0.1	0.1	0.1	0.1	0.1	—	—	—	—	—	—	—	—	—	—	—	—
200	1.0	0.6	0.4	0.3	0.2	0.1	0.1	0.1	—	—	—	—	—	—	—	—	—	—	—	—	—	—	—
100	0.5	0.3	0.2	0.1	0.1	0.1	—	—	—	—	—	—	—	—	—	—	—	—	—	—	—	—	—

* Solid Conductors. Other conductors are stranded.

Note 1—The above table gives voltage drops encountered in a single phase two-wire system. The voltage drops in other systems may be obtained through multiplication by appropriate factors listed below:

System for Which Voltage Drop is Desired	Multiplying Factors for Modification of Values in Table
Single Phase—3 Wire—Line to Line	1.00
Single Phase—3 Wire—Line to Neutral	0.50
Three Phase—3 Wire—Line to Line	0.866
Three Phase—4 Wire—Line to Line	0.866
Three Phase—4 Wire—Line to Neutral	0.50

Note 2—Allowable voltage drops for systems other than single phase, two wire cannot be used directly in the above table. Such drops should be modified through multiplication by the appropriate factor listed below. The voltage thus modified may then be used to obtain the proper wire size directly from the table.

System for Which Allowable Voltage Drop is Known	Multiplying Factor for Modification of Known Value to Permit Direct Use of Table
Single Phase—3 Wire—Line to Line	1.00
Single Phase—3 Wire—Line to Neutral	2.00
Three Phase—3 Wire—Line to Line	1.155
Three Phase—4 Wire—Line to Line	1.155
Three Phase—4 Wire—Line to Neutral	2.00

Note 3—The footage employed in the tabulated ampere feet refers to the length of run of the circuit rather than to the footage of individual conductor.

Note 4—The above table is figured at 60°C since this is an estimate of the average temperature which may be anticipated in service. The table may be used without significant error for conductor temperatures up to and including 75°C.

TABLE 3.70 Volts Drop for CU Conductor in Nonmagnetic Conduit—90 Percent PF

WIRE SIZE AWG or MCM → Ampere Feet ↓	1000	900	800	750	700	600	500	400	350	300	250	4/0	3/0	2/0	1/0	1	2	4	6	8*	10*	12*	14*
												Volts Drop											
500,000	27.0	28.0	30.0	31.0	32.0	34.0	39.0	44.0	49.0	55.0	62.0	69.0	85.0	103.0	127.0	155.0	191.0	295.0	456.0	—	—	—	—
400,000	21.6	22.4	24.0	24.8	25.6	26.2	31.2	35.2	39.2	44.0	49.6	55.2	68.0	82.4	102.0	124.0	153.0	236.0	364.0	—	—	—	—
300,000	16.2	16.8	18.0	18.6	19.2	20.4	23.4	26.4	29.4	33.0	37.2	41.4	51.0	61.8	76.2	93.0	115.0	177.0	273.0	417.0	—	—	—
200,000	10.8	11.2	12.0	12.4	12.8	13.6	15.6	17.6	19.6	22.0	24.8	27.6	34.0	41.2	50.8	62.0	76.4	118.0	182.0	278.0	440.0	—	—
100,000	5.4	5.6	6.0	6.2	6.4	6.8	7.8	8.8	9.8	11.0	12.4	13.8	17.0	20.6	25.4	31.0	38.2	59.0	91.2	139.0	220.0	350.0	—
90,000	4.8	4.9	5.4	5.5	5.8	6.2	7.0	7.9	8.8	9.9	11.3	12.4	15.3	18.5	22.9	27.9	34.4	53.2	81.7	125.0	198.0	315.0	497.0
80,000	4.3	4.4	4.8	4.9	5.2	5.5	6.2	7.0	7.8	8.8	10.0	11.0	13.6	16.5	20.3	24.8	30.6	47.4	72.7	111.0	176.0	280.0	442.0
70,000	3.8	3.9	4.2	4.3	4.5	4.8	5.5	6.2	6.9	7.7	8.7	9.6	11.9	14.4	17.8	21.7	26.8	41.4	63.7	97.3	154.0	245.0	386.0
60,000	3.2	3.4	3.6	3.7	3.8	4.1	4.7	5.3	5.9	6.6	7.4	8.2	10.2	12.4	15.2	18.6	22.9	35.4	54.7	83.5	132.0	210.0	331.0
50,000	2.7	2.8	3.0	3.1	3.2	3.4	3.9	4.4	4.9	5.5	6.2	6.9	8.5	10.3	12.7	15.5	19.1	29.5	45.6	69.6	110.0	175.0	276.0
40,000	2.2	2.2	2.4	2.5	2.6	2.7	3.1	3.5	3.9	4.4	4.9	5.5	6.8	8.2	10.2	12.4	15.3	23.6	36.4	55.6	88.0	140.0	221.0
30,000	1.6	1.7	1.8	1.8	1.9	2.0	2.3	2.6	2.9	3.3	3.7	4.1	5.1	6.2	7.6	9.3	11.5	17.7	27.3	41.7	66.0	105.0	166.0
20,000	1.1	1.1	1.2	1.2	1.3	1.4	1.5	1.8	2.0	2.2	2.5	2.8	3.4	4.1	5.1	6.2	7.6	11.8	18.2	27.8	44.0	70.0	110.0
10,000	0.5	0.6	0.6	0.6	0.6	0.7	0.8	0.9	1.0	1.1	1.2	1.4	1.7	2.1	2.5	3.1	3.8	5.9	9.1	13.9	22.0	35.0	55.2
9,000	0.5	0.5	0.5	0.6	0.6	0.6	0.7	0.8	0.9	1.0	1.1	1.2	1.5	1.9	2.3	2.8	3.4	5.3	8.2	12.5	19.8	31.5	49.7
8,000	0.4	0.4	0.5	0.5	0.5	0.6	0.6	0.7	0.8	0.9	1.0	1.1	1.4	1.7	2.0	2.5	3.1	4.7	7.3	11.1	17.6	28.0	44.2
7,000	0.4	0.4	0.4	0.4	0.5	0.5	0.6	0.6	0.7	0.8	0.9	1.0	1.2	1.4	1.8	2.2	2.7	4.1	6.4	9.7	15.4	24.5	38.6
6,000	0.3	0.3	0.4	0.4	0.4	0.5	0.5	0.5	0.6	0.7	0.7	0.8	1.0	1.2	1.5	1.9	2.3	3.5	5.5	8.4	13.2	21.0	33.1
5,000	0.3	0.3	0.3	0.3	0.3	0.4	0.4	0.4	0.5	0.6	0.6	0.7	0.9	1.0	1.3	1.6	1.9	2.9	4.6	6.9	11.0	17.5	27.6
4,000	0.2	0.2	0.2	0.2	0.3	0.3	0.3	0.3	0.4	0.4	0.5	0.6	0.7	0.8	1.0	1.2	1.5	2.4	3.6	5.6	8.8	14.0	22.1
3,000	0.2	0.2	0.2	0.2	0.2	0.2	0.2	0.2	0.3	0.3	0.3	0.4	0.5	0.6	0.8	0.9	1.1	1.8	2.7	4.2	6.6	10.5	16.6
2,000	0.1	0.1	0.1	0.1	0.1	0.1	0.2	0.2	0.2	0.2	0.2	0.3	0.3	0.4	0.5	0.6	0.8	1.2	1.8	2.8	4.4	7.0	11.0
1,000	0.1	0.1	0.1	0.1	0.1	0.1	0.1	0.1	0.1	0.1	0.1	0.1	0.2	0.2	0.3	0.3	0.4	0.6	0.9	1.4	2.2	3.5	5.5
900	0.1	0.1	0.1	0.1	0.1	0.1	0.1	0.1	0.1	0.1	0.1	0.1	0.2	0.2	0.2	0.3	0.3	0.5	0.8	1.3	1.9	3.2	4.9
800	—	—	0.1	0.1	0.1	0.1	0.1	0.1	0.1	0.1	0.1	0.1	0.1	0.2	0.2	0.2	0.3	0.5	0.7	1.1	1.8	2.8	4.4
700	—	—	—	—	—	—	0.1	0.1	0.1	0.1	0.1	0.1	0.1	0.1	0.2	0.2	0.3	0.4	0.6	1.0	1.5	2.5	3.9
600	—	—	—	—	—	—	—	0.1	0.1	0.1	0.1	0.1	0.1	0.1	0.2	0.2	0.2	0.4	0.5	0.8	1.3	2.1	3.3
500	—	—	—	—	—	—	—	—	0.1	0.1	0.1	0.1	0.1	0.1	0.1	0.2	0.2	0.3	0.3	0.7	1.1	1.8	2.8
400	—	—	—	—	—	—	—	—	—	—	0.1	0.1	0.1	0.1	0.1	0.1	0.2	0.2	0.2	0.6	0.9	1.4	2.2
300	—	—	—	—	—	—	—	—	—	—	—	—	—	0.1	0.1	0.1	0.1	0.2	0.1	0.4	0.7	1.1	1.7
200	—	—	—	—	—	—	—	—	—	—	—	—	—	—	—	0.1	0.1	0.1	0.1	0.3	0.4	0.7	1.1
100	—	—	—	—	—	—	—	—	—	—	—	—	—	—	—	—	0.1	0.1	0.1	0.1	0.2	0.4	0.6

* Solid Conductors. Other conductors are stranded.

Note 1—The above table gives voltage drops encountered in a single phase two-wire system. The voltage drops in other systems may be obtained through multiplication by appropriate factors listed below:

System for Which Voltage Drop is Desired	Multiplying Factors for Modification of Values in Table
Single Phase—3 Wire—Line to Line	1.00
Single Phase—3 Wire—Line to Neutral	0.50
Three Phase—3 Wire—Line to Line	0.866
Three Phase—4 Wire—Line to Line	0.866
Three Phase—4 Wire—Line to Neutral	0.50

Note 2—Allowable voltage drops for systems other than single phase, two wire cannot be used directly in the above table. Such drops should be modified through multiplication by the appropriate factor listed below. The voltage thus modified may then be used to obtain the proper wire size directly from the table.

System for Which Allowable Voltage Drop is Known	Multiplying Factor for Modification of Known Value to Permit Direct Use of Table
Single Phase—3 Wire—Line to Line	1.00
Single Phase—3 Wire—Line to Neutral	2.00
Three Phase—3 Wire—Line to Line	1.155
Three Phase—4 Wire—Line to Line	1.155
Three Phase—4 Wire—Line to Neutral	2.00

Note 3—The footage employed in the tabulated ampere feet refers to the length of run of the circuit rather than to the footage of individual conductor.

Note 4—The above table is figured at 60°C since this is an estimate of the average temperature which may be anticipated in service. The table may be used without significant error for conductor temperatures up to and including 75°C,

TABLE 3.71 Volts Drop for CU Conductor in Nonmagnetic Conduit—95 Percent PF

Wire Size (AWG or MCM) across top; Ampere Feet down the side. Values are Volts Drop.

Ampere Feet	1000	900	800	750	700	600	500	400	350	300	250	4/0	3/0	2/0	1/0	1	2	4	6	8*	10*	12*	14*
500,000	23.0	25.0	26.0	27.0	29.0	31.0	36.0	41.0	46.0	53.0	60.0	67.0	84.0	103.0	128.0	157.0	195.0	305.0	473.0	—	—	—	—
400,000	18.4	20.0	20.8	21.6	23.2	24.8	28.8	32.8	36.8	42.4	48.0	53.6	67.2	82.4	102.0	126.0	156.0	244.0	378.0	—	—	—	—
300,000	13.8	15.0	15.6	16.2	17.4	18.6	21.6	24.6	27.6	31.8	36.0	40.2	50.4	61.8	76.8	94.2	117.0	183.0	283.0	435.0	—	—	—
200,000	9.2	10.0	10.4	10.8	11.6	12.4	14.4	16.4	18.4	21.2	24.0	26.8	33.6	41.2	51.2	62.8	78.0	122.0	189.0	290.0	462.0	—	—
100,000	4.6	5.0	5.2	5.4	5.8	6.2	7.2	8.2	9.2	10.6	12.0	13.4	16.8	20.6	25.6	31.4	39.0	61.0	94.5	145.0	231.0	368.0	—
90,000	4.1	4.5	4.7	4.9	5.2	5.6	6.5	7.4	8.3	9.5	10.8	12.1	15.1	18.5	23.0	28.3	35.1	54.9	85.2	131.0	208.0	331.0	—
80,000	3.7	4.0	4.2	4.3	4.6	5.0	5.8	6.6	7.4	8.5	9.6	10.7	13.4	16.5	20.5	25.1	31.2	48.8	75.7	116.0	185.0	294.0	466.0
70,000	3.2	3.5	3.6	3.8	4.1	4.3	5.0	5.7	6.4	7.4	8.4	9.4	11.7	14.4	17.9	22.0	27.3	42.7	66.2	102.0	162.0	258.0	408.0
60,000	2.8	3.0	3.1	3.2	3.5	3.7	4.3	4.9	5.5	6.4	7.2	8.0	10.1	12.4	15.4	18.8	23.4	36.6	56.7	87.5	139.0	221.0	350.0
50,000	2.3	2.5	2.6	2.7	2.9	3.1	3.6	4.1	4.6	5.3	6.0	6.7	8.4	10.3	12.8	15.7	19.5	30.5	47.3	72.8	116.0	184.0	291.0
40,000	1.8	2.0	2.1	2.2	2.3	2.5	2.9	3.3	3.7	4.2	4.8	5.4	6.7	8.2	10.2	12.6	15.6	24.4	37.9	58.0	92.4	147.0	232.0
30,000	1.4	1.5	1.6	1.6	1.7	1.9	2.2	2.5	2.8	3.2	3.6	4.0	5.0	6.2	7.7	9.4	11.7	18.3	28.3	43.5	69.3	110.0	174.0
20,000	0.9	1.0	1.0	1.1	1.2	1.2	1.4	1.6	1.8	2.1	2.4	2.7	3.4	4.1	5.1	6.3	7.8	12.2	18.9	29.0	46.2	73.6	116.0
10,000	0.5	0.5	0.5	0.5	0.6	0.6	0.7	0.8	0.9	1.1	1.2	1.3	1.7	2.1	2.6	3.1	3.9	6.1	9.5	14.5	23.1	36.8	58.1
9,000	0.4	0.5	0.5	0.5	0.5	0.6	0.6	0.7	0.8	1.0	1.1	1.2	1.5	1.9	2.3	2.8	3.5	5.5	8.5	13.1	20.8	33.1	52.4
8,000	0.4	0.4	0.4	0.4	0.5	0.5	0.6	0.7	0.7	0.8	1.0	1.1	1.3	1.6	2.0	2.5	3.1	4.9	7.6	11.6	18.5	29.4	46.6
7,000	0.3	0.4	0.4	0.4	0.4	0.4	0.5	0.6	0.6	0.7	0.8	0.9	1.2	1.4	1.8	2.2	2.7	4.3	6.6	10.2	16.2	25.8	40.8
6,000	0.3	0.3	0.3	0.3	0.3	0.4	0.4	0.5	0.6	0.6	0.7	0.8	1.0	1.2	1.5	1.9	2.3	3.7	5.7	8.8	13.9	22.1	35.0
5,000	0.2	0.2	0.3	0.3	0.3	0.3	0.4	0.4	0.5	0.5	0.6	0.7	0.8	1.0	1.3	1.6	1.9	3.1	4.7	7.3	11.6	18.4	29.1
4,000	0.2	0.2	0.2	0.2	0.2	0.2	0.3	0.3	0.4	0.4	0.5	0.5	0.7	0.8	1.0	1.3	1.6	2.4	3.8	5.8	9.2	14.7	23.2
3,000	0.1	0.2	0.2	0.2	0.2	0.2	0.2	0.2	0.3	0.3	0.4	0.4	0.5	0.6	0.8	0.9	1.2	1.8	2.8	4.4	6.9	11.0	17.4
2,000	0.1	0.1	0.1	0.1	0.1	0.1	0.1	0.2	0.2	0.2	0.2	0.3	0.3	0.4	0.5	0.6	0.8	1.2	1.9	2.9	4.6	7.4	11.6
1,000	0.1	0.1	0.1	0.1	0.1	0.1	0.1	0.1	0.1	0.1	0.1	0.1	0.2	0.2	0.3	0.3	0.4	0.6	0.9	1.5	2.3	3.7	5.8
900	—	—	—	—	0.1	0.1	0.1	0.1	0.1	0.1	0.1	0.1	0.2	0.2	0.2	0.3	0.4	0.5	0.9	1.3	2.1	3.3	5.2
800	—	—	—	—	—	—	0.1	0.1	0.1	0.1	0.1	0.1	0.1	0.2	0.2	0.3	0.3	0.5	0.8	1.2	1.8	2.9	4.7
700	—	—	—	—	—	—	0.1	0.1	0.1	0.1	0.1	0.1	0.1	0.1	0.2	0.2	0.3	0.4	0.7	1.0	1.6	2.6	4.1
600	—	—	—	—	—	—	—	—	0.1	0.1	0.1	0.1	0.1	0.1	0.2	0.2	0.2	0.4	0.6	0.9	1.4	2.2	3.5
500	—	—	—	—	—	—	—	—	—	0.1	0.1	0.1	0.1	0.1	0.1	0.2	0.2	0.3	0.5	0.7	1.2	1.8	2.9
400	—	—	—	—	—	—	—	—	—	—	—	0.1	0.1	0.1	0.1	0.1	0.2	0.2	0.4	0.6	0.9	1.5	2.3
300	—	—	—	—	—	—	—	—	—	—	—	—	0.1	0.1	0.1	0.1	0.1	0.2	0.3	0.4	0.7	1.1	1.7
200	—	—	—	—	—	—	—	—	—	—	—	—	—	—	0.1	0.1	0.1	0.1	0.2	0.3	0.5	0.7	1.2
100	—	—	—	—	—	—	—	—	—	—	—	—	—	—	—	—	—	0.1	0.1	0.1	0.2	0.4	0.6

* Solid Conductors. Other conductors are stranded.

Note 1—The above table gives voltage drops encountered in a single phase two-wire system. The voltage drops in other systems may be obtained through multiplication by appropriate factors listed below:

System for Which Voltage Drop is Desired	Multiplying Factors for Modification of Values in Table
Single Phase—3 Wire—Line to Line	1.00
Single Phase—3 Wire—Line to Neutral	0.50
Three Phase—3 Wire—Line to Line	0.866
Three Phase—4 Wire—Line to Line	0.866
Three Phase—4 Wire—Line to Neutral	0.50

Note 2—Allowable voltage drops for systems either than single phase, two wire cannot be used directly in the above table. Such drops should be modified through multiplication by the appropriate factor listed below. The voltage thus modified may then be used to obtain the proper wire size directly from the table.

System for Which Allowable Voltage Drop is Known	Multiplying Factor for Modification of Known Value to Permit Direct Use of Table
Single Phase—3 Wire—Line to Line	1.00
Single Phase—3 Wire—Line to Neutral	2.00
Three Phase—3 Wire—Line to Line	1.155
Three Phase—4 Wire—Line to Line	1.155
Three Phase—4 Wire—Line to Neutral	2.00

Note 3—The footage employed in the tabulated ampere feet refers to the length of run of the circuit rather than to the footage of individual conductor.

Note 4—The above table is figured at 60°C since this is an estimate of the average temperature which may be anticipated in service. The table may be used without significant error for conductor temperatures up to and including 75°C.

TABLE 3.72 Volts Drop for CU Conductor in Nonmagnetic Conduit—100 Percent PF

WIRE SIZE AWG or MCM	1000	900	800	750	700	600	500	400	350	300	250	4/0	3/0	2/0	1/0	1	2	4	6	8*	10*	12*	14*
Ampere Feet	Volts Drop																						
500,000	13.0	15.0	16.0	17.0	19.0	22.0	26.0	32.0	36.0	42.0	51.0	59.0	75.0	95.0	121.0	152.0	192.0	306.0	483.0	—	—	—	—
400,000	10.4	12.0	12.8	13.6	15.2	17.6	20.8	25.6	28.8	33.6	40.8	47.2	60.0	76.0	96.8	122.0	154.0	244.0	386.0	—	—	—	—
300,000	7.8	9.0	9.6	10.2	11.4	13.2	15.6	19.2	21.6	25.2	30.6	35.4	45.0	57.0	72.6	91.2	115.0	184.0	290.0	450.0	—	—	—
200,000	5.2	6.0	6.4	6.8	7.6	8.8	10.4	12.8	14.4	16.8	20.4	23.6	30.0	38.0	48.4	60.8	76.8	122.0	193.0	300.0	480.0	—	—
100,000	2.6	3.0	3.2	3.4	3.8	4.4	5.2	6.4	7.2	8.4	10.2	11.8	15.0	19.0	24.2	30.4	38.4	61.2	96.6	150.0	240.0	384.0	—
90,000	2.3	2.7	2.9	3.0	3.4	4.0	4.6	5.8	6.4	7.6	9.2	10.6	13.5	17.1	21.8	27.4	34.5	55.1	87.0	135.0	216.0	345.0	—
80,000	2.1	2.4	2.5	2.7	3.0	3.5	4.1	5.2	5.7	6.7	8.2	9.4	12.0	15.2	19.4	24.3	30.7	49.0	77.3	120.0	192.0	307.0	487.0
70,000	1.8	2.1	2.2	2.4	2.7	3.1	3.6	4.5	5.0	5.9	7.1	8.3	10.5	13.3	17.0	21.2	26.8	42.8	67.6	105.0	168.0	269.0	426.0
60,000	1.6	1.8	1.9	2.0	2.3	2.6	3.1	3.8	4.3	5.0	6.1	7.1	9.0	11.4	14.5	18.2	23.0	36.7	58.0	90.0	144.0	230.0	365.0
50,000	1.3	1.5	1.6	1.7	1.9	2.2	2.6	3.2	3.6	4.2	5.1	5.9	7.5	9.5	12.1	15.2	19.2	30.6	48.3	74.9	120.0	192.0	304.0
40,000	1.0	1.2	1.3	1.3	1.5	1.7	2.1	2.6	2.9	3.4	4.1	4.7	6.0	7.6	9.7	12.2	15.4	24.4	38.6	60.0	96.0	154.0	243.0
30,000	0.8	0.9	1.0	1.0	1.1	1.3	1.6	1.9	2.2	2.5	3.1	3.5	4.5	5.7	7.3	9.1	11.5	18.4	29.0	45.0	72.0	115.0	182.0
20,000	0.5	0.6	0.6	0.7	0.8	0.9	1.0	1.3	1.4	1.7	2.0	2.4	3.0	3.8	4.8	6.1	7.7	12.2	19.3	30.0	48.0	76.8	122.0
10,000	0.3	0.3	0.3	0.3	0.4	0.4	0.5	0.6	0.7	0.8	1.0	1.2	1.5	1.9	2.4	3.0	3.8	6.1	9.7	15.0	24.0	38.4	60.8
9,000	0.2	0.3	0.3	0.3	0.3	0.4	0.5	0.6	0.6	0.8	0.9	1.1	1.4	1.7	2.2	2.7	3.5	5.5	8.7	13.5	21.6	34.5	54.7
8,000	0.2	0.2	0.3	0.3	0.3	0.4	0.4	0.5	0.6	0.7	0.8	0.9	1.2	1.5	1.9	2.4	3.1	4.9	7.7	12.0	19.2	30.7	48.7
7,000	0.2	0.2	0.2	0.2	0.3	0.3	0.4	0.5	0.5	0.6	0.7	0.8	1.1	1.3	1.7	2.1	2.7	4.3	6.8	10.5	16.8	26.9	42.6
6,000	0.2	0.2	0.2	0.2	0.2	0.3	0.3	0.4	0.4	0.5	0.6	0.7	0.9	1.1	1.5	1.8	2.3	3.7	5.8	9.0	14.4	23.0	36.5
5,000	0.1	0.2	0.2	0.2	0.2	0.2	0.3	0.3	0.4	0.4	0.5	0.6	0.8	1.0	1.2	1.5	1.9	3.1	4.8	7.5	12.0	19.2	30.4
4,000	0.1	0.1	0.1	0.1	0.2	0.2	0.2	0.3	0.3	0.3	0.4	0.5	0.6	0.8	1.0	1.2	1.5	2.4	3.9	6.0	9.6	15.4	24.3
3,000	0.1	0.1	0.1	0.1	0.1	0.1	0.2	0.2	0.2	0.3	0.3	0.4	0.5	0.6	0.7	0.9	1.2	1.8	2.9	4.5	7.2	11.5	18.2
2,000	0.1	0.1	0.1	0.1	0.1	0.1	0.1	0.1	0.1	0.2	0.2	0.2	0.3	0.4	0.5	0.6	0.8	1.2	1.9	3.0	4.8	7.7	12.2
1,000	—	—	—	—	—	—	0.1	0.1	0.1	0.1	0.1	0.1	0.2	0.2	0.2	0.3	0.4	0.6	1.0	1.5	2.4	3.8	6.1
900	—	—	—	—	—	—	0.1	0.1	0.1	0.1	0.1	0.1	0.1	0.2	0.2	0.3	0.4	0.6	0.9	1.4	2.2	3.5	5.5
800	—	—	—	—	—	—	—	0.1	0.1	0.1	0.1	0.1	0.1	0.2	0.2	0.2	0.3	0.5	0.8	1.2	1.9	3.1	4.9
700	—	—	—	—	—	—	—	0.1	0.1	0.1	0.1	0.1	0.1	0.1	0.2	0.2	0.3	0.4	0.7	1.1	1.7	2.7	4.3
600	—	—	—	—	—	—	—	—	—	0.1	0.1	0.1	0.1	0.1	0.2	0.2	0.2	0.4	0.6	0.9	1.4	2.3	3.7
500	—	—	—	—	—	—	—	—	—	—	0.1	0.1	0.1	0.1	0.1	0.2	0.2	0.3	0.5	0.8	1.2	1.9	3.0
400	—	—	—	—	—	—	—	—	—	—	—	0.1	0.1	0.1	0.1	0.1	0.2	0.2	0.4	0.6	1.0	1.5	2.4
300	—	—	—	—	—	—	—	—	—	—	—	—	0.1	0.1	0.1	0.1	0.1	0.2	0.3	0.5	0.7	1.2	1.8
200	—	—	—	—	—	—	—	—	—	—	—	—	—	—	0.1	0.1	0.1	0.1	0.2	0.3	0.5	0.8	1.2
100	—	—	—	—	—	—	—	—	—	—	—	—	—	—	—	—	—	0.1	0.1	0.2	0.2	0.4	0.6

* Solid Conductors. Other conductors are stranded.

Note 1—The above table gives voltage drops encountered in a single phase two-wire system. The voltage drops in other systems may be obtained through multiplication by appropriate factors listed below:

System for Which Voltage Drop is Desired	Multiplying Factors for Modification of Values in Table
Single Phase—3 Wire—Line to Line	1.00
Single Phase—3 Wire—Line to Neutral	0.50
Three Phase—3 Wire—Line to Line	0.866
Three Phase—4 Wire—Line to Line	0.866
Three Phase—4 Wire—Line to Neutral	0.50

Note 2—Allowable voltage drops for systems other than single phase, two wire cannot be used directly in the above table. Such drops should be modified through multiplication by the appropriate factor listed below. The voltage thus modified may then be used to obtain the proper wire size directly from the table.

System for Which Allowable Voltage Drop is Known	Multiplying Factor for Modification of Known Value to Permit Direct Use of Table
Single Phase—3 Wire—Line to Line	1.00
Single Phase—3 Wire—Line to Neutral	2.00
Three Phase—3 Wire—Line to Line	1.155
Three Phase—4 Wire—Line to Line	1.155
Three Phase—4 Wire—Line to Neutral	2.00

Note 3—The footage employed in the tabulated ampere feet refers to the length of run of the circuit rather than to the footage of individual conductor.

Note 4—The above table is figured at 60°C since this is an estimate of the average temperature which may be anticipated in service. The table may be used without significant error for conductor temperatures up to and including 75°C.

In busways, Tables 3.73 and 3.74 and Figures 3.39 through 3.41 show voltage drops per 100 feet at rated current (end loading) for the entire range of lagging power factors.

The voltage drop for a single-phase load connected to a three-phase system busway is 15.5 percent higher than the values shown in the tables. For a two-pole busway serving a single-phase load, the voltage drop values in Tables 3.73 and 3.74 should be multiplied by 1.08.

The tables show end-loaded conditions; that is, the entire load is concentrated at one end at rated capacity. Because plug-in types of busways are particularly adapted to serving the distributed blocks of load, care should be exercised to ensure proper handling of such voltage drop cal-

FIGURE 3.39   Voltage drop curves for typical interleaved construction of copper busway at rated load, assuming 70°C (158°F) as the operating temperature.

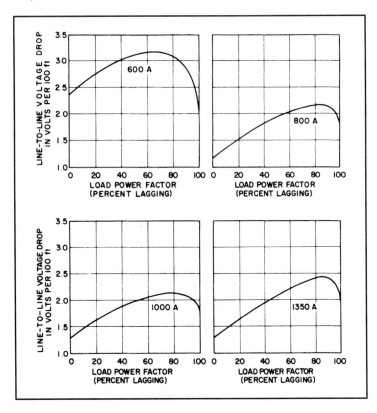

TABLE 3.73 Voltage Drop Values for Three-Phase Busways with Copper Bus Bars, in Volts per 100 Feet, Line-to-Line, at Rated Current with Balanced Entire Load at End

Rating (amperes)	Power Factor									
	20	30	40	50	60	70	80	90	95	100
Low-voltage-drop ventilated feeder										
800	3.66	3.88	4.04	4.14	4.20	4.20	4.16	3.92	3.60	2.72
1000	1.84	2.06	2.22	2.40	2.54	2.64	2.72	2.70	2.62	2.30
1350	2.24	2.44	2.62	2.74	2.86	2.94	2.96	2.90	2.78	2.30
1600	1.88	2.10	2.30	2.46	2.62	2.74	2.82	2.84	2.76	2.42
2000	2.16	2.34	2.52	2.66	2.78	2.84	2.90	2.80	2.68	2.30
2500	2.04	2.18	2.38	2.48	2.62	2.68	2.72	2.62	2.50	2.14
3000	1.96	2.12	2.28	2.40	2.52	2.58	2.60	2.52	2.40	2.06
4000	2.18	2.36	2.54	2.68	2.80	2.80	2.90	2.80	2.68	2.28
5000	2.00	2.16	2.30	2.40	2.50	2.60	2.68	2.60	2.40	2.10
Low-voltage-drop ventilated plug-in										
800	6.80	6.86	6.92	6.86	6.72	6.52	6.04	5.26	4.64	2.76
1000	2.26	2.56	2.70	2.86	2.96	3.00	3.00	2.92	2.80	2.28
1350	2.98	3.16	3.32	3.38	3.44	3.46	3.40	3.22	3.00	2.32
1600	2.28	2.44	2.62	2.78	2.90	3.00	2.96	2.94	2.88	2.44
2000	2.58	2.78	2.92	3.02	3.10	3.16	3.08	3.00	2.82	2.28
2500	2.32	2.50	2.66	2.76	2.86	2.90	2.86	2.78	2.66	2.18
3000	2.18	2.34	2.48	2.60	2.70	2.74	2.72	2.66	2.58	2.10
4000	2.42	2.56	2.76	2.88	3.00	3.02	3.00	2.96	2.84	2.36
5000	2.22	2.30	2.48	2.60	2.70	2.76	2.74	2.68	2.60	2.16
Plug-in										
225	2.82	2.94	3.04	3.12	3.18	3.18	3.10	2.86	2.70	2.04
400	4.94	5.08	5.16	5.18	5.16	5.02	4.98	4.30	3.94	2.64
600	5.24	5.34	5.40	5.40	5.36	5.00	4.50	2.10	3.62	2.92
800	5.06	5.12	5.16	5.06	5.00	4.74	4.50	3.84	3.32	1.94
1000	5.80	5.88	5.84	5.76	5.56	5.30	4.82	4.12	3.52	1.94
Trolley busway										
100	1.2	1.38	1.58	1.74	1.80	2.06	2.20	2.30	2.30	2.18
Current-limiting ventilated										
1000	12.3	12.5	12.3	12.2	11.8	11.1	10.1	8.65	7.45	3.8
1350	15.5	15.6	15.4	15.3	14.7	13.9	12.6	10.7	9.2	4.7
1600	18.2	18.2	18.0	17.5	16.6	15.6	14.1	11.5	9.5	4.0
2000	20.4	20.3	20.0	19.4	18.4	17.0	13.9	12.1	10.1	3.8
2500	23.8	23.6	23.0	22.2	21.0	19.2	17.2	13.5	10.7	3.8
3000	26.0	26.2	25.8	24.8	23.4	21.5	19.1	15.1	12.0	4.0
4000	29.1	28.8	28.2	27.2	25.6	25.2	21.0	16.6	13.0	4.1

TABLE 3.74 Voltage Drop Values for Three-Phase Busways with Aluminum Bus Bars, in Volts per 100 Feet, Line-to-Line, at Rated Current with Balanced Entire Load at End

Rating (amperes)	Power Factor									
	20	30	40	50	60	70	80	90	95	100
Low-voltage-drop ventilated feeder										
800	1.68	1.96	2.20	2.46	2.68	2.88	3.04	3.12	3.14	2.90
1000	1.90	2.16	2.38	2.60	2.80	2.96	3.06	3.14	3.12	2.82
1350	1.88	2.20	2.48	2.74	3.02	3.24	3.44	3.56	3.58	2.38
1600	1.66	1.92	2.18	2.42	2.64	2.84	3.02	3.12	3.16	2.94
2000	1.82	2.06	2.30	2.50	2.70	2.88	3.02	3.10	3.04	2.80
2500	1.86	2.10	2.34	2.56	2.74	2.90	3.04	3.06	3.08	2.78
3000	1.76	2.06	2.26	2.52	2.68	2.86	2.98	3.06	3.04	2.78
4000	1.74	1.98	2.24	2.48	2.70	2.88	3.04	3.08	3.12	2.88
5000	1.72	1.98	2.20	2.42	2.62	2.80	2.92	3.02	3.02	2.80
Low-voltage-drop ventilated plug-in										
800	2.12	2.38	2.58	2.80	3.00	3.16	3.26	3.30	3.24	2.90
1000	2.44	2.66	2.86	3.06	3.22	3.36	3.42	3.38	3.28	2.84
1350	2.22	2.48	2.78	3.00	3.24	3.46	3.60	3.68	3.64	3.30
1600	1.82	2.12	2.38	2.62	2.80	2.96	3.08	3.16	3.14	2.88
2000	2.00	2.30	2.50	2.76	2.92	3.06	3.12	3.18	3.12	2.80
2500	2.00	2.28	2.50	2.70	2.92	3.02	3.12	3.16	3.08	1.78
3000	1.98	2.26	2.44	2.66	2.86	3.00	3.10	3.18	3.14	2.82
4000	1.94	2.20	2.48	2.64	2.86	3.00	3.12	3.18	3.16	2.88
5000	1.90	2.16	2.38	2.58	2.76	2.92	3.06	3.10	3.08	2.52

Plug-in										
100	1.58	2.10	2.62	3.14	3.56	4.00	4.46	4.94	5.10	5.20
225	2.30	2.54	2.76	3.68	3.12	3.26	3.32	3.32	3.26	2.86
400	3.38	3.64	3.90	4.12	4.22	4.34	4.38	4.28	4.12	3.42
600	3.46	3.68	3.84	3.96	4.00	4.04	3.96	3.74	3.52	2.48
800	3.88	4.02	4.08	4.20	4.20	4.14	4.00	3.66	3.40	2.40
1000	3.30	3.48	3.62	3.72	3.78	3.80	3.72	3.50	3.30	2.50
Small plug-in										
50	2.2	2.6	3.0	3.5	3.8	4.1	4.5	4.7	4.8	4.6
Current-limiting ventilated										
1000	12.3	12.3	12.1	11.8	11.2	10.9	9.5	8.0	6.6	3.1
1350	16.3	16.3	16.1	15.6	14.7	13.7	12.1	8.1	8.0	3.1
1600	18.0	17.9	17.7	17.0	16.1	14.9	13.4	10.7	8.6	3.3
2000	22.5	22.4	21.8	21.1	19.9	18.2	16.0	12.7	9.9	3.1
2500	25.0	24.6	23.9	23.1	21.7	19.9	17.5	13.7	10.8	3.0
3000	26.2	25.8	25.1	24.1	22.7	20.8	18.2	14.2	10.9	2.9
4000	31.4	31.0	30.2	28.8	27.4	24.8	21.5	16.5	12.7	2.9

FIGURE 3.40   Voltage drop curves for typical plug-in-type busway at balanced rated load, assuming 70°C (158°F) as the operating temperature.

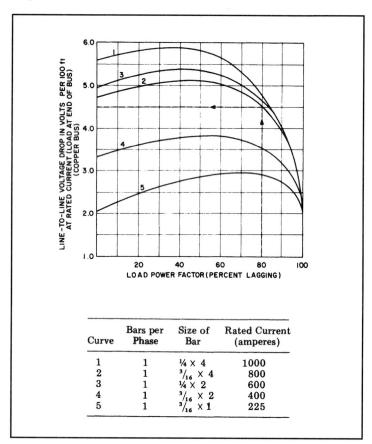

Curve	Bars per Phase	Size of Bar	Rated Current (amperes)
1	1	¼ × 4	1000
2	1	³/₁₆ × 4	800
3	1	¼ × 2	600
4	1	³/₁₆ × 2	400
5	1	³/₁₆ × 1	225

culations. Thus, with uniformly distributed loading, the values in the tables should be divided by 2. When several separate blocks of load are tapped off the run at various points, the voltage drop should be determined for the first section using the total load. The voltage drop in the next section is then calculated using the total load minus what was tapped off in the first section, and so on.

Figure 3.42 shows the voltage drop curve versus power factor for typical light-duty trolley busway carrying rated load.

FIGURE 3.41   Voltage drop curves for typical feeder busways at balanced rated load mounted flat horizontally, assuming 70°C (158°F) as the operating temperature.

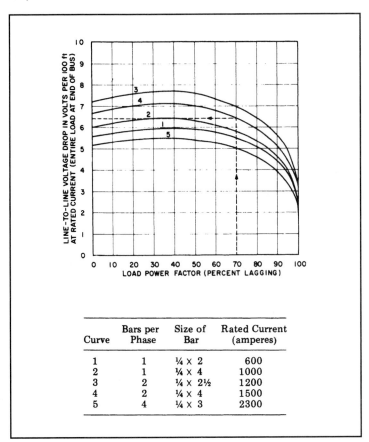

Curve	Bars per Phase	Size of Bar	Rated Current (amperes)
1	1	¼ × 2	600
2	1	¼ × 4	1000
3	2	¼ × 2½	1200
4	2	¼ × 4	1500
5	4	¼ × 3	2300

Figure 3.43 may be used to determine the approximate voltage drop in single-phase and three-phase 60-Hz liquid-filled, self-cooled transformers. The voltage drop through a single-phase transformer is found by entering the chart at a kilovolt-ampere value three times the rating of the single-phase transformer. Figure 3.43 covers transformers in the following ranges:

FIGURE 3.42 Voltage drop curve versus power factor for typical light-duty trolley busway carrying rated load, assuming 70°C (158°F) as the operating temperature.

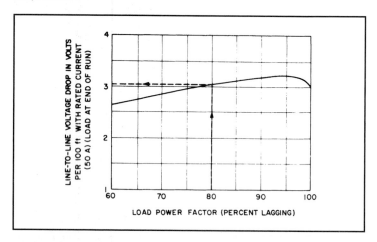

FIGURE 3.43 Voltage drop curves for three-phase transformers, 225 to 10,000 kVA, 5 to 25 kV. Note: This figure applies to 5.5 percent impedance transformers. For transformers of substantially different impedance, the information for the calculation should be obtained from the manufacturer.

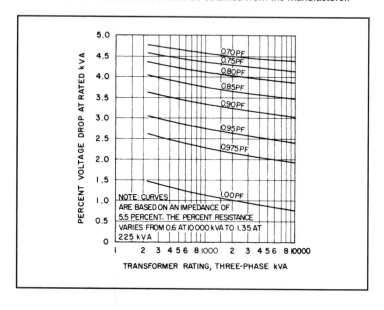

Single-phase
- 250 to 500 kVA, 8.6- to 15-kV insulation classes
- 833 to 1,250 kVA, 2.5- to 25-kV insulation classes

Three-phase
- 225 to 750 kVA, 8.6- to 15-kV insulation classes
- 1,000 to 10,000 kVA, 2.5- to 25-kV insulation classes

APPLICATION TIPS

1. Always locate the source of the low-voltage supply (service transformer and service equipment, distribution transformers, distribution panels, generators, and UPS systems) as close to the center of the load as possible.

2. When you oversize a feeder or branch circuit for voltage drop compensation, note it as such on the design drawings. This prevents confusion for the electrical contractor(s) bidding and/or installing the work.

3. *Rule of thumb:* When the distance in circuit feet equals the nominal system voltage (e.g., you are at 120 circuit feet and the nominal system voltage is 120 V), it serves as a "flag" that you should check the voltage drop. In practice, experience has generally shown that it is safe to go another 50 percent in circuit feet without a voltage drop problem (180 circuit feet for the example given).

4. As is the case with short-circuit calculations, the only significant circuit impedance parameters generally needed for the voltage drop calculations are those of transformers, busways, and conductors in conduit. Devices such as switches, circuit breakers, transfer switches, and so forth, contribute negligible impedance and generally can be ignored.

5. The NEC recommends (not mandatory) that the voltage drop from the point-of-service entrance to the farthest extremity of the electrical distribution system not exceed 5 percent. With this guideline, it is generally good practice to limit the voltage drop to distribution panels to a maximum of 2 to 3 percent, leaving the remaining 2 to 3 percent for the smaller branch circuits to the extremities of the system. For example, limiting the voltage drop to 2 percent to a distribution panel would allow up to 3 percent voltage drop for the branch circuits served by that panel.

Voltage Dips—Momentary Voltage Variations

The previous discussion covered relatively slow changes in voltage associated with steady-state voltage spreads and tolerance limits. However, sudden voltage changes should be given special consideration. Lighting equipment output is sensitive to applied voltage, and people are sensitive to sudden changes in light. Intermittently operated equipment, such as compressor motors, elevators, x-ray machines, and flashing signs, may produce a flicker when connected to lighting circuits. Care should be taken to design systems that will not irritate building occupants with flickering lights. In extreme cases, sudden voltage changes may even disrupt sensitive electronic equipment.

As little as a 0.5 percent voltage change produces a noticeable change in the output of an incandescent lamp. The problem is that individuals vary widely in their susceptibility to light flicker. Tests indicate that some individuals are irritated by a flicker that is barely noticeable to others. Studies show that sensitivity depends on how much illumination changes (magnitude), how often it occurs (frequency), and the type of work activity undertaken. The problem is further compounded by the fact that fluorescent and other lighting systems have different response characteristics to voltage changes (see previous parts of this section). Illumination flicker can be especially objectionable if it occurs often and is cyclical.

Figure 3.44 shows acceptable voltage dip limits for incandescent lights. Two curves show how the acceptable voltage flicker magnitude

FIGURE 3.44 Flicker of incandescent lamps caused by recurrent voltage dips.

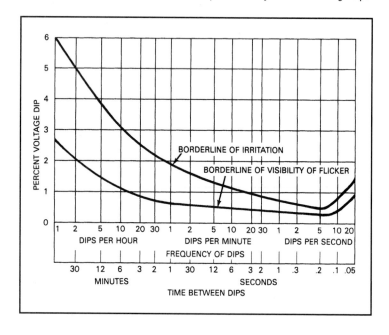

depends on the frequency of occurrence. The lower curve shows a borderline where people begin to detect the flicker. The upper curve is the borderline where some people will find the flicker objectionable. At 10 dips per hour, people begin to detect incandescent lamp flicker for voltage dips larger than 1 percent and begin to object when the magnitude exceeds 3 percent.

One source of voltage dips in commercial buildings is the inrush current while starting large motors on a distribution transformer that also supplies incandescent lights. A quick way to estimate flicker problems from motor starting is to multiply the motor locked-rotor starting kilovolt-ampere by the supply transformer impedance. A typical motor may draw 5 kVA/hp and a transformer impedance may be 6 percent. The equation below estimates flicker while starting a 15-hp motor on a 150-kVA transformer.

$$15 \text{ hp} \times 5 \text{ kVA/hp} \times 6\%/150 \text{ kVA} = 3\% \text{ flicker}$$

The estimated 3 percent dip associated with starting this motor reaches the borderline of irritation at 10 starts/hr. If the voltage dip combined with the starting frequency approaches the objectionable zone, more accurate calculations should be made using the actual locked-rotor current of the motor. Accurate locked-rotor kilovolt-amperes for motors are available from the motor manufacturer and from the starting code letter on the motor nameplate. The values for the code letters are listed in Table 3.39 of this handbook. More accurate methods for calculating motor-starting voltage dips are beyond the scope of this book.

One slightly more accurate method of quickly calculating voltage dip is to ratio the inrush current, or kilovolt-amperes, to the available short-circuit current, or kilovolt-amperes (if known), times 100 percent, to that point in the system of concern. This takes into account all impedance to the point in the system.

When the amount of the voltage dip in combination with the frequency falls within the objectionable range, then consideration should be given to methods of reducing the dip to acceptable values, such as using two or more smaller motors, providing a separate transformer for motors, separating motor feeders from other feeders, or using reduced-voltage motor starting.

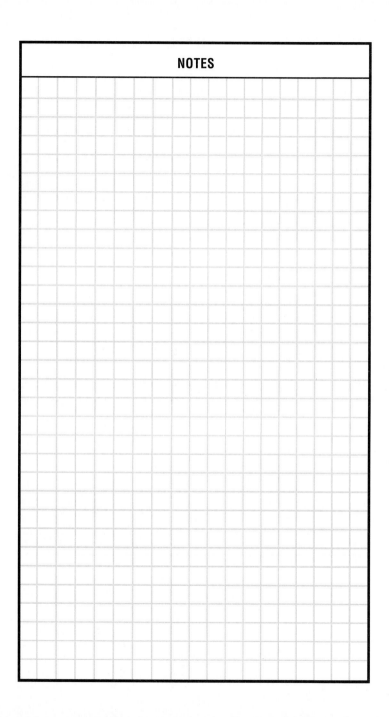

NOTES

CHAPTER FOUR
Grounding and Ground Fault Protection

4.0   GROUNDING

Introduction

Grounding encompasses several different but interrelated aspects of electrical distribution system design and construction, all of which are essential to the safety and proper operation of the system and equipment supplied by it. Among these are equipment grounding, system grounding, static and lightning protection, and connection to earth as a reference (zero) potential.

Equipment Grounding

Equipment grounding is essential to the safety of personnel. Its function is to ensure that all exposed noncurrent-carrying metallic parts of all structures and equipment in or near the electrical distribution system are at the same potential, and that is the zero reference potential of the earth. Grounding is required by both the National Electrical Code (Article 250) and the National Electrical Safety Code.

Equipment grounding also provides a return path for ground fault currents, permitting protective devices to operate effectively. Accidental contact of an energized conductor of the system with an improperly grounded noncurrent-carrying metallic part of the system (such as a motor frame or panelboard enclosure) would raise the potential of the metal object above ground potential. Any person coming in contact with such an object while grounded could be seriously injured or killed. In addition, current flow from the accidental grounding of an energized part of the system could generate sufficient heat (often with arcing) to start a fire.

To prevent the establishment of such an unsafe potential difference requires that (1) the equipment-grounding conductor provide a return path for the ground fault currents of sufficiently low impedance to prevent unsafe voltage drop (i.e., voltage rise due to the IZ drop), and (2) the equipment-grounding conductor be large enough to carry the maximum ground fault current, without burning off, for sufficient time to allow protective devices (ground fault relays, circuit breakers, fuses) to clear the fault. The grounded conductor of the system (usually the neutral conductor), although grounded at the source, must not be used for equipment grounding.

The equipment-grounding conductor may be the metallic conduit or raceway of the wiring system, or a separate equipment-grounding conductor, run with the circuit conductors, as permitted by the NEC. For minimum-size equipment-grounding conductors for grounding raceway and equipment, see Table 4.1. If a separate equipment-grounding conductor is used, it may be bare or insulated; if it is insulated, the insula-

TABLE 4.1 Minimum Size of Equipment Grounding Conductors for Grounding Raceway and Equipment (NEC Table 250-95)

Rating or Setting of Automatic Overcurrent Device in Circuit Ahead of Equipment, Conduit, etc., Not Exceeding (Amperes)	Size	
	Copper Wire No.	Aluminum or Copper-Clad Aluminum Wire No.*
15	14	12
20	12	10
30	10	8
40	10	8
60	10	8
100	8	6
200	6	4
300	4	2
400	3	1
500	2	1/0
600	1	2/0
800	1/0	3/0
1000	2/0	4/0
1200	3/0	250 kcmil
1600	4/0	350 "
2000	250 kcmil	400 "
2500	350 "	600 "
3000	400 "	600 "
4000	500 "	800 "
5000	700 "	1200 "
6000	800 "	1200 "

* See installation restrictions in Section 250-92(a).
Note: Equipment grounding conductors may need to be sized larger than specified in this table in order to comply with Section 250-51.

tion must be green. Conductors with green insulation may not be used for any purpose other than for equipment grounding. Where conductors are run in parallel in multiple raceways or cables, the equipment-grounding conductor, where used, shall be run in parallel. Each parallel equipment-grounding conductor shall be sized in accordance with Table 4.1 (NEC Table 250-95).

The equipment-grounding system must be bonded to the grounding electrode at the source or service; however, it may also be connected to ground at many other points. This will not cause problems with the safe operation of the electrical distribution system. Where computers, data processing, or microprocessor-based industrial process control systems are installed, the equipment-grounding system must be designed to minimize interference with their proper operation. Often, isolated grounding of this equipment, or completely isolated electrical supply systems are required to protect microprocessors from power system "noise" that does not in any way affect motors or other electrical equipment.

Low-Voltage System Grounding

System grounding connects the electrical supply, from the utility, from transformer secondary windings, or from a generator, to ground. A system can be solidly grounded (no intentional impedance to ground), impedance-grounded (through a resistance or reactance), or ungrounded (with no intentional connection to ground).

The most commonly used grounding point is the neutral of the system, or the neutral point, created by means of a zigzag-wye or an open-delta grounding transformer in a system that was operating as an ungrounded-delta system.

In general, it is a good practice that all source neutrals be grounded with the same grounding impedance. Where one of the medium-voltage sources is the utility, their consent for impedance grounding must be obtained.

The neutral impedance must have a voltage rating at least equal to the rated line-to-neutral voltage class of the system. It must have at least a 10-s rating equal to the maximum future line-to-ground fault current and a continuous rating to accommodate the triplen harmonics that may be present.

Solidly grounded three-phase systems (Figure 4.1) are usually wye-connected, with the neutral point grounded. Less common is the *red-leg,* or *high-leg,* delta, a 240-V system supplied by some utilities with one winding center-tapped to provide 120 V to ground for lighting and receptacles. This 240-V, three-phase, four-wire system is used where a 120-V lighting load is small compared with a 240-V power load, because the installation is low in cost to the utility. A corner-grounded, three-phase delta system is sometimes found, with one phase grounded to sta-

FIGURE 4.1 Solidly grounded systems.

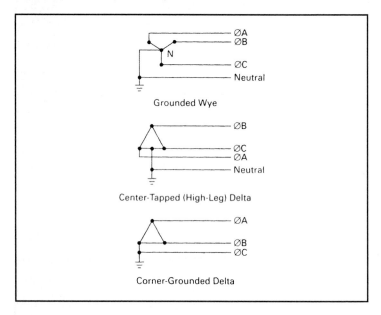

Grounded Wye

Center-Tapped (High-Leg) Delta

Corner-Grounded Delta

bilize all voltages to ground. Better solutions are available for new installations.

Ungrounded systems (Figure 4.2) can be either wye or delta, although the ungrounded delta system is far more common.

Resistance-grounded systems (Figure 4.3) are simplest with a wye connection, grounding the neutral point directly through the resistor.

FIGURE 4.2 Ungrounded systems.

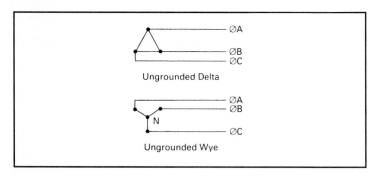

Ungrounded Delta

Ungrounded Wye

FIGURE 4.3 Resistance-grounded systems.

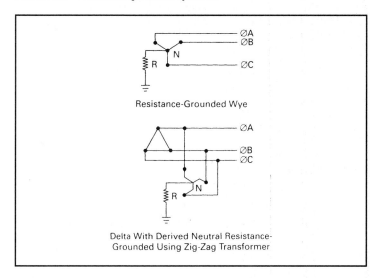

Resistance-Grounded Wye

Delta With Derived Neutral Resistance-
Grounded Using Zig-Zag Transformer

Delta systems can be grounded by means of a zigzag or other grounding transformer. Open-delta transformer banks may also be used.

This drives a neutral point, which can be either solidly or impedance-grounded. If the grounding transformer has sufficient capacity, the neutral created can be solidly grounded and used as a part of a three-phase, four-wire system. Most transformer-supplied systems are either solidly grounded or resistance-grounded. Generator neutrals are often grounded through a reactor, to limit ground fault (zero sequence) currents to values the generator can withstand. Generators that operate in parallel are sometimes resistance-grounded to suppress circulating harmonics.

Grounding-Electrode System

At some point, the equipment and system grounds must be connected to earth by means of a grounding-electrode system.

Outdoor substations usually use a ground grid, consisting of a number of ground rods driven into the earth and bonded together by buried copper conductors. The required grounding-electrode system for a building is spelled out in the NEC, Article 250-H. The preferred grounding electrode is a metal underground water pipe in direct contact with the earth for at least 10 ft. However, because underground water piping is often plastic outside of the building, or may later be replaced by plastic piping, the NEC requires this electrode to be supplemented by and bonded to at least one other grounding electrode, such as the effectively grounded metal frame of the building, a concrete-encased electrode, a copper conductor ground ring encircling the building, or a made electrode such as one or more driven ground rods or a buried plate. Where any of these electrodes are present, they must be bonded together into one grounding-electrode system.

One of the most effective grounding electrodes is the concrete-encased electrode, sometimes called the Ufer ground, after the man who developed it. It consists of at least 20 ft of steel reinforcing bars or rods not less than $\frac{1}{2}$ in diameter, or at least 20 ft of bare copper conductor, size #4 AWG or larger, encased in at least 2 in of concrete. It must be located within and near the bottom of a concrete foundation or footing that is in direct contact with earth. Tests have shown this electrode to provide a low-resistance earth ground even in poor soil conditions.

The electrical distribution system and equipment ground must be connected to this grounding-electrode system by a grounding-electrode conductor. All other grounding electrodes, such as those for the light-

FIGURE 4.4   Grounding-electrode system (NEC Articles 250-81 and 250-83).

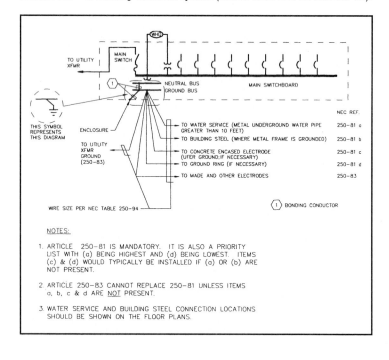

TABLE 4.2 Grounding Electrode Conductor for AC Systems
(NEC Table 250-94)

Size of Largest Service-Entrance Conductor or Equivalent Area for Parallel Conductors		Size of Grounding Electrode Conductor	
Copper	Aluminum or Copper-Clad Aluminum	Copper	Aluminum or Copper-Clad Aluminum*
2 or smaller	1/0 or smaller	8	6
1 or 1/0	2/0 or 3/0	6	4
2/0 or 3/0	4/0 or 250 kcmil	4	2
Over 3/0 thru 350 kcmil	Over 250 kcmil thru 500 kcmil	2	1/0
Over 350 kcmil thru 600 kcmil	Over 500 kcmil thru 900 kcmil	1/0	3/0
Over 600 kcmil thru 1100 kcmil	Over 900 kcmil thru 1750 kcmil	2/0	4/0
Over 1100 kcmil	Over 1750 kcmil	3/0	250 kcmil

Where multiple sets of service-entrance conductors are used as permitted in Section 230-40, Exception No. 2, the equivalent size of the largest service-entrance conductor shall be determined by the largest sum of the areas of the corresponding conductors of each set.

Where there are no service-entrance conductors, the grounding electrode conductor size shall be determined by the equivalent size of the largest service-entrance conductor required for the load to be served.

*See installation restrictions in Section 250-92(a).

(FPN): See Section 250-23(b) for size of alternating-current system grounded conductor brought to service equipment.

ning protection system, the telephone system, television antenna and cable TV system grounds, and computer systems, must be bonded to this grounding-electrode system.

The NEC (Article 250-81) requires a grounding-electrode system, illustrated by Figure 4.4 as an example, with the grounding-electrode conductor sized in accordance with Table 4.2 Grounding Electrode Conductor for AC Systems (NEC Table 250-94).

In general, where loads will be connected line to neutral, solidly grounded systems are used.

In commercial and institutional installations, such as office buildings, shopping centers, schools, and hospitals, lighting loads are often more than 50 percent of the total load. In addition, a feeder outage on the first ground fault is seldom crucial—even in hospitals, which have emergency power in critical areas. For these reasons, a solidly grounded wye distribution system, with the neutral used for lighting circuits, is usually the most economical, effective, and convenient design.

Medium-Voltage System Grounding

Because the method of grounding affects the voltage rise of the unfaulted phases above ground, ANSI C62.92 classifies systems from the point of view of grounding in terms of a coefficient of grounding (COG), which equals the highest power frequency rms line-to-ground voltage divided by the rms line-to-line voltage at the fault location with the fault removed.

This same standard also defines systems as effectively grounded when the COG is less than or equal to 0.8. Such a system would have X_0/X_1 less than or equal to 3.0 and R_0/X_1 less than or equal to 1.0. Any other grounding means that does not satisfy these conditions at any point in the system is not effectively grounded.

The aforementioned definition is of significance in medium-voltage distribution systems with long lines and with grounded sources removed during light-load periods so that in some locations in the system the X_0/X_1, R_0/X_1 ratios may exceed the defining limits.

Other standards (cable and lightning arrester) allow the use of 100 percent rated cables and arresters selected on the basis of an effectively grounded system only where the preceding criteria are met. In effectively grounded systems, the line-to-ground fault current is high, and there is no significant voltage rise in the unfaulted phases.

With selective ground fault isolation, the fault current will be at 60 percent of the three-phase current at the point of fault. Damage to cable shields must be checked. This fact is not a problem except in small cables. To prevent cable damage, it is a good idea to supplement cable shields as returns of ground fault current.

The burdens on the current transformers (CTs) must also be checked where residually connected ground relays are used and the CTs supply current to phase relays and meters. If ground sensor current transformers are used, they must also be of high-burden capacity.

Table 4.3 indicates the characteristics of the various methods of grounding.

Features of ungrounded and grounded systems are summarized in Table 4.4.

Reactance grounding is generally used in the grounding of generator neutrals, in which generators are directly connected to the distribution system bus, to limit the line-to-ground fault to somewhat less than the three-phase fault at the generator terminals. If the reactor is so sized, in all probability the system will remain effectively grounded.

When resistors are used in medium-voltage system grounding, they are generally low in resistance value. The fault is limited from 20 to 25 percent of the three-phase fault value down to about 400 A. With a properly sized resistor and relaying application, selective fault isolation is feasible. The fault limit provided has a bearing on whether residually connected relays are used or ground sensor current transformers are used for ground fault relaying.

In general, where residually connected relays are used, the fault current at each grounded source should not be limited to less than the current transformer's rating of the source. This rule will provide sensitive

TABLE 4.3 Characteristics of Grounding

Grounding Classes and Means	Ratios of Symmetrical Component Parameters[1]			Percent Fault Current [2]	Per Unit Transient LG Voltage [3]
	X_0/X_1	R_0/X_1	R_0/X_0		
A. Effectively [4]					
1. Effective	0-3	0-1	--	>60	≤2
2. Very effective	0-1	0-0.1	--	>95	<1.5
B. Noneffectively					
1. Inductance					
a. Low inductance	3-10	0-1		>25	<2.3
b. High inductance	>10		<2	<25	≤2.73
2. Resistance					
a. Low resistance	0-10		≥2	<25	<2.5
b. High resistance		>100	≤(-1)	<1	≤2.73
3. Inductance and resistance	>10	--	>2	<10	≤2.73
4. Resonant	[5]	--		<1	≤2.73
5. Ungrounded/capacitance					
a. Range A	-∞ to -40[6]	--	--	<8	≤3
b. Range B	-40 to 0	--	--	>8	>3 [7]

TABLE 4.4 Medium Voltage System Grounding Features of Ungrounded and Grounded Systems (from ANSI C62.92)

	A Ungrounded	B Solidly Grounded	C Reactance Grounded	D Resistance Grounded	E Resonant Grounded
(1) Apparatus Insulation	Fully insulated	Lowest	Partially graded	Partially graded	Partially graded
(2) Fault to Ground Current	Usually low	Maximum value rarely higher than three-phase short circuit current	Cannot satisfactorily be reduced below one-half or one-third of values for solid grounding	Low	Negligible except when Petersen coil is short circuited for relay purposes when it may compare with solidly-grounded systems
(3) Stability	Usually unimportant	Lower than with other methods but can be made satisfactory by use of high-speed breakers	Improved over solid grounding particularly if used at receiving end of system	Improved over solid grounding particularly if used at receiving end of system	Is eliminated from consideration during single line-to-ground faults unless neutralizer is short circuited to isolate fault by relays
(4) Relaying	Difficult	Satisfactory	Satisfactory	Satisfactory	Requires special provisions but can be made satisfactory
(5) Arcing Grounds	Likely	Unlikely	Possible if reactance is excessive	Unlikely	Unlikely
(6) Localizing Faults	Effect of fault transmitted as excess voltage on sound phases to all parts of conductively connected network	Effect of faults localized to system or part of system where they occur	Effect of faults localized to system or part of system where they occur unless reactance is quite high	Effect of faults transmitted as excess voltage on sound phases to all parts of conductively connected network	Effect of faults transmitted as excess voltage on sound phases to all parts of conductively connected network
(7) Double Faults	Likely	Likely	Unlikely unless reactance is quite high and insulation weak	Unlikely unless resistance is quite high and insulation weak	Seem to be more likely but conclusive information not available
(8) Lightning Protection	Ungrounded neutral service arresters must be applied at sacrifice in cost and efficiency	Highest efficiency and lowest cost	If resistance is very high arresters for ungrounded neutral service must be applied at sacrifice in cost and efficiency	Arresters for ungrounded, neutral service usually must be applied at sacrifice in cost and efficiency	Ungrounded neutral service arresters must be applied at sacrifice in cost and efficiency

(9) Telephone Interference	Will usually be low except in cases of double faults or electrostatic induction with neutral displaced but duration may be great	Will be greatest in magnitude due to higher fault currents but can be quickly cleared particularly with high speed breakers	Will be reduced from solidly grounded values	Will be reduced from solidly grounded values	Will be low in magnitude except in cases of double faults or series resonance at harmonic frequencies, but duration may be great
(10) Ratio Interference	May be quite high during faults or when neutral is displayed	Minimum	Greater than for solidly grounded, when faults occur	Greater than for solidly grounded, when faults occur	May be high during faults
(11) Line Availability	Will inherently clear themselves if total length of interconnected line is low and require isolation from system in increasing percentages as length becomes greater	Must be isolated for each fault	Must be isolated for each fault	Must be isolated for each fault	Need not be isolated but will inherently clear itself in about 60 to 80 percent of faults
(12) Adaptability to Interconnection	Cannot be interconnected unless interconnecting system is ungrounded or isolating transformers are used	Satisfactory indefinitely with reactance-grounded systems	Satisfactory indefinitely with solidly-grounded systems	Satisfactory with solidly- or reactance-grounded systems with proper attention to relaying	Cannot be interconnected unless interconnected system is resonant grounded or isolating transformers are used. Requires coordination between interconnected systems in neutralizer settings
(13) Circuit Breakers	Interrupting capacity determined by three-phase conditions	Same interrupting capacity as required for three-phase short circuit will practically always be satisfactory	Interrupting capacity determined by three-phase fault conditions	Interrupting capacity determined by three-phase fault conditions	Interrupting capacity determined by three-phase fault conditions
(14) Operating Procedure	Ordinarily simple but possibility of double faults introduces complication in times of trouble	Simple	Simple	Simple	Taps on neutralizers must be changed when major system switching is performed and difficulty may arise in interconnected systems. Difficult to tell where faults are located
(15) Total Cost	High, unless conditions are such that arc tends to extinguish itself, when transmission circuits may be eliminated, reducing total cost	Lowest	Intermediate	Intermediate	Highest unless the are suppressing characteristic is relied on to eliminate transmission circuits when it may be lowest for the particular types of service

differential protection for wye-connected generators and transformers against line-to-ground faults near the neutral. Of course, if the installation of ground fault differential protection is feasible, or ground sensor current transformers are used, sensitive differential relaying in a resistance-grounded system with greater fault limitation is possible. In general, ground sensor current transformers do not have high-burden capacity. Resistance-grounded systems limit the circulating currents of triplen harmonics and limit the damage at the point of fault. This method of grounding is not suitable for line-to-neutral connection of loads.

4.1 GROUND FAULT PROTECTION

Introduction

A ground fault normally occurs in one of two ways: by accidental contact of an energized conductor with normally grounded metal, or as a result of an insulation failure of an energized conductor. When an insulation failure occurs, the energized conductor contacts normally noncurrent-carrying metal, which is bonded to a part of the equipment-grounding conductor. In a solidly grounded system, the fault current returns to the source primarily along the equipment-grounding conductors, with a small part using parallel paths such as building steel or piping. If the ground return impedance were as low as that of the circuit conductors, ground fault currents would be high, and the normal phase-overcurrent protection would clear them with little damage. Unfortunately, the impedance of the ground return path is usually higher; the fault itself is usually arcing; and the impedance of the arc further reduces the fault current. In a 480Y/277-V system, the voltage drop across the arc can be from 70 to 140 V. The resulting ground fault current is rarely enough to cause the phase overcurrent protection device to open instantaneously and prevent damage. Sometimes, the ground fault is below the trip setting of the protective device and it does not trip at all until the fault escalates and extensive damage is done. For these reasons, low-level ground protection devices with minimum time-delay settings are required to rapidly clear ground faults. This is emphasized by the NEC requirement that a ground fault relay on a service shall have a maximum delay of 1 s for faults of 3000 A or more.

The NEC (Article 230-95) requires that ground fault protection, set at no more than 1200 A, be provided for each service-disconnecting means rated 1000 A or more on solidly grounded wye services of more than 150 V to ground, but not exceeding 600 V phase-to-phase. Practically, this makes ground fault protection mandatory on 480Y/277-V services, but not on 208Y/120-V services. On a 208-V system, the voltage to ground is 120 V. If a ground fault occurs, the arc will extinguish at current zero, and the voltage to ground is often too low to cause it to restrike. Therefore, arcing ground faults on 208-V systems tend to be self-extinguishing. On a 480-V system, with 277 V to ground, restrike usually takes place after current zero, and the arc tends to be self-sustaining, causing severe and increasing damage, until the fault is cleared by a protective device.

The NEC requires ground fault protection only on the service-disconnecting means. This protection works so fast that for ground faults on feeders, or even branch circuits, it will often open the service disconnect before the feeder or branch overcurrent device can operate. This is highly undesirable, and in the NEC (Article 230-95) a fine-print note (FPN) states that additional ground fault–protective equipment will be needed on feeders and branch circuits where maximum continuity of electric service is necessary. Unless it is acceptable to disconnect the entire service on a ground fault almost anywhere in the system, such additional stages of ground fault protection must be provided. At least two stages of ground fault protection are mandatory in health care facilities (NEC Article 517-14).

Overcurrent protection is designed to protect conductors and equipment against currents that exceed their ampacity or rating under prescribed time values. An overcurrent can result from an overload, short circuit, or high-level ground fault condition. When currents flow outside the normal current path to ground, supplementary ground fault protection equipment will be required to sense low-level ground fault currents and initiate the protection required. Normal phase-overcurrent protection devices provide no protection against low-level ground faults.

Basic Means of Sensing Ground Faults

There are three basic means of sensing ground faults. The most simple and direct method is the ground return method as illustrated in Figure 4.5. This sensing method is based on the fact that all currents supplied by a transformer must return to that transformer.

When an energized conductor faults to grounded metal, the fault current returns along the ground return path to the neutral of the source transformer. This path includes the grounding electrode conductor—sometimes called the ground strap—as shown in Figure 4.5. A current sensor on this conductor (which can be a conventional bar-type or window-type CT) will respond to ground fault currents only. Normal neutral currents resulting from unbalanced loads will return along the neutral conductor and will not be detected by the ground return sensor.

This is an inexpensive method of sensing ground faults in which only minimum protection per NEC Article 230-95 is desired. For it to operate properly, the neutral must be grounded in only one location, as indicated in Figure 4.5. In many installations, the servicing utility grounds the neutral at the transformer, and additional grounding is required in

FIGURE 4.5 Ground return sensing method.

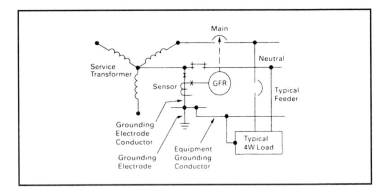

the service equipment per NEC Article 250-23a. In such cases and others, including multiple source with multiple interconnected neutral ground points, residual or zero-sequence sensing methods should be employed.

A second method of detecting ground faults is the use of a zero-sequence sensing method as illustrated in Figure 4.6. This sensing method requires a single, specially designed sensor, either of a toroidal- or rectangular-shaped configuration. This core balance current transformer surrounds all the phase and neutral conductors in a typical three-phase, four-wire distribution system.

The sensing method is based on the fact that the vectorial sum of the

FIGURE 4.6 Zero-sequence sensing method.

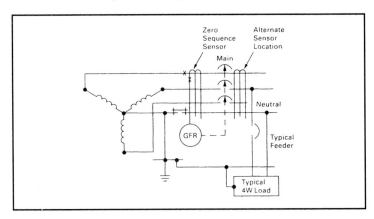

phase and neutral currents in any distribution circuit will equal zero unless a ground fault condition exists downstream from the sensor. All currents that flow only in the circuit conductors, including balanced or unbalanced phase-to-phase and phase-to-neutral normal or fault currents, and harmonic currents, will result in zero sensor output. However, should any conductor become grounded, the fault current will return along the ground path—not the normal circuit conductors—and the sensor will have an unbalanced magnetic flux condition, and a sensor output will be generated to actuate the ground fault relay.

Zero-sequence sensors are available with various window openings for circuits with small or large conductors, and even with large rectangular windows to fit over bus bars or multiple large-size conductors in parallel. Some sensors have split cores for installations over existing conductors without disturbing the connections.

This method of sensing ground faults can be employed on the main disconnect where minimum protection per NEC Article 230-95 is desired. It can also be employed in multitier systems where additional ground fault protection is desired for added service continuity. Additional grounding points may be employed upstream of the sensor, but not on the load side.

Ground fault protection employing ground return or zero-sequence sensing methods can be accomplished by the use of separate ground fault relays (GFRs) and disconnects equipped with standard shunt trip devices or by circuit breakers with integral ground fault protection with external connections arranged for these modes of sensing.

The third basic method of detecting ground faults involves the use of multiple current sensors connected in a residual sensing method, as illustrated in Figure 4.7. This is a very common sensing method used with circuit breakers equipped with electronic trip units and integral ground fault protection. The three-phase sensors are required for normal phase-overcurrent protection. Ground fault sensing is obtained with the addition of an identically rated sensor mounted on the neutral. In a residual sensing scheme, the relationship of the polarity markings—as noted by the X on each sensor—is critical. Because the vectorial sum of the currents in all of the conductors will total zero under normal, nonground-faulted conditions, it is imperative that proper polarity connections are employed to reflect this condition.

As with the zero-sequence sensing method, the resultant residual sensor output to the ground fault relay or integral ground fault tripping circuit will be zero if all currents flow only in the circuit conductors. Should a ground fault occur, the current from the faulted conductor will return along the ground path, rather than on the other circuit conductors, and the residual sum of the sensor outputs will not be zero. When the level of ground fault current exceeds the preset current and time-delay settings, a ground fault tripping action will be initiated.

This method of sensing ground faults can be economically applied on

FIGURE 4.7 Residual sensing method.

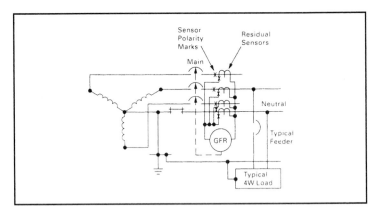

main-service disconnects, in which circuit breakers with integral ground fault protection are provided. It can be used in minimum-protection schemes per NEC Article 230-95 or in multitier schemes, in which additional levels of ground fault protection are desired for added service continuity. Additional grounding points may be employed upstream of the residual sensors, but not on the load side.

Both the zero-sequence and residual sensing methods have been commonly referred to as *vectorial summation* methods.

Most distribution systems can use any of the three sensing methods exclusively, or a combination of the sensing methods depending upon the complexity of the system and the degree of service continuity and selective coordination desired. Different methods will be required depending upon the number of supply sources and the number and location of system-grounding points.

As an example, one of the more frequently used systems in which continuity of service to critical loads is a factor is the dual-source system illustrated in Figure 4.8. This system uses tie-point grounding as permitted under NEC Article 230-23(a). The use of this grounding method is limited to services that are dual-fed (double-ended) in a common enclosure or grouped together in separate enclosures and employing a secondary tie.

This system uses individual sensors connected in ground-return fashion. Under tie breaker–closed operating conditions, either the M1 sensor or M2 sensor could see neutral unbalance current and possibly initiate an improper tripping operation. However, with the polarity arrangements of these two sensors, along with the tie breaker auxiliary switch (T/a) and the interconnections as shown, this possibility is eliminated. Selective ground fault tripping coordination between the tie breaker and

FIGURE 4.8 Dual-source system—single-point grounding.

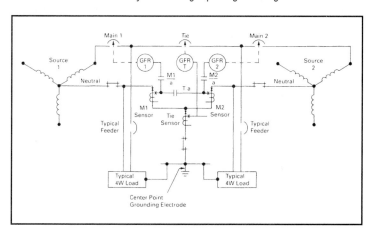

the two main circuit breakers is achieved by preset current pickup and time-delay settings between devices GFR/1, GFR/2, and GFR/T.

The advantages of increased service continuity offered by this system can only be effectively used if additional levels of ground fault protection are added on each downstream feeder. Some users prefer individual grounding of the transformer neutrals. In such cases, a partial differential ground fault scheme should be used for the mains and the tie breaker.

An infinite number of ground fault protection schemes can be developed depending upon the number of alternate sources, the number of grounding points, and system interconnections involved. Depending upon the individual system configuration, either mode of sensing or a combination of all types may be employed to accomplish the desired end results.

Because the NEC Article 230-95 limits the maximum setting of the ground fault protection used on service equipment to 1200 A (or 3000 A for 1 s), to prevent tripping of the main-service disconnect on a feeder ground fault, ground fault protection must be provided on all the feeders. To maintain maximum service continuity, more than two levels (zones) of ground fault protection will be required, so that ground fault outages can be localized and service interruption minimized. To retain selectivity between different levels of ground fault relays, time-delay settings should be employed with the GFR furthest downstream having the minimum time delay. This will allow the GFR nearest the fault to operate first. With several levels of protection, this will reduce the level of protection for faults within the GFR zones. Zone interlocking was developed for GFRs to overcome this problem.

Ground fault relays (or circuit breakers with integral ground fault protection) with zone interlocking are coordinated in a system to operate in a time-delayed mode for ground faults occurring most remote from the source. However, this time-delayed mode is only actuated when the GFR next upstream from the fault sends a restraining signal to the upstream GFRs. The absence of a restraining signal from a downstream GFR is an indication that any occurring ground fault is within the zone of the GFR next upstream from the fault and that device will operate instantaneously to clear the fault with minimum damage and maximum service continuity. This operating mode permits all GFRs to operate instantaneously for a fault within their zone and to still provide complete selectivity between zones. The National Electrical Manufacturers' Association (NEMA) states, in their application guide for ground fault protection, that zone interlocking is necessary to minimize damage from ground faults. A two-wire connection is required to carry the restraining signal from the GFRs in one zone to the GFRs in the next zone.

Circuit breakers with integral ground fault protection and standard circuit breakers with shunt trips activated by the ground fault relay are ideal for ground fault protection. Many fused switches over 1200 A, and some fusible switches in ratings from 400 to 1200 A, are listed by UL as suitable for ground fault protection. Fusible switches so listed must be equipped with a shunt trip and be able to open safely on faults up to 12 times their rating.

Power distribution systems differ widely from each other, depending on the requirements of each user, and total system overcurrent protection, including ground fault currents, must be individually designed to meet these needs. Experienced and knowledgeable engineers must consider the power sources (utility and on-site), the effects of outages and downtime, safety for people and equipment, initial and life-cycle costs, and many other factors. They must apply protective devices, analyzing the time-current characteristics, fault-interrupting capacity, and selectivity and coordination methods to provide the safest and most cost-effective distribution system.

4.2 LIGHTNING PROTECTION

Introduction

Lightning protection deals with the protection of buildings and other structures due to direct damage from lightning. Requirements will vary with geographic location, building type and environment, and many other factors. Any lightning protection system must be grounded, and the lightning protection ground must be bonded to the electrical equipment-grounding system.

Nature of Lightning

Lightning is an electric discharge between clouds or between clouds and earth. Charges of one polarity are accumulated in the clouds and of the opposite polarity in the earth. When the charge increases to the point that the insulation between can no longer contain it, a discharge takes place. This discharge is evidenced by a flow of current, usually great in magnitude, but extremely short in time.

Damage to buildings and structures is the result of heat and mechanical forces produced by the passage of current through resistance in the path of discharge. Although the discharge takes place at the point at which the potential difference exceeds the dielectric strength of the insulation, which implies low resistance relative to other paths, it is not uncommon for the current to follow the path of high resistance. This may be a tree, a masonry structure, or a porcelain insulator. Obviously, damage due to direct stroke can be minimized by providing a direct path of low resistance to earth.

Lightning can cause damage to structures by direct stroke and to equipment by surges coming in over exposed power lines. Surges may be the result of direct strokes to the line at some distance away, or they may be electrostatically induced voltages.

Need for Protection

Damage to structures and equipment due to surge effect is a subject in itself, and protection against this type of damage is not within the scope of this text except as grounding is involved.

It is not possible to positively protect a structure against damage from a direct stroke except by completely enclosing it with metal. The extent to which lightning protection should be provided is governed by weighing the cost of protection against the possible consequences of being struck. The following factors are to be considered:

1. Frequency and severity of thunderstorms
2. Value and nature of structure or content
3. Personnel hazards
4. Consequential loss, such as a loss of production, salaries of workers, damage suits, and other indirect losses
5. Effect on insurance premiums

The above factors are listed primarily to call attention to their importance. No general conclusions can be drawn as to the relative importance of each or to the necessity for or the extent of lightning protection for any given combination of conditions. As a matter of interest, a map showing the frequency of thunderstorm days for various areas of the United States is shown in Figure 4.9. It should be noted, however, that

FIGURE 4.9 Annual isokeraunic map showing number of thunderstorm days per year.

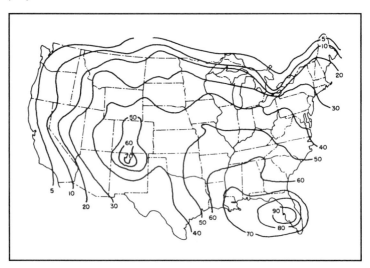

the severity of storms is much greater in some local areas than in others, and, therefore, the need for protection is not necessarily in direct proportion to the frequency.

Equipment and Structures That Should Be Considered for Protection

The nature of buildings and their content is important in deciding whether lightning protection is desirable. Some of the structures that should be considered are as follows:

- All-metal structures
- Metal-roofed and metal-clad buildings
- Metal-frame buildings with nonmetallic facings
- Buildings of wood, stone, brick, tile, and other nonconducting materials
- Spires, steeples, and flagpoles
- Buildings of historical value
- Buildings containing readily combustible or explosive materials
- Tanks and tank farms
- Transmission lines
- Power plants, substations, and water-pumping stations
- High stacks and chimneys

- Water towers, silos, and similar structures
- Buildings containing a significant amount of sensitive electronic equipment such as data centers
- Hospitals and health care facilities
- High-rise buildings

Metal buildings and structures offer a very satisfactory path to earth and require little in the way of additional protection. Metal-frame buildings with nonmetallic facings require more extensive measures. Buildings made entirely of nonconducting materials require complete lightning protection systems.

In special cases, buildings may have historical value out of proportion to their intrinsic value and may justify extensive protection systems. Power stations, substations, and water-pumping stations providing extremely important functions to outside facilities may demand protective measures far more extensive than would normally be warranted by the value of the structure. By the same token, structures containing combustible or explosive materials, liquids, and gases of a toxic nature or otherwise harmful to personnel or property if allowed to escape from their confining enclosures, may justify extensive protection systems.

Requirements for Good Protection

The fundamental theory of lightning protection of structures is to provide means by which a discharge may enter or leave the earth without passing through paths of high resistance. Such a condition is usually met by grounded steel-frame structures. Suitable protection is nearly always provided by the installation of lightning conductors.

A lightning conductor system consists of terminals projecting into the air above the uppermost parts of the structure, with interconnecting and ground conductors. Terminals should be placed so as to project above all points that are likely to be struck. Conductors should present the least possible impedance to earth. There should be no sharp bends or loops. Each projecting terminal above the structure should have at least two connecting paths to earth and more if practicable.

Each conductor running down from the terminals on top of the structure should have an earth connection. Properly made connections to earth are an essential feature of a lightning rod system for protection of buildings. It is more important to provide ample distribution of metallic contacts in the earth than to provide low-resistance connections. Low-resistance connections are desirable, however, and should be provided where practicable. Earth connections should be made at uniform intervals about the structure, avoiding as much as possible the grouping of connections on one side. Electrodes should be at least 2 ft (0.6 m) away from and should extend below building foundations (except when using reinforcing bars for grounds). They should make contact with the earth from the surface downward to avoid flashing at the surface.

Interior metal parts of buildings or structures should be grounded independently, and if they are within 6 ft (1.8 m) of metallic roofs, walls, or conductors running down from the terminals on top of the structure, they should be securely connected thereto.

Terminals projecting above the structure should be of ample length to bring the top point at least 10 in (0.25 m) above the object to be protected. In many cases, a greater height is desirable. Experiments have indicated that a vertical conductor, or point, will divert to itself direct hits that might otherwise fall within a cone-shaped space, of which the apex is the point and the base is a circle whose radius is approximately equal to the height of the point (only for single aerial terminals).

The foregoing outlines requirements for good protection of buildings. Good protection of electrical substations, power stations, tanks and tank farms, and other special applications is beyond the scope of this book. For further information, refer to IEEE Standard 142.

Application Tips

- As a practical matter, once it is decided that a lightning protection system is needed, consulting electrical engineers generally write a performance specification calling for a UL Master Label System. The system is actually designed and installed by a qualified lightning protection contractor.

- When considering a lightning protection system for a building, it is important to verify the history of frequency and severity of thunderstorms in the immediate area of the building being considered. This could be checked through the weather service and building owners in the local area.

- Experience has shown that adding a lightning protection system to a building increases its susceptibility to lightning strokes.

- If a lightning protection system is to be provided for a building addition, it must also be added to all existing contiguous buildings to obtain a UL Master Label. Even if the existing contiguous buildings already have a lightning protection system, their lightning protection system may have to be upgraded to obtain a UL Master Label.

NOTES

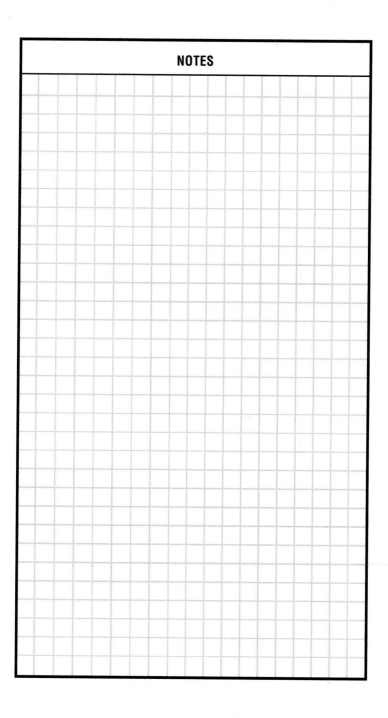

NOTES

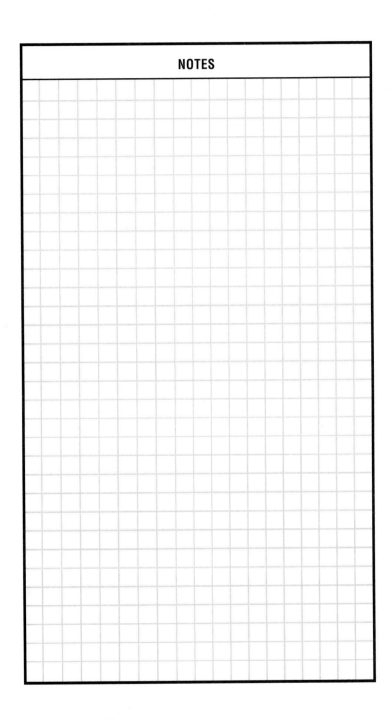

CHAPTER FIVE
Emergency and Standby Power Systems

5.0 GENERAL NEED FOR EMERGENCY AND STANDBY POWER SYSTEMS

Introduction

Emergency electric services are required for protection of life, property, or business where loss might be the result of an interruption of the electric service. The extent of the emergency services required depends on the type of occupancy, the consequences of a power interruption, and the frequency and duration of expected power interruptions.

Municipal, state, and federal codes define minimum requirements for emergency systems for some types of public buildings and institutions. These shall be adhered to, but economics or other advantages may result in making provisions beyond these minimums (see the NEC, Articles 517, 700, 701, and 702). The following presents some of the basic information on emergency and standby power systems. For additional information, design details, and maintenance requirements, see ANSI/IEEE Standard 446-1987 ("IEEE Recommended Practice for Emergency and Standby Power Systems for Industrial and Commercial Applications"), ANSI/NFPA 110-1988 ("Emergency and Standby Power Systems"), and ANSI/NFPA 110A-1989 ("Stored Energy Systems").

Emergency power systems should be separated from the normal power systems by using separate raceways and panelboards. The NEC requires that each item of emergency equipment be clearly marked as to its purpose. In large public buildings, physical separation of the emergency system from the normal system elements would enhance the reliability of the emergency system in the event of fire or other contingencies. Also, more and more states are requiring that the emergency systems not only be separated from the normal systems, but that they be enclosed in 2-h fire-rated construction.

Definitions

The following is intended to conveniently provide selected terms and definitions applicable to this chapter for the purpose of aiding in its overall understanding.

Automatic transfer switch: Self-acting equipment for transferring one or more load conductor connections from one power source to another.

Bypass/isolation switch: A manually operated device used in conjunction with an automatic transfer switch to provide a means of directly connecting load conductors to a power source and of disconnecting the automatic transfer switch.

Commercial power: Power furnished by an electric power utility company. When available, it is usually the prime power source; however, when economically feasible, it sometimes serves as an alternative or standby source.

Emergency power system: An independent reserve source of electric energy that, upon failure or outage of the normal source, automatically provides reliable electric power within a specified time to critical devices and equipment whose failure to operate satisfactorily would jeopardize the health and safety of personnel or result in damage to property.

Standby power system: An independent reserve source of electric energy that, upon failure or outage of the normal source, provides electric power of acceptable quality so that the user's facilities may continue in satisfactory operation.

Uninterruptible power supply (UPS): A system designed to automatically provide power, without delay or transients, during any period when the normal power supply is incapable of performing acceptably.

Lighting

Exit and emergency lights that are sufficient to permit safe exit from buildings in which the public may congregate should be supplied from an emergency power source (i.e., auditoriums, theaters, hotels, large stores and malls, sports arenas, and so on). Local regulations should always be referred to for more specific requirements. When the emergency lighting units are not used under normal conditions, power should be immediately available to them upon loss of the normal power supply. When the emergency lights are normally in service and served from the normal power supply, provisions should be made to transfer them automatically to the emergency power source when the normal power supply fails.

Sufficient lighting should be provided in stairs, exits, corridors, and halls so that the failure of any one unit will not leave any area dark or endanger persons leaving the building. Adequate lighting and rapid automatic transfer to prevent a period of darkness is important in public areas. Public safety is improved and the chance of pilfering or damage to property is minimized.

ANSI/NFPA 101-1988 ("Life Safety Code") requires that emergency power sources for lighting be capable of carrying their connected loads for at least 90 min. There are cases in which provisions should be made for providing emergency service for much longer periods of time, such as in health care facilities, communications, police, fire fighting, and emergency services. A 2- to 3-h capacity is more practical and, in many installations, a 5- to 6-h or even several-day capacity is provided. During a severe storm or catastrophe, the demands on hospitals, communications, police, fire fighting, and emergency service facilities will be increased. A third source of power to achieve the lighting reliability may be required.

When installation of a separate emergency power supply is not warranted but some added degree of continuity of service for exit lights is desired, they may be served from circuits connected ahead of the main service-entrance switch for some occupancies. This assures that load switching and tripping due to faults in the building's electric system will not cause loss of the exit lights. However, this arrangement does not protect against failures in the electric utility system.

ILLUMINATION OF MEANS OF EGRESS

In its occupancy chapter, ANSI/NFPA 101-1988 has illumination requirements for building egress, which includes stating the type of emergency lighting required.

Primary or normal illumination is required to be continuous during the time "the conditions of occupancy" require that the means of egress be available for use. ANSI/NFPA 101-1988 specifies the illuminances and equipment for providing this type of lighting.

Emergency power sources listed in the NEC, Article 700 include the following:

1. Storage batteries (rechargeable type) to supply the load for 90 min without the voltage at the load decreasing to 87.5 percent of normal

2. Generator sets that will accept the emergency lighting load within 10 s, unless an auxiliary lighting source is available

3. Uninterruptible power supplies

4. Separate electric utility service, which is widely separated electrically and physically from the normal service

5. Unit equipment (permanently installed) consisting of a rechargeable storage battery, automatic charger, lamp(s), and automatic transfer relay.

Refer to the ANSI/NFPA 101-1988, Sections 5-8 and 5-9 ("Illumination of Means of Egress" and "Emergency Lighting"), respectively.

Power Loads

An emergency source for supplying power loads is required when loss of such a load could cause extreme inconvenience or hazard to personnel, loss of product or material, or contamination of property. The size and type of the emergency system should be determined through consideration of the health and convenience factors involved and whether the utilization affects health care facilities, communication systems, alarm systems, police, fire fighting, and emergency services facilities. The installation should comply with any applicable codes and standards and be acceptable to the authority that has jurisdiction. For example, health care facilities may require conformance to ANSI/NFPA 99-1990 ("Health Care Facilities") and the NEC, Article 517. Fire pump installations may require conformance to ANSI/NFPA 20-1990 ("Centrifugal Fire Pumps").

In laboratories in which continuous processes are involved or in which chemical, biological, or nuclear experimentation is conducted, requirements are very demanding insofar as power and ventilating system requirements are concerned. Loss of adequate power for ventilation could permit the spread of poisonous gases, biological contamination, or radioactive contamination throughout the building, and can even cause loss of life. A building contaminated from radioactive waste could be a total loss or require extensive cleanup measures. Many processes or experiments cannot tolerate a power loss that could interrupt cooling, heating, agitation, and so forth.

Emergency power for fire pumps should be provided when water requirements cannot be met from other sources. Emergency power for elevators should also be considered when elevators are necessary to evacuate buildings or the cost seems warranted to avoid inconvenience to the public. This does not mean that the emergency power supply should have the full capacity for the demand of all elevators simultaneously.

Summary of Codes for Emergency Power in the United States

Table 5.1 is a guide to state codes and regulations for emergency power systems in the United States. All the latest codes and regulations for the area in which the industrial or commercial facility is located must be consulted and followed.

TABLE 5.1 Codes for Emergency Power by States and Major Cities (Completed September 1984)

State/City	Does State/City Have Legislation?	Legislation Code Type?	Hospitals	Nursing Homes	Schools	Theaters (Public Gathering Places)	Office Buildings	Hotels	Apartment Buildings	Airports	Fire and Police Stations	Water Treatment Plants	Sewage Treatment Plants	All Public Buildings, State	All Public Buildings, Commercial	Applicable Government Agency	Smokeproof Enclosures in High-Rise Buildings
Alabama	Yes	4,6	A,C,D	A,C,D	C,D	C,D	C,D	C,D	C,D	C,D	C	C	C	C,D	C,D	M,N,O	B
Birmingham **	Yes	4,6	A,C,D	A,C,D	C,D	C,D	C,D	C,D	C,D	C,D	C			C,D	C,D	Q,S	A,B
Mobile	Yes	1,4	A,C,D	A,C,D	C,D	A,C,D	A,C,D	A,C,D	A,C,D	C,D	C,D	C,D	C,D		C	S	B
Alaska	Yes	1,3	A,C,D	A,C,D	C,D	A,C,D	A,C,D	A,C,D	A,C,D	A,C,D				C,D	A,C,D	M,P	
Arizona **	No																
Phoenix	Yes	1	A,C,D	A,C,D	C,D	C,D	C,D	C,D	A,C,D	C,D	C,D	C,D	C,D	C,D	C,D	O	B
Arkansas	Yes	1,6	A,C,D	A,C,D	A,C,D	A,C,D	A,C,D	A,C,D	A,C,D	C,D				C,D	A,C,D	M	A
California	Yes	2,3,4	A,C,D	A,C,D	C,D	C,D	C,D	C,D	C,D	C,D				C,D	C,D	O,T	
Anaheim **	Yes	2,3,4,7	A,C,D	A,C,D	C,D	C,D	C,D	A,C,D	C,D	C,D	C,D		C,D	C,D	C,D	M,N,O,Q	
Berkeley	Yes	1,3	A,C,D	A,C,D	A,C,D	A,C,D	A,C,D	A,C,D	A,C,D	C,D	C,D	A,C,D		C,D	C,D	M,Q	B
Fresno	Yes	3,4	A,C,D	A,C,D	A,C,D	A,C,D	A,C,D	A,C,D	A,C,D	C,D	A,C,D			C,D	C,D	O,S	B
Glendale **	Yes	3,4	A,C,D	A,C,D	A,C,D	A,C,D	C,D	C,D	A,C,D	A,C,D	C,D			C,D	C,D	M,Q	
Long Beach	Yes	3	A,C,D	A,C,D	A,C,D	A,C,D	A,C,D	A,C,D	A,C,D	B,C,D	C,D	C,D	C,D	C,D	C,D	M,O	B
Los Angeles	Yes	3,4,8	A,C,D	A,C,D	A,C,D	A,C,D	A,C,D	A,C,D	A,C,D	C,D	C,D	C,D	C,D	C,D	C,D	M,Q,S	B
Oakland	Yes	1,3,4,8	A,C,D	A,C,D	A,C,D	A,C,D	A,C,D	A,C,D	A,C,D	C,D	C,D	C,D	C,D	C,D	C,D	M,O	
Pasadena	Yes	1,2,3,4	A,C,D	A,C,D	A,C,D	A,C,D	C,D	C,D	C,D	C,D	C,D	C,D		C,D	C,D	O,Q	B
San Diego	Yes	3,4	A,C,D	C,D	A,C,D	A,C,D	A,C,D	A,C,D	A,C,D	C,D	C,D	C,D	C,D		C,D	S	B
San Francisco	Yes	3,4	A,C,D	A,C,D	C,D	A,C,D	A,C,D	A,C,D	A,C,D	A,C,D	C,D		C,D	C,D	C,D	O	
San Jose **	Yes	4														M,S	B
Santa Ana **	Yes	3,4														M,Q	B
Colorado	Yes	3,4	A,C,D	A,C,D	C,D	C,D	A,C,D	A,C,D	A,C,D	C,D				C,D	C,D	Q	A
Denver	Yes	4,8	A,C,D	A,C,D	C,D	A,C,D	A,C,D	A,C,D	A,C,D	A,C,D	C,D		C,D	A,C,D	A,C,D	P	
Connecticut	Yes	2,4	A,C,D	A,C,D	A,C,D	A,C,D	A,C,D	A,C,D	C,D	C,D				C,D	C,D	M,R	
Hartford **	Yes	2	A,C,D	A,C,D	A,C,D	A,C,D	A,C,D	A,C,D	A,C,D	C,D	C,D	C,D	A,C,D	A,C,D	A,C,D	M,Q	
New Haven **	Yes	1,2,4	A,C,D	A,C,D	A,C,D	A,C,D	A,C,D	A,C,D	A,C,D	A,C,D	A,C,D	A,C,D	A,C,D	A,C,D	A,C,D		A
Delaware	Yes	1,2,4	A,C,D	A,C,D	C,D	C,D	C,D	C,D	C,D	C,D					C,D	M	A

NOTE: An explanation of the numbers and letters used is given at the end of the table.

TABLE 5.1 Codes for Emergency Power by States and Major Cities (Completed September 1984) *(Continued)*

State/City	Does State/City Have Legislation?	Legislation Code Type?	Hospitals	Nursing Homes	Schools	Theaters (Public Gathering Places)	Office Buildings	Hotels	Apartment Buildings	Airports	Fire and Police Stations	Water Treatment Plants	Sewage Treatment Plants	All Public Buildings, State	All Public Buildings, Commercial	Applicable Government Agency	Smokeproof Enclosures in High-Rise Buildings
District of Columbia	Yes	2,4	A,C,D	A,C,D	C,D	C,D	C,D	C,D	C,D	C,D	C,D	C,D	C,D	C,D	C,D	M,Q	A
Florida	Yes	1,2,4,6	A,C,D	A,C,D		C,D	C,D	C,D	C,D	C,D	C,D	C,D	C,D	C,D	C,D	M,O	A
Jacksonville	Yes	8	A,C,D	A,C,D	C,D	C,D	C,D	C,D	C,D	C,D					C,D	Q,S	
St Petersburg **	Yes	1,4,7	A,C,D	A,C,D	A,C,D	A,C,D	A,C,D	A,C,D	A,C,D	A,C,D	C,D				A,C,D	U	
Tampa	Yes	1,4,8	A,C,D	A,C,D	C,D	C,D	C,D	C,D	C,D	A,C,D				C,D	C,D	M	A
Georgia	Yes	1,4,7	A,C,D	A,C,D	C,D	C,D	C,D	C,D	C,D	C,D	C,D			C,D	C,D	M	
Atlanta	Yes	1,4,8	A,C,D	A,B,C,D	C,D	C,D	C,D	B,C,D	C,D	C,D	C,D		B	C,D	C,D	O	
Columbus **	Yes	4,6	A,C,D	B,C,D	B,C,D	B,C,D	C,D	C,D	C,D	C,D	C,D	B	B	C,D	C,D	S	B
Savannah	Yes	4	A,C,D	A,C,D	C,D	C,D	C,D	C,D	C,D	C,D	C,D			C,D	C,D	S	
Hawaii	Yes	1,3	A,C,D	A,C,D	C,D	C,D	C,D	C,D	C,D	C,D	A,C,D	C,D	C,D	C,D	C,D	M,Q	B
Honolulu	Yes	1,3,8	A,C,D	A,C,D	A,C,D	A,C,D	C,D	C,D	C,D	C,D	C,D	B	B	C,D	C,D	M,Q	B
Idaho	Yes	1,3,4	A,C,D	A,C,D	C,D	A,C,D	C,D	C,D	C,D	C,D	C,D	B	B	C,D	C,D	M,R	B
Illinois	No	2	A,C,D	A,C,D	C,D	A,C,D	A,C,D	A,C,D	A,C,D	A,C,D	A,C,D	A,C,D	A,C,D	A,C,D	A,C,D	M,N	
Chicago **	Yes	8	A,C,D	A,C,D	C,D	A,C,D	A,C,D	A,C,D	A,C,D	A,C,D	C,D	C,D	C,D	A,C,D	A,C,D	S	B
Rockford **	Yes	1,4	A,C,D	A,C,D	A,C,D	A,C,D	A,C,D	A,C,D	A,C,D	A,C,D	A,C,D	A,C,D	A,C,D	A,C,D	A,C,D	M,Q,R	B
Indiana	Yes	2,3,4	A,C,D	A,C,D	A,C,D	A,C,D	A,C,D	A,C,D	A,C,D	C,D	A,C,D	C,D	B,C,D	C,D	C,D	Q	B
Evansville	Yes	3,4	A,C,D	A,C,D	A,C,D	A,C,D	A,C,D	A,C,D	A,C,D	A,C,D	A,C,D	A,C,D	B,C,D	A,C,D	A,C,D	M,Q	B
Fort Wayne	Yes	1,4	A,C,D	A,C,D	A,C,D	A,C,D	C,D	C,D	A,C,D	C,D	C,D	C,D	C,D	C,D	C,D	S	B
Gary	Yes	2	A,C,D	A,C,D	A,C,D	A,C,D	A,C,D	A,C,D	C,D	A,C,D	A,C,D			A,C,D	A,C,D	S	B
Indianapolis **	Yes	1,3,4	A,C,D	A,C,D	C,D	C,D	A,C,D	C,D	C,D	C,D	A,C,D	C,D	C,D	C,D	C,D	M	B
South Bend	Yes	1,4	A,C,D	A,C,D	A,C,D	A,C,D	A,C,D	A,C,D	C,D	A,C,D	A,C,D			A,C,D	A,C,D	M,Q	B
Iowa	Yes	1,4	A,C,D	A,C,D	C,D	C,D	C,D	C,D	C,D	C,D	C,D			C,D	C,D	M	B
Des Moines	Yes	3,4	A,C,D	A,C,D	C,D	C,D	A,C,D	C,D	C,D	C,D	C,D			C,D	C,D	M,Q	B
Kansas	Yes	1,3,4	A,C,D	A,C,D	C,D											M,Q	B
Kansas City **	Yes	3,4	A,C,D	A,C,D	C,D											M,O	B
Wichita **	Yes	4	A,C,D	A,C,D	C,D											S	B
Kentucky	Yes	1,2,4,5	A,C,D	A,C,D	C,D	C,D	C,D	C,D	C,D	C,D				C,D	C,D	O,Q	B

NOTE: An explanation of the numbers and letters used is given at the end of the table.

TABLE 5.1 Codes for Emergency Power by States and Major Cities (Completed September 1984) (Continued)

State/City	Does State/City Have Legislation?	Legislation Code Type?	Hospitals	Nursing Homes	Schools	Theaters (Public Gathering Places)	Office Buildings	Hotels	Apartment Buildings	Airports	Fire and Police Stations	Water Treatment Plants	Sewage Treatment Plants	All Public Buildings, State	All Public Buildings, Commercial	Applicable Government Agency	Smokeproof Enclosures in High-Rise Buildings
Louisiana	Yes	1,4	A,C,D	A,C,D	C,D	C,D	C,D	A,C,D	C,D	C,D	A,C,D			C,D	C,D	M,Q	A
Baton Rouge **	Yes	4	A,C,D	A,C,D	B,C,D	A,C,D	B,C,D	A,C	C	A,C,D	C,D			C,D	C,D	S	A
New Orleans **	Yes	4	A,C,D	A,C,D	C,D	C,D	A,C,D*	A,C,D*	C,D	C,D	A,C,D		C,D	C,D	A,C,D	S	B
Shreveport	Yes	1,4	A,C,D	A,C,D	B,C,D	C,D	C,D	C,D	C,D	C,D	C,D		C,D	C,D	C,D	M,S	
Maine	Yes	1,2,4	A,C,D	C,D	C,D	C,D	C,D	C,D	C,D	A,C,D	A,C,D				C,D	M,N	
Maryland	Yes	1,2,4,5	A,C,D	A,C,D	C,D	A,C,D	C,D	A,C,D	C,D		A,C,D			C,D	C,D	S	
Baltimore **	Yes	4,8	A,C,D	A,C,D	C,D	C,D	C,D	C,D	C,D	C,D	A,C,D			C,D	C,D	P,Q	
Massachusetts	Yes	2,5	A,C,D	A,C,D	A,C,D	C,D	C,D	A,C,D	C,D	C,D	A,C,D		C,D	C,D	C,D	S	
Bedford **	Yes	2	A,C,D	C,D	C,D	C,D	C,D	C,D	C,D	C,D	A,C,D		C,D	C,D	C,D	S	
Boston **	Yes	2	A,C,D	A,C,D	A,C,D	A,C,D	A,C,D	A,C,D	C,D		A,C,D	B	B	A,C,D	A,C,D	S	B
Cambridge **	Yes	2	A,B	A,C,D	A,C,D	A,C,D	A,C,D	A,C,D	A,C,D		A,C,D			A,C,D	A,C,D	Q	
Springfield **	Yes	2	A,C,D	A,C,D	A,C,D	A,C,D	A,C,D	A,C,D	A,C,D	A,C	A,C			C,D	C,D	Q	
Worcester **	Yes	2	A,C	A,C	A,C	A,C	A,C,D	A,C	A,C	A,C,D	C,D				A,C,D	R	
Michigan	Yes	2,4,5	A,C,D	A,C,D	A,C,D	A,C,D	C,D	A,C,D	C,D	A,C,D	B,C,D	B,C,D	B,C,D	C,D	C,D	M,S	B
Detroit	Yes	1,4,5	A,C,D	A,C,D	A,C,D	A,C,D	A,C,D	A,C,D	A,C,D	A,C,D	A,C,D	B,C,D	B,C,D		D	M,S	B
Flint **	Yes	4,5	A,C,D	A,C,D	A,C,D	A,C,D		A,C,D	A,C,D		B,C,D				C,D	M,S	B
Grand Rapids	Yes	1,4,5	A,C,D	A,C,D	A,C,D	A,C,D	C,D	A,C,D	C,D	C,D	C,D	B	B	C,D	C,D	S	
Lansing **	Yes	2	A,C,D	A,C,D	A,C,D	A,C,D	C,D	A,C,D	C,D	A,C,D	C,D	C,D	C,D	C,D	C,D	M,N,O	
Minnesota	Yes	2,3,4,7	A,C,D	A,C,D	A,C,D	A,C,D	C,D	A,C,D	A,C,D	A,C,D	C,D	C,D	C,D	C,D	A,C,D	M,S	
Minneapolis	Yes	1,3,4	A,C,D	A,C,D	A,C,D	A,C,D	A,C,D	A,C,D	A,C,D	A,C,D	A,C,D	A,C,D	A,C,D	A,C,D	C,D	M,S	B
Saint Paul	Yes	1,3,4	A,C,D	A,C,D	A,C,D	A,C,D	A,C,D	A,C,D	A,C,D	C,D	A	A,C,D	A,C,D	C,D	C,D	M,S	
Mississippi	Yes	1,4,6	A,C,D	A,C,D	C,D	C,D	A,C,D	A,C,D	C,D	C,D	C,D	C,D	C,D	A,C,D	C,D	M	
Jackson	Yes	1,4	A,C,D	A,C,D	A,C,D	A,C,D	A,C,D	A,C,D	C,D	C,D				C,D	C,D	S	
Missouri	No																
Kansas City	Yes	4	A,C,D	A,C,D	C,D	C,D	C,D	C,D	C,D	C,D	C,D			C,D	C,D		A
Montana	Yes	2,3,4	A,C,D	A,C,D	C,D	C,D	C,D	C,D	C,D	C,D	C,D			C,D	C,D	M,Q	
Nebraska	Yes	1,4	A,C,D	A,C,D	C,D	C,D	C,D	C,D	C,D	C,D	C,D			C,D	C,D	M	

NOTE: An explanation of the numbers and letters used is given at the end of the table.

TABLE 5.1 Codes for Emergency Power by States and Major Cities (Completed September 1984) (*Continued*)

State/City	Does State/City Have Legislation?	Legislation Code Type?	Hospitals	Nursing Homes	Schools	Theaters (Public Gathering Places)	Office Buildings	Hotels	Apartment Buildings	Airports	Fire and Police Stations	Water Treatment Plants	Sewage Treatment Plants	All Public Buildings, State	All Public Buildings, Commercial	Applicable Government Agency	Smokeproof Enclosures in High-Rise Buildings
Lincoln	Yes	4,8	A,C,D	A,C,D	C,D	C,D	C,D	C,D	C,D	C,D					C,D	M,S	A*
Omaha	Yes	4,8	A,C,D	A,C,D	C,D	C,D	C,D	C,D	C,D	C,D					C,D	S	
Nevada	Yes	1,2,3,4	A,C,D	A,C,D	C,D	C,D	C,D	C,D	C,D	C,D					C,D	M,O,Q	A
New Hampshire	Yes	1,2,4	A,C,D	A,C,D	C,D	C,D	C,D	C,D	C,D	C,D				A,C,D	C,D	M	
New Jersey	Yes	2,3,4,5	A,C,D	A,C,D	C,D	C,D	C,D	C,D	C,D	C,D				C,D	C,D	Q	A
New Mexico	Yes	1,2,3,4	A,C,D	A,C,D	A,C,D	A,C,D	A,C,D	A,C,D	A,C,D	A,C,D		C,D	C,D	C,D	A,C,D	Q	
Albuquerque	Yes	3,4	A,C,D	A,C,D	C,D	C,D	C,D	C,D	C,D	C,D	C,D	C,D	C,D	C,D	C,D	S	
New York	Yes	2,4	A,C,D	A,C,D	C,D	C,D	C,D	C,D	C,D	C,D		B,C,D	B,C,D	C,D	C,D	O	
Albany	Yes	2,4	A,C,D	A,C,D	C,D	C,D	C,D	C,D	C,D	C,D		B,C,D	B,C,D	C,D	C,D	O	
Buffalo	Yes	2,8	A,C,D	A,C,D	C,D	C,D	C,D	C,D	C,D	C,D		B,C,D	B,C,D	C,D	C,D	Q	B
New York	Yes	2,8	A,C,D	A,C,D	C,D	C,D	C,D	C,D	C,D	C,D	C,D			C,D	C,D	N,O,R	A
Syracuse	Yes	3,8	A,C,D	A,C,D	C,D	C,D	C,D	C,D	C,D	C,D				C,D	C,D	Q	
North Carolina	Yes	2,4,6	A,C,D	A,C,D	C,D	C,D	C,D	C,D	C,D	C,D	C,D	C,D	C,D	C,D	C,D	M,Q	A
North Dakota	Yes	1,3,4	A,C,D	A,C,D	C,D	C,D	C,D	C,D	C,D	C,D	C,D	C,D	C,D	C,D	C,D	O,T	A
Ohio	Yes	2,4,5	A,C,D	A,C,D	C,D	C,D	C,D	C,D	C,D	C,D	C,D	C,D	C,D	C,D	C,D	M,O	A
Akron	Yes	2,4,5	A,C,D	A,C,D	C,D	A,C,D	A,C,D	A,C,D	A,C,D	C,D	C,D	C,D	C,D	C,D	C,D	Q,S	A
Cincinnati	Yes	2,4,5	A,C,D	A,C,D	C,D	A,C,D	A,C,D	A,C,D	A,C,D	C,D	C,D	C,D	C,D	C,D	C,D	O	A
Cleveland	Yes	2,4	A,C,D	C,D	C,D	C,D	C,D	C,D	C,D	C,D				C,D	C,D	Q,S	
Dayton	Yes	2,4,5	A,C,D	A,C,D	C,D	C,D	C,D	C,D	C,D	C,D				C,D	C,D	M,Q	A
Youngstown	Yes	2,4,5	A,C,D	A,C,D	C,D	C,D	C,D	C,D	C,D	C,D				C,D	C,D	Q	
Oklahoma	Yes†	1,2	A,C,D	C,D	C,D	C,D	C,D	C,D	C,D	C,D				C,D	C,D	M	
Oregon	Yes	1,2,3,4	A,C,D	A,C,D	C,D	C,D	C,D	C,D	C,D	A,C,D	A,C,D	C,D	C,D	C,D	C,D	M,Q	B
Portland **	Yes	1,2	A,C,D	C,D	A,C,D	A,C,D	A,C,D	A,C,D	A,C,D	C,D	A,C,D			C,D	C,D	M,Q	
Pennsylvania	Yes	4,8	A,C,D	A,C,D	A,C,D	A,C,D	A,C,D	A,C,D	A,C,D	A,C,D	C,D			A,C,D	A,C,D	R	
Philadelphia	Yes	2,4,5	A,C,D	A,C,D	A,C,D	A,C,D	A,C,D	A,C,D	A,C,D	A,C,D	C,D	C,D	C,D		A,C,D	M,Q	A
Rhode Island	Yes	1,4,6	A,C,D	A,C,D	C,D	C,D	C,D	C,D	C,D	C,D		C,D	C,D	C,D	C,D	M,Q	A
South Carolina	Yes	1,4,6	A,C,D	A,C,D	C,D	C,D	C,D	C,D	C,D	C,D				C,D	C,D	M,O	A

NOTE: An explanation of the numbers and letters used is given at the end of the table.

TABLE 5.1 Codes for Emergency Power by States and Major Cities (Completed September 1984) *(Continued)*

State/City	Does State/City Have Legislation?	Legislation Code Type?	Hospitals	Nursing Homes	Schools	Theaters (Public Gathering Places)	Office Buildings	Hotels	Apartment Buildings	Airports	Fire and Police Stations	Water Treatment Plants	Sewage Treatment Plants	All Public Buildings, State	All Public Buildings, Commercial	Applicable Government Agency	Smokeproof Enclosures in High-Rise Buildings
South Dakota	Yes	1,2,4	A,C,D	A,C,D	C,D	C,D	C,D	C,D	C,D	C,D				C,D	C,D	M	A
Tennessee	Yes	1,4,6	A,C,D	A,C,D	C,D	C,D	C,D	C,D	C,D	C,D	A,C			C,D	C,D	M,N,O	
Texas **	Yes	2	A,C,D	B,C,D					C	A,C	C,D				C	M,O,U	
Amarillo **	Yes	8	A,C	A,C	C,D	C,D	C,D	C,D	C,D	C,D		A	A	A	C	O	A
Austin	Yes	3	A,C,D	A,C,D	C,D	C,D	C,D	C,D	C,D	A,C,D	C,D			C,D	C,D	Q	
Corpus Christi **	Yes	3	A,C,D	A,C,D	C,D	A,C,D	C,D	A,C,D	A,C,D	C,D	C,D	C,D	C,D	C,D	C,D	Q	
Dallas	Yes	3,4	A,C,D	A,C,D	C,D	A,C,D	A,C,D	C,D	A,C,D	C,D	A,C,D	C,D	C,D	A,C,D	A,C,D	S	A
El Paso	Yes	4,6	A,C,D	A,C,D	C,D	C,D	C,D	C,D	C,D	B,C,D*	C,D	C,D	C,D	B,C,D*	B,C,D*	S	
Fort Worth **	Yes	3,4	A,C,D	C,D	C,D	C,D	B,C,D*	B,C,D*	B,C,D*	A,C,D	A,C,D			A,C,D	A,C,D	S	B
Houston	Yes	3,4,8	A,C,D	A,C,D	C,D	C,D	C,D	C,D	C,D	B,C,D*	C,D	C,D	C,D	B,C,D*	B,C,D*	M,Q	
Lubbock	Yes	1,3,4	A,C,D	A,C,D	C,D	A,C,D	C,D	C,D	C,D	A,C,D	C,D			C,D	A,C,D	M	
San Antonio **	Yes	4,8	A,C,D	A,C,D	C,D	C,D	C,D	C,D	C,D	C,D	C,D	C,D	C,D	C,D	C,D	S	B
Wichita Falls **	Yes	7	A,C,D	A,C,D	C,D	C,D	C,D	C,D	C,D	C,D	C,D			C,D	C,D	M,Q	A
Utah	Yes	1,2,3,4	A,C,D	A,C,D	C,D	C,D	C,D	C,D	C,D	C,D	C,D	C,D	C,D	C,D	C,D	M	B
Salt Lake City **	Yes	3,8	C,D	C,D	A,C,D	A,C,D	C,D	C,D	C,D	A,C,D	B,C,D			C,D	C,D	S	
Vermont	Yes	1,2,4,5	A,C,D	A,C,D	C,D	C,D	C,D	C,D	C,D	C,D	C,D	C,D	C,D	C,D	C,D	M,Q	B
Virginia	Yes	2,4,5	A,C,D	C,D	C,D	C,D	C,D	C,D	C,D	C,D	C,D	C,D	C,D	C,D	C,D	M,Q	
Richmond	Yes	1,4,5	A,C,D	A,C,D	C,D	C,D	C,D	C,D	C,D	A,C,D	C,D			C,D	C,D	R	
Virginia Beach	Yes	4,5	A,C,D	A,C,D	C,D	C,D	C,D	C,D	C,D	C,D	C,D			C,D	C,D	O,Q	
Washington	Yes	2,3,4	A,C,D	A,C,D	C,D	C,D	C,D	C,D	C,D	C,D	C,D	A,B	A,B	C,D	C,D	Q	
Seattle **	Yes	3,4	A,C,D	A,C,D	C,D	C,D	C,D	A,C,D	A,C,D	C,D	C,D	A,B	A,B	A,B	C,D	R	B
West Virginia	Yes	1,2,4	A,C,D	A,C,D	C,D	C,D	C,D	C,D	C,D	C,D	C,D	C,D	C,D	C,D	C,D	Q	
Wisconsin	Yes	2,4	A,C,D	A,C,D	C,D	C,D	C,D	C,D	C,D	C,D	C,D	C,D	C,D	C,D	C,D	M,O	
Madison	Yes	2,4	A,C,D	A,C,D	C,D	C,D	C,D	C,D	C,D	C,D	C,D	C,D	C,D	C,D	C,D	R,S	
Milwaukee	Yes	2,4,8	A,C,D	A,C,D	C,D	C,D	C,D	C,D	C,D	C,D	C,D	C,D	C,D	C,D	C,D	O	
Wyoming	Yes	1,2,3	A,C,D	A,C,D	A,C,D	C,D	C,D	C,D	C,D	C,D	C,D	C,D	C,D	C,D	C,D	M,N,P	

NOTE: An explanation of the numbers and letters used is given at the end of the table.

TABLE 5.1 Codes for Emergency Power by States and Major Cities (Completed September 1984) (*Continued*)

Explanation of Numbers and Letters Used in Table 1:

Legislation Code Type	Power Source	Governing Agency
1. Life Safety Code, ANSI/NFPA 101-1985 [11]	A. Emergency Power	M. Fire Marshal or Division of Fire
2. State	B. Standby Power	N. Department of Public Health
3. Uniform Building Code [24]	C. Exit Lighting	O. Local Government Units
4. National Electrical Code, ANSI/NFPA 70-1987 [9]	D. Egress Lighting	P. Public Safety
5. Building Officials and Code Administration (BOCA)		Q. Building Commission or Department
6. Standard Building Code [23]		R. Department of Labor
7. Health Care Facilities Code, ANSI/NFPA 99-1984 [10]		S. Inspection Department
8. City		T. Department of Insurance
		U. Various, but usually depends on occupancy

* High-rise building.
** No changes made since previous report.
† State buildings only.

Table 1 courtesy of the Electrical Generating Systems Marketing Association (April 1975).

Condensed General Need Criteria

Table 5.2 lists the needs in 13 general categories, with some breakdown under each, to indicate major requirements. Ranges under the columns "Maximum Tolerance Duration of Power Failure" and "Recommended Minimum Auxiliary Supply Time" are assigned based upon experience. Written standards have been referenced where applicable.

In some cases, under the columns "Type of Auxiliary Power System," both emergency and standby have been indicated as required. An emergency supply of limited time capacity may be used at a low cost for immediate or interruptible power until a standby supply can be brought on-line. An example would be the case in which battery lighting units come on until a standby generator can be started and transferred to critical loads.

Readers using this text may find that various combinations of general needs will require an in-depth system and cost analysis that will modify the recommended equipment and systems to best meet all requirements.

Small commercial establishments and manufacturing plants will usually find their requirements under two or three of the general need guidelines given in this chapter. Large manufacturers and commercial facilities will find that portions or all of the need guidelines given here apply to their operations and justify or require emergency and backup standby electric power.

Typical Emergency/Standby Lighting Recommendations

For short time durations, primarily lighting for personnel safety and evacuation purposes, battery units are satisfactory. Where longer service and heavier loads are required, an engine or turbine-driven generator is usually used, which starts automatically upon failure of the prime power source with the load applied by an automatic transfer switch. It is generally considered that an average level of 0.4 footcandles (fc) is adequate in which passage is required and no precise operations are expected.

Table 5.3 summarizes the user's needs for emergency and standby electric power for lighting by application and areas.

5.1 EMERGENCY/STANDBY POWER SOURCE OPTIONS

Power Sources

Sources of emergency power may include batteries, local generation, a separate source over separate lines from the electric utility, or various combinations of these. The quality of service required, the amount of load to be served, and the characteristics of the load will determine which type of emergency supply is required.

TABLE 5.2 Condensed General Criteria for Preliminary Consideration

General Need	Specific Need	Maximum Tolerance Duration of Power Failure	Recommended Minimum Auxiliary Supply Time	Type of Auxiliary Power System Emergency	Standby	System Justification
Lighting	Evacuation of personnel	Up to 10 s, preferably not more than 3 s	2 h	×		Prevention of panic, injury, loss of life Compliance with building codes and local, state, and federal laws Lower insurance rates Prevention of property damage Lessening of losses due to legal suits
	Perimeter and security	10 s	10–12 h during all dark hours	×	×	Lower losses from theft and property damage Lower insurance rates Prevention of injury
	Warning	From 10 s up to 2 or 3 min	To return to prime power source	×		Prevention or reduction of property loss Compliance with building codes and local, state, and federal laws Prevention of injury and loss of life
	Restoration of normal power system	1 s to indefinite depending on available light	Until repairs completed and power restored	×	×	Risk of extended power and light outage due to a longer repair time
	General lighting	Indefinite; depends on analysis and evaluation	Indefinite; depends on analysis and evaluation		×	Prevention of loss of sales Reduction of production losses Lower risk of theft Lower insurance rates

TABLE 5.2 Condensed General Criteria for Preliminary Consideration (*Continued*)

General Need	Specific Need	Maximum Tolerance Duration of Power Failure	Recommended Minimum Auxiliary Supply Time	Type of Auxiliary Power System Emergency	Standby	System Justification
	Hospitals and medical areas	Uninterruptible to 10 s ANSI/NFPA 99-1984 [10], 101-1985 [11] allow 10 s for alternate power source to start and transfer	To return of prime power	×	×	Facilitate continuous patient care by surgeons, medical doctors, nurses, and aids Compliance with all codes, standards, and laws Prevention of injury or loss of life Lessening of losses due to legal suits
	Orderly shutdown time	0.1 s to 1 h	10 min to several hours	×		Prevention of injury or loss of life Prevention of property loss by a more orderly and rapid shutdown of critical systems Lower risk of theft Lower insurance rates
Startup power	Boilers	3 s	To return of prime power	×	×	Return to production Prevention of property damage due to freezing Provision of required electric power
	Air compressors	1 min	To return of prime power		×	Return to production Provision for instrument control
Transportation	Elevators	15 s to 1 min	1 h to return of prime power		×	Personnel safety Building evacuation Continuation of normal activity
	Material handling	15 s to 1 min	1 h to return of prime power		×	Completion of production run Orderly shutdown Continuation of normal activity
	Escalators	15 s to no requirement for power	Zero to return of prime power		×	Orderly evacuation Continuation of normal activity

TABLE 5.2 Condensed General Criteria for Preliminary Consideration (*Continued*)

General Need	Specific Need	Maximum Tolerance Duration of Power Failure	Recommended Minimum Auxiliary Supply Time	Type of Auxiliary Power System		System Justification
				Emergency	Standby	
	Conveyors	15 s to 1 min	As analyzed and economically justified		×	Completion of production run Completion of customer order Orderly shutdown Continuation of normal activity
Mechanical utility systems	Water (cooling and general use)	15 s	½ h to return of prime power		×	Continuation of production Prevention of damage to equipment Supply of fire protection
	Water (drinking and sanitary)	1 min to no requirement	Indefinite until evaluated		×	Providing of customer service Maintaining personnel performance
	Boiler power	0.1 s	1 h to return of prime power	×	×	Prevention of loss of electric generation and steam Maintaining production Prevention of damage to equipment
	Pumps for water, sanitation, and production fluids	10 s to no requirement	Indefinite until evaluated		×	Prevention of flooding Maintaining cooling facilities Providing sanitary needs Continuation of production Maintaining boiler operation
	Fans and blowers for ventilation and heating	0.1 s to return of normal power	Indefinite until evaluated	×	×	Maintaining boiler operation Providing for gas-fired unit venting and purging Maintaining cooling and heating functions for buildings and production
Heating	Food preparation	5 min	To return of prime power		×	Prevention of loss of sales and profit Prevention of spoilage of in-process preparation

TABLE 5.2 Condensed General Criteria for Preliminary Consideration (*Continued*)

General Need	Specific Need	Maximum Tolerance Duration of Power Failure	Recommended Minimum Auxiliary Supply Time	Type of Auxiliary Power System		System Justification
				Emergency	Standby	
	Process	5 min	Indefinite until evaluated; normally for time for orderly shutdown, or to return of prime power		×	Prevention of in-process product damage Prevention of property damage Continued production Prevention of payment to workers during no production Lower insurance rates
Refrigeration	Special equipment or devices which have critical warmup (cryogenics)	5 min	To return of prime power		×	Prevention of equipment or product damage
	Depositories of critical nature (blood banks, etc)	5 min (10 s per ANSI/NFPA 99-1984 [10]	To return of prime power		×	Prevention of loss of material stored
	Depositories of noncritical nature (meat, produce, etc)	2 h	Indefinite until evaluated		×	Prevention of loss of material stored Lower insurance rates
Production	Critical process power (sugar factory, steel mills, chemical processes, glass products, etc)	1 min	To return of prime power or until orderly shutdown		×	Prevention of product and equipment damage Continued normal production Reduction of payment to workers on guaranteed wages during nonproductive period Lower insurance rates Prevention of prolonged shutdown due to nonorderly shutdown

TABLE 5.2 Condensed General Criteria for Preliminary Consideration (*Continued*)

General Need	Specific Need	Maximum Tolerance Duration of Power Failure	Recommended Minimum Auxiliary Supply Time	Type of Auxiliary Power System		System Justification
				Emergency	Standby	
	Process control power	Uninterruptible (UPS) to 1 min	To return of prime power	×	×	Prevention of loss of machine and process computer control program Maintaining production Prevention of safety hazards from developing Prevention of out-of-tolerance products
Space conditioning	Temperature (critical application)	10 s	1 min to return of prime power	×	×	Prevention of personnel hazards Prevention of product or property damage Lower insurance rates Continuation of normal activities Prevention of loss of computer function
	Pressure (critical) pos/neg atmosphere	1 min	1 min to return of prime power	×	×	Prevention of personnel hazards Continuation of normal activities Prevention of product or property damage Lower insurance rates Compliance with local, state, and federal codes, standards, and laws
	Humidity (critical)	1 min	To return of prime power	×	×	Prevention of loss of computer functions Maintenance of normal operations and tests Prevention of explosions or other hazards

TABLE 5.2 Condensed General Criteria for Preliminary Consideration (*Continued*)

General Need	Specified Need	Maximum Tolerance Duration of Power Failure	Recommended Minimum Auxiliary Supply Time	Type of Auxiliary Power System		System Justification
				Emergency	Standby	
	Static charge	10 s or less	To return of prime power	×	×	Prevention of static electric charge and associated hazards Continuation of normal production (printing press operation, painting spray operations)
	Building heating and cooling	30 min	To return of prime power	×	×	Prevention of loss due to freezing Maintenance of personnel efficiency Continuation of normal activities
	Ventilation (toxic fumes)	15 s	To return of prime power or orderly shutdown	×	×	Reduction of health hazards Compliance with local, state, and federal codes, standards, and laws Reduction of pollution
	Ventilation (explosive atmosphere)	10 s	To return of prime power or orderly shutdown	×	×	Reduction of explosion hazard Prevention of property damage Lower insurance rates Compliance with local, state, and federal codes, standards, and laws Lower hazard of fire Reduce hazards to personnel
	Ventilation (building general)	1 min	To return of prime power	×	×	Maintaining of personnel efficiency Providing make-up air in building

TABLE 5.2 Condensed General Criteria for Preliminary Consideration (*Continued*)

General Need	Specified Need	Maximum Tolerance Duration of Power Failure	Recommended Minimum Auxiliary Supply Time	Type of Auxiliary Power System Emergency	Type of Auxiliary Power System Standby	System Justification
	Ventilation (special equipment)	15 s	To return of prime power or orderly shutdown	×	×	Purging operation to provide safe shutdown or startup Lowering of hazards to personnel and property Meeting requirements of insurance company Compliance with local, state, and federal codes, standards, and laws Continuation of normal operation
	Ventilation (all categories noncritical)	1 min	Optional		×	Maintaining comfort Preventing loss of tests
	Air pollution control	1 min	Indefinite until evaluated; compliance or shutdowns are options	×	×	Continuation of normal operation Compliance with local, state, and federal codes, standards, and laws
Fire protection	Annunciator alarms	1 s	To return of prime power	×		Compliance with local, state, and federal codes, standards, and laws Lower insurance rates Minimizing life and property damage
	Fire pumps	10 s	To return of prime power		×	Compliance with local, state, and federal codes, standards, and laws Lower insurance rates Minimizing life and property damage

TABLE 5.2 Condensed General Criteria for Preliminary Consideration (*Continued*)

General Need	Specified Need	Maximum Tolerance Duration of Power Failure	Recommended Minimum Auxiliary Supply Time	Type of Auxiliary Power System — Emergency	Standby	System Justification
	Auxiliary lighting	10 s	5 min to return of prime power	×	×	Servicing of fire pump engine should it fail to start Providing visual guidance for fire-fighting personnel
Data processing	CPU memory tape/disk storage, peripherals	½ cycle	To return of prime power or orderly shutdown	×	×	Prevention of program loss Maintaining normal operations for payroll, process control, machine control, warehousing, reservations, etc
	Humidity and temperature control	5 to 15 min (1 min for water-cooled equipment)	To return of prime power or orderly shutdown		×	Maintenance of conditions to prevent malfunctions in data processing system Prevention of damage to equipment Continuation of normal activity
Life support and life safety systems (medical field, hospitals, clinics, etc)	X-ray	Milliseconds to several hours	From no requirement to return of prime power, as evaluated		×	Maintenance of exposure quality Availability for emergencies
	Light	Milliseconds to several hours	To return of prime power	×	×	Compliance with local, state, and federal codes, standards, and laws Preventing interruption to operation and operating needs
	Critical to life, machines, and services	½ cycle to 10 s	To return of prime power	×	×	Maintenance of life Prevention of interruption of treatment or surgery Continuation of normal activity Compliance with local, state, and federal codes, standards, and laws

TABLE 5.2 Condensed General Criteria for Preliminary Consideration (*Continued*)

General Need	Specified Need	Maximum Tolerance Duration of Power Failure	Recommended Minimum Auxiliary Supply Time	Type of Auxiliary Power System Emergency	Type of Auxiliary Power System Standby	System Justification
	Refrigeration	5 min	To return of prime power		×	Maintaining blood, plasma, and related stored material at recommended temperature and in prime condition
Communication systems	Teletypewriter	5 min	To return of prime power		×	Maintenance of customer services Maintenance of production control and warehousing Continuation of normal communication to prevent economic loss
	Inner building telephone	10 s	To return of prime power	×		Continuation of normal activity and control
	Television (closed circuit and commercial)	10 s	To return of prime power		×	Continuation of sales Meeting of contracts Maintenance of security Continuation of production
	Radio systems	10 s	To return of prime power	×		Maintenance of security and fire alarms Providing evacuation instructions Continuation of service to customers Prevention of economic loss Directing vehicles normally
	Intercommunication systems	10 s	To return of prime power	×		Providing evacuation instructions Directing activities during emergency Providing for continuation of normal activities Maintaining security

TABLE 5.2 Condensed General Criteria for Preliminary Consideration (*Continued*)

General Need	Specific Need	Maximum Tolerance Duration of Power Failure	Recommended Minimum Auxiliary Supply Time	Type of Auxiliary Power System — Emergency	Type of Auxiliary Power System — Standby	System Justification
	Paging systems	10 s	½ h	×	×	Locating of responsible persons concerned with power outage Providing evacuation instructions Prevention of panic
Signal circuits	Alarms and annunciation	1 to 10 s	To return of prime power	×	×	Prevention of loss from theft, arson, or riot Maintaining security systems Compliance with codes, standards, and laws Lower insurance rates Alarm for critical out-of-tolerance temperature, pressure, water level, and other hazardous or dangerous conditions Prevention of economic loss
	Land-based aircraft, railroad, and ship warning systems	1 s to 1 min	To return of prime power	×	×	Compliance with local, state, and federal codes, standards, and laws Prevention of personnel injury Prevention of property and economic loss

TABLE 5.3 Typical Emergency and Standby Lighting Recommendations

Standby*	Immediate, Short-Term†	Immediate, Long-Term‡
Security lighting	Evacuation lighting	Hazardous areas
Outdoor perimeters	Exit signs	Laboratories
Closed circuit TV	Exit lights	Warning lights
Night lights	Stairwells	Storage areas
Guard stations	Open areas	Process areas
Entrance gates	Tunnels	
	Halls	Warning lights
Production lighting		Beacons
Machine areas	Miscellaneous	Hazardous areas
Raw materials storage	Standby generator areas	Traffic signals
Packaging	Hazardous machines	
Inspection		Health care facilities
Warehousing		Operating rooms
Offices		Delivery rooms
		Intensive care areas
Commercial lighting		Emergency treatment areas
Displays		
Product shelves		Miscellaneous
Sales counters		Switchgear rooms
Offices		Elevators
		Boiler rooms
Miscellaneous		Control rooms
Switchgear rooms		
Landscape lighting		
Boiler rooms		
Computer rooms		

* An example of a standby lighting system is an engine-driven generator.
† An example of an immediate short-term lighting system is the common unit battery equipment.
‡ An example of an immediate long-term lighting system is a central battery bank rated to handle the required lighting load only until a standby engine-driven generator is placed on-line.

Batteries

Batteries are the fundamental and most commonly used standby power source. They are typically in the form of unitized equipment (wall-packs) consisting of a rechargeable storage battery, automatic charger, floodlight-type lamps, and automatic transfer relay. They sometimes have remote lighting heads and usually have exit lights connected to them. Operation is typically at 12 VDC. These constitute decentralized systems.

There are also centralized systems that power remote lighting heads and exit lights that typically operate at 24 or 32 VDC. A variation of this is centralized inverter systems, which operate regular light fixtures and exit lights on their normal AC voltage of 120 or 277 VAC. Another variation is decentralized, self-contained, emergency lighting inverter units.

Batteries are also used as a backup power source for communications, security systems, telephone, and fire alarm systems.

Batteries provide a low first-cost option as an emergency source, but have a relatively high maintenance cost. They also have limited capacity, thereby restricting the equipment loads that they are suitable for supplying; their low-voltage operation presents voltage drop limitations.

Local Generation

Local generation is advisable when service is absolutely essential for lighting or power loads, or both, and when these loads are relatively large and are distributed over large areas. Several choices are available in the type of prime mover, voltage of the generator, and method of connection to the system. Various alternates should be considered. The prime mover supply may be steam, natural gas, gasoline, diesel fuel, or liquefied petroleum gas (LPG).

For generators over 500 kW, gas turbine–driven units may be a favorable choice. This type of unit has acceptable efficiency at full load but is much less efficient than other types of drives at partial load. Gas turbine–driven units do not start as rapidly as other drives, but they are reliable and require a minimum of attention. They generally will not meet NEC requirements for emergency systems. Generator sets requiring more than 10 s to develop power require that an auxiliary system supply power until the generator can pick up the load. Of all the prime mover supply choices, diesel fuel is probably the most widely used for commercial and institutional applications.

Fuel storage requirements should be determined after considering the frequency and duration of power outages, the types of emergency loads to be served, and the ease of replenishing fuel supplies. Some installations may require a supply sufficient for 3 months be maintained, whereas a 1-day supply may be adequate for others. Code requirements [see ANSI/NFPA 37-1990 ("Stationary Combustion Engines and Gas Turbines")] severely limit the amount of fuel that can be stored in buildings, so that fuel may have to be piped to a small local (day) tank adjacent to the generator. The NEC and other codes [e.g., EGSA 109C-1984 ("Codes for Emergency Power by States and Major Cities")] require an on-site fuel supply capable of operating the prime mover at full-demand load for at least 2 h.

A significant additional consideration germane to the fuel source is its emissions. The federal and state Environmental Protection Agencies have strict and complicated regulations for which compliance is mandatory. It is generally advisable to engage the services of an environmental consultant to ensure compliance with these laws and regulations. What it means to the electrical design professional is determining the total hours of operation for the engine-driven generator on an annual basis, including time for emergency operation, exercise, peak-shaving or load-shedding, parallel operation with the electric utility, and so on. The emissions resulting from the hours of operation are taken in concert with any other source of emissions from the site, such as boilers, for total site emissions as a source. It is customary to estimate the hours of operation using your best judgment with a conservative margin of safety. There is close monitoring and stiff penalties for noncompliance.

Generator selection can only be made after a careful study of the system to which it is connected and the loads to be carried by it. The voltage, frequency, and phase relationships of the generator should be the same as in the normal system. The size of the generator will be determined by the load to be carried, with consideration given to the size of the individual motors to be started. Another consideration is the distortion created by the loads that the system will be supplying. The speed and voltage regulation required will determine the accuracy and sensitivity of regulating devices. When a generator is required to carry emergency loads only during power outages and should not operate in parallel with the normal system, the simplest type of regulating equipment is usually adequate. For parallel operation, good-quality voltage regulators and governors are needed to ensure proper and active and reactive power loading of the generator. When the generator is small in relation to the system, it is usually preferable to have a large drooping characteristic in the governor and considerable compensation in the voltage regulator so that the local generator will follow the larger system rather than try to regulate it. Automatic synchronizing packages for paralleling generators are available that may include all the protective features required for paralleling generators. The design of this equipment should be coordinated with the characteristics of the generator.

Multiple Service Connections

When the local utility company can provide two or more service connections over separate lines from separate generation points so that system disturbances or storms are not apt to affect both supplies simultaneously, local generation or batteries may not be justified. A second line for emergency power should not be relied upon, however, unless total loss of power can be tolerated on rare occasions. The alternate feeder can either serve as a standby with primary switching or have its own transformer with secondary switching.

Often, an alternate primary service feeder can be run physically separate from the normal service feeder but is not from a separate generation source. Because of this, it is common for critical load facilities such as hospitals and data centers to have multiple service connections in combination with local generation to ensure reliability and, thus, service continuity.

5.2 TYPICAL EMERGENCY/STANDBY SYSTEM ARRANGEMENTS

Some arrangements commonly found for multiple utility services and/or engine-driven local generation are as follows:

Multiple Utility Services

Multiple utility services may be used as an emergency or standby source of power. Required is an additional utility service from a separate source and the required switching equipment. Figure 5.1 shows automatic transfer between two low-voltage utility supplies. Utility source 1 is the normal power line and utility source 2 is a separate utility supply providing emergency power. Both circuit breakers are normally closed. The load must be able to tolerate the few cycles of interruption while the automatic transfer device operates.

Automatic switching equipment may consist of three circuit breakers with suitable control and interlocks, as shown in Figure 5.2. Circuit breakers are generally used for primary switching in which the voltage exceeds 600 V. They are more expensive but safer to operate, and the use of fuses for overcurrent protection is avoided.

Relaying is provided to transfer the load automatically to either source if the other one fails, provided that circuit is energized. The supplying utility will normally designate which source is for normal use and which is for emergency. If either supply is not able to carry the entire load, provisions must be made to drop noncritical loads before the

FIGURE 5.1 Two-utility source system using one automatic transfer switch.

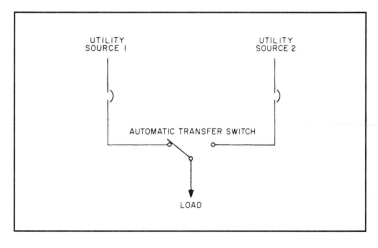

UTILITY
SOURCE 1

UTILITY
SOURCE 2

AUTOMATIC TRANSFER SWITCH

LOAD

FIGURE 5.2 Two-utility source system in which any two circuit breakers can be closed.

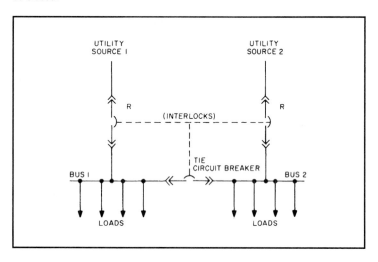

transfer takes place. If the load can be taken from both services, the two R circuit breakers are closed and the tie circuit breaker is open. This mode of operation is generally preferred by the supplying utility and the customer. The three circuit breakers are interlocked to permit any two to be closed but prevent all three from being closed. The advantages of this arrangement are that the momentary transfer outage will occur only on the load supplied from the circuit that is lost, the loads can be balanced between the two buses, and the supplying utility doesn't have to keep track of reserve capacity for the emergency feeder. However, the supplying utility may not allow the load to be taken from both sources, especially because a more expensive totalizing meter may be required. A manual override of the interlock system should be provided so that a closed transition transfer can be made if the supplying utility wants to take either line out of service for maintenance or repair and a momentary tie is permitted.

If the supplying utility will not permit power to be taken from both sources, the control system must be arranged so that the circuit breaker on the normal source is closed, the tie circuit breaker is closed, and the emergency-source circuit breaker is open. If the utility will not permit dual or totalized metering, the two sources must be connected together to provide a common metering point and then connected to the distribution switchboard. In this case, the tie circuit breaker can be eliminated and the two circuit breakers act as a transfer device (sometimes called a transfer pair). Under these conditions, the cost of an extra circuit breaker can rarely be justified.

The arrangement shown in Figure 5.2 only provides protection against failure of the normal utility service. Continuity of power to critical loads can also be disrupted by

1. An open circuit within the building (load side of the incoming service)
2. An overload or fault tripping out a circuit
3. An electrical or mechanical failure of the electric power distribution system within the building

It may be desirable to locate transfer devices close to the load and have the operation of the transfer devices independent of overcurrent protection. Multiple transfer devices of lower current rating, each supplying a part of the load, may be used rather than one transfer device for the entire load.

The arrangement shown in Figure 5.2 can represent the secondary of a double-ended substation configuration or a primary service. It is sometimes referred to as a "main-tie-main" configuration.

Availability of multiple utility service systems can be improved by adding a standby engine-generator set capable of supplying the more critical load. Such an arrangement, using multiple automatic transfer switches, is shown in Figure 5.3.

Transfer Methods

Figure 5.4, panel *a,* shows a typical switching arrangement in which a local emergency generator is used to supply the entire load upon loss of the normal power supply. All emergency loads are normally supplied through device A. Device B is open and the generator is at rest. When the normal supply fails, the transfer switch undervoltage relay is de-energized and, after a predetermined time delay, closes its engine-starting contacts. The time delay is introduced so that the generator will not be started unnecessarily during transient voltage dips and momentary outages. When the alternate source is a generator, sufficient time or speed monitoring should be allowed to permit the generator to reach acceptable speed (thus frequency and voltage) before transfer and application of load. It should be noted that the arrangement shown in Figure 5.4 (*a*) does not provide complete protection against power disruption within the building.

Panel *b* of Figure 5.4 shows a typical switching arrangement in which only the critical loads are transferred to the emergency source—in this case, an emergency generator. For maximum protection, the transfer switch is located close to the critical loads.

FIGURE 5.3 Diagram illustrating multiple automatic double-throw transfer switches providing varying degrees of emergency and standby power.

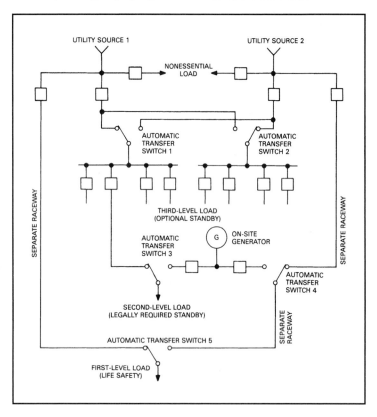

Other transfer methods are illustrated in the foregoing discussion of multiple utility services.

Parallel Generation

Enhanced reliability can be provided in large measure through redundancy, and engine-driven emergency generators are no exception. If, for example, a single 300-kW generator can accommodate all of the critical emergency load of a building and it is the only generator, should it fail to start for any reason or be out of service for routine maintenance at the time it is needed, you have no emergency service. To preclude this situation, good practice dictates that you have two generators, each

FIGURE 5.4 Typical transfer-switching methods. (*a*) Total transfer.
(*b*) Critical load transfer.

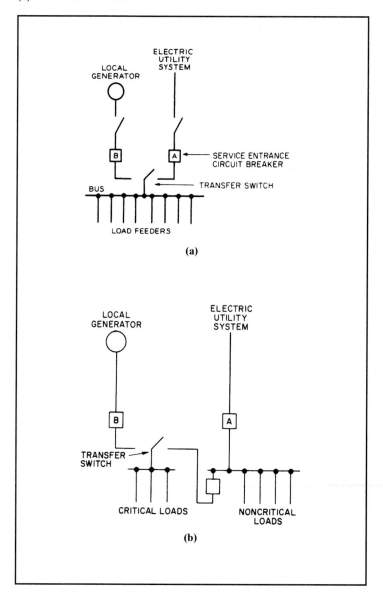

sized to accommodate the entire load and automatically synchronized, thus ensuring that at least one generator is available at all times. This concept can be extended to any situation (i.e., any two out of three units, three out of four, and so on). A good general philosophy is multiple small, rather than singular large, generating units.

To illustrate the operation of a typical multiengine automatic paralleling system and its sequence of operation, Figure 5.5 shows four engine generators that comprise an emergency source.

The operation is for a random-access paralleling system, and the loads are connected to the bus in random order, as they become available.

The loads, however, are always connected to the emergency bus in ascending order of priority beginning with priority one. For load shedding, the loads are disconnected in descending order of priority beginning with the last priority of load to be connected.

Upon a loss of normal-source voltage as determined by any one or more of the automatic transfer switches shown, a signal initiates starting of all engine-generator sets. The first set to come up to 90 percent of nominal voltage and frequency is connected to the alternate source bus. Critical and life safety loads are then transferred via ATS No. 1 and No. 2 to the bus upon sensing availability of power on the bus. As the remaining engine-generator sets achieve 90 percent of the nominal voltage and frequency, their respective synchronizing monitors will control the voltage and frequency of these oncoming units to produce synchronism with the bus. Once the oncoming unit is matched in voltage, frequency, and the phase angle with the bus, its synchronizer will initiate paralleling. Upon connection to the bus, the governor will cause the engine-generator set to share the connected load with the other on-line sets.

Each time an additional set is added to the emergency bus, the next load is transferred in a numbered sequence via additional transfer switches, such as ATS No. 3, until all sets and essential loads are connected to the bus. Control circuitry should prevent the automatic transfer or connection of loads to the bus until there is sufficient capacity to carry these loads. Provision is made for manual override of the load addition circuits for supervised operation.

Upon the restoration of the normal source of supply as determined by the automatic transfer switches, the engines are run for a period of up to 15 min for cooling down and then for shutdown. All controls automatically reset in readiness for the next automatic operation.

The system is designed so that reduced operation is automatically initiated upon failure of any plant through load dumping. This mode overrides any previous manual controls to prevent overloading the emergency bus. Upon sensing a failure mode on an engine, the controls automatically initiate disconnect, shutdown, and lockout of the failed engine, and reduction of the connected load to within the capacity of

FIGURE 5.5 Typical multiengine automatic paralleling system.

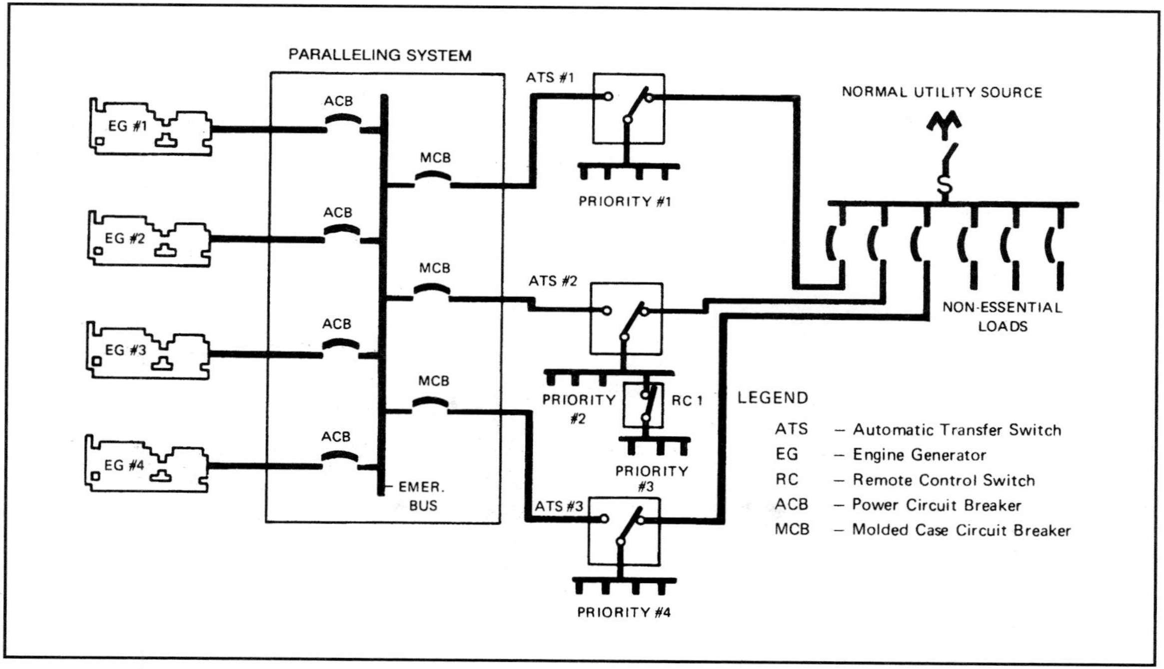

PARALLELING SYSTEM
EG #1
EG #2
EG #3
EG #4
ACB
ACB
ACB
ACB
MCB
MCB
MCB
EMER. BUS
ATS #1
ATS #2
ATS #3
PRIORITY #1
PRIORITY #2
RC 1
PRIORITY #3
PRIORITY #4
NORMAL UTILITY SOURCE
NON-ESSENTIAL LOADS
LEGEND
ATS – Automatic Transfer Switch
EG – Engine Generator
RC – Remote Control Switch
ACB – Power Circuit Breaker
MCB – Molded Case Circuit Breaker

the remaining plants. Controls should require manual reset under these conditions.

Protection of the engine and generator against motorization is provided. A reverse-power monitor, upon sensing a motorizing condition on any plant, will initiate load shedding, disconnect the failing plant, and shut it down.

Sometimes a higher level of reliability is economically justifiable in a parallel generation arrangement for critical loads such as hospitals and data centers. This is known as providing an $(N + 1)$ level of reliability (redundancy) (i.e., providing one more generator than is needed to serve the emergency load). Thus, if one of the emergency generators fails to start or is out of service for any reason, the remaining plants can serve the entire emergency load. This precludes the need for automatic load shedding, which can be expensive in itself. Thus, this provides for two levels of contingency operation, the first being loss of the normal source of power, and the second being loss of one of the emergency/ standby generators. Providing an even higher level of reliability is rarely justifiable.

Elevator Emergency Power Transfer System

Elevators present a unique emergency power situation. Where elevator service is critical for personnel and patients, it is desirable to have automatic power transfer with manual supervision. Operators and maintenance personnel may not be available in time if the power failure occurs on a weekend or at night.

1. *Typical elevator system:* Figure 5.6 shows an elevator emergency power transfer system whereby one preferred elevator is fed from a vital load bus through an emergency riser, while the rest of the elevators are fed from the normal service. By providing an automatic transfer switch for each elevator and a remote selector station, it is possible to select individual elevators, thus permitting complete evacuation in the event of power failure. The engine-generator set and emergency riser need only be sized for one elevator, thus minimizing the installation cost. The controls for the remote selector, automatic transfer switches, and engine starting are independent of the elevator controls, thereby simplifying installation.

2. *Regenerated power:* Regenerated power is a concern for motor-generator-type elevator applications. In some elevator applications, the motor is used as a brake when the elevator is descending and generates electricity. Electric power is then pumped back into the power source. If the source is commercial utility power, it can easily be absorbed. If the power source is an engine-driven generator, the regenerated power can cause the generating set and the

elevator to overspeed. To prevent overspeeding of the elevator, the maximum amount of power that can be pumped back into the generating set must be known. The permissible amount of absorption is approximately 20 percent of the generating set's rating in kilowatts. If the amount pumped back is greater than 20 percent, other loads must be connected to the generating set, such as emergency lights or "dummy" (parasitic) load resistances. Emergency lighting should be permanently connected to the generating set for maximum safety. A dummy (parasitic) load can also be automatically switched on the line whenever the elevator is operating from an engine-driven generator.

FIGURE 5.6   Elevator emergency power transfer system.

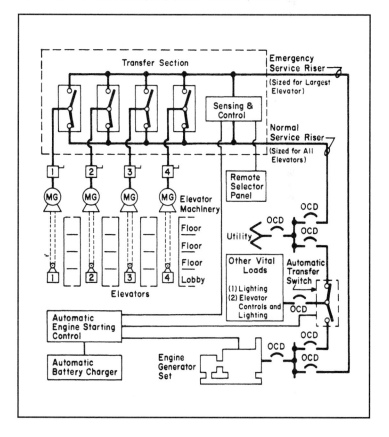

Hospitals/Health Care Facilities

Hospital/health care facilities present a unique situation. ANSI/NFPA 99-1984 mandates that emergency loads be broken into three distinct branches, namely critical, life safety, and equipment. This concept is illustrated in Figure 5.7.

This arrangement provides a very high level of reliability and integrity. Critical, life safety, and essential equipment loads are transferred automatically and immediately (i.e., with no intentional delay), to the emergency source upon loss of commercial power. Lower-priority nonessential loads are transferred manually via nonautomatic transfer switches when the system has stabilized in the emergency mode and available capacity has been verified.

FIGURE 5.7   Typical hospital installation with a nonautomatic transfer switch and several automatic transfer switches.

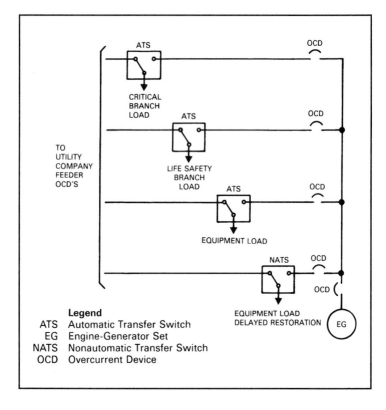

5.3 GENERATOR AND GENERATOR SET SIZING

Introduction

Proper sizing of a generator is an important task. The following guidelines represent the general and specific considerations that must be taken into account in properly sizing a generator for a specific application. These guidelines are based on Caterpillar Generator Sets as an industry leader. A common practice in the industry is to base a given design around a specific manufacturer of a major piece of equipment, such as a generator, and to make allowances for idiosyncratic differences that allow competitive bids and supply to the purchaser. It is in this context that the Caterpillar guidelines are offered.

I. APPLICATION DATA RATINGS

Diesel-Electric Power Generation
All ratings shown and thermal ratings are subject to manufacturing tolerances of ±3 percent.

When using a generator set, use the following guidelines to determine whether standby, prime, prime plus 10 percent, or continuous rating applies.

STANDBY RATING:

Typical load factor = 60 percent or less

Typical hours/year = 100 h

Typical peak demand = 80 percent of standby-rated kilowatts with 100 percent of rating available for the duration of an emergency outage

Enclosure/sheltered environment

PRIME + 10 PERCENT RATING:

Typical load factor = 60 percent or less

Typical hours/year = 500 h

Typical demand = 80 percent of standby-rated kilowatts with 100 percent of rating available for the duration of an emergency outage

Typical application = Standby, rental, power module, unreliable utility, or interruptible rates

PRIME RATING:

Typical load factor = 60 to 70 percent

Typical hours/year = No limit

Typical peak demand = 100 percent of prime-rated kilowatts used occasionally, but for less than 10 percent of operating hours

Typical application = Industrial, pumping, construction, peak shaving, or cogeneration

CONTINUOUS RATING:

    Typical load factor = 70 to 100 percent

    Typical hours/year = No limit

    Typical peak demand = 100 percent of continuous-rated kilowatts for 100 percent of operating hours

    Typical application = Base load, utility, cogeneration, or peak shaving

For conditions outside the above limits, refer to the manufacturer.

Operating units above these rating definitions will result in a shorter life until overhaul.

Gas-Electric Power Generation

All ratings shown and thermal ratings are subject to manufacturing tolerances of ±3 percent.

When using a generator set, use the following guidelines to determine whether standby or continuous rating applies.

Remember the typical load factor is the sum of the loads a generator set experiences while it is running under load divided by the number of hours it operates under those loads. Extended idling time and the time when the generator is not operating do not enter into the calculation for load factor.

STANDBY RATING:

Adds 5 percent to continuous rating when using natural gas. When using other fuels, contact manufacturer. Applies to all gas engine-generator sets.

    Typical load factor = 60 percent or less

    Maximum hours/year = 100 h

    Typical peak demand = 80 percent of standby-rated kilowatts with 100 percent of rating available for the duration of the emergency outage

    Typical application = Building service standby and enclosure/sheltered environment

CONTINUOUS RATING:

    Typical load factor = 70 to 100 percent

    Typical hours/year = No limit

    Typical peak demand = 100 percent of continuous-rated kilowatts for 100 percent of operating hours

    Typical application = Base load, utility, cogeneration, or peak shaving

For conditions outside the above limits, refer to the manufacturer.

Operating units above these rating definitions will result in shorter life until overhaul and possible catastrophic failure.

Power for gas engines is based on fuel having a low heating value (LHV) of 33.74 kJ/L (905 Btu/ft^3) for pipeline natural gas.

Propane ratings are based on having an LHV of 85.75 kJ/L (2300 Btu/ft^3). Landfill gas ratings are based on fuel having an LHV of 16.78 kJ/L (450 Btu/ft^3). Digester gas ratings are based on fuel having an LHV of 22.37 kJ/L (600 Btu/ft^3). The gas volume is based on conditions of 101 kPa (29.88 in Hg) and 15.5°C (60°F). Variations in altitude, temperature, and gas composition from standard conditions may require a reduction in engine horsepower.

II. LOADS

All resistive and inductive loads are summarized. Information from motor nameplates are as noted whenever possible. Table 5.7 approximates motor efficiencies.

III. ENGINE SIZING

Total engine load is determined by calculating effects of motor efficiencies and adding to resistive loads.

IV. ENGINE SELECTION

Consideration of load (kW), frequency (Hz), speed (rpm), and engine configuration (gas, diesel, turbocharged, aftercooled, naturally aspirated) allow engine selection from Table 5.4.

V. GENERATOR SIZING

Generator capacity (kVA) is determined not only by total load but by motor size, configuration, starting sequence, and possible motor-starting aids. Minimize motor-starting requirements by starting largest motors first. Random-starting sequence requires worst-case application by starting smallest motors first. Use Table 5.5 to calculate starting kVA (SKVA) or full-load amperes.

Effective SKVA
Motors on-line diminish generator capability (SKVA) to start additional motors (Figure 5.8). Reduced-voltage starting decreases demand on the generator (Table 5.6), but also reduces the torque capability of the motor.

Select a generator that provides motor-starting requirements (SKVA) with acceptable voltage dip (Table 5.4).

Voltage dip is measured on an oscilloscope as SKVA, noted in Table 5.4, while driven by a synchronized motor.

TABLE 5.4 Motor Starting Data Diesel and Gas Generator Sets

PRIME POWER — 60 Hz-1200 RPM*

Engine Model	Generator Frame	Rating w/Fan kW	Starting kV·A at Voltage Dip**		
			10%	20%	30%
3516 TA	809	1100	788	1773	3039
3516 TA	806	900	444	1000	1714
3512 TA	806	830	444	1000	1714
3512 TA	687	650	411	925	1587
3508 TA	686	550	277	625	1071
3508 TA	683	425	231	520	892
3412 TA	587	325	214	481	824
3408 TA	585	225	161	362	621
3406 TA	583	170	142	321	549

* ISO power with 10% overload capability except as noted by ***.
** Noted SKVA values are for low voltage (below 600V) generators. Consult Caterpillar for medium voltage generator capabilities.
NOTE: SCR rectifiers and variable speed motor controls require detailed analysis. Contact Caterpillar and the SCR supplier.

TABLE 5.4 Motor Starting Data Diesel and Gas Generator Sets (Continued)

PRIME POWER — 60 Hz-1800 RPM*

Engine Model	Generator Frame	Rating w/Fan kW	Starting kV·A at Voltage Dip** 10%	20%	30%
3516 TA	807	1600	1234	2777	4761
	806	1360	1010	2272	3896
3512 TA	889	1135	966	2173	3726
	887	1000	888	2000	3428
	685	910	584	1315	2255
3508 TA	685	820	584	1315	2255
	681	725	419	943	1617
	681	680	419	943	1617
	589	650	396	892	1530
3412 TA	589	545	444	1000	1714
T	588	455	317	714	1224
T	586	425	278	625	1071
3408 TA	584	365	242	543	932
3406 TA #0	450	320	188	424	726
TA #1	449	275	171	385	659
TA	448	250	159	357	612
3306 ATAAC	447	225	142	321	549
TA	446	205	139	313	536
TA	446	180	139	313	536
3208 T	443	160	111	250	428

* ISO power with 10% overload capability except as noted by ***.
** Noted SKVA valves are for low voltage (below 600V) generators. Consult Caterpillar for medium voltage generator capabilities.

TABLE 5.4 Motor Starting Data Diesel and Gas Generator Sets (*Continued*)

STANDBY POWER — 60 Hz-1200 RPM

Engine Model	Generator Frame	Rating w/Fan kW	Starting kV·A at Voltage Dip**		
			10%	20%	30%
3516 TA	809	1250	788	1773	3039
3516 TA	806	975	444	1000	1714
3512 TA	806	925	444	1000	1714
3512 TA	687	700	411	925	1587
3508 TA	686	615	277	625	1071
3508 TA	683	465	231	520	892
3412 TA	587	355	214	481	824
3408 TA	585	245	161	362	621
3406 TA	583	185	142	321	549

** Noted SKVA valves are for low voltage (below 600V) generators. Consult Caterpillar for medium voltage generator capabilities.
NOTE: SCR rectifiers and variable speed motor controls require detailed analysis. Contact Caterpillar and the SCR supplier.

TABLE 5.4 Motor Starting Data Diesel and Gas Generator Sets (*Continued*)

STANDBY POWER — 60 Hz-1800 RPM

Engine Model	Generator Frame	Rating w/Fan kW	Starting kV·A at Voltage Dip		
			10%	20%	30%
3516 TA	808	2000	1355	3048	5226
	807	1750	1234	2777	4761
3512 TA	806	1500	1010	2272	3896
	805	1400	793	1785	3061
	689	1250	966	2173	3726
	687	1100	888	2000	3428
3508 TA	685	1000	584	1315	2255
	685	900	584	1315	2255
	681	800	419	943	1617
	681	750	419	943	1617
	589	700	396	892	1530
3412 TA	589	600	444	1000	1714
T	588	500	317	714	1224
T	586	475	278	625	1071
3408 TA	584	400	242	543	932
3406 TA	449	350	218	490	840
	448	300	202	455	779
	447	275	161	362	821
3306 ATAAC	446	250	156	352	604
T	445	225	146	329	564
3208 ATAAC	444	200	123	278	476
T	443	175	93	208	357

** Noted SKVA valves are for low voltage (below 600V) generators. Consult Caterpillar for medium voltage generator capabilities.
NOTE: SCR rectifiers and variable speed motor controls require detailed analysis. Contact Caterpillar and the SCR supplier.

TABLE 5.4 Motor Starting Data Diesel and Gas Generator Sets (*Continued*)

CONTINUOUS POWER — 60 Hz-1800 RPM

Engine Model	Type °C (°F)/Ratio	Generator Frame	Rating w/o Fan kW	Starting kV·A at Voltage Dip		
				10%	20%	30%
3412 TA	32 (90)	588	460	317	714	1224
	54 (130)	586	410	278	625	1071
3408 TA	32 (90)	582	300	171	385	659
	54 (130)	582	270	171	385	659
3306 TA	HCR	445	150	111	250	428
3306 TA	LCR	444	135	74	167	286
3306 NA	HCR	444	100	74	167	286
3306 NA	LCR	444	85	74	167	286

CONTINUOUS POWER — 60 Hz-1200 RPM

Engine Model	Type °C (°F)/Ratio	Generator Frame	Rating w/o Fan kW	Starting kV·A at Voltage Dip		
				10%	20%	30%
G3516 LE	32 (90)	807	820	444	1000	1714
G3516 LE	54 (130)	807	770	444	1000	1714
G3516 NA	—	686	465	231	521	893
G3512 LE	32 (90)	686	600	278	625	1071
G3512 LE	54 (130)	686	570	278	625	1071
G3512 NA	—	683	365	231	521	893
G3508 LE	32	683	395			
G3508 LE	54	683	375			
G3508 NA	—	683	210			

** Noted SKVA valves are for low voltage (below 600V) generators. Consult Caterpillar for medium voltage generator capabilities.
NOTE: SCR rectifiers and variable speed motor controls require detailed analysis. Contact Caterpillar and the SCR supplier.
10% overload of TA engines can be factory demonstrated.

TABLE 5.5   Code Letters on AC Motors

NEMA Code Letter	SKVA per hp	Mid-Value
A	0.00- 3.14	1.57
B	3.15- 3.54	3.34
C	3.55- 3.99	3.77
D	4.00- 4.49	4.24
E	4.50- 4.99	4.74
F	5.00- 5.59	5.30
G	5.60- 6.29	5.94
H	6.30- 7.09	6.70
J	7.10- 7.99	7.54
K	8.00- 8.99	8.50
L	9.00- 9.99	9.50
M	10.00-11.19	10.60
N	11.20-12.49	11.84
P	12.50-13.99	13.24
R	14.00-15.99	15.00
S	16.00-17.99	17.00
T	18.00-19.99	19.00
U	20.00-22.39	21.20
V	22.40-	

Use 6.0 if code letter unknown

Wound Rotor Motor has no code letter

VI. GENERATOR SET SIZING

Match engine-running load (kW) with generator motor-starting requirements (SKVA) to satisfy application. Table 5.7 will assist in determining running load kW for squirrel cage induction motors. Engines and generators may be interchanged with model configurations, but mechanical considerations should be reviewed with the manufacturer.

Silicon-controlled rectifiers (SCRs) and variable-speed motor controls require detailed analysis. These should be reviewed with the respective manufacturers.

In Figure 5.9, panel *a* shows a sample generator sizing calculation, and panel *b* provides a blank form for the reader's use.

Critical Installation Considerations

The following summary contains important points to remember for a successful generator installation:

1. The generator set must be sized properly for the installation. Determine the duty cycle: continuous, prime, standby, or peak shaving or sharing (paralleled or not paralleled with the utility).

FIGURE 5.8 Motor preload multiplier.

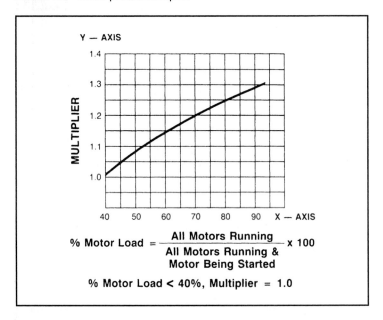

$$\% \text{ Motor Load} = \frac{\text{All Motors Running}}{\text{All Motors Running \&}} \times 100$$
$$\text{Motor Being Started}$$

% Motor Load < 40%, Multiplier = 1.0

Continuous: Output available without varying load for an unlimited time.

Prime: Output available with varying load for an unlimited time.

Standby: Output available with varying load for the duration of the interruption of the normal source of power. The standby duty cycle is usually sized initially for 60 percent of actual load, because loads tend to increase during the 30-year life of the unit. Normal hours of operation are less than 100 h per year.

Peak shaving/sharing: Prime if paralleled with the utility, standby if not paralleled with the utility and if the load meets the definition of prime or standby. Normally peak shaving/sharing is less than 200 h per year of operation.

Loads that are too light cause engine slobber. Overloading causes excessive piston loading and high exhaust temperatures.

Standby engines that must be exercised regularly but cannot be loaded should only be run long enough to achieve normal oil pressure and then shut off—less than 5 min of running time. Good practice dic-

TABLE 5.6   Reduced-Voltage Starting Factors

Type	Multiply SKVA By
Resistor, Reactor, Impedance	
80% Tap	0.80
65% Tap	0.65
50% Tap	0.50
45% Tap	0.45
Autotransformer	
80% Tap	0.68
65% Tap	0.46
50% Tap	0.29
Y Start, Run	0.33

Solid State: Adjustable, consult manufacturer or estimate 300% of full load kV•A
(Use 1 if no reduced voltage starting aids used)

tates that this be done weekly and that once a month the generators be run under load for a half hour or so, then unloaded briefly for cooldown. The load should be at least two-thirds of capacity, either using a dummy resistive load bank, or preferably under actual building load. The latter requirement is mandatory for hospitals under NFPA 99.

2. The generator set must be properly installed in an atmosphere that allows it to achieve the required life.

TABLE 5.7   Approximate Efficiencies—Squirrel Cage Induction Motors

hp	kW	Full-Load Efficiency
5-7½	4-6	0.83
10	7.5	0.85
15	11	0.86
20-25	15-19	0.89
30-50	22-37	0.90
60-75	45-56	0.91
100-300	74.6-224	0.92
350-600	261-448	0.93

FIGURE 5.9 Generator sizing chart. (a) Filled-out sample. (b) Blank.

Customer _____ Project _____ Analyst _____ Date _____

I. APPLICATION DATA
Prime/Standby Power   Gas/Diesel Fuel   _480_ Volts   _3_ Phase   _60_ Hz

II. LOADS
A. Lighting Loads .... kW
B. Other Non-Motor Loads .... kW
C. Motors

Starting Sequence	hp	Nema Code	Reduced Voltage Starting Type	Nameplate Data Acceptable Voltage Dip Percent
1	200	F	Res. 65%	.12
2	75	G	A-T 80%	.91
3	60	F	Across the line	.91
4	50	F	Across the line	.90
5				

III. ENGINE SIZING
A. Lighting Loads _____ kW
B. Non-Motor Loads _____ kW
C. Motor Loads (hp) _____ kW (Motor) × 0.746

Motor Efficiency (Chart 5)
162 → 163 kW
61 → 49 kW
49 → 41 kW
41 → kW
→ kW
313 → kW

Total Motor Load _____ kW
Total Engine Load (A + B + C) _425_ kW   _60_ Hz   _1800_ rpm

IV. ENGINE SELECTION Model: _34/2 J S_   Frame: _586_

V. GENERATOR SIZING Start Sequence

	Motor(s) 1	Motor(s) 2	Motor(s) 3	Motor(s) 4	Motor (s) 5
A. Starting KV•A (SKVA)		Rating (With Fan):			
1. Motor Ratings	200 hp	75 hp	60 hp	50 hp	hp
2. NEMA Code	F	G	F	F	
3. SKVA/hp (Use 6.0 if Code Letter Unknown)	5.30	5.94	5.30	5.30	
4. SKVA/hp × Motor hp (A.1 × A.3)	1040 SKVA	446 SKVA	318 SKVA	265 SKVA	SKVA
B. Effective SKVA					
1. All Motors Running	0 kW	162 kW	223 kW	272 kW	kW
2. All Motors Running & Motor Being Started	162 kW	223 kW	272 kW	313 kW	kW
3. B.1 × 100 / B.2	0 %	73 %	82 %	87 %	%
4. Compensation for Motors Already Started (Chart 2)	1.0	1.21	1.26	1.28	
5. Step A.4. × Step B.4.	1040 SKVA	540 SKVA	401 SKVA	339 SKVA	SKVA
6. Reduced Voltage Factor (Chart 3) (use 1.0 if no starting aid used)	.65	.68	1.0	1.0	
7. Effective SKVA = Step B.5. × B.6.	677 SKVA	362 SKVA	401 SKVA	339 SKVA	SKVA
8. Acceptable Voltage Dip (10, 20, 30%)	20 %	20 %	20 %	20 %	%
C. Generator Selection (Chart 1)					
1. Frame	588	448	584	448	
2. Rating	455 kW	255 kW	265 kW	225 kW	kW
3. SKVA at Selected Voltage Dip	714	385	543	357	

VI. GENERATOR SET SIZING
Select Largest Generator Set Model of Step IV and Step V.C.1.
Model: _34/2 J S_   Frame: _588_   Rating: _455_ kW   Prime/Standby   _60_ Hz   _1800_ rpm

NEMA

380

FIGURE 5.9 Generator sizing chart. (*a*) Filled-out sample. (*b*) Blank. (*Continued*)

Customer _____ Project _____ Analyst _____ Date _____

I. APPLICATION DATA
Prime/Standby Power _____ Gas/Diesel Fuel _____ Volts _____ Phase _____ Hz _____

III. ENGINE SIZING
_____ kW
_____ kW
$kW\ (Engine) = \dfrac{hp\ (Motor) \times 0.746}{Motor\ Efficiency}$ (Chart 5)

II. LOADS
A. Lighting Loads .... _____ kW
B. Other Non-Motor Loads .... _____ kW
C. Motors

Nameplate Data

Starting Sequence	hp	Nema Code	Reduced Voltage Starting Type	Acceptable Voltage Dip Percent	Motor Eff. (Chart 5)	
1						___ kW
2						___ kW
3						___ kW
4						___ kW
5						___ kW

Total Motor Load _____ kW
Total Engine Load (A + B + C) _____ kW

IV. ENGINE SELECTION Model: _____ Frame: _____ Rating (With Fan): _____ Hz _____ rpm

V. GENERATOR SIZING Start Sequence

	Motor(s) 1	Motor(s) 2	Motor(s) 3	Motor(s) 4	Motor(s) 5
A. Starting kV•A (SKVA)					
1. Motor Ratings	___ hp	___ hp	___ hp	___ hp	___ hp
2. NEMA Code					
3. SKVA/hp (Use 6.0 if Code Letter Unknown)					
4. SKVA/hp x Motor hp (A.1 x A.3)	___ SKVA	___ SKVA	___ SKVA	___ SKVA	___ SKVA
B. Effective SKVA					
1. All Motors Running	0 kW	___ kW	___ kW	___ kW	___ kW
2. All Motors Running & Motor Being Started	___ kW	___ kW	___ kW	___ kW	___ kW
3. B.1 x 100 / B.2	0 %	___ %	___ %	___ %	___ %
4. Compensation for Motors Already Started (Chart 2)	1.0				
5. Step A.4. x Step B.4.	___ SKVA	___ SKVA	___ SKVA	___ SKVA	___ SKVA
6. Reduced Voltage Factor (Chart 3) (use 1.0 if no starting aid used)					
7. Effective SKVA = Step B.5. x B.6.	___ SKVA	___ SKVA	___ SKVA	___ SKVA	___ SKVA
8. Acceptable Voltage Dip (10, 20, 30%)	___ %	___ %	___ %	___ %	___ %
C. Generator Selection (Chart 1)					
1. Frame					
2. Rating	___ kW	___ kW	___ kW	___ kW	___ kW
3. SKVA at Selected Voltage Dip					

VI. GENERATOR SET SIZING
Select Largest Generator Set Model of Step IV and Step V.C.1.
Model: _____ Frame: _____ Rating: _____ kW Prime/Standby _____ Hz _____ rpm

NEMA

Air flow: Provide adequate clean, cool air for cooling and combustion. High engine room temperatures may require ducting cooler outside air to the engine intake to avoid power derating. Restriction of radiator air reduces its cooling capability.

Exhaust: Isolate exhaust piping from the engine with flexible connections. Wrap the piping with a thermal blanket to keep exhaust heat out of the engine room. The exhaust stack and muffler need to be sized so that the exhaust back pressure at the turbocharger outlet does not exceed 6.7 kPa (27 in) of water. Excessive back pressure raises exhaust temperatures and reduces engine life.

Fuel: Use clean fuel. Fuel day tanks should be below the level of the injectors.

Mounting: The generator sets must have a flat and secure mounting surface. The generator set mounting must allow adequate space around the generator set for maintenance and repairs.

Starting: Batteries should be close to the starter and protected from very cold temperatures. Do not disconnect batteries from a running engine or a plugged-in battery charger.

3. SCR loads can affect generator output waveform. Make sure the SCR supplier is aware of the possible problems.

Every generator set installation is unique and requires careful consideration of the particular application and site-specific conditions. It is therefore best to determine the foundation, ventilation, exhaust, fuel, vibration isolation, and other requirements in conjunction with the generator set manufacturer for the specific application and site conditions.

5.4   UNINTERRUPTIBLE POWER SUPPLY (UPS) SYSTEMS

A UPS is a device or system that provides quality and continuity of an AC power source. Every UPS should maintain some specified degree of continuity of load for a specified stored-energy time upon AC input failure [see NEMA PE1-1990 ("Uninterruptible Power Systems")]. The term *UPS* commonly includes equipment, backup power source(s), environmental equipment (enclosure, heating and ventilating equipment), switchgear, and controls, which, together, provide a reliable, continuous-quality electric power system.

The following definitions are given for clarification:

1. *Critical load:* That part of the load that requires continuous-quality electric power for its successful operation.

2. *Uninterruptible power supply (UPS) system:* Consists of one or more UPS modules, energy storage battery (per module or com-

mon battery), and accessories (as required) to provide a reliable and high-quality power supply. The UPS isolates the load from the primary and emergency sources, and, in the event of a power interruption, provides regulated power to the critical load for a specified period depending on the battery capacity. (The battery is normally sized to provide a capacity of 15 min when operating at full load.)

3. *UPS module:* The power conversion portion of the UPS system. A UPS module may be made entirely of solid-state electronic construction, or a hybrid combining rotary equipment (motor-generator) and solid-state electronic equipment. A solid-state electronic UPS consists of a rectifier, an inverter, and associated controls along with synchronizing, protective, and auxiliary devices. UPS modules may be designed to operate either individually or in parallel. A rotary UPS consists of a pony motor, a motor-generator, or, alternatively, a synchronous machine in which the synchronous motor and generator have been combined into a single unit. This comprises a stator whose slots carry alternate motor and generator windings, and a rotor with DC excitation, a rectifier, an inverter, a solid-state transfer switch, and associated controls along with synchronizing, protective, and auxiliary devices.

4. *Nonredundant UPS configuration:* Consists of one or more UPS modules operating in parallel with a bypass circuit transfer switch and a battery (see Figure 5.10). The rating and number of UPS modules are chosen to supply the critical load with no intentional excess capacity. Upon the failure of any UPS module, the bypass circuit automatically transfers the critical load to the bypass source without an interruption. The solid-state electronic UPS configuration relies upon a static transfer switch for transfer within 4.17 milliseconds (ms). The rotary UPS configuration relies upon the stored energy of the flywheel to propel the generator and maintain normal voltage and frequency for the time that the electromechanical circuit breakers are transferring the critical load to the alternate source. All operational transfers are "make before break."

5. *"Cold" standby redundant UPS configuration:* Consists of two independent, nonredundant modules with either individual module batteries or a common battery (see Figure 5.11). One UPS module operates on the line, and the other UPS module is turned off. Should the operating UPS module fail, its static bypass circuit will automatically transfer the critical load to the bypass source without an interruption to the critical load. The second UPS module is then manually energized and placed on the bypass mode of operation. To transfer the critical load, external make-before-break nonautomatic circuit breakers are operated to place the

FIGURE 5.10 Nonredundant UPS system configuration.

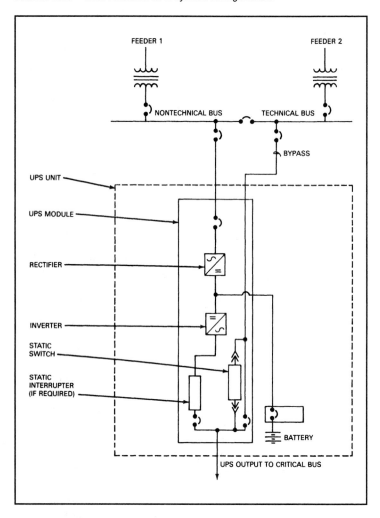

load on the second UPS bypass circuit. Finally, the critical load is returned from the bypass to the second UPS module via the bypass transfer switch. The two UPS modules cannot operate in parallel; therefore, a safety interlock circuit should be provided to prevent this condition. This configuration is rarely used.

FIGURE 5.11 "Cold" standby redundant UPS system.

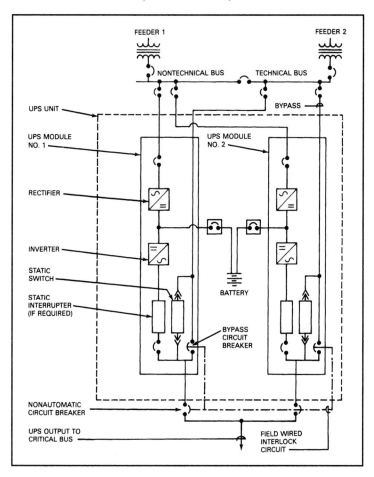

6. *Parallel redundant UPS configuration:* Consists of two or more UPS modules with static inverter turnoff(s), a system control cabinet, and either individual module batteries or a common battery (see Figure 5.12). The UPS modules operate in parallel and normally share the load, and the system is capable of supplying the rated critical load upon failure of any one UPS module. A static interrupter will disconnect the failed UPS module from the other UPS modules without an interruption to the critical load. A system bypass is usually included to permit system maintenance.

FIGURE 5.12   Parallel redundant UPS system.

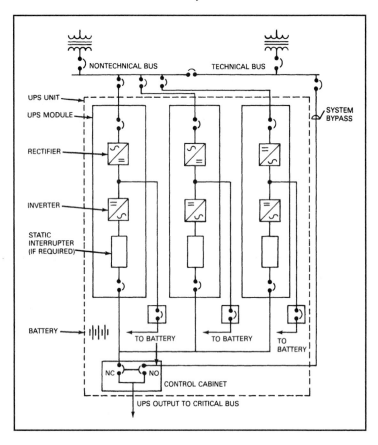

7. *Isolated redundant UPS configuration:* Uses a combination of automatic transfer switches and a reserve system to serve as the bypass source for any of the active systems (in this case, a system consists of a single module with its own system switchgear). This is shown in Figure 5.13. The use of this configuration requires each active system to serve an isolated/independent load. The advantage of this type of configuration minimizes single-point failure modes (i.e., systems do not communicate via logic connections with each other; the systems operate independently of one another). The disadvantage of this type of system is that each system requires its own separate feeder to its dedicated load.

FIGURE 5.13   Isolated redundant UPS system.

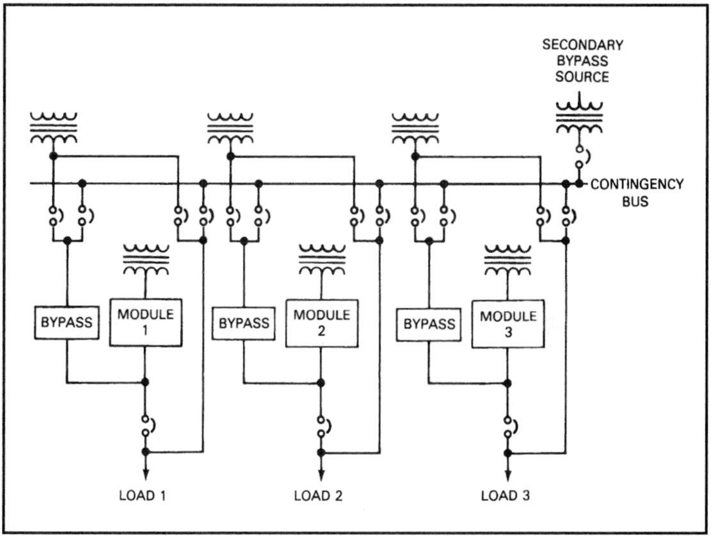

Application of UPS

1. The nonredundant UPS may be satisfactory for many critical load applications.
2. The installation of a parallel redundant UPS system is justified when the criticality of the load demands the greatest protection and the load cannot be divided into suitable blocks.

Power System Configuration for 60-Hz Distribution

In 60-Hz power distribution systems, the following basic concepts are used:

1. *Single-module UPS system:* A single unit that is capable of supplying power to the total load (see Figure 5.14). In the event of an overload or if the unit fails, the critical bus is transferred to the bypass source via the bypass transfer switch. Transfer is uninterrupted.
2. *Parallel capacity UPS system:* Two or more units capable of supplying power to the total load (see Figure 5.15). In the event of overload, or if either unit fails, the critical load bus is transferred to the bypass source via the bypass transfer switch. Transfer is uninterrupted. The battery may be common or separate.

FIGURE 5.14 Single-module UPS system.

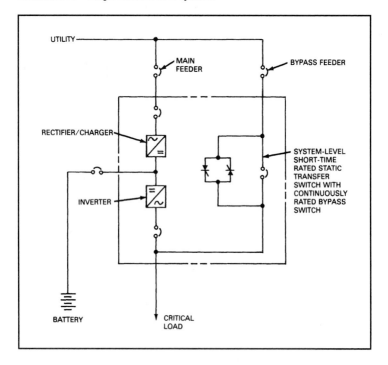

3. *Parallel redundant UPS system:* Two or more units with more capacity than is required by the total load (see Figure 5.16). If any unit fails, the remaining units should be capable of carrying the total load. If more than one unit fails, the critical bus will be transferred to the bypass source via the bypass transfer switch. The battery may be common or separate per module.

4. *Dual redundant UPS systems:* One UPS module is standing by, running unloaded (see Figure 5.17). If the loaded module fails, the load is transferred to the standby module. Each rating is limited to the size of the largest available module.

5. *Isolated redundant UPS system:* Multiple UPS modules, usually three, are individually supplied from transformer sources (see Figure 5.18). Each UPS module supplies a critical load and is available to supply a common contingency bus. The common contingency bus supplies the bypass circuit for each UPS module. In addition to being supplied from the common contingency bus, the bypass switch of each module is supplied from an individual trans-

FIGURE 5.15   Parallel capacity UPS system.

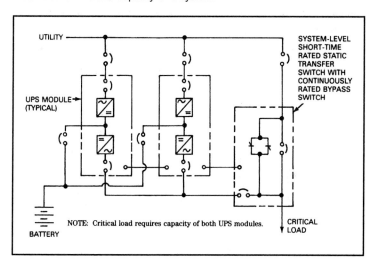

NOTE: Critical load requires capacity of both UPS modules.

FIGURE 5.16   Parallel redundant UPS system.

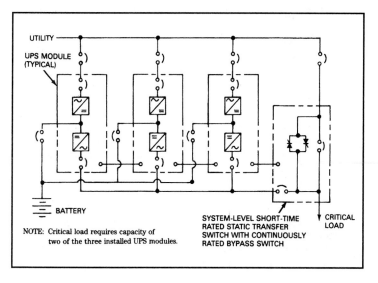

NOTE: Critical load requires capacity of
        two of the three installed UPS modules.

FIGURE 5.17 Dual redundant UPS system.

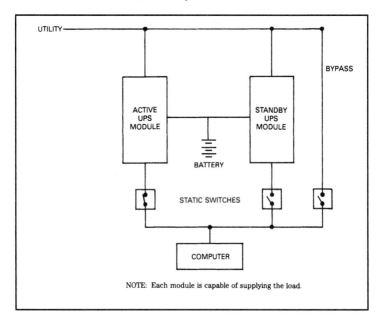

NOTE: Each module is capable of supplying the load.

former source. Furthermore, the common contingency bus is also supplied from a separate standby transformer called a *secondary bypass source.* The arrangement includes one UPS module in reserve as a "hot" standby. When a primary UPS module fails, the reserve UPS module is transferred to the load.

6. *Parallel tandem UPS system:* The tandem configuration is a special case of two modules in parallel redundancy (see Figure 5.19). In this arrangement, both modules have rectifier/chargers, DC links, and inverters; also, one of the modules houses the system-level static transfer switch. Either module can support full system load while the other has scheduled or corrective maintenance performed.

7. *Hot tied-bus UPS system:* The UPS tied-bus arrangement consists of two individual UPS systems (single module, parallel capacity, or redundant), with each one supplying a critical load bus (see Figure 5.20). The two critical load buses can be paralleled via a tie breaker (normally open) while remaining on inverter power, which allows greater user flexibility for scheduled maintenance or damage control due to various failures.

FIGURE 5.18   Isolated redundant UPS system.

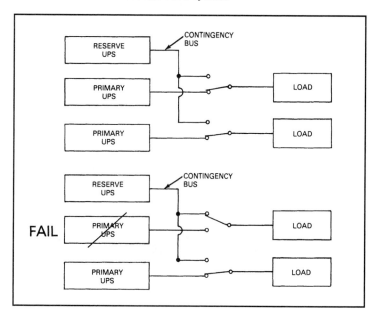

FIGURE 5.19   Parallel tandem UPS system.

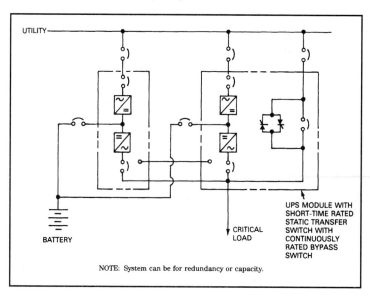

FIGURE 5.20 Hot tied-bus UPS system.

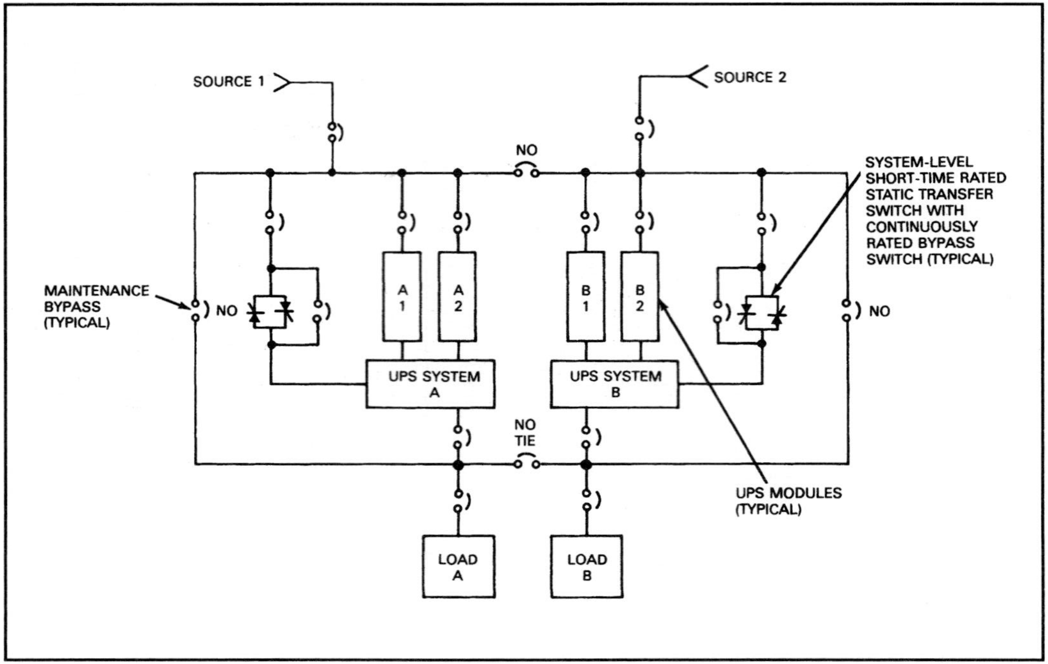

8. *Superredundant parallel system–hot tied-bus UPS system:* The superredundant UPS arrangement consists of *n* UPS modules (limited by a 4000-A bus). Each UPS module is supplied from dual sources (either/or) to supply two critical paralleling buses. Each paralleling bus is connected via a circuit breaker to a common bus in parallel with the output feeder of one of the system static bypass switches. This junction is connected via a breaker to a system critical load bus. A tie enables the two system critical load buses to be paralleled. Bypass sources for each system supply their own respective static bypass switches and maintenance bypasses. The superredundant UPS arrangement normally operates with the tie breaker open between the two system critical load buses. When all UPS modules are supplying one paralleling bus, then the tie breaker is closed. All operations are preselected, automatic, and allow the user to do module- and system-level reconfigurations without submitting either critical load to utility power. See Figure 5.21.

9. *Uninterruptible power with dual utility sources and static transfer switches:* Essentially, uninterruptible electric power to the critical load may be achieved by the installation of dual utility sources, preferably from two separate substations, supplying secondary buses via step-down transformers as required (see Figure 5.22).

FIGURE 5.21 Superredundant parallel system–hot tied-bus UPS system.

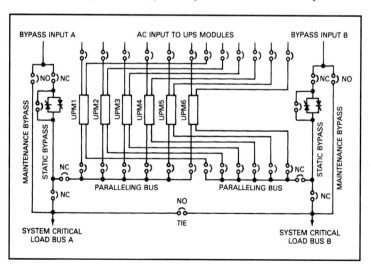

FIGURE 5.22 Uninterruptible power with dual utility sources and static transfer switches.

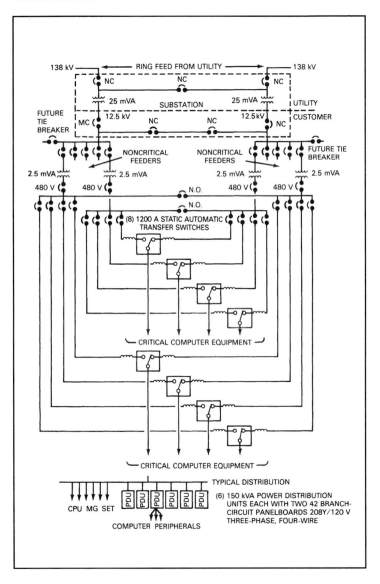

Feeders from each of two source buses are connected to static transfer switches as sources 1 and 2. A feeder from the load connection of the static transfer switch supplies a power line conditioner, if needed. The power line conditioner filters transients and provides voltage regulation. Filtered and regulated power is then supplied from the power line conditioner to the critical load distribution switchgear. This system eliminates the need for energy storage batteries, emergency generators, and other equipment. The reliability of this system is dependent upon the two utility sources and power conditioners.

Power System Configuration with 60-Hz UPS

1. *Electric service and bypass connectors:* Two separate electric sources, one to the UPS rectifier circuit and the other to the UPS bypass circuit, should be provided. When possible, they should emanate from two separate buses with the UPS bypass connected to the noncyclical load bus (also called the *technical bus*). This connection provides for the isolation of sensitive technical loads from the effects of UPS rectifier harmonic distortion and motor start-up current inrush.

2. *Maintenance bypass provisions:* To provide for the maintenance of equipment, bypass provisions are necessary to isolate each UPS module or system.

UPS Distribution Systems

The UPS serves critical loads only. Noncritical loads are served by separate distribution systems that are supplied from either noncyclical load bus (technical bus) or the cyclical load bus (nontechnical bus), as appropriate.

1. *Critical load protection:* Critical load overcurrent devices equipped with fast-acting fuses to shorten the transient effects of undervoltage caused by short circuits will result in a reliable system. Solid-state transient suppression (metal-oxide type) should also be supplied to lessen the overvoltage transients caused by reactive load switching.

2. *Critical motor loads:* Due to the energy losses and the starting current inrush inherent in motors, the connection of motors to the UPS bus should be limited to frequency conversion applications, that is, motor-generator sets. Generally, due to the current inrush, motor-generator sets are started on the UPS bypass circuit. Motor-generator sets may be started on the rectifier/inverter mode of operation under the following conditions:

    a. When the rating of the motor-generator set is less than 10 percent of the UPS rating.

    b. When reduced-voltage and peak current starters, such as the wye-delta closed transition type, are used for each motor load.

    c. When more than one motor-generator set is connected to the critical bus, each set should be energized sequentially rather than simultaneously.

Refer all applications requiring connection of induction and synchronous motor loads to the UPS manufacturer. Application rules differ depending on the design and rating of the UPS.

Power System Configuration for 400-Hz Distribution

In 400-Hz power distribution systems, the following basic concepts are used:

1. *Direct utility supply to dual-rotary frequency converters parallel at the output critical load bus:* Each frequency converter is sized for 100 percent load or the arrangement has redundant capacity. The frequency converters may be equipped with an inverter/charger and battery upon utility failure. Transfer from the utility line to the inverter occurs by synchronizing the inverter to the residual voltage of the motor.

2. *Dual-utility supply:* Dual-utility feeders supply an automatic transfer switch. The automatic transfer switch supplies multiple-rotary frequency converters (flywheel equipped). The frequency converters are parallel at the critical load bus. Transfer from one utility line to another occurs within the ride-through capability of the rotary frequency converters.

3. *UPS:* A static or rotary UPS supplies multiple-frequency converters and other 60-Hz loads.

4. *UPS with local generation backup:* Both the utility feeder (connected to the normal terminals) and the feeder from the backup generation (connected to the emergency terminals) supply the automatic transfer switch. The automatic transfer switch in turn supplies the UPS. Critical load distribution is as described above.

5. *Parallel 400-Hz single-CPU configuration:* Two or more 60- to 400-Hz frequency converters are normally connected in a redundant configuration to supply the critical load. There is no static switch or bypass breaker. Note that, on static converters, it is possible to use a 400-Hz motor-generator as a bypass source.

6. *Common UPS for single-mainframe computer site:* Two 60- to 400-Hz frequency converters are normally connected in a redundant configuration supplying the mainframe computer, while a 60-Hz UPS supplies the peripherals.

7. *Alternative combination UPS for single-mainframe computer site:* A 60-Hz UPS supplies a critical load bus that, in turn, supplies the peripherals plus the input to a motor-generator set frequency converter (60 to 400 Hz).

8. *Combination UPS for multiple-mainframe computer site:* A utility source supplies a redundant 400-Hz UPS system. This paralleled system supplies a 400-Hz critical load distribution bus. Feeders from the 400-Hz distribution bus, equipped with line drop compensators (LDCs) to reactive voltage drop, supply computer mainframes. A utility source also supplies a parallel redundant 60-Hz UPS system. This system supplies the critical peripheral load.

9. *Remote redundant 400-Hz UPS:* A 60-Hz UPS and a downstream parallel redundant 400-Hz motor-generator frequency conversion system with paralleling and distribution switchgear and line drop compensators, which are all installed in the facility power equipment room with 60- and 400-Hz feeders distributed into the computer room.

10. *Point-of-use redundant 400-Hz UPS:* A 60-Hz UPS and a parallel redundant frequency conversion system as in item 9, except that the motor-generators are equipped with silencing enclosures and are installed in the computer room near the mainframes.

11. *Point-of-use 400-Hz UPS:* A 60-Hz UPS and a nonparalleled point-of-use static or rotary 400-Hz frequency converter installed in the computer room adjacent to each mainframe.

12. *Remote 400-Hz UPS:* A 60-Hz UPS and a separate parallel redundant 400-Hz UPS installed in the power equipment room, which is similar to item 8.

13. *Wiring:* For 400-Hz circuits, the reactance of circuit conductors may produce unacceptable voltage drops. Multiple conductor cables and use of conductors in parallel, if necessary, should be installed in accordance with the *NEC,* Article 310-4. Also, use of a nonmagnetic conduit will help in reducing voltage drop.

It should be noted that 400-Hz (actually 415-Hz) mainframe computers are rarely used today. Most mainframe computers are now 60 Hz.

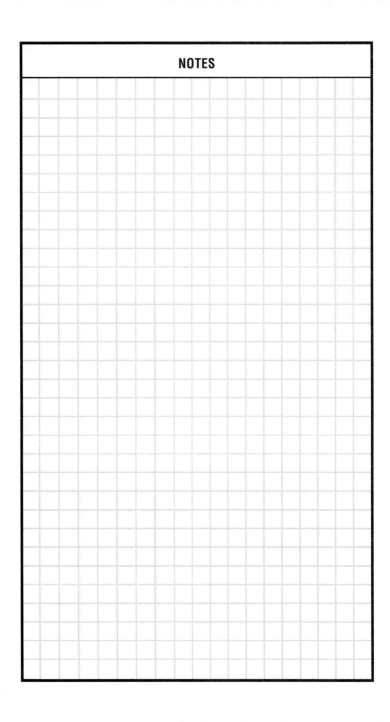

NOTES

CHAPTER SIX
Lighting

6.0 MEASURING LIGHT AND ILLUMINATION TERMS

Definitions

Luminous intensity, I, is the solid angular flux density in a given direction measured in candlepower in American National Standards Institute (ANSI) units and candela (cd) in SI units. The candela and candlepower have the same magnitude. See Figure 6.1.

Lumen (lm) is the unit of luminous flux equal to the flux in a unit solid angle of 1 steradian (sr) from a uniform point source of 1 cd. On a unit sphere, an area of 1 ft^2 (or 1 m^2) will subtend an angle of 1 sr. Because the area of a unit sphere is $4 \times$ pi, a source of 1 candlepower (1 cd) produces 12.57 lm.

Illuminance (E) is the density of luminous flux *incident* on a surface in lumens per unit area. One lumen uniformly incident on 1 ft^2 of area produces an illuminance of one footcandle. The unit of measurement, therefore, is the footcandle (fc) in ANSI units. In SI units, the measurement is lux (lx), or lumens per square meter.

$$1 \text{ footcandle} = 10.764 \text{ lux}$$
$$\text{fc} = \text{lm/ft}^2$$
$$\text{fc} = \text{lm/m}^2$$

As a rule of thumb, 10 lx is taken as being approximately equal to 1 fc.

Luminance, L, is the luminous flux per unit of projected area (apparent) area and a unit solid angle *leaving* a surface, either reflected or transmitted. The unit is the footlambert (fL), in which 1 fL = $1/\pi$ candelas per square foot. In SI units, it is candela per square meter. Luminance takes into account the reflectance and transmittance properties of materials and the direction in which they are viewed (the apparent area). Thus, 100 fc striking a surface with 50 percent reflectance would result in a luminance of 50 fL.

Another way to view illuminance is to say that a surface emitting, transmitting, or reflecting 1 lm/ft^2 in the direction being viewed has a

FIGURE 6.1 Relationship of light source, illumination, transmittance, and reflectance. (*Source:* GE Lighting Business Group)

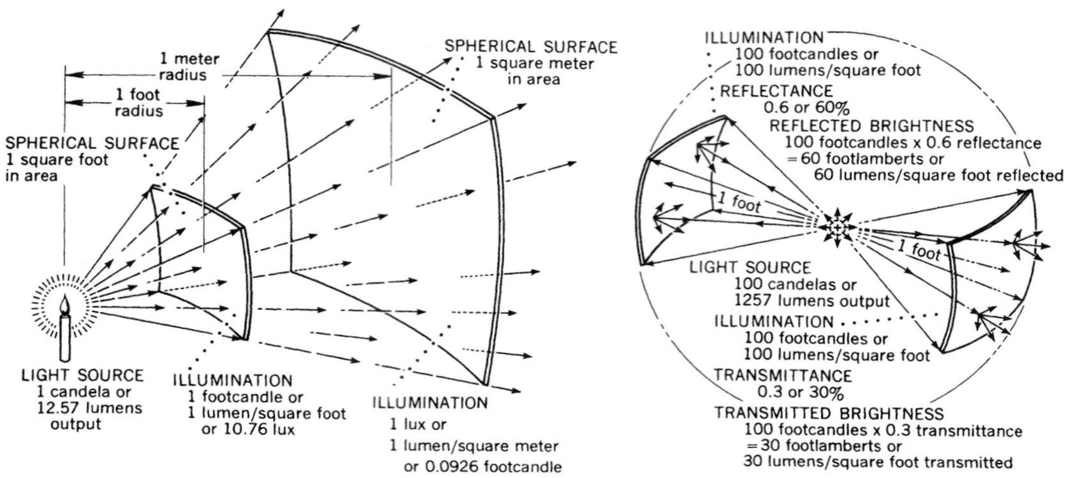

TABLE 6.1 Conversion Factors of Units of Illumination

Given	Multiply by	to obtain
Illuminance (E) in lux	0.0929	footcandles
Illuminance (E) in footcandles	10.764	lux
Luminance (L) in cd/sq. m	0.2919	footlamberts
Luminance (L) in footlamberts	3.4263	cd/sq. m
Intensity (I) candelas	1.0	candlepower

luminance of 1 fL. For more information about conversion factors of units of illumination, see Table 6.1.

Inverse Square Law

The illumination at a point on a surface when the surface is perpendicular to the direction of the source varies directly with the luminous intensity of the source and inversely with the square of the distance between the source and the point:

$$E = \frac{I}{d^2}$$

where: E = illumination in footcandles (or lux)
I = luminous intensity in candlepower (or candela)
d = distance in feet (or meters)

This equation assumes the source is a *point source.* Because a point source is only theoretical, the formula is applicable when the maximum dimension of the source is less than five times the distance to the point at which the illumination is being calculated.

The value for I at various angles can be obtained from the candlepower distribution curves or tables supplied by the manufacturer of the luminaire under consideration.

Cosine Law

The illumination of any surface varies as the cosine of the angle of incidence, θ, where the angle of incidence is the angle between the normal to the surface and the direction of the incident light. See Figure 6.2.

Combined with the equation just given, the formula becomes:

$$E = \frac{I}{d^2} \cos \theta$$

FIGURE 6.2 Cosine law of illumination.

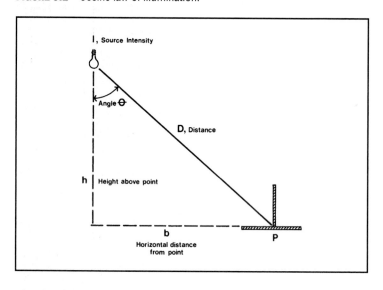

This is the illumination on the horizontal surface at point P. For illumination on a vertical surface at point P, the equation becomes:

$$E(v) = \frac{I}{d^2} \cos \theta$$

Because

$$\cos \theta = \frac{h}{d}$$

and

$$\sin \theta = \frac{b}{d}$$

the equations for horizontal and vertical illumination can be rewritten as follows:

$$E(h) = \frac{I}{h^2} \cos^3 \theta$$

$$E(v) = \frac{I}{b^2} \sin^3 \theta$$

Example: What is the vertical surface illumination on a wall 6 ft down from the ceiling that is illuminated by a downlight placed 3 ft from the wall? The candlepower distribution curve for the fixture indicates an intensity of 2550 fc at 25° from vertical.

The angle, θ, is arctan 3/6, or 26.6°. Because this is very close to the reading at 25°, use $I = 2550$ fc. Thus:

$$E (v) = 2550/3^2 \sin^3 26.6°$$
$$E (v) = 25 \text{ fc}$$

If the reflectance of the wall is 55 percent, the luminance, L, is 25×0.55, or about 14 fL.

6.1 HOW TO SELECT THE RECOMMENDED ILLUMINANCE LEVEL

Different tasks under different conditions require different levels of illumination. The variables include the task itself, the age of the person performing the task, the reflectances of the room, and the demand for speed and/or accuracy in performing the task. The Illuminating Engineering Society of North America (IESNA) has established a range of illumination levels for various tasks, areas, and activities to take into account these variables.

To determine the required illumination level in footcandles (or lux), first determine the illuminance category for the task under consideration from Table 6.2. This table lists representative activities for common occupancies. For a detailed listing, refer to the complete table in the *IESNA Lighting Handbook.* Illuminance categories are given a letter

TABLE 6.2　Illuminance Categories for Selected Activities

Area/Activity	Illuminance Category
Auditoriums	
Assembly	C
Social activity	B
Banks	
General lobby area	C
Lobby writing area	D
Tellers' stations	E
Barber shops and beauty parlors	E
Conference rooms—conferring	D
For critical seeing, refer to individual task	
Drafting	
High contrast	E
Low contrast	F

TABLE 6.2 Illuminance Categories for Selected Activities (*Continued*)

Area/Activity	Illuminance Category
Educational facilities	
General classrooms (see Reading)	
Science laboratories	E
Lecture rooms—audience (see Reading)	
Lecture rooms—demonstrations	F
Cafeterias (see Food Service)	
Food service facilities	
Cashier	D
Cleaning	C
Dining	B
Kitchen	E
Hotels	
Bathrooms, for grooming	D
Bedrooms, for reading	D
Corridors, elevators, and stairs	C
Front desk	E
Lobby, general lighting	C
Libraries	
Reading areas (see Reading)	
Active stacks	D
Inactive stacks	B
Card files	E
Circulation desks	D
Merchandising spaces	
Dressing areas	D
Fitting areas	F
Wrapping and packaging	D
Sales transaction area	E
Offices	
Conference (see Conference rooms)	
General and private offices (see Reading)	
Lobbies, lounges, and reception areas	C
Mail sorting	E
Reading	
Copied tasks	
Microfiche reader	B
Photographs, moderate detail	E
Xerograph	D
Electronic data-processing tasks	
CRT screens	B
Impact printer, good ribbon	D
Keyboard reading	D
Machine rooms, active operations	D
Handwritten tasks	
# 3 pencil and softer leads	E
# 4 pencil and harder leads	F
Felt-tip pen	D
Chalkboards	E

TABLE 6.2 Illuminance Categories for Selected Activities (*Continued*)

Area/Activity	Illuminance Category
Printed tasks	
6 point type	E
8 and 10 point type	D
Maps	E
Typed originals	D
Telephone books	E
Residences	
General lighting	B
Dining	C
Grooming	D
Kitchen duties, critical seeing	E
Kitchen duties, non-critical	D
Reading, normal	D
Reading, prolonged	E
Service spaces	
Stairways, corridors	C
Elevators	C
Toilets and washrooms	C

Note: Refer to the *IES Lighting Handbook* for a detailed list of requirements for individual spaces and for industrial, transportation, and outdoor activities.

Source: Data extracted from IES Lighting Handbook, *1981 Reference Volume.*

from A to I: A represents the lowest values for general lighting in noncritical areas, and I represents requirements for specialized and difficult visual tasks.

Table 6.3 gives the corresponding range of illuminances for each category.

With the illuminance category and the knowledge of the age of the occupant, the approximate (or assumed) surface reflectances, and the importance of the task, find which of the three values should be used by referring to Table 6.4. Note that the values in this table are in lux. For recommended footcandle levels, divide the values by 10.

The following caveats apply to selecting illumination levels and using them in lighting calculations:

1. All aspects of a quality design must be considered—control of glare, contrast ratios, color-rendering properties, and so on—not just raw illumination levels.

2. The values determined in the illumination categories are *maintained* values in the space, not initial values.

3. The values in categories A through C are average maintained illuminances and are most appropriate for lighting calculations using

TABLE 6.3 Illuminance Categories and Illuminance Values for Generic Types of Activities in Interiors

Type of Activity	Illuminance Category	Range of Illuminances (in footcandles)
General lighting throughout space:		
Public spaces with dark surroundings	A	2-3-5
Simple orientation for short temporary visits	B	5-7.5-10
Working spaces where visual tasks are only occasionally performed	C	10-15-20
Illuminance on task:		
Performance of visual tasks of high contrast or large size	D	20-30-50
Performance of visual tasks of medium contrast of small size	E	50-75-100
Performance of visual tasks of low contrast or very small size	F	100-150-200
Illuminance on task, obtained by a combination of general and local (supplementary) lighting:		
Performance of visual tasks of low contrast and very small size over a prolonged period	G	200-300-500
Performance of very prolonged and exacting visual tasks	H	500-750-1000
Performance of very special visual tasks of extremely low contrast and small size	I	1000-1500-2000

Source: IES Lighting Handbook.

the zonal cavity method, as described in the next section and for daylighting calculations. The values in categories D through I are illumination levels on the task. Point calculation methods, as described in the previous section, are more appropriate for these categories, although achieving the recommended illumination level may be accomplished with a combination of general and task lighting.

4. Special analysis and design is required for lighting for visual tasks in categories G through I.

6.2 ZONAL CAVITY METHOD OF CALCULATING ILLUMINATION

The number of luminaires required to light a space to a desired illumination level (footcandles) can be calculated knowing certain characteristics of the room and light source. The following method is the zonal cavity method of calculating illumination.

$$\frac{\text{Area}}{\text{luminaire}} = \frac{N \times \text{lumens per lamp} \times CU \times LLF}{\text{footcandles required } (E)}$$

TABLE 6.4 Illuminance Values, Maintained, in Lux, for a Combination of Illuminance Categories and User, Room, and Task Characteristics (for Illuminance in Footcandles, Divide by 10)

a. General Lighting Throughout Room

	Weighting Factors	Illuminance Categories		
Average of Occupants Ages	Average Room Surface Reflectance (per cent)	A	B	C
Under 40	Over 70	20	50	100
	30–70	20	50	100
	Under 30	20	50	100
40–55	Over 70	20	50	100
	30–70	30	75	150
	Under 30	50	100	200
Over 55	Over 70	30	75	150
	30–70	50	100	200
	Under 30	50	100	200

b. Illuminance on Task

	Weighting Factors		Illuminance Categories					
Average of Workers Ages	Demand for Speed and/or Accuracy*	Task Background Reflectance (per cent)	D	E	F	G**	H**	I**
Under 40	NI	Over 70	200	500	1000	2000	5000	10000
		30–70	200	500	1000	2000	5000	10000
		Under 30	300	750	1500	3000	7500	15000
	I	Over 70	200	500	1000	2000	5000	10000
		30–70	300	750	1500	3000	7500	15000
		Under 30	300	750	1500	3000	7500	15000
	C	Over 70	300	750	1500	3000	7500	15000
		30–70	300	750	1500	3000	7500	15000
		Under 30	300	750	1500	3000	7500	15000
40–55	NI	Over 70	200	500	1000	2000	5000	10000
		30–70	300	750	1500	3000	7500	15000
		Under 30	300	750	1500	3000	7500	15000
	I	Over 70	300	750	1500	3000	7500	15000
		30–70	300	750	1500	3000	7500	15000
		Under 30	300	750	1500	3000	7500	15000
	C	Over 70	300	750	1500	3000	7500	15000
		30–70	300	750	1500	3000	7500	15000
		Under 30	500	1000	2000	5000	10000	20000
Over 55	NI	Over 70	300	750	1500	3000	7500	15000
		30–70	300	750	1500	3000	7500	15000
		Under 30	300	750	1500	3000	7500	15000
	I	Over 70	300	750	1500	3000	7500	15000
		30–70	300	750	1500	3000	7500	15000
		Under 30	500	1000	2000	5000	10000	20000
	C	Over 70	300	750	1500	3000	7500	15000
		30–70	500	1000	2000	5000	10000	20000
		Under 30	500	1000	2000	5000	10000	20000

* NI = not important, I = important, and C = critical
** Obtained by a combination of general and supplementary lighting.

Source: IES Lighting Handbook.

where: N = number of lamps
              CU = coefficient of utilization
             LLF = light loss factor
                E = recommended illumination (maintained)

The formula can be rewritten to find the number of luminaires or to determine the maintained footcandle level.

$$\text{Number of luminaires} = \frac{\text{footcandles required} \times \text{area of room}}{N \times \text{lumens per lamp} \times CU \times LLF}$$

$$\text{Footcandles} = \frac{N \times \text{lumens per lamp} \times CU \times LLF}{\text{area per luminaire}}$$

The coefficient of utilization (CU) is a factor that reflects the fact that not all of the lumens produced by a luminaire reach the work surface. It depends on the particular light fixture used as well as the characteristics of the room in which it is placed, including the room size and the surface reflectances of the room. If you know the specific luminaire you want to use, obtain coefficient of utilization factors from the manufacturer and use those. They are usually included in product catalogs.

If you do not know specifically what fixture you will be selecting, you can use general coefficient of utilization tables based on luminaire types (see Table 6.5).

Light Loss Factor (*LLF*)

The light loss factor is a fraction that represents the amount of light that will be lost due to things such as dirt on lamps, reduction of light output of a lamp over time, and similar factors. The following items are the individual components of the *LLF.* The total *LLF* is calculated by multiplying all of the individual factors together.

Ambient temperature: For normal indoor temperatures, use 1. For air-handling luminaires, use 1.10.

Voltage: Use 1 for luminaire operation at rated temperature.

Luminaire surface depreciation: Over time, the various surfaces of a light fixture will change (some plastic lenses yellow, for example). In the absence of data, use a value of 1.

Nonstandard components: Use of different components such as ballasts, louvers, and so on can affect light output. Use a value of 1 if no other information is available.

In the absence of other data, use a factor of 0.9 for the combination of the four factors just mentioned. This is usually adequate for most situations.

TABLE 6.5 Coefficients of Utilization

Typical Luminaire	Typical Intensity Distribution and Per Cent Lamp Lumens	Maint. Cat.	SC	RCR ↓	ρ_{cc} → 80			70			50			30			10			0
					ρ_w → 50	30	10	50	30	10	50	30	10	50	30	10	50	30	10	0
					Coefficients of Utilization for 20 Per Cent Effective Floor Cavity Reflectance (ρ_{fc} = 20)															
1 Pendant diffusing sphere with incandescent lamp	V · 35½%↑ · 45%↓		1.5	0	.87	.87	.87	.81	.81	.81	.70	.70	.70	.59	.59	.59	.49	.49	.49	.45
				1	.71	.67	.63	.66	.62	.59	.56	.53	.50	.47	.45	.42	.38	.37	.35	.31
				2	.60	.54	.49	.56	.50	.45	.47	.43	.39	.39	.36	.33	.32	.29	.27	.23
				3	.52	.45	.39	.48	.42	.37	.41	.36	.31	.34	.30	.26	.27	.24	.22	.18
				4	.46	.38	.33	.42	.36	.30	.36	.30	.26	.30	.26	.22	.24	.21	.18	.15
				5	.40	.33	.27	.37	.30	.25	.31	.26	.22	.26	.22	.18	.21	.18	.15	.12
				6	.36	.28	.23	.33	.26	.21	.28	.23	.19	.23	.19	.16	.19	.15	.13	.10
				7	.32	.25	.20	.29	.23	.18	.25	.20	.16	.21	.16	.13	.17	.13	.11	.09
				8	.29	.22	.17	.26	.20	.16	.23	.17	.14	.19	.15	.12	.15	.12	.09	.07
				9	.26	.19	.15	.24	.18	.14	.20	.15	.12	.17	.13	.10	.14	.11	.08	.06
				10	.23	.17	.13	.22	.16	.12	.19	.14	.10	.16	.12	.09	.13	.09	.07	.05
2 Concentric ring unit with incandescent silvered-bowl lamp	II · 83%↓ · 3½%↑		N.A.	0	.83	.83	.83	.72	.72	.72	.50	.50	.50	.30	.30	.30	.12	.12	.12	.03
				1	.72	.69	.66	.62	.60	.57	.43	.42	.40	.26	.25	.25	.10	.10	.10	.03
				2	.63	.58	.54	.54	.50	.47	.38	.36	.33	.23	.22	.21	.09	.09	.08	.02
				3	.55	.49	.45	.48	.43	.39	.33	.30	.28	.20	.19	.17	.08	.08	.07	.02
				4	.48	.42	.37	.42	.37	.33	.29	.26	.24	.18	.16	.15	.07	.07	.06	.02
				5	.43	.36	.32	.37	.32	.28	.26	.23	.20	.16	.14	.13	.06	.06	.05	.01
				6	.38	.32	.27	.33	.28	.24	.23	.20	.17	.14	.12	.11	.06	.05	.04	.01
				7	.34	.28	.23	.30	.24	.21	.21	.17	.15	.13	.11	.09	.05	.04	.04	.01
				8	.31	.25	.20	.27	.21	.18	.19	.15	.13	.12	.10	.08	.05	.04	.03	.01
				9	.28	.22	.18	.24	.19	.16	.17	.14	.11	.10	.09	.07	.04	.03	.03	.01
				10	.25	.20	.16	.22	.17	.14	.16	.12	.10	.10	.08	.06	.04	.03	.03	.01
3 Porcelain-enameled ventilated standard dome with incandescent lamp	IV · 0%↑ · 83½%↓		1.3	0	.99	.99	.99	.97	.97	.97	.93	.93	.93	.89	.89	.89	.85	.85	.85	.83
				1	.88	.85	.82	.86	.83	.81	.83	.80	.78	.79	.78	.76	.77	.75	.73	.72
				2	.78	.73	.68	.76	.72	.67	.73	.69	.66	.71	.67	.64	.68	.65	.63	.61
				3	.69	.62	.57	.67	.61	.57	.65	.60	.56	.63	.58	.55	.61	.57	.54	.52
				4	.61	.54	.49	.60	.53	.48	.58	.52	.48	.56	.51	.47	.54	.50	.46	.45
				5	.54	.47	.41	.53	.46	.41	.51	.45	.41	.50	.44	.40	.48	.43	.40	.38
				6	.48	.41	.35	.47	.40	.35	.46	.39	.34	.44	.39	.34	.43	.38	.34	.32
				7	.43	.35	.30	.42	.35	.30	.41	.34	.30	.39	.34	.30	.38	.33	.29	.28
				8	.38	.31	.26	.38	.31	.26	.37	.30	.26	.36	.30	.26	.35	.30	.26	.24
				9	.35	.28	.23	.34	.27	.23	.33	.27	.23	.32	.27	.23	.31	.26	.22	.21
				10	.31	.25	.20	.31	.24	.20	.30	.24	.20	.29	.24	.20	.29	.23	.20	.18
4 Prismatic square surface drum	V · 18½%↑ · 60½%↓		1.3	0	.89	.89	.89	.85	.85	.85	.77	.77	.77	.70	.70	.70	.63	.63	.63	60
				1	.78	.75	.72	.74	.72	.69	.68	.66	.64	.62	.60	.58	.56	.55	.54	.51
				2	.69	.65	.61	.66	.62	.58	.61	.57	.54	.56	.53	.50	.51	.49	.47	.44
				3	.62	.57	.52	.60	.55	.50	.55	.51	.47	.50	.47	.44	.46	.43	.41	.39
				4	.56	.50	.46	.54	.49	.44	.50	.45	.42	.46	.42	.39	.42	.39	.37	.35
				5	.51	.45	.40	.49	.43	.39	.45	.41	.37	.42	.38	.35	.39	.36	.33	.31
				6	.46	.40	.36	.45	.39	.35	.42	.37	.33	.39	.35	.31	.36	.32	.30	.28
				7	.42	.36	.32	.41	.35	.31	.38	.33	.29	.35	.31	.28	.33	.29	.27	.25
				8	.38	.32	.28	.37	.32	.28	.35	.30	.26	.32	.28	.25	.30	.27	.24	.22
				9	.35	.29	.25	.34	.29	.25	.32	.27	.24	.30	.26	.23	.28	.24	.22	.20
				10	.32	.27	.23	.31	.26	.22	.29	.25	.21	.27	.23	.20	.26	.22	.20	.18
5 R-40 flood without shielding	IV · 0%↑ · 100%↓		0.8	0	1.19	1.19	1.19	1.16	1.16	1.16	1.11	1.11	1.11	1.06	1.06	1.06	1.02	1.02	1.02	00
				1	1.09	1.07	1.04	1.07	1.05	1.02	1.03	1.01	.99	.99	.98	.96	.96	.95	.93	.92
				2	1.01	.97	.93	.99	.95	.92	.96	.93	.90	.93	.90	.88	.90	.88	.86	.84
				3	.93	.88	.84	.92	.87	.83	.89	.85	.81	.87	.83	.80	.84	.81	.79	.77
				4	.87	.81	.76	.85	.80	.75	.83	.78	.75	.81	.77	.74	.79	.76	.73	.71
				5	.80	.74	.69	.79	.73	.69	.77	.72	.68	.76	.71	.67	.74	.70	.67	.65
				6	.74	.68	.63	.73	.67	.63	.72	.66	.62	.70	.66	.62	.69	.65	.61	.60
				7	.69	.62	.57	.68	.62	.57	.67	.61	.57	.65	.60	.56	.64	.60	.56	.55
				8	.64	.57	.53	.63	.57	.52	.62	.56	.52	.61	.56	.52	.60	.55	.52	.50
				9	.59	.52	.48	.59	.52	.48	.58	.52	.48	.57	.51	.48	.56	.51	.47	.46
				10	.55	.49	.44	.55	.48	.44	.54	.48	.44	.53	.48	.44	.52	.47	.44	.42
6 R-40 flood with specular anodized reflector skirt; 45° cutoff	IV · 0%↑ · 85%↓		0.7	0	1.01	1.01	1.01	.99	.99	.99	.94	.94	.94	.90	.90	.90	.87	.87	.87	.85
				1	.96	.94	.92	.94	.92	.91	.90	.89	.88	.88	.86	.85	.85	.84	.83	.82
				2	.91	.88	.86	.90	.87	.85	.87	.85	.83	.84	.83	.82	.82	.81	.80	.79
				3	.87	.84	.81	.86	.83	.81	.84	.81	.79	.82	.80	.78	.80	.78	.77	.76
				4	.83	.80	.77	.82	.79	.77	.81	.78	.76	.79	.77	.75	.78	.76	.74	.73
				5	.79	.76	.73	.79	.75	.73	.77	.74	.72	.76	.73	.71	.75	.73	.71	.70
				6	.76	.73	.70	.76	.72	.70	.75	.72	.69	.74	.71	.69	.73	.70	.68	.67
				7	.73	.69	.66	.72	.68	.66	.72	.68	.66	.71	.68	.66	.70	.67	.66	.64
				8	.70	.66	.63	.70	.66	.63	.69	.65	.63	.68	.65	.63	.67	.65	.63	.62
				9	.67	.63	.60	.67	.63	.60	.66	.62	.60	.65	.62	.60	.65	.62	.60	.59
				10	.64	.60	.58	.64	.60	.58	.63	.60	.58	.63	.60	.57	.62	.59	.57	.56

TABLE 6.5   Coefficients of Utilization (*Continued*)

Typical Luminaire	Typical Intensity Distribution and Per Cent Lamp Lumens	Maint. Cat.	SC	RCR	ρ_{cc} → 80			70			50			30			10			0
				ρ_w →	50	30	10	50	30	10	50	30	10	50	30	10	50	30	10	0
					Coefficients of Utilization for 20 Per Cent Effective Floor Cavity Reflectance (ρ_{FC} = 20)															
7 EAR-38 lamp above 51 mm (2") diameter aperture (increase efficiency to 54 ½% for 76 mm (3") diameter aperture)*	IV, 0%↑, 43½%↓	IV	0.7	0	.52	.52	.52	.51	.51	.51	.48	.48	.48	.46	.46	.46	.45	.45	.45	.44
				1	.49	.48	.48	.48	.48	.47	.47	.46	.46	.45	.45	.44	.44	.43	.43	.42
				2	.47	.46	.45	.46	.45	.44	.45	.44	.43	.44	.43	.42	.43	.42	.42	.41
				3	.45	.44	.43	.45	.43	.42	.44	.42	.42	.43	.42	.41	.42	.41	.40	.40
				4	.43	.42	.41	.43	.41	.40	.42	.41	.40	.41	.40	.39	.41	.40	.39	.38
				5	.42	.40	.39	.41	.40	.38	.41	.39	.38	.40	.39	.38	.39	.38	.38	.37
				6	.40	.39	.37	.40	.38	.37	.39	.38	.37	.39	.38	.37	.38	.37	.36	.36
				7	.39	.37	.36	.39	.37	.36	.38	.37	.35	.38	.36	.35	.37	.36	.35	.35
				8	.37	.36	.34	.37	.35	.34	.37	.35	.34	.36	.35	.34	.36	.35	.34	.33
				9	.36	.34	.33	.36	.34	.33	.35	.34	.33	.35	.34	.33	.35	.33	.33	.32
				10	.35	.33	.32	.35	.33	.32	.34	.33	.32	.34	.33	.32	.34	.32	.31	.31
8 Medium distribution unit with lens plate and inside frost lamp	V, 0%↑, 54½%↓	V	1.0	0	.65	.65	.65	.63	.63	.63	.60	.60	.60	.58	.58	.58	.55	.55	.55	.54
				1	.60	.58	.57	.58	.57	.56	.56	.55	.54	.54	.53	.52	.52	.51	.50	.50
				2	.55	.53	.51	.54	.52	.50	.52	.50	.49	.51	.49	.48	.49	.48	.47	.46
				3	.51	.48	.46	.50	.47	.45	.49	.46	.44	.47	.45	.44	.46	.44	.43	.42
				4	.47	.44	.41	.47	.44	.41	.45	.43	.41	.44	.42	.40	.43	.41	.40	.39
				5	.44	.40	.38	.43	.40	.38	.42	.39	.37	.41	.39	.37	.41	.38	.37	.36
				6	.41	.37	.35	.40	.37	.35	.39	.36	.34	.39	.36	.34	.38	.36	.34	.33
				7	.38	.34	.32	.37	.34	.32	.37	.34	.31	.36	.33	.31	.35	.33	.31	.30
				8	.35	.32	.29	.35	.31	.29	.34	.31	.29	.34	.31	.29	.33	.30	.29	.28
				9	.33	.29	.27	.32	.29	.27	.32	.29	.26	.31	.28	.26	.31	.28	.26	.25
				10	.30	.27	.25	.30	.27	.24	.30	.27	.24	.29	.26	.24	.29	.26	.24	.23
9 Recessed baffled downlight, 140 mm (5 ½") diameter aperture—150-PAR/FL lamp	IV, 0%↑, 6a½↓	IV	0.5	0	.82	.82	.82	.80	.80	.80	.76	.76	.76	.73	.73	.73	.70	.70	.70	.69
				1	.78	.77	.76	.77	.76	.75	.74	.74	.73	.72	.71	.71	.69	.69	.68	.68
				2	.76	.74	.73	.75	.73	.72	.73	.71	.70	.71	.70	.69	.69	.68	.67	.67
				3	.74	.72	.70	.73	.71	.70	.71	.70	.69	.70	.69	.68	.68	.67	.67	.66
				4	.72	.70	.68	.71	.69	.68	.70	.68	.67	.69	.67	.66	.67	.66	.66	.65
				5	.70	.68	.66	.69	.67	.66	.68	.67	.65	.67	.66	.64	.66	.65	.64	.64
				6	.69	.66	.65	.68	.66	.65	.67	.66	.64	.66	.65	.64	.66	.65	.64	.63
				7	.67	.65	.63	.67	.65	.63	.66	.64	.63	.65	.64	.63	.65	.64	.62	.62
				8	.66	.64	.62	.65	.63	.62	.65	.63	.62	.64	.63	.62	.64	.62	.61	.61
				9	.65	.63	.61	.64	.62	.61	.64	.62	.61	.63	.62	.61	.63	.62	.60	.60
				10	.63	.61	.60	.63	.61	.60	.63	.61	.60	.62	.61	.60	.62	.61	.60	.59
10 Recessed baffled downlight, 140 mm (5 ½") diameter aperture—75ER30 lamp	IV, 0%↑, 85%↓	IV	0.5	0	1.01	1.01	1.01	.99	.99	.99	.95	.95	.95	.91	.91	.91	.87	.87	.87	.85
				1	.97	.95	.94	.95	.94	.92	.92	.91	.90	.88	.88	.87	.86	.85	.84	.83
				2	.93	.91	.89	.91	.89	.88	.89	.87	.86	.86	.85	.83	.84	.83	.82	.81
				3	.90	.87	.85	.89	.86	.84	.87	.85	.83	.85	.83	.82	.83	.82	.81	.79
				4	.87	.84	.82	.86	.83	.81	.84	.82	.80	.83	.81	.79	.81	.80	.79	.78
				5	.84	.81	.79	.83	.80	.78	.82	.79	.78	.81	.79	.77	.80	.78	.76	.75
				6	.82	.79	.76	.81	.78	.76	.80	.78	.76	.79	.77	.75	.78	.76	.75	.74
				7	.79	.76	.74	.79	.76	.74	.78	.75	.73	.77	.75	.73	.76	.74	.73	.72
				8	.77	.74	.72	.77	.74	.72	.76	.73	.71	.75	.73	.71	.75	.73	.71	.70
				9	.75	.72	.70	.75	.72	.70	.74	.71	.69	.73	.71	.69	.73	.71	.69	.68
				10	.73	.70	.68	.73	.70	.68	.72	.69	.68	.72	.69	.67	.71	.69	.67	.67
11 Wide distribution unit with lens plate and inside frost lamp	V, 0%↑, 53½%↓	V	1.4	0	.63	.63	.63	.62	.62	.62	.59	.59	.59	.57	.57	.57	.54	.54	.54	.53
				1	.58	.56	.54	.57	.55	.54	.54	.53	.52	.52	.51	.50	.50	.50	.49	.48
				2	.53	.50	.48	.52	.49	.47	.50	.48	.46	.48	.47	.45	.47	.45	.44	.43
				3	.48	.45	.42	.47	.44	.42	.46	.43	.41	.44	.42	.40	.43	.41	.40	.39
				4	.44	.40	.37	.43	.40	.37	.42	.39	.37	.41	.38	.36	.40	.37	.36	.35
				5	.40	.36	.33	.39	.36	.33	.38	.35	.33	.37	.35	.32	.36	.34	.32	.31
				6	.36	.32	.30	.36	.32	.29	.35	.32	.29	.34	.31	.29	.33	.31	.29	.28
				7	.33	.29	.26	.33	.29	.26	.32	.28	.26	.31	.28	.26	.30	.28	.26	.25
				8	.30	.26	.23	.30	.26	.23	.29	.26	.23	.28	.25	.23	.28	.25	.23	.22
				9	.27	.23	.21	.27	.23	.21	.26	.23	.21	.26	.23	.20	.25	.22	.20	.19
				10	.25	.21	.18	.25	.21	.18	.24	.21	.18	.24	.20	.18	.24	.20	.18	.17
12 Recessed unit with dropped diffusing glass	V, 1½↑, 50½%↓	V	1.3	0	.62	.62	.62	.60	.60	.60	.57	.57	.57	.54	.54	.54	.52	.52	.52	.51
				1	.53	.51	.48	.52	.49	.47	.49	.47	.46	.47	.45	.44	.45	.43	.42	.41
				2	.46	.42	.39	.45	.42	.39	.43	.40	.38	.41	.39	.36	.39	.37	.35	.34
				3	.40	.36	.33	.40	.35	.32	.38	.34	.31	.36	.33	.31	.35	.32	.30	.29
				4	.36	.31	.28	.35	.31	.28	.34	.30	.27	.32	.29	.26	.31	.28	.26	.25
				5	.32	.27	.24	.31	.27	.24	.30	.26	.23	.29	.25	.23	.28	.25	.22	.21
				6	.29	.24	.20	.28	.24	.20	.27	.23	.20	.26	.22	.20	.25	.22	.19	.18
				7	.26	.21	.18	.25	.21	.18	.24	.20	.17	.23	.20	.17	.22	.19	.17	.16
				8	.23	.19	.15	.22	.18	.15	.22	.18	.15	.21	.18	.15	.20	.17	.15	.14
				9	.21	.17	.14	.21	.16	.14	.20	.16	.13	.19	.16	.13	.19	.15	.13	.12
				10	.19	.15	.12	.19	.15	.12	.18	.14	.12	.18	.14	.12	.17	.14	.12	.11

* Also, reflector downlight with baffles and inside frosted lamp.

TABLE 6.5 Coefficients of Utilization (*Continued*)

Typical Luminaire	Typical Intensity Distribution and Per Cent Lamp Lumens	Maint. Cat.	SC	RCR	ρCC 80 ρW 50	30	10	70 50	30	10	50 50	30	10	30 50	30	10	10 50	30	10	0
31 — 150 mm × 150 mm (6" × 6") cell parabolic wedge louver (multiply by 1.1 for 250 × 250 mm (10 × 10") cells)	IV — 0%↑ 58%↓	IV	1.5/1.2	0	.69	.69	.69	67	67	67	64	64	64	62	62	62	59	59	59	58
				1	.63	.61	.59	62	60	58	59	58	57	57	56	55	55	54	53	52
				2	.57	.54	.52	56	53	51	54	52	50	52	50	49	51	49	48	47
				3	.52	.48	.45	51	47	45	49	46	44	48	45	43	46	44	42	41
				4	.47	.42	.39	46	42	39	44	41	38	43	40	38	42	40	38	36
				5	.42	.37	.34	41	37	34	40	36	34	39	36	33	38	35	33	32
				6	.38	.33	.30	37	33	30	36	32	29	35	32	29	34	31	29	28
				7	.34	.29	.26	33	29	26	32	29	26	32	28	26	31	28	25	24
				8	.30	.26	.22	30	25	22	29	25	22	28	25	22	28	24	22	21
				9	.27	.22	.19	27	22	19	26	22	19	25	22	19	25	21	19	18
				10	.24	.20	.17	24	20	17	23	19	17	23	19	17	22	19	17	16
32 — 2 lamp, surface mounted, bare lamp unit—Photometry with 460 mm (18") wide panel above luminaire (lamps on 150 mm (6") centers)	I — 9½%↑ 78%↓	I	1.3	0	1.02	1.02	1.02	99	99	99	92	92	92	86	86	86	81	81	81	78
				1	.86	.82	.78	83	79	75	78	74	71	73	70	67	68	66	64	61
				2	.74	.67	.61	71	65	60	66	61	57	62	58	54	58	55	52	49
				3	.64	.56	.50	62	55	49	58	52	47	54	49	45	51	47	43	41
				4	.56	.48	.42	55	47	41	51	45	39	48	42	38	45	40	36	34
				5	.49	.41	.35	48	40	34	45	38	33	42	36	32	39	34	30	28
				6	.44	.36	.30	43	35	29	40	33	28	38	32	27	35	30	26	24
				7	.39	.31	.25	38	30	25	36	29	24	34	28	23	32	27	23	21
				8	.35	.27	.22	34	27	22	32	26	21	30	24	20	29	23	19	18
				9	.32	.24	.19	31	23	18	29	22	18	27	21	17	26	20	17	15
				10	.29	.21	.17	28	21	16	26	20	16	25	19	15	23	18	15	13
33 — Luminous bottom suspended unit with extra-high output lamp	VI — 66%↑ 12%↓	VI	N.A.	0	.77	.77	.77	68	68	68	50	50	50	34	34	34	19	19	19	12
				1	.67	.64	.62	59	57	54	44	42	41	30	29	28	17	16	16	10
				2	.59	.54	.50	52	48	45	38	36	34	26	25	23	15	14	13	09
				3	.51	.46	.42	45	41	37	34	31	28	23	21	20	13	12	12	07
				4	.45	.40	.35	40	35	31	30	27	24	20	18	17	12	11	10	06
				5	.40	.34	.30	35	30	27	26	23	20	18	16	14	10	09	08	05
				6	.36	.30	.26	32	27	23	24	20	18	16	14	12	09	08	07	05
				7	.32	.26	.22	28	23	20	21	18	15	15	12	11	08	07	06	04
				8	.29	.23	.19	25	21	17	19	16	13	13	11	09	08	06	06	03
				9	.26	.20	.17	23	18	15	17	14	12	12	10	08	07	06	05	03
				10	.24	.18	.15	21	16	13	16	12	10	11	09	07	06	05	04	03
34 — Prismatic bottom and sides, open top, 4 lamp suspended unit—see note 7	VI — 33%↑ 50%↓	VI	1.4/1.2	0	.91	.91	.91	85	85	85	74	74	74	64	64	64	54	54	54	50
				1	.80	.77	.74	75	73	70	66	64	62	57	56	54	49	48	47	43
				2	.71	.66	.62	67	63	59	59	56	53	51	49	47	44	43	41	38
				3	.63	.58	.53	60	55	50	53	49	45	46	43	41	40	38	36	33
				4	.57	.50	.45	53	48	43	47	43	39	41	38	35	36	34	32	29
				5	.50	.44	.39	48	42	37	42	38	34	37	34	31	33	30	28	25
				6	.45	.39	.34	43	37	33	38	33	30	34	30	27	30	27	24	22
				7	.41	.34	.30	39	33	28	34	29	26	32	28	24	27	24	21	19
				8	.37	.30	.26	35	29	25	31	26	23	27	24	21	24	21	19	17
				9	.33	.27	.22	31	26	22	28	23	20	25	21	18	22	19	16	15
				10	.30	.24	.20	28	23	19	25	21	18	23	19	16	20	17	14	13
35 — 2 lamp prismatic wraparound—see note 7	V — 11½%↑ 58½%↓	V	1.5/1.2	0	.81	.81	.81	78	78	78	72	72	72	66	66	66	61	61	61	59
				1	.71	.69	.66	69	66	64	64	62	60	59	58	56	55	54	53	50
				2	.64	.59	.56	61	58	54	57	54	51	53	51	49	49	48	46	44
				3	.57	.52	.48	55	50	47	51	48	45	48	45	42	45	42	40	38
				4	.51	.46	.41	49	44	41	46	42	39	43	40	37	41	38	35	34
				5	.46	.40	.36	44	39	35	41	37	34	39	35	32	37	33	31	29
				6	.41	.35	.31	40	35	31	38	33	30	35	31	28	33	30	27	26
				7	.37	.31	.27	36	31	27	34	29	26	32	28	25	30	27	24	23
				8	.33	.28	.24	32	27	23	30	26	22	29	25	22	27	24	21	19
				9	.30	.24	.20	29	24	20	27	23	19	26	22	19	24	21	18	17
				10	.27	.22	.18	26	21	18	25	20	17	23	19	16	22	18	16	15
36 — 2 lamp prismatic wraparound—see note 7	V — 24%↑ 50%↓	V	1.2	0	.82	.82	.82	77	77	77	69	69	69	61	61	61	53	53	53	50
				1	.71	.68	.65	67	65	62	60	58	56	53	51	50	47	45	44	41
				2	.63	.58	.54	59	55	52	53	50	47	47	45	42	42	40	38	35
				3	.56	.50	.46	53	48	44	47	44	40	42	39	37	38	35	33	31
				4	.50	.44	.40	48	42	38	43	39	35	38	35	32	34	32	29	27
				5	.45	.39	.34	43	37	33	38	34	31	35	31	28	31	28	26	24
				6	.41	.35	.30	39	33	29	35	30	27	32	28	25	28	25	23	21
				7	.37	.31	.27	35	30	26	32	27	24	29	25	22	26	23	20	19
				8	.33	.27	.23	32	26	23	29	24	21	26	22	20	23	20	18	16
				9	.30	.24	.20	29	23	20	26	22	19	24	20	17	21	18	16	14
				10	.27	.22	.18	26	21	16	24	19	16	22	18	15	19	16	14	13

Coefficients of Utilization for 20 Per Cent Effective Floor Cavity Reflectance (ρFC = 20)

TABLE 6.5 Coefficients of Utilization (*Continued*)

Typical Luminaire	Maint. Cat.	SC	RCR ↓	$\rho_{CC}\to$ 80, ρ_W 50	30	10	70, 50	30	10	50, 50	30	10	30, 50	30	10	10, 50	30	10	0
37 2 lamp diffuse wraparound—see note 7	V	1.3	0	52	52	52	50	50	50	46	46	46	43	43	43	39	39	39	38
			1	45	43	41	43	41	39	40	38	37	36	35	34	34	33	32	30
			2	39	35	33	37	34	32	34	32	30	32	30	28	29	28	26	25
			3	34	30	27	33	29	26	30	27	25	28	26	24	26	24	22	21
			4	30	26	23	29	25	22	27	24	21	25	22	20	23	21	19	18
			5	26	22	19	25	21	19	23	20	18	22	19	17	20	18	16	15
			6	23	19	16	23	19	16	21	18	15	19	17	14	18	16	14	13
			7	21	17	14	20	16	14	19	16	13	18	15	13	16	14	12	11
			8	19	15	12	18	14	12	17	14	11	16	13	11	15	12	10	09
			9	17	13	10	16	13	10	15	12	10	14	11	09	13	11	09	08
			10	15	12	09	15	11	09	14	11	09	13	10	08	12	10	08	07
38 4 lamp, 610 mm (2') wide troffer with 45° plastic louver—see note 7	IV	1.0	0	60	60	60	58	58	58	56	56	56	53	53	53	51	51	51	50
			1	54	52	50	52	51	49	50	49	48	48	47	46	47	46	45	44
			2	48	45	43	47	44	42	45	43	41	44	42	40	42	41	39	39
			3	43	40	37	42	39	37	41	38	36	40	37	36	39	37	35	34
			4	39	35	32	38	35	32	37	34	32	36	33	31	35	33	31	30
			5	35	31	28	35	31	28	34	30	28	33	30	28	32	29	27	26
			6	32	28	25	32	28	25	31	27	25	29	27	25	29	26	24	23
			7	29	25	22	29	25	22	28	25	22	27	24	22	27	24	22	21
			8	26	22	20	26	22	20	25	22	20	25	22	19	24	21	19	18
			9	24	20	17	24	20	17	23	20	17	23	19	17	22	19	17	16
			10	22	18	16	22	18	16	21	18	16	21	18	15	20	17	15	15
39 4 lamp, 610 mm (2') wide troffer with 45° white metal louver—see note 7	IV	0.9	0	55	55	55	54	54	54	51	51	51	49	49	49	47	47	47	46
			1	50	48	47	49	47	46	47	46	45	45	44	43	43	43	42	41
			2	45	43	41	44	42	40	43	41	39	41	40	38	40	39	37	37
			3	41	38	36	40	38	35	39	37	35	38	36	34	37	35	34	33
			4	37	34	32	37	34	31	36	33	31	35	32	30	34	32	30	29
			5	34	30	28	33	30	28	32	30	27	32	29	27	31	29	27	26
			6	31	27	25	31	27	25	30	27	25	29	27	25	28	26	24	24
			7	29	25	23	28	25	23	28	25	22	27	24	22	26	24	22	21
			8	26	23	20	26	23	20	25	22	20	25	22	20	24	22	20	19
			9	24	20	18	24	20	18	23	20	18	23	20	18	22	20	18	17
			10	22	19	16	22	19	16	21	18	16	21	18	16	20	18	16	15
40 Fluorescent unit dropped diffuser, 4 lamp 610 mm (2') wide—see note 7	V	1.2	0	73	73	73	71	71	71	68	68	68	65	65	65	62	62	62	60
			1	64	61	59	62	60	58	60	58	56	57	56	54	55	54	52	51
			2	56	52	49	55	51	48	52	49	47	50	48	46	48	46	44	43
			3	50	45	41	49	44	41	47	43	40	45	42	39	43	41	38	37
			4	44	39	35	43	38	35	42	37	34	40	36	33	39	36	33	32
			5	39	34	30	38	33	29	37	32	29	36	32	29	34	31	28	27
			6	35	30	26	34	29	25	33	29	25	32	28	25	31	27	25	23
			7	31	26	22	31	26	22	30	25	22	29	25	22	28	24	22	20
			8	28	23	19	28	23	19	27	22	19	26	22	19	26	22	19	18
			9	25	20	17	25	20	17	24	20	17	23	19	16	23	19	16	15
			10	22	18	15	22	18	15	21	17	15	21	17	15	21	17	14	13
41 Fluorescent unit with flat bottom diffuser, 4 lamp 610 mm (2') wide—see note 7	V	1.2	0	69	69	69	67	67	67	64	64	64	61	61	61	59	59	59	58
			1	61	58	56	59	57	56	57	55	54	55	53	52	53	52	51	49
			2	53	50	47	52	49	46	50	48	45	49	46	44	47	45	43	42
			3	47	43	40	47	42	39	45	41	38	43	40	38	42	39	37	36
			4	42	37	34	41	37	33	40	36	33	39	35	33	37	35	32	31
			5	37	32	29	37	32	28	35	31	28	35	31	28	33	30	27	26
			6	33	28	25	33	28	25	32	28	24	31	27	24	30	27	24	23
			7	29	25	22	29	24	21	29	24	21	28	24	21	27	24	21	20
			8	27	22	19	27	22	19	26	22	18	25	21	18	24	21	18	17
			9	24	19	16	24	19	16	23	19	16	23	19	16	22	18	16	15
			10	22	17	14	22	17	14	21	17	14	21	17	14	21	17	13	13
42 Fluorescent unit with flat prismatic lens, 4 lamp 610 mm (2') wide—see note 7	V	1.4/1.2	0	75	75	75	73	73	73	70	70	70	67	67	67	64	64	64	63
			1	67	65	63	66	64	62	63	62	60	61	60	58	59	58	57	55
			2	60	57	54	59	56	53	57	54	52	55	53	51	53	51	50	49
			3	54	50	47	53	49	46	52	48	45	50	47	45	48	46	44	43
			4	49	44	40	48	44	40	47	43	40	45	42	39	44	41	39	37
			5	44	39	35	43	38	35	42	38	35	41	37	34	40	36	34	33
			6	40	34	31	39	34	31	38	34	30	37	33	30	36	32	30	29
			7	36	30	27	35	30	27	35	30	27	34	29	27	32	29	26	25
			8	32	27	23	32	27	23	31	26	23	30	26	23	29	26	23	22
			9	29	24	20	28	23	20	28	23	20	27	23	20	26	23	20	19
			10	26	21	18	26	21	18	25	21	18	24	20	18	24	20	18	16

Coefficients of Utilization for 20 Per Cent Effective Floor Cavity Reflectance (ρ_{FC} = 20)

TABLE 6.5 Coefficients of Utilization (*Continued*)

Typical Luminaire	Typical Intensity Distribution and Per Cent Lamp Lumens	Maint. Cat.	SC	ρ_{CC} →	80			70			50			30			10			0
				ρ_W →	50	30	10	50	30	10	50	30	10	50	30	10	50	30	10	0
				RCR ↓	Coefficients of Utilization for 20 Per Cent Effective Floor Cavity Reflectance (ρ_{FC} = 20)															
43  4 lamp, 610 mm (2') wide unit with sharp cutoff (high angle—low luminance) flat prismatic lens—see note 7	0 ▲ 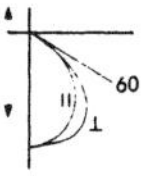 65½ ▼	V	1.4/1.3	0	.78	.78	.78	.76	.76	.76	.73	.73	.73	.70	.70	.70	.67	.67	.67	.66
				1	.71	.69	.67	.70	.68	.66	.67	.65	.64	.64	.63	.62	.62	.61	.60	.59
				2	.64	.61	.58	.63	.60	.58	.61	.59	.56	.59	.57	.55	.57	.55	.54	.53
				3	.58	.54	.51	.58	.54	.51	.56	.52	.50	.54	.51	.49	.52	.50	.48	.47
				4	.53	.48	.45	.52	.48	.44	.51	.47	.44	.49	.46	.43	.48	.45	.43	.42
				5	.48	.43	.39	.47	.42	.39	.46	.42	.39	.45	.41	.38	.43	.40	.38	.37
				6	.43	.38	.35	.43	.38	.34	.42	.37	.34	.40	.37	.34	.40	.36	.34	.32
				7	.39	.34	.30	.38	.34	.30	.38	.33	.30	.37	.33	.30	.36	.32	.30	.28
				8	.35	.30	.26	.35	.30	.26	.34	.29	.26	.33	.29	.26	.32	.29	.26	.25
				9	.31	.26	.23	.31	.26	.23	.30	.26	.23	.30	.26	.23	.29	.25	.23	.21
				10	.28	.24	.20	.28	.23	.20	.28	.23	.20	.27	.23	.20	.26	.23	.20	.19
44 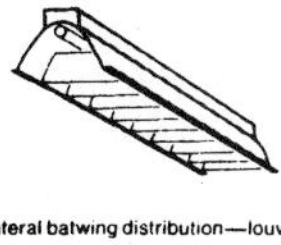 Bilateral batwing distribution—louvered fluorescent unit	0° ▲ 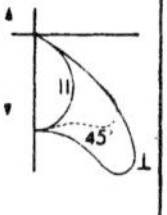 60 ▼	IV	N.A.	0	.71	.71	.71	.70	.70	.70	.66	.66	.66	.64	.64	.64	.61	.61	.61	.60
				1	.65	.63	.61	.63	.62	.60	.61	.59	.58	.59	.57	.56	.57	.56	.55	.54
				2	.59	.55	.53	.58	.55	.52	.55	.53	.51	.54	.52	.50	.52	.50	.49	.48
				3	.53	.49	.46	.52	.48	.45	.50	.47	.45	.49	.46	.44	.47	.45	.43	.42
				4	.47	.43	.40	.47	.43	.40	.45	.42	.39	.44	.41	.39	.43	.40	.38	.37
				5	.42	.38	.34	.42	.37	.34	.41	.37	.34	.40	.36	.34	.39	.36	.33	.32
				6	.38	.33	.30	.38	.33	.30	.37	.33	.30	.36	.32	.29	.35	.32	.29	.28
				7	.34	.29	.26	.33	.29	.26	.33	.28	.25	.32	.28	.25	.31	.28	.25	.24
				8	.30	.25	.22	.30	.25	.22	.29	.25	.22	.28	.24	.22	.27	.24	.21	.20
				9	.27	.22	.18	.26	.22	.18	.26	.21	.18	.25	.21	.18	.24	.21	.18	.17
				10	.24	.19	.16	.24	.19	.16	.23	.19	.16	.22	.19	.16	.22	.18	.16	.15
45  Bilateral batwing distribution—4 lamp, 610 mm (2') wide fluorescent unit with flat prismatic lens and overlay—see note 7	0° ▲ 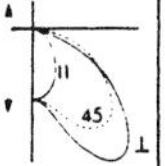 48 ▼	V	N.A.	0	.57	.57	.57	.56	.56	.56	.53	.53	.53	.51	.51	.51	.49	.49	.49	.48
				1	.50	.48	.47	.49	.47	.46	.47	.46	.44	.45	.44	.43	.44	.43	.42	.41
				2	.44	.41	.38	.43	.40	.38	.41	.39	.37	.40	.38	.36	.38	.37	.35	.34
				3	.39	.35	.32	.38	.34	.31	.37	.33	.31	.35	.33	.30	.34	.32	.30	.29
				4	.34	.30	.27	.33	.29	.26	.32	.29	.26	.31	.28	.26	.30	.27	.25	.24
				5	.30	.25	.22	.29	.25	.22	.28	.24	.22	.27	.24	.21	.26	.23	.21	.20
				6	.26	.22	.19	.26	.22	.18	.25	.21	.18	.24	.21	.18	.23	.20	.18	.17
				7	.23	.19	.16	.23	.19	.16	.22	.18	.16	.21	.18	.15	.21	.18	.15	.14
				8	.21	.16	.13	.20	.16	.13	.19	.16	.13	.19	.15	.13	.18	.15	.13	.12
				9	.18	.14	.11	.18	.14	.11	.17	.14	.11	.17	.13	.11	.16	.13	.11	.10
				10	.16	.12	.09	.16	.12	.09	.16	.12	.09	.15	.12	.09	.15	.12	.09	.08
46 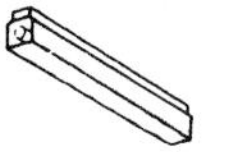 Bilateral batwing distribution—one lamp, surface mounted fluorescent with prismatic wraparound lens	12° ▲ 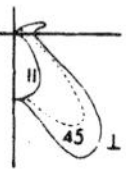 63½° ▼	V	N.A.	0	.87	.87	.87	.84	.84	.84	.77	.77	.77	.72	.72	.72	.66	.66	.66	.64
				1	.76	.73	.70	.73	.70	.67	.67	.65	.63	.63	.61	.59	.58	.57	.55	.53
				2	.66	.61	.57	.64	.59	.56	.59	.56	.52	.55	.52	.49	.51	.49	.47	.44
				3	.59	.53	.48	.56	.51	.47	.53	.48	.44	.49	.45	.42	.46	.43	.40	.38
				4	.52	.45	.40	.50	.44	.40	.47	.42	.38	.44	.39	.36	.41	.37	.34	.32
				5	.46	.39	.34	.44	.38	.33	.41	.36	.32	.38	.34	.31	.36	.32	.29	.27
				6	.41	.34	.29	.39	.33	.29	.37	.31	.27	.34	.30	.26	.32	.28	.25	.23
				7	.36	.30	.25	.35	.29	.24	.33	.27	.23	.31	.26	.23	.29	.25	.22	.20
				8	.32	.26	.21	.31	.25	.21	.29	.24	.20	.27	.23	.19	.26	.21	.18	.17
				9	.29	.22	.18	.28	.22	.18	.26	.21	.17	.24	.20	.16	.23	.19	.15	.14
				10	.26	.20	.16	.25	.19	.15	.23	.18	.15	.22	.17	.14	.20	.16	.13	.12
47  Radial batwing distribution—4 lamp, 610 mm (2') wide fluorescent unit with flat prismatic lens—see note 7	0 ▲  59½ ▼	V	1.7	0	.71	.71	.71	.69	.69	.69	.66	.66	.66	.63	.63	.63	.61	.61	.61	.60
				1	.62	.60	.58	.61	.59	.57	.59	.57	.55	.56	.55	.53	.54	.53	.52	.51
				2	.55	.51	.47	.53	.50	.47	.51	.48	.46	.49	.47	.45	.48	.45	.44	.42
				3	.48	.43	.39	.47	.43	.39	.45	.41	.38	.44	.40	.38	.42	.39	.37	.36
				4	.42	.37	.33	.41	.37	.33	.40	.36	.32	.39	.35	.32	.37	.34	.31	.30
				5	.37	.32	.27	.36	.31	.27	.35	.30	.27	.34	.30	.27	.33	.29	.26	.25
				6	.33	.27	.23	.32	.27	.23	.31	.26	.23	.30	.26	.23	.29	.25	.23	.21
				7	.29	.24	.20	.29	.24	.20	.28	.23	.20	.27	.23	.20	.26	.22	.19	.18
				8	.26	.21	.17	.25	.20	.17	.25	.20	.17	.24	.20	.17	.23	.19	.16	.15
				9	.23	.18	.14	.23	.18	.14	.22	.17	.14	.21	.17	.14	.21	.17	.14	.13
				10	.21	.16	.12	.20	.16	.12	.20	.15	.12	.19	.15	.12	.19	.15	.12	.11
48  2 lamp fluorescent strip unit	20½ ▲ 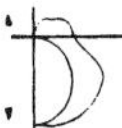 68 ▼	I	1.6/1.2	0	1.01	1.01	1.01	.96	.96	.96	.87	.87	.87	.79	.79	.79	.72	.72	.72	.68
				1	.85	.81	.77	.81	.77	.73	.73	.70	.67	.66	.64	.62	.60	.58	.56	.53
				2	.73	.66	.61	.69	.63	.58	.63	.58	.54	.57	.53	.50	.51	.48	.45	.42
				3	.63	.56	.50	.60	.53	.48	.55	.49	.44	.50	.45	.41	.45	.41	.38	.35
				4	.56	.47	.41	.53	.46	.40	.48	.42	.37	.44	.39	.34	.40	.35	.32	.29
				5	.49	.40	.34	.46	.39	.33	.42	.36	.31	.38	.33	.29	.35	.30	.26	.24
				6	.43	.35	.29	.41	.34	.28	.38	.31	.26	.34	.29	.24	.31	.26	.23	.20
				7	.39	.31	.25	.37	.29	.24	.34	.27	.23	.31	.25	.21	.28	.23	.19	.17
				8	.34	.27	.21	.33	.26	.21	.30	.24	.19	.27	.22	.18	.25	.20	.17	.15
				9	.31	.23	.18	.30	.23	.18	.27	.21	.17	.25	.19	.15	.22	.18	.14	.12
				10	.28	.21	.16	.27	.20	.16	.25	.19	.15	.22	.17	.14	.20	.16	.13	.11

TABLE 6.6 Lamp Group and Burnout Replacement Factors

Lamp Type	Group Replacement	Burnout Replacement
Fluorescent	0.90	0.85
Incandescent	0.94	0.88
Metal-halide	0.87	0.80
Mercury	0.82	0.74
Tungsten-halogen	0.94	0.88
High-pressure sodium	0.94	0.88

Lamp burnouts: If lamps are replaced as they burn out, use a factor of 0.95. If a group replacement maintenance program is employed, use a factor of 1.

Lamp lumen depreciation: All lamps put out less light as they age. Specific information is available from each manufacturer, or you can use the figures in Table 6.14. For preliminary calculations the factors in Table 6.6 can also be used.

Luminaire Dirt Depreciation (LDD)

This factor depends on the type of luminaire, its design, the maintenance schedule of cleaning, and the cleanliness of the room in which the luminaire is used. The manufacturer's literature should give the maintenance category to which an individual fixture belongs. If not, follow the procedure given in Table 6.7 to find the maintenance category to which a fixture belongs.

Next, determine the degree of dirt conditions from the following examples:

Very clean: High-grade offices, not near production; laboratories; clean rooms

Clean: Offices in older buildings or near production, light assembly, inspection

Medium: Mill offices, paper processing, light machine

Dirty: Heat treating, high-speed printing, rubber processing

Very dirty: Similar to dirty but luminaires within immediate area of contamination

Finally, estimate the expected cleaning cycle. With these three factors, use Table 6.8 to determine the LDD factor.

TABLE 6.7 Procedure for Determining Luminaire Maintenance Categories

To assist in determining Luminaire Dirt Depreciation (LDD) factors, luminaires are separated into six maintenance categories (I through VI). To arrive at categories, luminaires are arbitrarily divided into sections, a Top Enclosure and a Bottom Enclosure, by drawing a horizontal line through the light center of the lamp or lamps. The characteristics listed for the enclosures are then selected as best describing the luminaire. Only one characteristic for the top enclosure and one for the bottom enclosure should be used in determining the category of a luminaire. Percentage of uplight is based on 100% for the luminaire.

The maintenance category is determined when there are characteristics in both enclosure columns. If a luminaire falls into more than one category, the lower numbered category is used.

Maintenance Category	Top Enclosure	Bottom Enclosure
I	1. None	1. None
II	1. None 2. Transparent with 15% or more uplight through apertures. 3. Translucent with 15% or more uplight through apertures. 4. Opaque with 15% or more uplight through apertures.	1. None 2. Louvers or baffles
III	1. Transparent with less than 15% upward light through apertures. 2. Translucent with less than 15% upward light through apertures. 3. Opaque with less than 15% upward light through apertures.	1. None 2. Louvers or baffles
IV	1. Transparent unapertured. 2. Translucent unapertured. 3. Opaque unapertured.	1. None 2. Louvers
V	1. Transparent unapertured. 2. Translucent unapertured. 3. Opaque unapertured.	1. Transparent unapertured 2. Translucent unapertured
VI	1. None 2. Transparent unapertured. 3. Translucent unapertured. 4. Opaque unapertured.	1. Transparent unapertured 2. Translucent unapertured 3. Opaque unapertured

Source: IES Lighting Handbook *1981 Reference Volume.*

Room Surface Dirt

This factor depends on the type of luminaire (how much it depends on surface reflectances), the type of use conditions, and the maintenance schedule. There are detailed ways of calculating this factor, but for preliminary design purposes, use the factors given in Table 6.9.

TABLE 6.8 Luminaire Dirt Depreciation Factors

Dirt Conditions	Cleaning Cycle in Years	Luminaire Maintenance Categories					
		I	II	III	IV	V	VI
Very clean	1.0	0.96	0.97	0.92	0.93	0.92	0.93
	1.5	0.95	0.96	0.90	0.91	0.91	0.90
	2.0	0.94	0.95	0.88	0.89	0.89	0.87
	3.0	0.92	0.94	0.84	0.86	0.87	0.82
Clean	1.0	0.93	0.93	0.90	0.88	0.88	0.87
	1.5	0.91	0.92	0.87	0.84	0.85	0.81
	2.0	0.89	0.90	0.84	0.81	0.83	0.77
	3.0	0.86	0.87	0.80	0.75	0.80	0.68
Medium	1.0	0.89	0.90	0.87	0.81	0.83	0.80
	1.5	0.86	0.88	0.83	0.75	0.79	0.73
	2.0	0.84	0.85	0.79	0.70	0.76	0.67
	3.0	0.79	0.82	0.73	0.62	0.71	0.56
Dirty	1.0	0.85	0.86	0.83	0.73	0.78	0.75
	1.5	0.81	0.83	0.78	0.66	0.73	0.67
	2.0	0.77	0.80	0.74	0.60	0.70	0.59
	3.0	0.71	0.75	0.67	0.50	0.64	0.47
Very dirty	1.0	0.74	0.83	0.79	0.64	0.73	0.67
	1.5	0.67	0.79	0.73	0.55	0.67	0.57
	2.0	0.62	0.75	0.68	0.47	0.63	0.48
	3.0	0.53	0.69	0.60	0.37	0.56	0.35

Source: IES Lighting Handbook *1981 Reference Volume.*

In lieu of combining all of the factors just given, the *LLF* can be estimated by using the following combination of task and area types:

Clean	0.70
Light dirt	0.65
Medium dirt	0.60
Dirty	0.55
Very dirty	0.50

TABLE 6.9 Approximate Room Surface Dirt Depreciation Factors

Room Cleanliness	Luminaire Distribution Types				
	Direct	Semidirect	Direct-indirect	Semi-indirect	Indirect
Very clean	0.97	0.95	0.94	0.94	0.89
Clean	0.95	0.91	0.87	0.85	0.80
Medium	0.94	0.88	0.83	0.81	0.73
Dirty	0.92	0.85	0.79	0.78	0.67
Very dirty	0.91	0.83	0.76	0.74	0.61

Source: IES Lighting Handbook *1981 Reference Volume.*

Step-by-Step Calculations for the Number of Luminaires Required for a Particular Room

1. Compile the following information:
  - Length and width of room.
  - Height of floor cavity—the distance from the floor to the work surface (usually taken as 2.5 ft).
  - Height of the ceiling cavity—the distance from the ceiling to the light fixture. If the fixture is recessed or ceiling-(surface-) mounted, the value is zero.
  - Height of the room cavity—the distance from the work surface to the light fixture.
  - Surface reflectances—of the ceiling, the walls, and the floor. If the wall surface of the floor cavity is different from the room cavity wall surface (as with a wainscot, for example) obtain both figures. Surface reflectances are usually available from paint companies, ceiling tile manufacturers, and manufacturers of other finishes. If these are not readily available, use the values in Table 6.10.

TABLE 6.10 Reflectance Values of Various Materials and Colors

Material	Approximate Reflectance (in %)
Acoustical ceiling tile	75–85
Aluminum, brushed	55–58
Aluminum, polished	60–70
Clear glass	8–10
Granite	20–25
Marble	30–70
Stainless steel	55–65
Wood	
Light oak	25–35
Dark oak	10–15
Mahogany	6–12
Walnut	5–10
Color	
White	80–85
Light gray	45–70
Dark gray	20–25
Ivory white	70–80
Ivory	60–70
Pearl gray	70–75
Buff	40–70
Tan	30–50
Brown	20–40
Green	25–50
Azure blue	50–60
Sky blue	35–40
Pink	50–70
Cardinal red	20–25
Red	20–40

2. Determine cavity ratios:

$$CR = 2.5 \times \frac{\text{area of cavity wall}}{\text{area of base of cavity}}$$

For rectangular spaces the formula becomes

$$CR = 5h \times \frac{l + w}{l \times w}$$

where: h = height of the cavity
l = length of the room
w = width of the room

Note that if the work surface is the floor or if the luminaires are surface-mounted, the floor cavity ratio or ceiling cavity ratio, respectively, are zero. Also, because the three cavity ratios are related, after finding one you can find the other two by ratios:

$$CCR = RCR \left(\frac{h_{cc}}{h_{rc}} \right)$$

$$FCR = RCR \left(\frac{h_{fc}}{h_{rc}} \right)$$

where: CCR = ceiling cavity ratio
FCR = floor cavity ratio
RCR = room cavity ratio
h_{cc} = height of ceiling cavity
h_{fc} = height of floor cavity
h_{rc} = height of room cavity

You can find the cavity ratios by calculation or use the values given in Table 6.11. First find the RCR and then use the ratios to find the values of the CCR and FCR.

3. Determine the effective ceiling cavity reflectance and the effective floor cavity reflectance. These are values of the imaginary planes at the height of the luminaire and the work surface that will be used in finding the coefficient of utilization of a particular light fixture. If the luminaires are recessed or surface-mounted, the effective ceiling cavity reflectance is the same as the reflectance of the ceiling itself. Use Table 6.12 to find the effective reflectances, knowing the cavity ratios you determined in step 2.

4. Determine the coefficient of utilization of the fixture under consideration by using the CU tables from the manufacturer's literature or from Table 6.5. Straight-line interpolation will probably be necessary. Most tables are set up for a floor reflectance of 20 percent. If the effective floor reflectance varies significantly from this, use the correction factors given in Table 6.13 and multiply by the *CU* for the fixture.

5. Determine the recommended illumination for the space being designed. Follow the procedure outlined in Section 6.1 ("How to Select the Recommended Illuminance Level").

6. Determine the lumen output of the lamps that will be used in the luminaire you have selected. Values for lumen output for some representative lamps are given in Table 6.14. More accurate data can be obtained from the fixture manufacturer or a lamp manufacturer. Determine the number of lamps that will be used in each luminaire.

7. With the information compiled in the previous steps and with the light loss factor (*LLF*), use the following formula.

$$\text{Number of luminaires} = \frac{\text{footcandles required} \times \text{area of room}}{N \times \text{lumens per lamp} \times CU \times LLF}$$

You can also determine the area per luminaire using the formula given at the beginning of this section.

6.3 LAMP CHARACTERISTICS AND SELECTION GUIDE (TABLES 6.14 THROUGH 6.19)

6.4 HOW LIGHT AFFECTS COLOR (TABLE 6.20)

Relationship of Light and Color

Light is the radiant energy produced by a light source. It may come to your eye directly from the source, or be reflected or transmitted by some object.

Color is the interaction of the light source, the reflector or transmitter, and our own ability to detect the color of light. Remember, you cannot perceive color without light. Different light sources radiate different wavelengths of light, influencing the appearance of colored objects or surfaces.

Color Temperature

Color temperature describes how the lamp itself appears when lit. Color temperature is measured by *Kelvin degrees,* ranging from 9000K (which appears blue) down to 1500K (which appears orange-red). Light sources lie somewhere between the two, with those of higher color temperature (4000K or more) being "cool," and those of lower color temperature (3100K or less) being "warm." Certain fluorescent lamps are "intermediate" types, lying somewhere between cool and warm.

TABLE 6.11 Room Cavity Ratios

Room W	Room L	Cavity Depth													
		2.5	5.5	6.0	6.5	7.0	7.5	8.0	8.5	9.0	10.0	12.0	14.0	16.0	18.0
10	10	2.5	5.5	6.0	6.5	7.0	7.5	8.0	8.5	9.0					
	12	2.3	5.0	5.5	6.0	6.4	6.9	7.3	7.8	8.3					
	14	2.1	4.7	5.1	5.6	6.0	6.4	6.9	7.3	7.7	8.6				
	15	2.1	4.6	5.0	5.4	5.8	6.3	6.7	7.1	7.5	8.3				
	16	2.0	4.5	4.9	5.3	5.7	6.1	6.5	6.9	7.3	8.1				
12	12	2.1	4.6	5.0	5.4	5.8	6.3	6.7	7.1	7.5	8.3				
	14	1.9	4.3	4.6	5.0	5.4	5.8	6.2	6.6	7.0	7.7				
	16	1.8	4.0	4.4	4.7	5.1	5.5	5.8	6.2	6.6	7.3				
	18	1.7	3.8	4.2	4.5	4.9	5.2	5.6	5.9	6.3	6.9				
	20	1.7	3.7	4.0	4.3	4.7	5.0	5.3	5.7	6.0	6.7				
14	14	1.8	3.9	4.3	4.6	5.0	5.4	5.7	6.1	6.4	7.1	8.6			
	16	1.7	3.7	4.0	4.4	4.7	5.0	5.4	5.7	6.0	6.7	8.0			
	18	1.6	3.5	3.8	4.1	4.4	4.8	5.1	5.4	5.7	6.3	7.6			
	20	1.5	3.3	3.6	3.9	4.3	4.6	4.9	5.2	5.5	6.1	7.3			
	22	1.5	3.2	3.5	3.8	4.1	4.4	4.7	5.0	5.3	5.8	7.0			
16	16	1.6	3.4	3.8	4.1	4.4	4.7	5.0	5.3	5.6	6.3	7.5	8.8		
	18	1.5	3.2	3.5	3.8	4.1	4.4	4.7	5.0	5.3	5.9	7.1	8.3		
	20	1.4	3.1	3.4	3.7	3.9	4.2	4.5	4.8	5.1	5.6	6.8	7.9		
	22	1.3	3.0	3.2	3.5	3.8	4.0	4.3	4.6	4.9	5.4	6.5	7.6		
	24	1.3	2.9	3.1	3.4	3.6	3.9	4.2	4.4	4.7	5.2	6.3	7.3		
18	18	1.4	3.1	3.3	3.6	3.9	4.2	4.4	4.7	5.0	5.6	6.7	7.8	8.9	
	22	1.3	2.8	3.0	3.3	3.5	3.8	4.0	4.3	4.5	5.1	6.1	7.1	8.1	
	26	1.2	2.6	2.8	3.1	3.3	3.5	3.8	4.0	4.2	4.7	5.6	6.6	7.5	
	30	1.1	2.4	2.7	2.9	3.1	3.3	3.6	3.8	4.0	4.4	5.3	6.2	7.1	
	34	1.1	2.3	2.5	2.8	3.0	3.2	3.4	3.6	3.8	4.2	5.1	5.9	6.8	

20	20	1.3	2.8	3.0	3.3	3.5	3.8	4.0	4.3	4.5	5.0	6.0	7.0	8.0	9.0
	24	1.1	2.5	2.8	3.0	3.2	3.4	3.7	3.9	4.1	4.6	5.5	6.4	7.3	8.3
	28	1.1	2.4	2.6	2.8	3.0	3.2	3.4	3.6	3.9	4.3	5.1	6.0	6.9	7.7
	32	1.0	2.2	2.4	2.6	2.8	3.0	3.3	3.5	3.7	4.1	4.9	5.7	6.5	7.3
	36	1.0	2.1	2.3	2.5	2.7	2.9	3.1	3.3	3.5	3.9	4.7	5.4	6.2	7.0
24	24	1.0	2.3	2.5	2.7	2.9	3.1	3.3	3.5	3.8	4.2	5.0	5.8	6.7	7.5
	28	1.0	2.1	2.3	2.5	2.7	2.9	3.1	3.3	3.5	3.9	4.6	5.4	6.2	7.0
	32	0.9	2.0	2.2	2.4	2.6	2.7	2.9	3.1	3.3	3.6	4.4	5.1	5.8	6.6
	36	0.9	1.9	2.1	2.3	2.4	2.6	2.8	3.0	3.1	3.5	4.2	4.9	5.6	6.3
	40	0.8	1.8	2.0	2.2	2.3	2.5	2.7	2.8	3.0	3.3	4.0	4.7	5.3	6.0
28	34	0.8			2.1	2.3	2.4	2.6	2.8	2.9	3.3	3.9	4.6	5.2	5.9
	40	0.8			2.0	2.1	2.3	2.4	2.6	2.7	3.0	3.6	4.3	4.9	5.5
	46	0.7			1.9	2.0	2.2	2.3	2.4	2.6	2.9	3.4	4.0	4.6	5.2
	52	0.7			1.8	1.9	2.1	2.2	2.3	2.5	2.7	3.3	3.8	4.4	4.9
32	38	0.7					2.2	2.3	2.4	2.6	2.9	3.5	4.0	4.6	5.2
	44	0.7					2.0	2.2	2.3	2.4	2.7	3.2	3.8	4.3	4.9
	50	0.6					1.9	2.1	2.2	2.3	2.6	3.1	3.6	4.1	4.6
	56	0.6					1.8	2.0	2.1	2.2	2.5	2.9	3.4	3.9	4.4
38	46	0.6					1.8	1.9	2.0	2.2	2.4	2.9	3.4	3.8	4.3
	54	0.6					1.7	1.8	1.9	2.0	2.2	2.7	3.1	3.6	4.0
	62	0.5					1.6	1.7	1.8	1.9	2.1	2.5	3.0	3.4	3.8
	70	0.5							1.7	1.8	2.0	2.4	2.8	3.2	3.7
44	50	0.5							1.8	1.9	2.1	2.6	3.0	3.4	3.8
	60	0.5							1.7	1.8	2.0	2.4	2.8	3.2	3.5
	70	0.5							1.6	1.7	1.9	2.2	2.6	3.0	3.3
	80	0.4							1.5	1.6	1.8	2.1	2.5	2.8	3.2

TABLE 6.12 Percent Effective Ceiling or Floor Cavity Reflectances for Various Reflectance Combinations

Per Cent Base* Reflectance	90										80										70										60										50									
Per Cent Wall Reflectance → Cavity Ratio ↓	90	80	70	60	50	40	30	20	10	0	90	80	70	60	50	40	30	20	10	0	90	80	70	60	50	40	30	20	10	0	90	80	70	60	50	40	30	20	10	0	90	80	70	60	50	40	30	20	10	0
0.2	89	88	88	87	87	86	85	84	84	82	79	78	77	77	76	76	75	74	73	72	70	69	69	68	68	67	66	66	65	64	60	60	59	59	58	58	57	56	56	53	50	50	49	48	48	47	46	46	45	44
0.4	88	86	85	85	84	83	81	80	79	76	79	77	76	75	74	74	73	72	71	68	70	68	67	66	65	64	63	61	60	58	60	58	57	56	55	55	54	53	52	50	50	48	48	47	46	45	44	43	42	42
0.6	87	86	84	83	82	80	79	77	76	72	78	76	75	73	72	71	70	69	67	65	70	67	66	64	63	62	61	60	58	54	60	57	56	55	54	53	52	50	49	46	50	48	47	46	45	44	43	42	41	38
0.8	87	85	83	82	80	77	75	73	71	69	78	75	73	71	69	67	66	65	63	63	69	66	64	62	60	58	56	55	53	50	59	57	55	53	51	48	47	46	45	44	50	47	46	46	44	43	42	41	39	36
1.0	86	85	82	80	77	75	72	69	66	64	77	74	72	69	66	64	62	60	58	57	68	65	62	60	58	55	52	50	49	47	59	57	55	53	51	49	46	42	40	38	50	47	46	45	43	41	40	38	37	34
1.2	85	82	79	78	77	75	72	69	69	57	67	63	59	57	54	51	48	46	45	44	66	60	57	54	51	48	45	43	41	38	59	56	54	51	49	46	44	42	39	38	50	47	45	43	41	39	36	35	34	29
1.4	85	80	77	73	69	64	62	56	54	52	73	67	61	56	50	45	41	38	36	34	65	60	54	51	47	43	39	37	35	32	56	53	51	49	47	45	42	39	37	36	50	47	45	42	40	36	35	34	32	27
1.6	84	79	75	71	67	60	59	53	50	50	71	64	59	54	48	45	40	36	34	31	65	59	53	49	45	41	37	35	33	30	55	53	50	47	44	42	40	37	35	33	50	46	44	41	39	36	33	32	30	26
1.8	83	78	73	69	64	58	53	50	48	47	70	63	58	53	47	42	38	34	31	28	64	58	52	47	43	38	34	31	29	26	55	52	48	44	40	37	35	33	31	31	50	46	43	40	37	34	33	30	28	25
2.0	83	77	72	67	62	56	50	47	43	43	69	62	56	52	45	40	36	34	30	27	64	56	50	45	42	38	34	30	27	24	54	50	46	43	39	35	33	31	29	29	50	46	43	40	37	34	30	28	26	24
2.2	82	76	70	65	59	54	47	44	40	40	66	60	55	51	48	43	38	34	32	32	66	60	54	48	43	37	35	33	29	26	58	53	49	45	42	37	34	31	29	28	46	46	42	38	36	33	29	27	24	22
2.4	82	75	69	64	58	53	45	42	38	37	65	59	54	50	47	42	36	35	32	30	65	59	53	47	43	37	35	32	28	24	58	52	48	44	41	36	32	29	27	26	46	46	41	37	36	32	26	24	21	21
2.6	81	74	67	62	56	51	46	42	38	35	64	58	53	49	45	40	35	33	30	28	65	59	53	47	42	36	33	30	26	24	58	53	47	43	39	35	31	28	26	24	46	46	41	37	33	30	23	21	20	20
2.8	81	73	66	60	54	49	40	36	34	34	63	57	53	48	44	38	33	30	28	26	64	57	52	46	41	34	31	28	25	22	58	51	47	42	38	34	29	26	24	22	46	45	41	36	33	29	22	21	19	17
3.0	80	72	64	58	52	47	42	38	34	30	62	56	52	47	43	37	32	29	27	24	64	56	52	45	42	37	29	27	23	20	57	52	46	42	37	32	28	25	23	20	46	45	40	36	32	28	21	19	17	17
3.2	79	71	63	56	50	45	40	36	32	28	65	52	45	40	35	33	28	25	23	18	72	65	58	51	46	36	32	27	25	23	57	51	45	41	36	31	26	23	20	18	50	39	35	31	27	23	20	17	14	14
3.4	79	70	62	54	48	43	38	34	30	27	64	54	46	39	34	29	25	22	20	17	71	64	57	50	44	35	31	27	24	22	56	51	44	40	35	30	25	22	19	16	44	35	30	26	21	18	15	14	13	13
3.6	78	69	61	53	47	42	36	32	28	25	63	53	47	38	32	28	24	21	18	15	71	63	56	49	43	33	29	25	22	20	56	50	44	39	34	29	24	21	18	15	44	34	30	25	21	16	14	13	12	12
3.8	78	60	40	40	45	40	35	31	27	23	70	61	53	37	31	28	23	20	17	14	70	61	54	48	42	32	28	24	20	18	55	49	43	38	33	29	23	20	17	14	48	33	29	24	19	15	13	12	11	11
4.0	77	59	39	39	44	39	33	29	25	22	69	58	51	46	35	30	26	23	20	14	70	61	53	47	41	31	26	23	20	17	54	49	42	37	32	28	23	20	17	11	37	32	28	24	18	14	12	12	11	11
4.2	77	62	43	42	41	36	32	28	24	21	69	57	50	43	34	30	25	22	19	16	77	62	55	48	41	30	25	22	19	16	56	49	42	36	32	27	22	19	17	14	43	33	32	27	24	20	17	14	11	12
4.4	76	61	42	41	40	36	31	27	23	20	68	56	49	43	37	27	24	21	18	13	76	62	54	47	40	29	25	21	18	13	56	49	41	36	31	27	21	18	16	13	43	32	27	23	23	18	16	13	11	12
4.6	76	60	40	39	38	34	29	26	22	18	67	55	47	41	36	28	23	20	17	12	75	61	54	47	39	28	25	20	18	13	55	49	41	35	31	26	21	18	15	11	42	31	26	23	22	16	15	13	10	11
4.8	75	59	40	38	36	33	28	25	20	18	66	54	46	40	35	27	22	19	15	10	75	60	53	45	38	27	23	20	16	12	54	48	40	34	30	24	20	17	15	10	42	30	25	22	21	15	13	10	08	10
5.0	75	59	39	38	35	32	27	24	20	14	66	53	45	40	34	26	21	18	14	09	75	59	52	44	36	26	22	19	16	12	54	47	40	34	29	23	20	17	14	11	42	30	25	21	21	14	12	10	08	09
6.0	73	65	53	44	34	29	24	20	16	11	60	51	44	37	31	24	20	16	13	09	63	55	48	41	35	24	21	17	14	11	55	45	37	31	25	21	17	14	11	07	42	29	23	19	19	15	13	10	06	06
7.0	70	58	45	38	27	23	18	14	11	08	58	48	41	35	28	22	17	14	11	05	59	52	46	39	32	22	19	16	13	11	54	43	35	28	24	21	17	14	12	09	41	32	21	18	14	14	11	09	08	05
8.0	68	55	40	35	27	21	15	12	06	05	57	46	38	32	23	19	16	13	10	04	57	46	38	36	19	16	13	11	10	08	53	42	33	26	18	15	13	12	10	07	40	25	18	16	13	12	10	07	07	04
9.0	66	52	38	30	26	16	14	11	05	04	56	45	36	31	27	18	14	12	07	03	52	45	38	31	18	14	12	10	08	03	52	40	31	26	16	14	12	10	07	03	29	24	15	11	15	10	07	07	03	03
10.0	65	51	36	29	22	15	11	09	04	03	55	43	33	28	21	16	12	10	08	03	51	39	29	24	16	12	10	08	06	02	51	39	29	24	14	11	09	07	05	02	27	22	14	08	14	08	06	02	...	02

* Ceiling, floor or floor of cavity.

TABLE 6.12 Percent Effective Ceiling or Floor Cavity Reflectances for Various Reflectance Combinations (*Continued*)

Per Cent Base Reflectance	40										30										20										10										0									
Per Cent Wall Reflectance / Cavity Ratio	90	80	70	60	50	40	30	20	10	0	90	80	70	60	50	40	30	20	10	0	90	80	70	60	50	40	30	20	10	0	90	80	70	60	50	40	30	20	10	0	90	80	70	60	50	40	30	20	10	0
0.2	40	40	39	39	39	38	38	37	37	36	31	31	30	30	29	29	29	28	28	27	21	20	20	20	20	20	19	19	19	17	11	11	11	10	10	10	10	09	09	09	02	02	02	02	01	01	01	00	00	00
0.4	41	40	39	39	38	38	37	36	35	34	31	31	30	30	29	28	28	27	26	25	22	20	20	20	20	19	19	18	18	16	12	11	11	10	10	10	10	09	09	08	04	03	03	03	02	02	01	01	00	00
0.6	41	40	39	38	38	36	35	34	33	32	32	31	30	30	29	28	27	26	25	23	23	21	21	20	19	19	18	17	17	15	13	13	12	11	11	11	10	09	09	08	05	05	04	03	03	02	02	01	01	00
0.8	41	40	38	37	36	35	33	32	31	29	32	31	30	29	28	27	26	25	23	22	24	22	21	20	19	18	17	16	16	14	15	14	13	12	11	11	10	10	09	08	07	06	05	04	04	03	02	02	01	00
1.0	42	40	38	37	35	33	32	31	29	27	33	32	30	29	29	27	25	24	23	20	25	22	22	20	19	18	17	16	15	13	16	14	13	12	12	11	11	10	08	07	08	07	06	05	04	03	02	02	01	00
1.2	42	40	38	37	34	32	30	29	27	25	33	32	30	28	27	25	23	22	21	19	25	23	22	20	19	17	16	15	13	12	17	15	14	13	12	12	11	10	09	06	10	08	07	06	05	04	03	02	01	00
1.4	42	39	37	35	33	31	29	27	25	23	34	32	30	28	26	24	22	21	19	18	26	24	22	20	18	17	16	15	13	11	18	16	15	13	12	11	11	10	08	06	11	09	08	07	06	04	03	02	01	00
1.6	42	39	37	35	32	30	27	25	23	22	34	33	30	28	25	23	22	20	18	17	26	24	23	20	18	17	15	14	12	11	19	17	15	14	13	12	11	09	08	06	12	10	09	07	06	05	04	03	01	00
1.8	42	39	36	34	31	29	26	24	22	21	35	33	29	27	25	23	21	19	17	16	27	25	23	20	18	17	15	13	11	10	19	18	16	14	13	12	11	09	08	05	13	11	09	08	07	05	04	03	01	00
2.0	42	39	36	34	31	28	25	23	21	19	35	33	29	27	24	22	20	18	16	14	28	25	23	20	18	16	15	13	11	09	20	18	16	14	13	13	11	09	08	05	14	12	10	09	07	05	04	03	01	00
2.2	42	39	36	33	30	27	24	22	19	18	36	32	29	26	24	22	19	17	15	13	28	26	23	20	18	16	14	12	10	09	21	19	16	14	13	13	11	09	07	05	15	13	11	09	08	06	04	03	01	00
2.4	43	39	36	33	29	26	24	21	18	17	36	32	29	26	24	21	19	16	14	12	29	26	23	20	18	16	14	12	10	08	22	20	17	15	13	13	11	09	07	05	16	13	11	10	08	06	04	03	01	00
2.6	43	39	35	32	29	26	23	20	17	15	36	32	29	26	23	20	18	16	14	12	29	26	23	20	18	16	14	12	10	08	23	20	17	15	14	13	11	09	07	04	17	14	12	10	09	07	05	03	02	00
2.8	43	39	35	32	28	25	22	19	16	14	36	33	29	25	22	20	17	15	13	11	30	27	23	20	17	15	13	11	09	07	23	20	18	16	14	13	10	09	06	04	17	15	12	10	09	07	05	04	02	00
3.0	43	39	35	31	27	24	21	18	15	13	37	33	29	25	21	20	17	15	13	10	30	27	23	20	17	15	13	11	09	07	24	21	18	16	15	13	10	09	06	03	18	16	13	11	09	07	05	04	02	00
3.2	43	39	35	31	27	23	20	17	15	13	37	33	29	25	21	19	16	14	12	10	31	27	23	20	17	15	12	11	09	06	25	21	18	16	15	13	10	09	06	03	19	16	14	11	10	07	05	04	02	00
3.4	43	39	34	30	26	23	20	17	14	11	37	33	29	25	21	19	16	14	11	09	31	27	23	20	17	15	12	10	08	06	26	22	18	16	15	13	10	09	06	03	20	17	14	12	10	07	05	04	02	00
3.6	44	39	34	30	26	22	19	16	13	11	38	33	28	24	21	18	16	13	10	09	32	27	23	20	17	15	12	10	08	05	26	22	19	17	15	12	10	08	06	03	20	17	15	12	10	08	06	04	02	00
3.8	44	38	33	29	25	22	18	15	13	10	38	33	28	24	20	18	15	13	10	08	32	28	23	20	17	15	12	10	08	05	27	23	19	17	15	12	10	08	06	02	21	18	15	12	11	08	06	04	02	00
4.0	44	38	33	29	25	21	18	15	12	10	38	33	28	24	20	18	15	12	10	07	33	28	23	20	17	15	12	10	07	05	27	23	20	17	16	12	10	08	06	02	22	18	16	13	11	08	06	04	02	00
4.2	44	38	33	29	24	21	17	14	12	10	38	33	28	24	19	17	14	12	09	07	33	28	23	20	17	14	11	09	06	04	28	24	20	17	16	11	09	08	06	02	22	19	16	13	11	08	06	04	02	00
4.4	44	38	33	28	24	20	17	14	11	09	39	33	28	23	19	17	14	11	09	06	34	28	24	20	17	14	11	09	06	04	28	24	20	17	16	11	09	08	06	02	23	19	16	13	11	08	06	04	02	00
4.6	44	38	32	28	23	20	16	14	11	09	39	33	28	23	19	17	13	11	08	06	34	29	24	20	17	14	11	08	06	04	29	25	20	17	16	11	09	08	06	02	23	20	17	13	11	08	06	04	02	00
4.8	44	38	32	28	23	19	16	13	10	08	39	33	28	23	19	16	13	10	08	05	35	29	24	20	16	13	10	08	06	04	29	25	20	17	16	11	09	08	06	02	24	20	17	14	11	08	06	04	02	00
5.0	45	38	32	27	22	19	15	13	10	07	39	33	27	22	18	16	13	10	07	05	35	29	24	20	16	13	10	08	06	04	29	25	20	17	16	11	09	08	05	02	25	21	18	14	11	08	06	04	02	00
6.0	44	37	30	25	20	17	13	11	08	05	39	32	26	21	16	14	11	09	06	04	36	30	24	20	16	12	10	07	04	02	31	26	20	16	13	10	08	06	04	01	27	23	18	15	12	09	06	04	02	00
7.0	44	36	29	24	19	16	12	10	07	04	40	32	25	20	15	14	10	08	06	03	36	30	24	20	15	12	09	07	04	02	32	27	21	17	13	10	08	06	04	01	28	24	19	15	12	09	06	04	02	00
8.0	44	35	28	23	18	15	11	09	06	03	40	33	25	20	15	13	09	07	05	02	37	30	23	19	15	11	08	06	04	01	33	27	21	17	13	10	08	06	04	01	29	25	20	15	12	09	06	04	02	00
9.0	44	35	26	21	16	13	10	09	05	02	40	33	25	20	15	12	09	07	04	02	37	29	23	19	14	11	08	06	04	01	34	28	21	17	13	10	07	05	02	01	31	25	20	15	12	09	06	04	02	00
10.0	43	34	25	20	15	12	08	07	05	02	40	32	24	19	14	11	08	06	03	01	37	29	22	18	13	10	07	05	03	01	34	28	21	17	12	10	07	05	02	01	31	25	20	15	12	09	06	04	02	00

* Ceiling, floor or floor of cavity

TABLE 6.13 Multiplying Factors for Other than 20 Percent Effective Floor Cavity Reflectance

% Effective Ceiling Cavity Reflectance, ρ_{cc}	80				70				50			30			10		
% Wall Reflectance, ρ_w	70	50	30	10	70	50	30	10	50	30	10	50	30	10	50	30	10
Room Cavity Ratio	For 30 Per Cent Effective Floor Cavity Reflectance (20 Per Cent = 1.00)																
1	1.092	1.082	1.075	1.068	1.077	1.070	1.064	1.059	1.049	1.044	1.040	1.028	1.026	1.023	1.012	1.010	1.008
2	1.079	1.066	1.055	1.047	1.068	1.057	1.048	1.039	1.041	1.033	1.027	1.026	1.021	1.017	1.013	1.010	1.006
3	1.070	1.054	1.042	1.033	1.061	1.048	1.037	1.028	1.034	1.027	1.020	1.024	1.017	1.012	1.014	1.009	1.005
4	1.062	1.045	1.033	1.024	1.055	1.040	1.029	1.021	1.030	1.022	1.015	1.022	1.015	1.010	1.014	1.009	1.004
5	1.056	1.038	1.026	1.018	1.050	1.034	1.024	1.015	1.027	1.018	1.012	1.020	1.013	1.008	1.014	1.009	1.004
6	1.052	1.033	1.021	1.014	1.047	1.030	1.020	1.012	1.024	1.015	1.009	1.019	1.012	1.006	1.014	1.008	1.003
7	1.047	1.029	1.018	1.011	1.043	1.026	1.017	1.009	1.022	1.013	1.007	1.018	1.010	1.005	1.014	1.008	1.003
8	1.044	1.026	1.015	1.009	1.040	1.024	1.015	1.007	1.020	1.012	1.006	1.017	1.009	1.004	1.013	1.007	1.003
9	1.040	1.024	1.014	1.007	1.037	1.022	1.014	1.006	1.019	1.011	1.005	1.016	1.009	1.004	1.013	1.007	1.002
10	1.037	1.022	1.012	1.006	1.034	1.020	1.012	1.005	1.017	1.010	1.004	1.015	1.009	1.003	1.013	1.007	1.002

For 10 Per Cent Effective Floor Cavity Reflectance (20 Per Cent = 1.00)

Room Cavity Ratio																	
1	.923	.929	.935	.940	.933	.939	.943	.948	.956	.960	.963	.973	.976	.979	.989	.991	.993
2	.931	.942	.950	.958	.940	.949	.957	.963	.962	.968	.974	.976	.980	.985	.988	.991	.995
3	.939	.951	.961	.969	.945	.957	.966	.973	.967	.975	.981	.978	.983	.988	.988	.992	.996
4	.944	.958	.969	.978	.950	.963	.973	.980	.972	.980	.986	.980	.986	.991	.987	.992	.996
5	.949	.964	.976	.983	.954	.968	.978	.985	.975	.983	.989	.981	.988	.993	.987	.992	.997
6	.953	.969	.980	.986	.958	.972	.982	.989	.977	.985	.992	.982	.989	.995	.987	.993	.997
7	.957	.973	.983	.991	.961	.975	.985	.991	.979	.987	.994	.983	.990	.996	.987	.993	.998
8	.960	.976	.986	.993	.963	.977	.987	.993	.981	.988	.995	.984	.991	.997	.987	.994	.998
9	.963	.978	.987	.994	.965	.979	.989	.994	.983	.990	.996	.985	.992	.998	.988	.994	.999
10	.965	.980	.989	.995	.967	.981	.990	.995	.984	.991	.997	.986	.993	.998	.988	.994	.999

For 0 Per Cent Effective Floor Cavity Reflectance (20 Per Cent = 1.00)

Room Cavity Ratio																	
1	.859	.870	.879	.886	.873	.884	.893	.901	.916	.923	.929	.948	.954	.960	.979	.983	.987
2	.871	.887	.903	.919	.886	.902	.916	.928	.926	.938	.949	.954	.963	.971	.978	.983	.991
3	.882	.904	.915	.942	.898	.918	.934	.947	.936	.950	.964	.958	.969	.979	.976	.984	.993
4	.893	.919	.941	.958	.908	.930	.948	.961	.945	.961	.974	.961	.974	.984	.975	.985	.994
5	.903	.931	.953	.969	.914	.939	.958	.970	.951	.967	.980	.964	.977	.988	.975	.985	.995
6	.911	.940	.961	.976	.920	.945	.965	.977	.955	.972	.985	.966	.979	.991	.975	.986	.996
7	.917	.947	.967	.981	.924	.950	.970	.982	.959	.975	.988	.968	.981	.993	.975	.987	.997
8	.922	.953	.971	.985	.929	.955	.975	.986	.963	.978	.991	.970	.983	.995	.976	.988	.998
9	.928	.958	.975	.988	.933	.959	.980	.989	.966	.980	.993	.971	.985	.996	.976	.988	.998
10	.933	.962	.979	.991	.937	.963	.983	.992	.969	.982	.995	.973	.987	.997	.977	.989	.999

TABLE 6.14 Characteristics of Typical Lamps

Standard Incandescent						
Bulb Description	Watts	Length/ Size (in in.)	Lamp Life (in hours) (1)	Color Temp. °K (1)	Initial Lumens (1)	Lamp Lumen Depreciation (1)
A-19	60		1000	2790	860	0.93
A-19	75		750	2840	1180	0.92
A-19	100		750	2900	1740	0.91
A-19	100		2500		1490	0.93
A-21	100		750	2880	1690	0.90
A-21	150		750	2960	2880	0.89
A-23	150		2500		2350	0.89
PS-25	150		750	2900	2660	0.88
A-23	200		750	2980	4000	0.90
A-23	200		2500		3400	0.88
PS-25	300		750	3010	6360	0.88
PS-30	300		2500		5200	0.79
PS-35	500		1000	3050	10600	0.89

R, PAR, and ER Lamps						
Bulb Description	Watts	Length/ Size (in in.)	Lamp Life (in hours) (1)	Color Temp. (1)	Initial Lumens (1,2)	Lamp Lumen Depreciation (1)
R-30 Spot/Flood	75		2000		850	
R-40 Spot/Flood	150		2000		1825	
R-40 Spot/Flood	300		2000		3600	
PAR-38 Spot/Flood	100		2000		1250	
PAR-38 Spot/Flood	150		2000		1730	
ER-30	50		2000		525	
ER-30	75		2000		850	
ER-30	90		5000		950	
ER-40	120		2000		1475	

Fluorescent						
Bulb Description	Watts	Length/ Size (in in.)	Lamp Life (in hours) (1,3)	Color Temp. (1,4)	Initial Lumens (1,5)	Lamp Lumen Depreciation (1)
F40T12CW/RS	40	48	20000	4300	3150	0.84
F40T12WW/RS	40	48	20000	3100	3170	0.84
F40T12CWX/RS	40	48	20000	4100	2200	0.84
F40T12WWX/RS	40	48	20000	3000	2170	0.84
F40T12D/RS	40	48	20000	6500	2600	0.84
F40T12W/RS	40	48	20000	3600	3180	0.84
F96T12CW	75	96	12000	4300	6300	0.89
F96T12WW	75	96	12000	3100	6335	0.89
F96T12CWX	75	96	12000	4100	4465	0.89
F96T12WWX	75	96	12000	3000	4365	

TABLE 6.14   Characteristics of Typical Lamps (*Continued*)

Tungsten-Halogen (Quartz-Iodine) Bulb Description	Watts	Length/ Size (in in.)	Lamp Life (in hours) (1)	Color Temp. (1)	Initial Lumens (1)	Lamp Lumen Depreciation (1)
T-4	100		1000		1800	0.93
T-4	150		1500	3000	2900	0.93
T-4	250		2000	2950	5000	0.97
PAR-38	250		6000		3500	0.95

Mercury Bulb Description	Watts	Length/ Size (in in.)	Lamp Life (in hours) (1)	Color Temp. (1)	Initial Lumens (1)	Lamp Lumen Depreciation (1)
H45AY-40/50 DX	50		16000		1680	
H43AY-75/DX	75		24000		3000	
H38BP-100/DX	100		24000+		2865	
H38JA-100/WDX	100		24000+		4000	
H38MP-100/DX	100		24000		4275	
H39BN-175/DX	175		24000		5800	
H39KC-175/DX	175		24000+		8600	
H37KC-250/DX	250		24000+		12775	

Metal-Halide Bulb Description	Watts	Length/ Size (in in.)	Lamp Life (in hours) (1)	Color Temp. (1)	Initial Lumens (1)	Lamp Lumen Depreciation (1)
M57PF-175	175		7500	3600	14000	
M58PH-250	250		10000		20500	
M59PK-400	400		1500	3800	34000	

High-Pressure Sodium Bulb Description	Watts	Length/ Size (in in.)	Lamp Life (in hours) (1)	Color Temp. (1)	Initial Lumens (1)	Lamp Lumen Depreciation (1)
S68MT-50	50		24000		3800	
S54MC-100	100		24000		8800	
S55MD-150	150		24000		15000	

(1) Figures listed are approximate. Exact values vary with manufacturer.

(2) Initial lumens for R, PAR, and ER lamps is for total lumens.

(3) Lamp life for fluorescent depends on number of hours per start; figures given are for approximately 10 hours per start.

(4) Technically, "color temperature" applies only to incandescent sources, but it is often used to describe the degree of whiteness of other light sources.

(5) Lumens at 40% of rated life.

TABLE 6.15 Guide to Lamp Selection

Lamp Type and Efficacy (1)	Lamp Appearance Effect on Neutral Surfaces	Effect on "Atmosphere"	Colors Strengthened	Colors Grayed	Effect on Complexions	Remarks
Fluorescent						
Cool white (#4) (2)	White	Neutral to moderately cool	Orange, blue, yellow	Red	Pale pink	Blends with natural daylight—good color acceptance
Deluxe cool white (#2) (2)	White	Neutral to moderately cool	All nearly equal	None appreciably	Most natural	Best overall color rendition, simulates natural daylight
Warm white (#4) (3)	Yellowish white	Warm	Orange, yellow	Red, green, blue	Sallow	Blends with incandescent light—poor color acceptance
Deluxe warm white (#2) (3)	Yellowish white	Warm	Red, orange, yellow, green	Blue	Ruddy	Good color rendition; simulates incandescent light
Daylight (#3)	Bluish white	Very cool	Green, blue	Red, orange	Grayed	Usually replaceable with cool white
White (#4)	Pale yellowish white	Moderately warm	Orange, yellow	Red, green, blue	Pale	Usually replaceable with cool white or warm white

(1) Efficacy (lumens/watt): #1 = low; #2 = medium; #3 = medium high; #4 = high.
(2) Greater preference at higher levels.
(3) Greater preference at lower levels.

Source: GE Lighting Business Group.

TABLE 6.15 Guide to Lamp Selection (*Continued*)

Lamp Type and Efficacy (1)	Lamp Appearance Effect on Neutral Surfaces	Effect on "Atmosphere"	Colors Strengthened	Colors Grayed	Effect on Complexions	Remarks
Incandescent, Tungsten-Halogen						
Filament (#1) (3)	Yellowish white	Warm	Red, orange, yellow	Blue	Ruddiest	Good color rendering
High-Intensity Discharge						
Clear mercury (#2)	Greenish blue-white	Very cool, greenish	Yellow, green, blue	Red, orange	Greenish	Very poor color rendering
White mercury (#2)	Greenish white	Moderately cool, greenish	Yellow, green, blue	Red, orange	Very pale	Moderate color rendering
Deluxe white mercury (#2)	Purplish white	Warm, purplish	Red, yellow, blue	Green	Ruddy	Color acceptance similar to cool white fluorescent
Metal-Halide (#4) (2)	Greenish white	Moderately cool greenish	Yellow, blue, green	Red	Grayed	Color acceptance similar to cool white
High-pressure sodium (#4)	Yellowish	Warm, yellowish	Yellow, green, orange	Red, blue	Yellowish	Color acceptance approaches warm white fluorescent

Source: GE Lighting Business Group.

(1) Efficacy (lumens/watt); #1 = low; #2 = medium; #3 = medium high; #4 = high.
(2) Greater preference at higher levels.
(3) Greater preference at lower levels.

TABLE 6.16 Recommended Reflectances of Interior Surfaces

	Recommended Reflectances in Percent					
	Ceilings	Walls	Floors	Furniture	Other	
Offices	80+	50–70	20–40	25–45	40–70	Partitions
Schools	70–90	40–60	30–50	35–50	up to 20	Chalkboards
Industrial	80–90	40–60	20+		25–45	Benchtops, machines, etc.
Residential	60–90	35–60 (1)	15–35 (1)		45–85	Large drapery areas

(1) Where specific visual tasks are more important than lighting for environment, minimum reflectances should be 40% for walls and 25% for floors.

Source: Data extracted from IES Lighting Handbook, *1981 Applications Volume.*

Color Rendition

Color rendition describes the effect a light source has on the appearance of colored objects. The color-rendering capability of a lamp is measured as the color-rendering index (CRI). In general, the higher the CRI, the less distortion of the object's color by the lamp's light output. The scale used ranges from 0 to 100. A CRI of 100 indicates that there is no color shift as compared with a reference source, and the lower the CRI, the more pronounced the shift may be.

It is important to recognize that the reference source (and thus the

TABLE 6.17 Recommended Luminance Ratios

		Recommended Ratios (1)				
		Between task and immediate darker surroundings		Between task and immediate lighter surroundings	Between task and general surroundings	
Use	**Task**	**Minimum**	**Desired**	**Maximum**	**Minimum**	**Desired**
Residential	1	1/5	1/3		0.1–10	0.2–5
Office	1		1/3			0.1–10
Classroom	1	1/3		5 (2)	1/3	
Merchandising	1	1/3	1/5			
Industrial	1		1/3	3	0.5–20	0.1–10

(1) These are recommended guidelines for most applications. Ratios higher or lower are acceptable if they do not exceed a significant portion of the visual field.

(2) Any significant surface normally viewed directly should be no greater than five times the luminance of the task.

Source: IES Lighting Handbook, *1981 Applications Volume.*

TABLE 6.18 Compact Fluorescent Fixture Operation Data

LAMP QUANTITY AND SIZE	120 VOLT NPF FIXT. TOTAL AMPS/WATTS	120 VOLT HPF FIXT. TOTAL AMPS/WATTS	277 VOLT HPF FIXT. TOTAL AMPS/WATTS	FIXTURE LUMEN LAMP LUMENS Per WATT	EQUIVALENT INCANDESCENT WATTAGE	STANDARD COLOR TEMP. C.R.I.**	MIN. LAMP START TEMP
2 x 9W	.36/25	.20/25	.13/32	$\frac{1200}{67}$	75W	$\frac{2700° \text{ K}}{82}$	25° F
2 x13W	.60/34	.28/34	.17/42	$\frac{1800}{69}$	120W	$\frac{2700° \text{ K}}{86}$	32° F
2 x 18W	.70/47	.44/47	.18/49	$\frac{2500}{69}$	150W	$\frac{2700° \text{ K}}{86}$	23°F
2 x 26W	1.0/64	.63/64	.26/54	$\frac{3600}{69}$	200W	$\frac{2700° \text{ K}}{86}$	23° F
1 x 9W	.18/13	.10/13	.065/13	$\frac{600}{67}$	40W	$\frac{2700° \text{ K}}{82}$	25° F
1 x 13W	.30/17	.14/17	.085/17	$\frac{900}{69}$	60W	$\frac{2700° \text{ K}}{82}$	32° F
1 x 18W	.35/24	.22/24	.09/24	$\frac{1250}{69}$	75W	$\frac{2700° \text{ K}}{86}$	23° F
1 x 26W	.50/32	.32/32	.13/32	$\frac{1800}{69}$	100W	$\frac{2700° \text{ K}}{86}$	23° F

* *Consult Lamp Manufacturers*
For Other Color Temp. Ratings

** *Color Rendering Index*

TABLE 6.19 Compact Fluorescent Socket Style/Lamp Information

NL Corp Socket Code	A	B	C	D	E	F	G	H
Socket Style and Base Design Number	G23	G23 2	GX23	GX23 2	G24 d-1	G24 d-2	G24 d-3	G24 q-3
Lamp Wattage and Style	9WTT	9WDTT	13WTT	13WDTT	13WDTT	18WDTT	26WDTT	26WD/E DTT
Manuf. Code	1,2,3,4	2,4	1,2,3,4	2,4	1,2	1,2,3,4	1,2,3,4	2

Lamp Manuf. Code System

1-G.E. Lighting
2-Osram
3-Philips Lighting
4-Sylvania/GTE

TABLE 6.20 Summary of Light Source Characteristics and Effect on Color

Light Source	Characteristics	Effect on Color
Incandescent Color Temperatures from 2750K to 3400K. CRI: 95 +	• Warm, inviting light • Standard light source • Relatively inefficient	• Brightens reds, oranges, yellows • Darkens blues and greens
Tungsten Halogen Color temperatures from 2850K to 3000K. CRI: 95 +	• Brighter, whiter light than standard incandescent • More efficient than regular incandescent	• Brightens reds, oranges, yellows • Darkens blues and greens
Fluorescent Color temperatures from 2700K to 6300K. CRIs from 48 to 90	• Wide selection of phosphor colors—select warm to cool light-ing atmosphere • Generally high efficiency • Much longer life	• Wide range of color temperatures and CRIs to light effectively any (basically indoor) area with a "warm" to "cool" environment as decor or task dictates
High Intensity Discharge Metal Halide (Metalarc®) High Pressure Sodium (Lumalux® and Unalux®) Mercury	• Different gases and phosphor colors create a variety of atmospheres • High efficiency • Very long life	• Sylvania Metalarc® (metal halide) lamps provide excellent color rendering. Mercury and High Pres-sure Sodium provide poor color rendering. Mercury gives a blue-green coloration and High Pres-sure Sodium imparts an orange-yellow color

CRI scale) is different at different color temperatures. As a result, CRI values should only be compared between lamps of similar color temperatures.

Additional Factors Affect Color Appearance

The color-rendering properties of a lamp are an important influence on the color appearance of an object. However, many other factors will affect color appearance, such as the finishes used on walls, floors, and furnishings; the intensity level of the lighting; and the presence of daylight in the room. All these factors should be considered in selecting the appropriate light source. Additionally, the room decor is a critical consideration in selecting a light source. If colors such as reds and oranges are the main element, a warm light source (color temperature below 3200 K) would be the best choice. Conversely, if blues and violets are being used, cool lamps (color temperature above 4000 K) should be used. For areas with mixed cool and warm elements, or where neutral colors such as gray predominate, an intermediate color temperature source (3400 to 3600 K) should be considered.

NOTES

NOTES

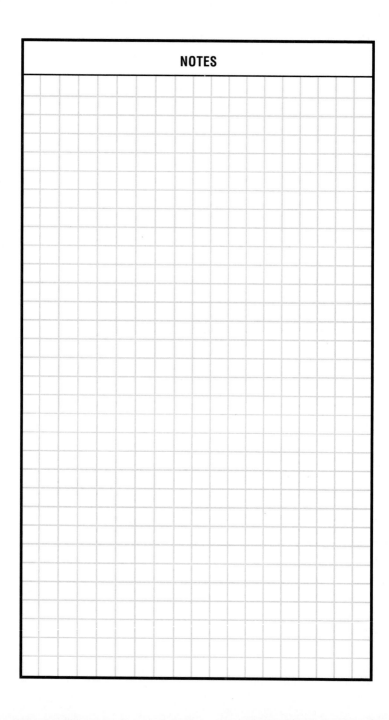

CHAPTER SEVEN
Special Systems

7.0 FIRE ALARM SYSTEMS

Introduction

Fire alarm systems have become increasingly sophisticated, and functionally more capable and reliable in recent years. They are designed to fulfill two general requirements: protection of property and assets, and protection of life. As a result of state and local codes, the life safety aspect of fire protection has become a major factor in the last two decades.

There are a number of reasons for the substantial increases in the life safety form of fire protection during recent years, foremost of which are:

1. The proliferation of high-rise construction and the concern for life safety within these buildings.

2. A growing awareness of the life safety hazard in residential, institutional, and educational occupancies.

3. Increased hazards caused by new building materials and furnishings, which create large amounts of toxic combustion products (i.e., plastics, synthetic fabrics, and so on).

4. Vast improvements in smoke detection and related technology made possible through quantum advances in electronic technology.

5. The passing of the Americans with Disabilities Act (ADA), signed into law on July 26, 1990, providing comprehensive civil rights protection for individuals with disabilities. With an effective date of January 26, 1992, these requirements included detailed accessability standards for both new construction and renovation toward the goal of equal usability of buildings for everyone, regardless of limitations of sight, hearing, and mobility. This had a significant impact on fire alarm system signaling devices, power requirements, and device locations.

Common Code Requirements

The following codes apply to fire alarm systems:

NFPA 70: National Electrical Code.

NFPA 72: National Fire Alarm Code.

NFPA 90A: Standard for the Installation of Air Conditioning and Ventilation Systems.

NFPA 101: Life Safety Code.

BOCA, SBCCI, ICBO: The National Basic Building Code and National Fire Prevention Code, published by the Building Officials Code Administrators International (BOCA), the Uniform Building Code and Uniform Fire Code of the International Conference of Building Officials (ICBO), and the Standard Building Code and the Standard Fire Prevention Code of the Southern Building Code Congress International (SBCCI) all have reference to fire alarm requirements.

Many states and municipalities have adopted these model building codes in full or in part.

You should consult with the local authority having jurisdiction (AHJ) to verify the requirements in your area.

Fire Alarm System Classifications

NFPA 72 classifies fire alarm systems as follows:

Household fire alarm system: A system of devices that produces an alarm signal in the household for the purpose of notifying the occupants of the presence of fire so that they will evacuate the premises.

Protected-premises (local) fire alarm system: A protected-premises system that sounds an alarm at the protected premises as the result of the manual operation of a fire alarm box or the operation of protection equipment or systems, such as water flowing in a sprinkler system, the discharge of carbon dioxide, the detection of smoke, or the detection of heat.

Auxiliary fire alarm system: A system connected to a municipal fire alarm system for transmitting a fire alarm to the public fire service communications center. Fire alarms from an auxiliary fire alarm system are received at the public fire service communications center on the same equipment and by the same methods as alarms transmitted manually from municipal fire alarm boxes located on streets. There are three subtypes of this system: local energy, parallel telephone, and shunt type.

Remote supervising-station fire alarm system: A system installed in accordance with NFPA 72 to transmit alarm, supervisory, and trouble signals from one or more protected premises to a remote location at which appropriate action is taken.

Proprietary supervising-station fire alarm system: An installation of fire alarm systems that serves contiguous and noncontiguous properties, under one ownership, from a proprietary supervising station located at the protected property, at which trained, competent personnel are in constant attendance. This includes the proprietary supervising station; power supplies; signal-initiating devices; initiating-device circuits; signal notification appliances; equipment for the automatic, permanent visual recording of signals; and equipment for initiating the operation of emergency building control services.

Central-station fire alarm system: A system or group of systems in which the operations of circuits and devices are transmitted automatically to, recorded in, maintained by, and supervised from a listed central station having competent and experienced servers and operators who, upon receipt of a signal, take such action as required by NFPA 72. Such service is to be controlled and operated by a person, firm, or corporation whose business is the furnishing, maintaining, or monitoring of supervised fire alarm systems.

Municipal fire alarm system: A system of alarm-initiating devices, receiving equipment, and connecting circuits (other than a public telephone network) used to transmit alarms from street locations to the public fire service communications center.

Fire Alarm Fundamentals—Basic Elements

Regardless of type, application, complexity, or technology level, any fire alarm system is comprised of four basic elements:

1. Initiating devices
2. Control panel
3. Signaling devices
4. Power supply

These components must be electrically compatible and are interconnected by means of suitable wiring circuits to form a complete functional system as illustrated in Figure 7.1.

Figure 7.1 shows a conventional version of a protected-premises (local) fire alarm system, which is the most widely used classification type in commercial and institutional buildings. The requirements for this type of system are detailed in Chapter 3 of NFPA 72. Some highlights of this chapter's requirements are worthy of note and are given in abridged form hereafter.

FIGURE 7.1 Typical local protective signaling system.

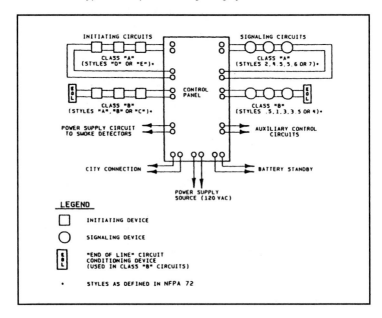

Circuit Designations

Initiating device, notification appliance, and signaling line circuits shall be designated by class or style, or both, depending on the circuits' capability to operate during specified fault conditions.

Class

Initiating device, notification appliance, and signaling line circuits shall be permitted to be designated as either Class A or Class B, depending on the capability of the circuit to transmit alarm and trouble signals during nonsimultaneous single-circuit fault conditions as specified by the following:

1. Circuits capable of transmitting an alarm signal during a single open or a nonsimultaneous single ground fault on a circuit conductor shall be designated as Class A.

2. Circuits not capable of transmitting an alarm beyond the location of the fault conditions specified in item 1 shall be designated as Class B.

Faults on both Class A and Class B circuits shall result in a trouble condition on the system in accordance with the requirements of NFPA 72, Article 1-5.8.

Style

Initiating device, notification appliance, and signaling line circuits shall be permitted to be designated by style also, depending on the capability of the circuit to transmit alarm and trouble signals during specified simultaneous multiple-circuit fault conditions in addition to the single-circuit fault conditions considered in the designation of the circuits by class.

1. An initiating device circuit shall be permitted to be designated as either Style A, B, C, D, or E, depending on its ability to meet the alarm and trouble performance requirements shown in Table 7.1, during a single open, single ground, wire-to-wire short, and loss-of-carrier fault condition.

2. A notification appliance circuit shall be permitted to be designated as either Style W, X, Y, or Z, depending on its ability to meet the alarm and trouble performance requirements shown in Table 7.2, during a single open, single ground, and wire-to-wire short fault condition.

3. A signaling line circuit shall be permitted to be designated as either Style 0.5, 1, 2, 3, 3.5, 4, 4.5, 5, 6, or 7, depending on its ability to meet the alarm and trouble performance requirements shown in Table 7.3, during a single open, single ground, wire-to-wire short, simultaneous wire-to-wire short and open, simultaneous wire-to-wire short and ground, simultaneous open and ground, and loss-of-carrier fault conditions.

Installation of Class A Circuits

All styles of Class A circuits using physical conductors (e.g., metallic, optical fiber) shall be installed such that the outgoing and return conductors, exiting from and returning to the control unit, respectively, are routed separately. The outgoing and return (redundant) circuit conductors shall not be run in the same cable assembly (i.e., multiconductor cable), enclosure, or raceway.

Exception No. 1: For a distance not to exceed 10 ft (3 m) where the outgoing and return conductors enter or exit the initiating device, notification appliance, or control unit enclosures; or

Exception No. 2: Where the vertically run conductors are contained in a 2-h rated cable assembly or enclosed (installed) in a 2-h rated enclosure other than a stairwell; or

TABLE 7.1 Performance of Initiating Device Circuits (IDC)

R = Required capability
X = Indication required at protected premises and as required by Chapter 4
α = Style exceeds minimum requirements for Class A

Class →	B			B			B			A			A		
Style →	A			B			C			D			Eα		
	1	2	3	4	5	6	7	8	9	10	11	12	13	14	15
Abnormal Condition	Alarm	Trouble	Alarm receipt capability during abnormal condition	Alarm	Trouble	Alarm receipt capability during abnormal condition	Alarm	Trouble	Alarm receipt capability during abnormal condition	Alarm	Trouble	Alarm receipt capability during abnormal condition	Alarm	Trouble	Alarm receipt capability during abnormal condition
A. Single open		X			X			X			X	X		X	X
B. Single ground		R			X	R		X	R		X	R		X	R
C. Wire-to-wire short	X			X				X		X				X	
D. Loss of carrier (if used)/channel interface								X						X	

442

TABLE 7.2 Notification Appliance Circuits (NAC)

Class	B		B		B		A	
Style	W		X		Y		Z	
X = Indication required at protected premises	Trouble indication at protected premises	Alarm capability during abnormal conditions	Trouble indication at protected premises	Alarm capability during abnormal conditions	Trouble indication at protected premises	Alarm capability during abnormal conditions	Trouble indication at protected premises	Alarm capability during abnormal condition
Abnormal condition	1	2	3	4	5	6	7	8
Single open	X		X	X	X		X	X
Single ground	X		X		X	X	X	X
Wire-to-wire short	X		X		X		X	

Exception No. 3: Where permitted and where the vertically run conductors are enclosed (installed) in a 2-h rated stairwell in a building fully sprinklered in accordance with NFPA 13, Standard for the Installation of Sprinkler Systems.

Exception No. 4: Where looped conduit/raceway systems are provided, single conduit/raceway drops to individual devices or appliances shall be permitted.

Exception No. 5: Where looped conduit/raceway systems are provided, single conduit/raceway drops to multiple devices or appliances installed within a single room not exceeding 1000 ft² (92.9 m²) in area shall be permitted.

Performance of Initiating Device Circuits (IDC)

The assignment of class designations or style designations, or both, to initiating circuits shall be based on their performance capabilities under

TABLE 7.3 Performance of Signaling Line Circuits (SLC)

M = May be capable of alarm with wire-to-wire short
R = Required capability
X = Indication required at protected premises and as required by Chapter 4
α = Style exceeds minimum requirements for Class A

Class	B			B			A			B			B			B			B			A			A			A		
Style	0.5			1			2α			3			3.5			4			4.5			5α			6α			7α		
	Alarm	Trouble	Alarm receipt capability during abnormal condition	Alarm	Trouble	Alarm receipt capability during abnormal condition	Alarm	Trouble	Alarm receipt capability during abnormal condition	Alarm	Trouble	Alarm receipt capability during abnormal condition	Alarm	Trouble	Alarm receipt capability during abnormal condition	Alarm	Trouble	Alarm receipt capability during abnormal condition	Alarm	Trouble	Alarm receipt capability during abnormal condition	Alarm	Trouble	Alarm receipt capability during abnormal condition	Alarm	Trouble	Alarm receipt capability during abnormal condition	Alarm	Trouble	Alarm receipt capability during abnormal condition
	1	2	3	4	5	6	7	8	9	10	11	12	13	14	15	16	17	18	19	20	21	22	23	24	25	26	27	28	29	30
Abnormal Condition																														
A. Single open		X			X	R		X	R		X	R		X			X	R		X	R		X	R		X	R		X	R
B. Single ground		X			X	R		X	R		X	R		X			X	R		X	R		X	R		X	R		X	R
C. Wire-to-wire short									M		X			X			X			X			X			X			X	R
D. Wire-to-wire short & open									M		X			X			X			X			X			X			X	
E. Wire-to-wire short & ground								X	M		X			X			X			X			X			X	X		X	
F. Open and ground								X	R		X			X			X			X			X			X			X	R
G. Loss of carrier (if used)/channel interface														X			X			X			X			X			X	

444

abnormal (fault) conditions in accordance with the requirements of Table 7.1.

Performance of Signaling Line Circuits (SLC)

The assignment of class designations or style designations, or both, to signaling line circuits shall be based on their performance capabilities under abnormal (fault) conditions in accordance with the requirements of Table 7.2.

Notification Appliance Circuits (NAC)

The assignment of class designations or style designations, or both, to notification appliance circuits shall be based on their performance capabilities under abnormal (fault) conditions in accordance with the requirements of Table 7.3.

Secondary Supply Capacity and Sources

From NFPA 72, Chapter 1 ("Fundamentals"), the secondary source for a protected-premises system should have a secondary supply source capacity of 24 h, and at the end of that period, shall be capable of operating all alarm notification appliances used for evacuation or to direct aid to the location of an emergency for 5 min. The secondary power supply for emergency voice/alarm communications service shall be capable of operating the system under maximum load for 24 h and then shall be capable of operating the system during a fire or other emergency condition for a period of 2 h. Fifteen minutes of evacuation alarm operation at maximum connected load shall be considered the equivalent of 2 h of emergency operation.

Audible/Visual Notification Appliance Requirements

Tables 7.4 and 7.5 summarize the audible and visual notification appliance requirements to comply with the Americans with Disabilities Act Accessibility Guidelines (ADAAG), NFPA 72—1993 and BOCA—1993. Also, refer to Figure 7.2 for the mounting heights for manual pull stations.

Application Tips

A very general rule of thumb for spacing automatic fire detectors is to allow 900 ft^2 per head. This is good for very rough estimating in preliminary stages of design. There are many factors to consider for each specific application (e.g., architectural and structural features such as beams and coves, special-use spaces, ambient temperature and other environmental considerations, and so on). It is therefore prudent to

TABLE 7.4 Audible Notification Appliances to Meet the Requirements of: ADA, NFPA 72 (1993), BOCA

ADA	NFPA	BOCA
• Intensity and frequency that can attract individuals who have partial hearing loss • Periodic element to its signal such as: • Single stroke bell • Hi-Low • Fast whoop • Avoid continuous or reverberating tones. Select a signal which has a sound characterized by three or four clear tones without a great deal of noise in between.	• To insure that audible public mode signals are clearly heard, it shall be required that their sound level be at least 15 dBA above the average ambient sound level, or 5 dBA above the maximum sound level having a duration of at least 60 seconds, whichever is greater, measured at 5' above the floor in the occupiable area • Mechanical Equipment Rooms • Design for a minimum of 85 dBA for all type occupancies • Sleeping Areas • Design for a minimum of 70 dBA at any point in the sleeping area • Mounting location • Wall mounted appliances -not less than 90" AFF -not less than 6" BFC • Combination A/V Units -Bottoms 80"- 96" AFF • Effective July 1, 1996, the fire alarm signal used to notify building occupants shall be in accordance with ANSI S3.41 (NFPA 3-7.2) • Temporal Slow Whoop or • Temporal Code 3-3, 1 second bursts of signal with 2 seconds quiet before repeating the 3 bursts	• Minimum of 15 dBA over average ambient • Every occupied space within the building • Minimum of 70 dBA in use groups R, I-1 • Minimum of 90 dBA in Mechanical Rooms • Minimum of 60 dBA in all other use groups • Maximum of 130 dBA at minimum hearing distance from audible appliance

Design Criteria	Design Comments	Available Devices
• Ratings/listings - most devices are rated for dBA output at 10' from device; • Doubling the distance from the device - drop of 6 dBA • A device with an output of 96 dBA at 10' will have 90 dBa at 20', 84 dBA at 40', 78 dBA at 80', etc. • Acustic tile ceiling causes approximately a 3 dBA drop in sound levels; • Rug on floor - causes approximately 3 dBA drop in sound levels; • An open door: 8-12 dBA drop; • Closed hollow core door: 12-20 dBA drop; • Closed solid core, rated door: 20-30 dBA drop; • 4" Partition: 40-45 dBA drop; • Multiple signals effect: add approximately 3 dBA at mid-point of signals; Typical ambient sound levels: • High School Office: 60 dBA • Corridor with back- ground music: 60 dBA • Classroom with students "Under Control": 62 dBA • Classroom with TV set turned on: 65 dBA • Classroom with students "out of control" end of day: 70 dBA • Corridor with students at end of day: 80 dBA • Normal Business Office: 55 to 60 dBA (air diff., computer on, 1 person talking on phone) • Hotel Room with A/C unit running in room and TV turned on: 65 dBA	• It is good fire alarm system design engineering to provide audible devices that allow for adjusting their sound level output to accommodate the sound level environment they are installed in; • "OVER KILL" in dBA output can be a disaster for the END USER (installing horns, mini-horns in all spaces) • No more than one type of Fire Alarm Signaling Device may be used in an area (PA Labor & Industry). All audible alarm notification appliance devices in a facility shall be distinguishable from all other audible devices in the building (BOCA): • Horns or bells in the corridor and buzzers in the rooms may not meet this criteria; • Under most circumstances, the only practical way to achieve the required sound level to meet the ADA and applicable codes, is to install an audible notification appliance in every room and occupied space in the facility • Presently, the only practical approved audible device available, with a wide range of dBA adjustments to meet these requirements is the Fire Alarm Speaker. • Present technology allows tones to be generated on the speakers to meet the desired sound characteristics	Fire Alarm Horn • Ratings from 88 dBA to 110 dBA • Settings of "loud to louder" • Normally mid to high frequency • Multi-tone settings in field available • Relatively low current draw • Low profile - standard mounting • Low to moderate price Fire Alarm Bell • Ratings of 87 dBA to 92 dBA • Output not adjustable • Low to mid range frequency • Low current draw • Approximate same cost as a horn • Surface mounting • Large in size than a horn Fire Alarm Speakers • Ratings from 75 dBA to 120 dBA • Wide range of adjustment • Frequency of low to high • Flush and surface mount • Slightly higher cost when supplied with variable taps Speakers • Speakers are available with outputs adjustable from 75 dBA to 120 dBA • A common tone can be generated at the main control and amplified and distributed to all speaker circuits • Emergency paging can normally be added as an option • Speakers can be re-taped if changes in ambient sound level occur in the area they are installed in • Design circuits to a maximum of 75% to 80% of rated capacity to allow for ambient sound level changes

TABLE 7.5 Visual Notification Appliances to Meet the Requirements of: ADA, NFPA 72 (1993), BOCA

ADA

- Xenon strobe or equivalent
- Clear or nominal white lens color
- Minimum of 75 caldela or equivalent facilitation
- 1 to 3 Hz flash rate
- 80" AFF or 6" BFC
- No place in any room or space required to have a visual signal shall be mote than 50' from the visual signal
- In large open spaces, such as auditoriums exceeding 100' across, mount 6' AFF, spaced a maximum of 100' apart
- No place in corridors or hallways shall be more than 50' from a visual signal
- Install in restrooms, general use areas, meeting rooms, hallways, lobbies and other common use area
- ADA does not mandate emergency alarm systems
- In existing buildings, the update of the fire alarm system requires ADA compliance
- Common Use areas include:
  - Meeting and conference rooms
  - Employee break rooms
  - Classrooms
  - Cafeterias
  - Filing and photocopy rooms
  - Dressing rooms
  - Examination rooms
  - Treatment rooms
- Similar space not used solely as employee work areas

ADA (continued)

- Not required in individual offices and work stations
- Visual units not required in:
  - Mechanical, electrical, telephone rooms
  - Janitor's closets
  - Similar non-occupiable spaces
  - Non-assigned work areas
- Lamps tested at 1/3 Hz were judged ineffective. Requires a flash rate of from1 to 3 Hz
- Mounting heights from 80" to 96" AFF are considered equivalent
- Recommend 100' spacing in corridors and installed on alternate walls
- Maximize lamp intensity to minimize number of fixtures
- Lesser intensity may be sufficient as an equivalent facilitation
- Equivalent facilitation permits alternate designs
- Where a single lamp can provide the necessary intensity and coverage, multiple lamps should not be installed because of their potential effect on persons with photosensitivity
- Health Care Facilities: modify to suit industry-accepted practices (NFPA 101).

Mounting Heights
- Forward Reach: 15"-48" AFF
- Side Reach: 9"-54" AFF

UL 1971
- 1/3 Hz rate
- Allows ceiling mount.
- 15 cd corridor units

ADA
- 1 to 3 Hz rate
- No ceiling mounting
- Equivalent facilitation

NFPA	BOCA
• NFPA accepts the requirements of UL 1971 to determine compliance for visual units; • It is important to determine if the system is designed to meet the ADA or UL 1971 Guidelines Mounting Heights • Minimum of 42" - Maximum of 54"	• Required in public and common areas of all buildings housing the hearing impaired. • In Use Groups I-1 and R-1, in required accessible sleeping rooms and suites. • Sleeping room visual units shall be activated by the in-room smoke detector and building fire alarm system Mounting Heights • Minimum of 42" - Maximum of 54"

Design Criteria	Design Comments
• Synchronization of strobes when more than two strobes are installed in the same room • Keep tuned: ADA is considering changes Mounting Heights • PAL&I - Minimum of 36"-Maximum of 44"	• Check with the strobe manufacturer's data sheets to determine coverage and compliance with the ADA for corridor strobes. • Some manufacturer's 15 cd strobes may be spaced 100' apart in corridors; others require closer spacing.

FIGURE 7.2   (*a*) High forward reach limit. (*b*) High and low side reach limits.

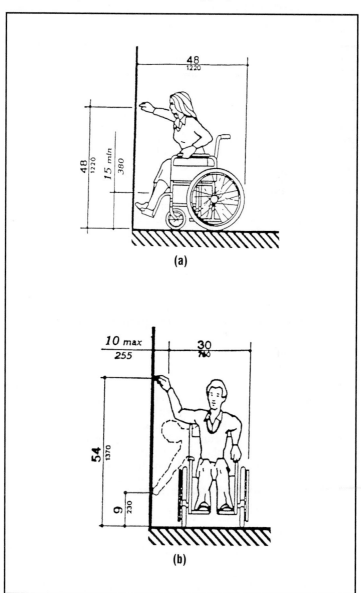

(a)

(b)

refer to and become familiar with NFPA 72, Appendix B ("Application Guide for Automatic Fire Detector Spacing") coupled with your own experience.

In the design of any fire alarm system, it is necessary to determine what codes and other requirements are applicable to the project site, what editions of the same have been adopted and are in effect at the time of design (sometimes states and/or municipalities don't adopt the latest edition of codes until several years later). It is also good practice to review the design with the AHJ periodically throughout the design process. This latter step will also be beneficial in resolving any conflicts between codes and the ADAAG (these do occur) through equivalent facilitation, thus achieving compliance with all codes and regulations that apply.

It is also essential to coordinate with the architect, structural engineer, and other trade disciplines (e.g., sprinkler systems), to determine their effects on fire alarm system requirements.

Fire alarm system technology today has reached a profoundly high level, with multiplexed digital communication, 100 percent addressable systems, and even "smart" automatic fire detectors that can be programmed with profiles of their ambient environmental conditions, thus preventing nuisance alarms by being able to discriminate between "normal" and "abnormal" conditions for their specific environment. These capabilities provide the designer with a lot of flexibility to design safe and effective fire alarm systems.

7.1 TELECOMMUNICATION-STRUCTURED CABLING SYSTEMS

Structured Cabling Design

Structured cabling is a term widely used to describe a generic voice, data, and video (telecommunications) cabling system design that supports a multiproduct, multivendor, and multimedia environment. It is an information technology (IT) infrastructure that provides direction for the cabling system design based on the end user's requirements, and it enables installation of structured cabling where there is little or no knowledge of the active equipment to be installed.

Suited to both campus and single-building installations, structured cabling consists of up to three subsystems that can be joined to form a complete network in a star topology.

The three subsystems of a structured cabling system are:

1. *Horizontal cabling:* Cabling on each building floor that connects the telecommunications outlets at the work area to a horizontal cross-connect located in a telecommunications closet; the cross-connect must be located on the same floor (the rack to the jack).

2. *Interbuilding backbone cabling:* Cabling that links the buildings in a campus environment. Each interbuilding backbone cable runs

from the main cross-connect (usually situated in the main building) to an intermediate cross-connect (closet to closet).

3. *Intrabuilding backbone cabling:* Cabling that connects each horizontal cross-connect within the same building to either a main or intermediate cross-connect.

Figure 7.3 illustrates these three subsystems.

FIGURE 7.3   (*a*) Typical commercial interbuilding and building wiring topology. (*b*) Typical commercial building wiring topology.

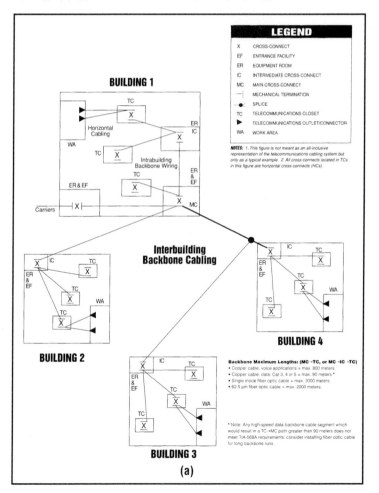

(a)

FIGURE 7.3 (*a*) Typical commercial interbuilding and building wiring topology. (*b*) Typical commercial building wiring topology. (*Continued*)

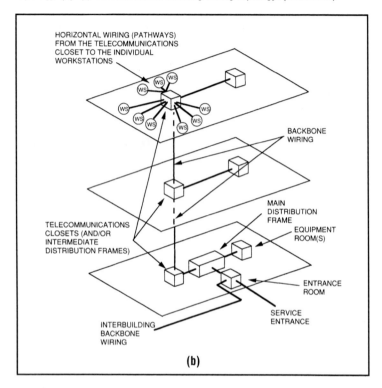

HORIZONTAL WIRING (PATHWAYS) FROM THE TELECOMMUNICATIONS CLOSET TO THE INDIVIDUAL WORKSTATIONS

BACKBONE WIRING

MAIN DISTRIBUTION FRAME

EQUIPMENT ROOM(S)

TELECOMMUNICATIONS CLOSETS (AND/OR INTERMEDIATE DISTRIBUTION FRAMES)

ENTRANCE ROOM

SERVICE ENTRANCE

INTERBUILDING BACKBONE WIRING

(b)

Industry Standards

The Electronic Industries Association (EIA) and the Telecommunications Industry Association (TIA) organized technical committees to develop a uniform set of standards for telecommunications cabling.

This resulted in the first comprehensive standards on telecommunications cabling, pathways, and spaces. These standards have been applied to many nations and have spawned additional specifications on administration, bonding, grounding/earthing, and testing, as well as the universally accepted cable and connector categories for 100-Ω unshielded twisted-pair (UTP). These standards were adopted by the American National Standards Institute (ANSI) and are known as ANSI/TIA/EIA-568-A, which is the current industry standard.

Work on cabling system standards continues to progress with new editions of ANSI/TIA/EIA-568-A and the pending ANSI/TIA/EIA-569-A, along with the introduction of an international generic cabling standard (ISO/IEC 11801) and the European generic cabling standard (CENELEC EN 50173).

Levels and Categories

To provide a common definition of types and to simplify choosing telecommunications media for various application requirements, component performance can be ascertained by its level or category rating.

THE LEVEL RATING SYSTEM

The level system was developed by "Anixter" distributors to classify cabling system components such as wire and jacks into grades of performance. The level system, given specific definition and testing by "Anixter," sparked industry demand for a single cabling system for all communications, and for better classification of product performance.

The only levels that may still be in use are Levels 1 and 2. These levels are defined by Underwriters Laboratories (UL) as follows:

Level 1: Plain old telephone service (POTS)
Level 2: IBM Type 3 cabling system

THE CATEGORY RATING SYSTEM FOR UTP

The category rating system was developed by the TIA in response to industry demands for higher data rate specifications on applications over unshielded twisted-pair.

The category information was released in two Technical Service Bulletins, TSB-36 and TSB-40. The TSBs recommended changes and additions to EIA/TIA-568 and added the category rating system, replacing the old level system. TSB-36 covered additional specifications for UTP cables. TSB-40 added specifications for connecting devices, such as jacks, cross-connect blocks, and patch panels. These TSBs have now been fully integrated into the main body of the ANSI/EIA/TIA-568-A standard document.

The categories characterize commercial building wiring systems as follows:

Category 3: 16 MHz (10 Mbps) 100-Ω UTP
Category 4: 20 MHz (16 Mbps) 100-Ω UTP
Category 5: 100 MHz (100 Mbps) 100-Ω UTP

You will notice that the category rating system only applies to 100-Ω UTP wiring systems. However, ANSI/EIA/TIA-568-A does allow 150-Ω STP and 62.5/125-μm multimode optical fiber. TSB-53, "Extended

Specifications for 150 Ohm STP Cables and Data Connectors," extends the 150-Ω cabling system from 20 MHz to 300 MHz.

Interpreting Compliance and Performance Test Data

To determine compliance with category specifications, cable and connecting hardware (jacks) must fulfill certain parameters as defined by ANSI/EIA/TIA-568A. For full category compliance, jacks and cable must meet ANSI/EIA/TIA-568A electrical and mechanical specifications and transmission requirements.

Attenuation, near-end cross talk (NEXT), and return loss are signal-degrading factors that greatly affect a system's transmission capability. The proportions of these factors determine a cable or connector's category of compliance.

Attenuation is the loss of signal strength during transmission, in which the received signal is lower in strength than the transmitted signal due to losses in the transmission medium (such as caused by resistance in the cable). *NEXT* is a distortion of the incoming signal, caused by the coupling of noise from one pair to another. *Return loss* is the measure of the similarity of the impedance of a transmission line and the impedance at its termination.

Consider the following when reading test data for category-rated products:

1. Theoretically, there is no advantage or disadvantage in one product's test results being a few decibels higher or lower than another, as long as the value is within the specifications of a given category. However, a few decibels higher is better.

2. Also, it is important to distinguish between full category *compliance* and category *performance.* Category compliance is the case when a cable or jack fully meets all of the specifications of a given category and fully complies with TIA-568A. Category performance is the case when the jack meets the transmission requirements of a category, but may not meet all the specifications of EIA/TIA-568A (such as mechanical or wiring specifications).

Overview Specs for Levels, Categories, and Emerging Standards
LEVEL 1

Level 1 wire types:

• One-hundred-ohm UTP is preferable, and required for any multiline application.

• "Quad" wire is not recommended for data or network installations, but can function adequately in certain limited situations (i.e., single-line analog voice applications in which it has already been installed).

Level 1 technical requirements defined by:

- FCC Part 68
- ICEA S-80-576
- Bellcore 48007

Level 1 performance criteria:

- None specified

Level 1 safety requirements defined by:

- UL 1459 (telephone)
- UL 1863 (wire and jacks)
- NEC Article 800-4

LEVEL 2

Level 2 wiring types:

- 100-Ω UTP

Level 2 technical requirements defined by:

- FCC Part 68
- GA27-3773-1, IBM Cabling System Technical Interface

Level 2 safety requirements:

- UL 1459 (telephone)
- UL 1863 (wire and jacks)
- NEC Article 800-4

Typical applications include voice and ISDN.

CATEGORY 3

Category 3 is characterized to 16 MHz, to support applications up to 10 Mbps.

CATEGORY 3 WIRING TYPES:

- 100-Ω UTP rated, Category 3

CATEGORY 3 TECHNICAL SPECIFICATIONS DEFINED BY:

- FCC Part 68
- ANSI/EIA/TIA-568A

CATEGORY 3 SAFETY REQUIREMENTS DEFINED BY:
- UL 1459 (telephone)
- UL 1863 (wire and jacks)
- NEC Article 800-4

Typical applications are voice, ISDN, 4-Mbps Token Ring, and 10Base-T.

CATEGORY 4

Category 4 defines cabling system requirements to support 20 MHz.

CATEGORY 4 WIRING TYPES:
- 100-Ω UTP, rated Category 4

CATEGORY 4 TECHNICAL REQUIREMENTS DEFINED BY:
- FCC Part 68
- ANSI/EIA/TIA-568A

CATEGORY 4 SAFETY REQUIREMENTS:
- UL 1459 (telephone)
- UL 1863 (wire and jacks)
- NEC Article 800-4

Typical applications are from voice to 16-Mbps Token Ring.

CATEGORY 5

Category 5 is a further extension of the ANSI/EIA/TIA-568A cabling system to 100 MHz.

CATEGORY 5 WIRING TYPES:
- 100-Ω UTP, rated Category 5

CATEGORY 5 TECHNICAL SPECIFICATIONS:
- UL 1459 (telephone)
- UL 1863 (wire and jacks)
- NEC Article 800-4

Other Network Applications

ONE-HUNDRED-FIFTY-OHM STP-WIRING SYSTEMS

STP is used extensively in Europe and may become the standard for frequencies over 100 MHz due to several factors, including radiation over FCC limits and EMI protection.

One-hundred-fifty-ohm wiring systems specifications are defined by ANSI/EIA/TIA-568A.

TP-PMD

Twisted pair–physical medium dependent (TP-PMD) is an application standard from an EIA/TIA committee called the X3T9.5 working group. It is a proposed implementation of fiber-distributed data interface (FDDI) on 100-Ω UTP. The TP-PMD cabling system provides a way to achieve Category 5 performance (100-Mbps transmission) using only the outside two pairs of an eight-position jack (see Figure 7.4).

The performance on the outside two pairs will be more than sufficient to meet the requirements of 100-Mbps applications (with lower performance on the two unused inside pairs). Note, however, that it cannot be considered an EIA/TIA-568A-compliant system.

The physical separation of the two outer pairs reduces near-end cross talk to a level that meets the TIA-568A transmission specifications for 100 Mbps. This allows jacks that otherwise do not comply with Category 5 requirements to support 100-Mbps applications.

As is typical with application standards, X3T9.5 is not defining the cabling system, but is defining a minimum set of requirements of the cabling system. TIA's Category 5 cabling system exceeds these requirements. TP-PMD wiring system is 100-Ω Category 5 UTP.

100VG-ANYLAN

100VG-ANYLAN is now an approved standard that was proposed by Hewlett-Packard and AT&T Microsystems to the IEEE 802.12 committee. (It is so named because it is based on 100 Mbps, voice-grade cable; it's called "ANYLAN" for its ability to support both Ethernet and Token Ring).

The 100VG-ANYLAN protocol is for a 100-Mbps half-duplex transmission, which allows 100 Mbps on a four-pair Category 3 cabling system, but is not based on the 802.12 Ethernet CSMA/CD (Carrier Sense Multiple Access with Collision Detection) protocol.

FIGURE 7.4 TP-PMD jack wiring.

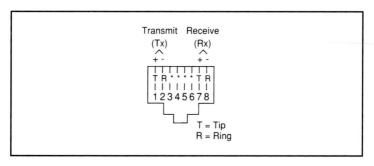

100BASE-T (FAST ETHERNET)

The IEEE 802.3 committee has also approved this standard for a 100-Mbps full-duplex Ethernet application on Category 5 and possible TP-PMD-style wiring systems. This protocol goes by the name 100Base-T (meaning, "based on 100 Mbps, TP-PMD wiring").

There are three variations: 100Base-T4 for Categories 3, 4, and 5; 100Base-TX specifically for Category 5 applications; and 100Base-FX for fiber. Note that backbone cable distances for category-rated UTP are restrictive—for example, if Category 5 cable is used for a 100Base-T backbone, it cannot exceed 90 m.

ATM

When first introduced, Asynchronous Transfer Mode (ATM) was a proposed application for a copper network capable of 155-Mbps data rates and possibly higher (300 and 600 Mbps). Since then, low-speed ATM transmission specifications have also been proposed: IBM recommends a 25-Mbps transmission, and Hewlett-Packard and AT&T recommend a 51-Mbps transmission specification.

The ATM proposals are being developed by a forum of over 120 application suppliers. Physical specifications for ATM links have been defined (but not yet released) in a document called *AF-PHY-0015.000*. Basically, ATM applications must meet all requirements for Category 5 as specified by TIA-568A.

ONE-HUNDRED-OHM UTP CABLING SYSTEM APPLICATIONS SUMMARY

Table 7.6 shows the minimum 100-Ω UTP cabling system required for several applications. Some require an external impedance-matching device, called a Balun, to correctly couple the application signal to the cabling system and are identified in Table 7.6.

7.2 HORIZONTAL DISTRIBUTION PRACTICES (COPPER CABLING)

Horizontal Cabling

Horizontal cabling in ANSI/EIA/TIA-568A is a generic cabling system. If horizontal cabling is installed according to the practices in the standard, the system will be sufficient for 100 MHz at 100-Mbps (minimum) applications.

For the purposes of this section, horizontal cabling is considered the cabling from the work area to the telecommunications closet. It includes cross-connects in the telecommunications closet, horizontal cable, and the outlet at the work areas.

TABLE 7.6   Summary of Applications for 100-Ohm UTP Cabling Systems

DESIRED → APPLICATION	Level 1 components / Level 1 cable	Level 2 components / Level 2 cable	Cat 3 jack / Cat 3 (or higher) cable	TP-PMD-wired Cat 3 Jack / Category 5 cable	Cat 4 Jack / Cat 4 (or higher) cable	Cat 4 jack, TP-PMD wiring / Cat 5 cable	TP-PMD jack, no Cat rating / Cat 5 cable	Cat 5 jack / Cat 5 cable
Voice	●	●	●	●	●	●		●
Modems	●	●	●	●	●	●		●
RS-232	●	●	●	●	●	●		●
Apple Talk	●	●	●	●	●	●		●
5250 w/baluns		●	●	●	●	●		●
3270 w/baluns		●	●	●	●	●		●
ISDN			●	●	●	●		●

MINIMUM REQUIRED CONNECTING DEVICE & CABLE:

TABLE 7.6 Summary of Applications for 100-Ohm UTP Cabling Systems (*Continued*)

T1	•	•	•	•			•
Wang w baluns	•	•	•	•			•
4 mbps Token ring *	•	•	•	•			•
10BaseT	•	•	•	•			•
100 Base-VG	• (1)	• (1)	• (1)	• (1)			• (1)
16 mbps token ring *		•	•			•	•
TP-PMD						•	•
100BaseT		• (1)	• (1)	• (1)	• (1)	• (1)	• (1)
ATM (155 mbps)		• (1)	• (1)	• (1)	• (1)	• (1)	• (1)

(1) 100BASE-VG, 100BASE-T and ATM are defined but unreleased. 100BASE-VG proposes to use all 4 pairs of a Category 3 cabling system in a 4 channel, half duplex transmission scheme. 100BASE-T and ATM are evaluating the TP-PMD cabling scheme. * May require an external impedance-matching device.

Leviton Telcom • 2222 - 222nd St. SE • Bothell, WA U.S.A 98021-4422 • Phone: 1 (206) 486-2222 • U.S.A. Fax: 1 (206) 485-3373 • Int'l Fax: 1 (206) 485-9170   1-9

Note that the scope of ANSI/EIA/TIA-568A covers only the building's cabling systems, not the pathways such as conduit or raceways. Commercial building pathways and spaces for telecommunications wiring are covered in a separate standard, TIA-569. Grounding is covered in TIA-607.

Benefits of ANSI/EIA/TIA-568A Compliance

Commercial building horizontal cabling that is installed in accordance with this standard is like having one type of foundation that can support any type of structure that is built upon it—even if the structure keeps changing. The benefits are that you have one system that will:

- Simplify ongoing maintenance, relocation, and addition
- Accommodate future equipment and service changes
- Accommodate a diversity of user applications, including voice, data, LAN, video, switching, and other building services
- Large base of vendors whose equipment is based on Category 5.

Star Topology

ANSI/EIA/TIA-568A specifies a star topology—a hierarchical series of distribution levels. In the backbone are the main distribution frame (MDF) and the optional intermediate distribution frame (IDF), as shown in Figure 7.3. Only one IDF is allowed between the MDF and the telecommunications closet.

The first-level backbone, the MDF, links to other hubs via the backbone cabling. The MDF may link to the third and final level of backbone, the telecommunications closet (TC) directly, or in large installations it may link to some TCs via an optional second-level backbone, the IDF. The TC terminates the backbone cable and cross-connects to the horizontal cabling. The horizontal cabling terminates in the work area. The TC and the work area must be on the same floor.

EQUIPMENT LOCATIONS

Communications equipment (phones, fax machines, computers, and so on) may be located in any space—work areas, TCs, distribution frames, or a separate space called an equipment room.

NETWORK TOPOLOGIES

Bus, tree, and ring topologies are implemented in the TC or other cross-connects rather than directly between the work areas. Therefore, the network can be determined or changed without affecting horizontal cabling. Application distance limitations must be checked.

Maximum Horizontal Distances

- Maximum cable length from the mechanical termination of the media in the TC to the telecommunications outlet is 90 m (295 ft), independent of media type.

- Splices and bridged taps are not allowed as part of the horizontal cabling.

- Only one transition point (TSB-75, Consolidation Point) is allowed between flat undercarpet cables and one of the recognized horizontal cables.

- It is suggested that the maximum equipment cable length from the telecommunications outlet to the work area equipment be limited to 3 m (10 ft). Work area cords, however, are outside the scope of the standard.

- In addition, the maximum cable length for jumpers and patch cords in the TCs shall be limited to 7 m (23 ft), with no single cord exceeding 6 m (see Figure 7.5). A maximum of two patch cords is allowed per horizontal run.

FIGURE 7.5   Maximum horizontal distances.

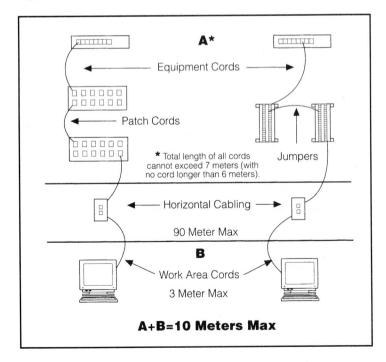

- It should be noted that the ANSI/TIA/EIA-568A performance is based on 90-m maximum horizontal cabling plus 10-m maximum of patch cords, jumper cables, and work area cords.

- It is suggested that equipment cables meet or exceed patch cable performance requirements.

Horizontal Cabling Systems

1. There must be a minimum of two cabling runs from the TC to each individual work area. The requirement of having two cabling runs is due to the importance of both voice and data telecommunications in a commercial building, and to allow implementation of bus and ring topologies. *TIA-568A realizes that most work areas will require both voice and data telecommunications within the lifetime of the cabling system, and so requires that all work areas be wired with a minimum of two outlets.*

2. The required two horizontal cabling runs to each work area shall be as follows:

    a. One shall be a four-pair 100-Ω UTP.

    b. The other/second shall be one of:

        (1) Four-pair, 24-AWG, 100-Ω UTP

        (2) Two-pair, 150-Ω STP

        (3) Two-strand, 62.5/125-μm optical fiber

3. Hybrid cables (consisting of more than one of the above recognized cables under a common sheath) may be used in the horizontal cabling *provided* that they meet the hybrid cable requirements in TIA-568A.

4. Coaxial cable (allowed, but not part of a compliant system).

Horizontal Cabling System Selection

There are three types of cable recognized for standard-compliant installation: 100-Ω UTP, 150-Ω STP, and 62.5/125-μm optical fiber.

One-hundred-ohm UTP is the most universal cabling system and generally the least expensive. It covers all applications up to 100 MHz with a minimum of cost. The user must decide which category of 100-Ω UTP cabling system is needed for the application. For voice cabling systems, Category 3 is sufficient; for data cabling systems, Category 5 is highly recommended.

One-hundred-fifty-ohm STP is usually installed as a hybrid system, consisting of one 150-Ω STP data cabling system and one 100-Ω UTP Category 3 voice cabling system under one sheath. One-hundred-fifty-ohm STP is usually used for Token Ring applications, but the extended-bandwidth 150-Ω STP has application hardware for broadband video up to 300-MHz and 155-Mbps ATM.

Optical fiber is typically the most expensive cabling system to install, but it has the widest bandwidth (in excess of 1 GHz). Although optical fiber is not practical for voice and other low-bandwidth applications in the horizontal, optical fiber should be the cabling system choice for high-bandwidth applications, such as FDDI, ATM, broadband video, and multiplexed signals. Fiber is also best used for minimizing EMI interference and grounding isolation.

Cabling System Components

Each cabling system is composed of four main components:

- Telecommunications outlet [work area outlet (WAO)]
- Horizontal cable
- Cross-connect hardware
- Patch cables, equipment cables, and jumpers

  Telecommunications outlet: In the work area and provides access to the building telecommunications cabling system

  Horizontal cables: Connect the work area outlet to the cross-connect system in the telecommunications closet

  Cross-connect hardware: Terminates the horizontal cable, backbone cable, and equipment in the telecommunications closet

  Patch cables and jumpers: Connect cross-connect hardware in the TCs

  Equipment cables: Connect telecommunications equipment to the outlet in the work area or to the cross-connects in the TC

- Per the NEC, each component must be listed for the purpose.
- All components must comply with FCC Part 68 (CS-03 in Canada).

Other Cabling Systems

- If you need to include other cabling systems but still wish to have a TIA-568A-compliant installation, other cabling systems may be included in the work area, as long as they are in addition to the two TIA-568A-required cabling systems.

Grounding and Bonding Considerations

Grounding and bonding systems are normally an integral part of the specific application or telecommunications cabling system that they protect. They protect personnel and equipment from hazardous voltage, and they reduce the effect of electromagnetic interference (EMI) to and from the telecommunications cabling system. Improper grounding and bonding can induce voltages that disrupt telecommunications circuits.

- Grounding and bonding shall meet the requirements of the NEC as a minimum and comply with more stringent requirements if imposed by authorities and/or codes having jurisdiction.
- Additionally, grounding and bonding shall conform with TIA-607 requirements.
- Grounding and bonding instructions and requirements of the equipment manufacturer should also be followed.

One-Hundred-Ohm UTP Cabling Systems

One-hundred-ohm UTP cabling systems currently are the most versatile and often the most cost-effective. They cover almost all applications up to 100 MHz with a minimum of cost.

One-hundred-ohm UTP cabling system components have been categorized into performance groups. Each performance group characterizes the performance of the components up to a specific frequency:

Category 3: Up to 16 MHz (10 Mbps)
Category 4: Up to 20 MHz (16 Mbps)
Category 5: Up to 100 MHz (100 Mbps)

A general rule of thumb is to use Category 3 for voice cabling systems, and Category 5 for data cabling systems. As a minimum for any category-rated installation, make sure *all* components are at least of the minimum category required.

WORK AREA OUTLETS

- Screw terminations for cable outlets generally limit an outlet to Category 3 performance.
- Category 4 outlets usually terminate on an insulation displacement connector (IDC).
- Category 5 outlets always use an IDC for cable terminations.

WORK AREA OUTLET INSTALLATION

- Each four-pair cable shall have all pairs terminated on an eight-position jack.
- Pin/pair assignments shall be as per T568A or T568B (Figure 7.6).
- Maintain the twists of the cable as close to the termination on the outlet as possible, to maintain the transmission characteristics of the category. Category specifications require that pair twisting be maintained to within the following distances from the outlet termination:

Category 3 maximum allowed untwisting: 3 in

Category 4 maximum allowed untwisting: 1 in

Category 5 maximum allowed untwisting: ½ in

FIGURE 7.6 ANSI/EIA/TIA-568A-compliant wiring configurations for pin/pair assignments.

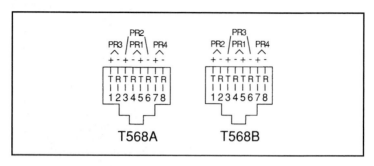

T568A   T568B

- Leave a sufficient service loop of the horizontal cable for future adds, moves, and changes. Usually ⅓ to 1 m (1 to 3 ft).
- Each telecommunications outlet must comply with pair color codes or have a conversion chart shipped with each outlet. Pair color codes are shown in Table 7.7.
- The bend radius of the cable must be no tighter than four times (4×) the cable's outside diameter.
- For multipair cable (more than four pairs; typically 25 pairs), the minimum-allowed bend radius is 10 times the outside diameter.

OUTLETS FOR CABLING SYSTEMS BEYOND THE MINIMUM TIA-568A

- Any 100-Ω UTP modular jack or adapter may be added to the minimum cabling system outlet, as long as the minimum cabling system for TIA-568A is met. Baluns, six-position USOC, or MMJ outlets may be added to the work area telecommunications outlet as required by the specific site.

ONE-HUNDRED-OHM UTP CABLING SYSTEMS

- The unshielded inside cable used in the horizontal cabling system is 24 AWG.
- Four-pair 22-AWG cables may be used if they meet the physical transmission requirements of the desired category.
- Four-pair screened twisted-pair (100-Ω STP) cables may be used if they meet the physical and transmission requirements of the desired category.
- Undercarpet cables may be used for certain applications, but only one transition point (TSB-75, Consolidation Point) from the round cable to flat undercarpet cable is permitted on any horizontal run.

TABLE 7.7 Telecommunications Wiring Color Codes

Wire Pair # and Lead Functions	Banded Colors	Semi-Solid Colors	Cat 5 Solid Colors (tightly twisted together)	8-Position T568A Jack Pin #	8-Position T568B Jack Pin #	6-Position Jack Solid Colors*	6-Position RJ25 USOC Jack Pin *
1 Tip	White-Blue	White-Blue	White	5	5	Green	4
1 Ring	Blue-White	Blue	Blue	4	4	Red	3
2 Tip	White-Orange	White-Orange	White	3	1	Black	2
2 Ring	Orange-White	Orange	Orange	6	2	Yellow	5
3 Tip	White-Green	White-Green	White	1	3	White	1
3 Ring	Green-White	Green	Green	2	6	Blue	6
4 Tip	White-Brown	White-Brown	White	7	7		
4 Ring	Brown-White	Brown	Brown	8	8		

Note: The wire insulation is white, and a colored marking is added for identification (see Book 2, Section 2.1.1 on band-striped color marking). For cables with tightly twisted pairs (all less than 38.1mm [1.5 inches]) the mate conductor may serve as the marking for the white conductor. A white marking is optional. *Added for informational purposes only; does not comply with EIA/TIA-568A.

467

Undercarpet cables shall meet ANSI/IPC-FC-21 and must be listed for that purpose.

CABLE INSTALLATION

- For TIA-568A-compliant installations, do not exceed 25-lb pulling tension on the cable (four-pair, 24-gauge).
- Do not chafe or damage the outer jacket of the cable.
- Installation in colder climates may require cables with special jackets.
- The wire color code shall be as shown in Table 7.7.

TC CONNECTING HARDWARE

The TC is where connecting hardware for 100-Ω UTP cable is installed as a means of connecting the horizontal cabling to the backbone cabling or equipment.

Two types of cross-connects are common: *patch panels* and *cross-connect blocks.*

Patch panels often have the backbone cable, horizontal cable, or electronic equipment cord directly terminated on the cable terminations. Cross-connecting is achieved by patch cords.

Cross-connect blocks are usually IDC connections with the electronic equipment cords, horizontal cables, and backbone cables terminated on one side. The cross-connect jumpers terminate to the other side of the block, and between blocks to complete the cross-connect.

- It is desirable that hardware used to terminate cables be of the IDC type (see Table 7.8).

TABLE 7.8   Termination Hardware for Category-Rated Cabling Systems

Termination Hardware	Category 3	Category 4	Category 5
Screw terminals	(1)	-	-
25 pair connector	(2)	(2)	(2)
66-clip	Yes	Yes	(2)
110	Yes	Yes	Yes
Krone®	Yes	Yes	(2)
BIX®	Yes	Yes	(2)

Note (1): If the application specifically requires it.
Note (2): Some versions comply; check with the manufacturer.

CONNECTING HARDWARE INSTALLATION

- Install connecting hardware in a neat, well-organized manner.
- Organize connecting hardware into connecting fields for ease of administration.
- Document the installation, and use color coding and labeling.
- Preserve wire pair twists as closely as possible to the point of mechanical termination, to minimize signal impairment.

CROSS-CONNECT JUMPERS, PATCH AND EQUIPMENT CORDS

- The summed lengths of the jumpers, patch cords, and equipment cords should not exceed 23 ft (7 m) in length in the TC.
- It may be preferable to buy premanufactured patch and equipment cords made to required lengths.
- The twists of the individual pairs must be maintained up to and into the plug. This is especially crucial for Category 5 applications (see Table 7.9).
- Modular plugs for solid wire provide the best connection on TIA-compliant patch cord cable.
- Use only the modular plug crimping tool recommended by the plug manufacturer.

TABLE 7.9   Patch Cord Wire Color Codes

Conductor Identification (1)	Wire Color
Pair 1 + Pair 1 -	White (2) Blue (3)
Pair 2 + Pair 2 -	White (2) Orange (3)
Pair 3 + Pair 3 -	White (2) Green (3)
Pair 4 + Pair 4 -	White (2) Brown (3)

Notes: (1) + = Tip, - = Ring
(2) Mostly white wire may have the associate color as a band or stripe.
(3) Mostly colored wire may have white as a band or stripe.

- Patch cord cable must be category-compliant. Tinsel cordage (silver satin) is not acceptable.

CROSS-CONNECT JUMPERS

- TIA-568A requires that all jumper cables comply with category transmission requirements.

PATCH CORDS

- TIA-568A requires all patch cords to comply with category transmission requirements. However, patch cords are allowed additional attenuation so that a more "lossy" (less stringent attenuation characteristics) flexible cable may be used.
- It is recommended that stranded, twisted conductor patch cords be used. Ordinary flat, solid, silver satin patch cords are allowed by TIA-568A, but they do not comply with any category.
- Patch cords do not reverse the wires with the plugs. Pin 1 of end 1 connects to pin 1 of end 2. See Figure 7.7, panel *b*.

EQUIPMENT CORDS

- Although equipment cords are supplied by the equipment vendor, TIA-568A does require that they meet the same performance criteria as patch cords, and comply with category transmission requirements. The only exception is for analog telephones.

FIGURE 7.7   Patch cords. (*a*) Reversed. (*b*) Straight-through pin-outs.

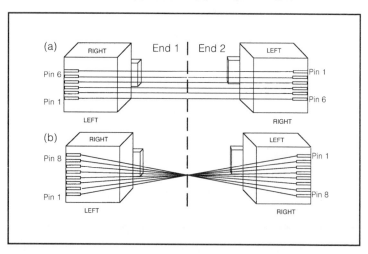

- Maximum length for work area equipment cords is 10 ft (3 m).
- In the telecommunications closet, the summed lengths of the equipment and patch cables and jumpers together should not exceed 7 m (23 ft) total.
- Equipment cords for data applications usually do *not* reverse the wires with the plugs. Pin 1 of end 1 connects to pin 1 of end 2. See Figure 7.7, panel *b*.
- Equipment cords for analog telephones usually reverse the wires with the plugs (i.e., pin 1 of end 1 connects to pin 6 of end 2). See Figure 7.7, panel *a*.

One-Hundred-Fifty-Ohm STP Cabling Systems

Two 150-Ω cabling systems exist today, which are used extensively in Europe. The initial (and unnamed) version is characterized to 20 MHz and is applicable for Token Ring applications up to 16 Mbps. The new version is the "Extended" 150-Ω STP, sometimes called "1A" (in reference to the extended-version designator for the new Type 1 cable).

The Extended 150-Ω system is characterized to 300 MHz and is not only applicable for Token Ring applications up to 100 Mbps, but is also being suggested for broadband video. TSB-53 defines the Extended 150-Ω STP. In TIA-568A, the Extended cabling system replaces the initial version.

Extended components will be designated either by an "A" following the type designation on the cables, or an "E" or the word "EXTENDED" on the cable and connector.

Upgrading existing 150-Ω STP cabling systems to the Extended 150-Ω cabling system will not usually require replacing the cable. This should be verified with the vendor. The connector, however, will still need to be replaced, as the initial connector falls short of meeting requirements for Extended systems.

OUTLETS

- The telecommunications connector to be used for terminating the 150-Ω STP cable shall be that specified by ANSI/IEEE 802.5 for the media interface connector.
- The new Extended version (specified in TSB-53) will be mateable with the old version. It is recommended that the Extended version be used in all new installations.

OUTLET INSTALLATION

This connector generally is installed directly on the horizontal cable at the work area or in the TC.

- It is suggested that a 1- to 3-ft ($\frac{1}{3}$- to 1-m) service loop be added at both locations for adds, moves, and changes.

CABLE

• One-Hundred-Fifty-ohm STP cable must meet the requirements of EIA Interim Standard Omnibus Specification, NQ-EIA/IS-43, and the Detail Specifications listed in the standard.

CABLE INSTALLATION

• Do not exceed the manufacturer's recommendations for pulling tension.
• Do not chafe or damage the outer jacket of the cable.
• Installation in colder climates may require cables with special jackets.

TC CONNECTING HARDWARE

Patch panels and passive or electronic hubs are the usual cross-connect hardware. Cross-connect blocks are rarely used and are not recommended. Termination of the backbone and horizontal cables is usually to either a 150-Ω STP media interface connector, or to an IDC on the patch panel.

CROSS-CONNECT PANELS

Patch panels are generally one of two types: an open panel with a hole for the 150-Ω STP media interface connector to snap into, or a panel with IDCs for termination of the building cables.

• In either case, follow the manufacturer's recommendations for termination.
• Allow 1 to 3 ft ($\frac{1}{3}$ to 1 m) of service loop for future adds, moves, and changes.
• For 19-in (483-mm) rack-mounted cross-connect panel installations, allow room on the rack for possible telecommunications equipment associated with the 150-Ω STP cable.
• Racks should have at least the following clearances for access and cable dressing space:

  30 in (762 mm) in the rear

  36 in (915 mm) in the front

  14 in (356 mm) on the side

HUBS

Passive or active hubs usually are connected via the 150-Ω STP media interface connectors and patch cords to the horizontal cabling. Backbone cables may be optical fiber or 150-Ω STP cables, and are usually connectorized and connected directly to the hub to minimize connections. In either case, follow the recommendations for 150-Ω patch panels.

PATCH AND EQUIPMENT CORDS

- One-hundred-fifty-ohm STP patch and equipment cords are usually purchased items and are not normally constructed in the field.
- If field construction is required, follow the equipment, patch panel, or hub vendor's recommendation.
- Patch cord length should be limited to 23 ft (7 m).
- Equipment cord length should be limited to 10 ft (3 m).

7.3 HORIZONTAL DISTRIBUTION PRACTICES (FIBER-OPTIC CABLING)

Fiber-Optic Technology Overview

Fiber optics is a technology in which electrical video/data signals are converted into light, and then the light is beamed through an optical fiber to transport the information from one point to another. The fiber is a thin filament of glass through which the light travels.

Because one optical fiber can do the job of hundreds of copper cables, optical fiber is increasingly being used for communications signaling systems. Applications such as FDDI, ATM, and broadband video are currently optical fiber–based. Also, fiber provides many advantages over copper wiring.

FIBER-OPTIC ADVANTAGES

Benefits of fiber over copper include the following:

- *Noise immunity:* Fiber is unaffected by EMI because it is a dielectric, whereas copper wire requires shielding.
- *Low attenuation (signal loss):* Current single-mode fibers have losses as low as 0.2 dB/km. Multimode losses can be less than 1 dB/km. This permits longer spans without the use of repeaters.
- *No possibility of short circuits:* The fiber is glass and does not carry electrical current, radiate energy, or produce heat or sparks.
- *Greater transmission security:* The fiber does not radiate electromagnetic pulses, radiation, or other energy that can be detected. It is not possible to tap into fiber-optic cable without creating a loss that is easily detected at the receiver.
- *Fiber is becoming more price competitive to copper:* Currently, it is only about a 10 percent premium (cable and electronic connections are more expensive) over the cost of copper.
- *Fiber carries more data much faster and in less space:* The current state-of-the-art optical fiber system is known as SONET/SDH, level

OC 192. It uses a single-mode fiber that transmits 129,024 voice channels at 9.953 Gbit/s—which is more than 120,000 times the information of electronic signals over copper wire. Besides having plenty of capacity (bandwidth) for current trends, fiber's large bandwidth also leaves room for capabilities beyond current technologies. Also, fiber takes up much less space and is more lightweight than copper cable of comparable performance, eliminating crowded conduits and other raceways.

• *Fiber is a mature, established, standardized industry:* Whereas copper applications keep pushing the limits of UTP (and thus constantly redefining and revising its relatively young TIA standards), the fiber industry has been around for decades and has long since established its standards and installation practices, including testing.

FIBER-OPTIC DISADVANTAGES

Time is sometimes an enemy to fiber installations because fiber takes longer to connectorize than copper, and requires more skilled labor and specialized tools.

Until recently, fiber's main disadvantage has been a much higher price. Lately, however, increased competition, larger manufacturing volumes, and standardization of common products have eroded the price to a margin that is very competitive with copper.

Fiber Cable Construction

An optical fiber is composed of three united layers. In the center is the *core* filament of glass, which transmits the light signals. This is surrounded by *cladding* to keep the optical signal within the core. Covering these is a plastic *inner coating,* which protects the glass from abrasion and adds strength. These layers make up what is called the *fiber* (see Figure 7.8).

To make the fiber into fiber-optic cable (Figure 7.9), certain materials are added. A variety of *buffers,* or *outer coatings,* are added by the cable manufacturer. Common fiber-coating sizes are 250 and 900 μm. The 250-μm size is used in outdoor cables, whereas the 900-μm size is used for indoor cable designs.

Kevlar®, a very strong aramid yarn made by DuPont, is used in cables as a strength member to both protect the fiber and facilitate pulling the cable through ducts or building riser sleeves. It should be noted that strength members can be metallic or nonmetallic.

Note that the cores can consist of plastic or glass, as can the cladding. Most common fibers used today have a silica glass core and cladding; however, the use of plastic fibers for short-distance links (100 m, or 328 ft) is increasing.

FIGURE 7.8 Optical fibers.

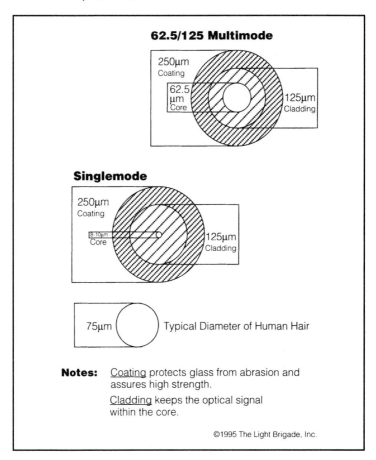

CABLE TYPES

Whereas all fiber cables share the same basic construction, there are multimode versions that differ in their proportion of core to inner cladding. This proportion is designated in the cable name by the two numbers separated by a slash. See Figure 7.8.

The TIA-568A standard recognizes two fibers: the 62.5/125-μm for multimode, and the 9/125-μm for single-mode applications. Other types are considered alternate fibers acceptable for installation. Note that 50/125-μm is gaining popularity to replace 62.5/125-μm cable.

FIGURE 7.9 Fiber cable construction.

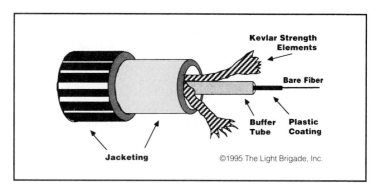

Kevlar Strength
Elements

Bare Fiber

Buffer   Plastic
Tube   Coating

Jacketing   ©1995 The Light Brigade, Inc.

SINGLE-MODE AND MULTIMODE

- A *mode* is the light path in the optical fiber.
- A *single-mode fiber* has a small core that allows only one light path, resulting in a very focused, high-quality signal. However, the core is so small that it usually requires a laser source to "launch" light into the cable—and laser sources are usually more expensive than the LED sources used for multimode fibers.
- A *multimode fiber* has a larger core diameter. The larger the core, the more modes it transmits. In other words, more light can be launched into a multimode core. However, the increase in the amount of modes actually decreases the optical bandwidth of a fiber, as the signal is less concentrated. Multimode fibers are very good at coupling light from inexpensive LEDs.

DISTRIBUTION AND BREAKOUT STYLES

- *Distribution cables* have smaller diameters and tend to work better in confined spaces. They are typically used in vertical backbone.
- *Breakout-style cables* tend to be used for horizontal runs because they have internal Kevlar®-strength members, which allow for direct connectorization onto the cable.

LOOSE-TUBE AND TIGHT-TUBE STYLES

- A plastic *buffer tube* surrounds the fiber, providing mechanical isolation to help protect it from crushing and impact loads, and to some extent from the macrobending induced during cabling operations. Buffer tubes can be either tight-tube or loose-tube/gel-filled.

- In a *tight-tube* cable construction, the buffer tube is an extrusion of plastic applied directly over the basic fiber coating. This type is used for indoor applications.
- In a *loose-tube* cable construction, the buffer tube's inner diameter is much larger than the fiber itself, leaving a gap between the tube and the fiber. This extra space is often filled with a gel to cushion the fiber, further isolating it from any exterior mechanical forces acting on the cable, and preventing moisture intrusion. This cable structure handles stress and environmental changes better than indoor-rated cable, and is primarily used for outdoor applications.

PERFORMANCE CRITERIA FOR COMPARING CABLES

The performance characteristics of fiber cables vary depending upon the materials used and the process of manufacturing. The following criteria are useful in verifying performance when choosing a fiber cable:

- *Attenuation:* A lower number means less signal loss.
- *Bandwidth:* A higher number means more capacity.
- *Numerical aperture (N.A.):* A lower number means more bandwidth. But note that less power will be coupled into the fiber.
- *Core size:* Follow these three rules of thumb:

  The smaller the core, the lower the attenuation

  The smaller the core, the higher the bandwidth

  The smaller the core, the lower the cable cost

- However, as the core size decreases, the price of the required connectors and equipment increases.
- *Environmental factors:* Temperature ratings, tensile stress, bend radius requirements, and so forth.

Optical Fiber Transmission

Starting with the source of a transmission signal, when a data signal travels through a fiber-optic system, it first goes into a *transmitter,* which converts the electrical signal to light pulses. Note that on schematic drawings, this transmitter is often designated the E/O (electrical/optical) device.

The optical fiber carries the signal from source to receiver. Along the way, it will be routed through various jumpers, cross-connect panels, and distribution panels located in various sites throughout a building or complex, until it gets to the receiver (see Figure 7.10).

The signal may also pass through a *multiplexer (MUX),* an electronic device that allows two or more signals to pass over one communications

FIGURE 7.10 The signal path through a fiber-optic system.

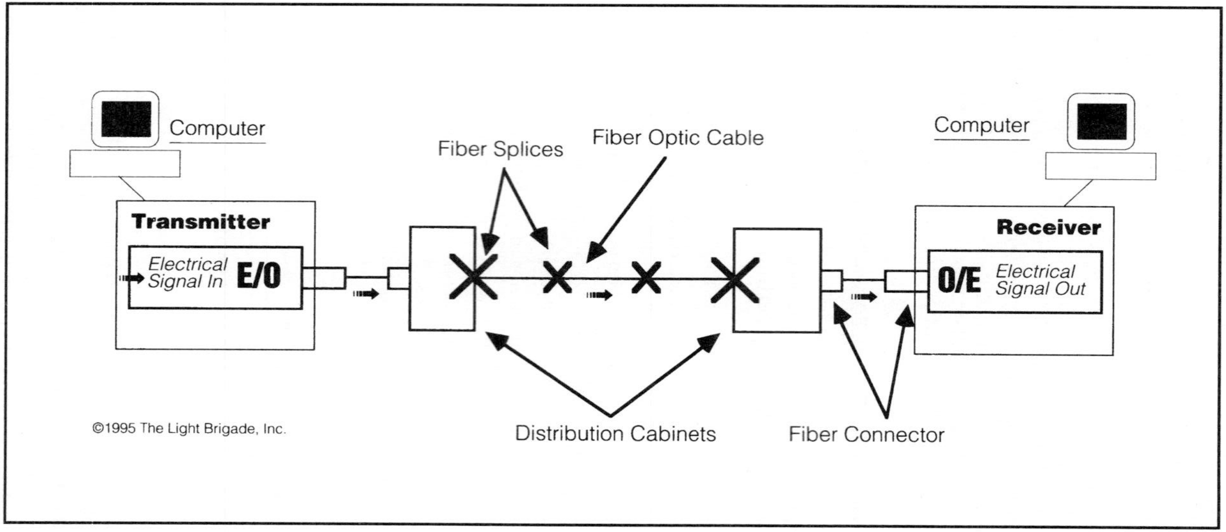

©1995 The Light Brigade, Inc.

circuit. The MUX combines multiple signals into one bit stream, and converts analog signals to digital, and vice versa.

At the receiving end, a *receiver* converts the light pulses back into electronic data signals. (In a schematic, the receiver is called the O/E, for optical/electrical).

TIP: To minimize core and numerical aperature mismatch, use the same fiber manufacturer for the transmitter-to-receiver cable segments.

Installing Optical Cable—Fiber System Topology

The horizontal topology for fiber is the same as for copper—a star topology (see Figure 7.11). Cable is pulled from the main distribution frame (MDF) to intermediate distribution frames (IDFs), and from IDFs individual home runs go out to each workstation termination point.

FIGURE 7.11 Fiber system topology.

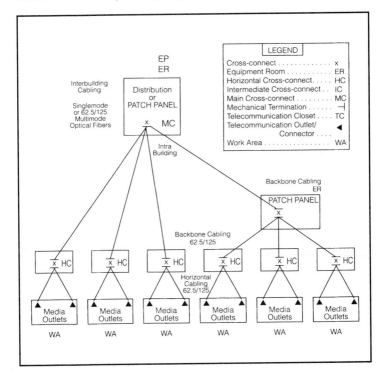

PRESERVING SIGNAL DIRECTION

The TIA building wiring committees recommend that each fiber-optic cable—horizontal, backbone, or patch cord—be reversed for one end to the other. In other words, side one of the transmit end will be coded as the receive side of the other end (see Figure 7.12). The same applies to connectors. If everything is reversed in this manner, then the signal direction is preserved, no matter how many connections are made.

Reducing Loss by Proper Handling

Like all cable, fiber-optic cable must be handled properly to minimize any installation-induced signal losses (the attenuation, or deterioration, of the light signal during transmission). Of course, some losses are *intrinsic,* or out of the installer's control; these include effects from internal absorption, scattering, fiber core variations, and microbends. But *extrinsic* losses related to workmanship can and should be controlled; these include connections, splices, end finishes, and macrobends.

Installing optical cables can be done effectively and with a minimal amount of effort if certain basic guidelines are followed. Details of installation are beyond the scope of this book, but there are many good manufacturers' references covering this topic, such as Leviton's® *Installation Strategies for Long-Term Cabling System Success.*

Fiber-Optic Cable Termination

Once the fiber cables are in, they are ready to be terminated. Fiber-optic connectors offer a mechanical means to terminate optical fibers to other fibers and to active devices, thereby connecting transmitters, receivers, and cables into working links.

Connectors are terminated onto the fiber via splicing or connectorizing. To *connectorize,* the connector is attached to the end of a raw fiber. Connectors can be plugged in and out of patch panels, or station outlets.

In splicing, two bare fibers are joined together with a mechanical or fusion splice. A *splice* is a permanent joining method, used to connect two cable runs together to make the run longer, or to add a pigtail connector onto the cable. A *pigtail* is a short piece of cable with a connector factory-attached to one end. Because of the great potential for misalignment, most single-mode terminations are factory-made onto pigtails.

The TIA-568A standard does allow for fusion or mechanical splicing methods using pigtails, allowing for a maximum loss of 0.3 dB.

A general rule of thumb is that single-mode fibers are spliced, and multimode fibers are connectorized. In most multimode applications, cable will be directly terminated with a connector.

FIGURE 7.12 Preserving signal direction.

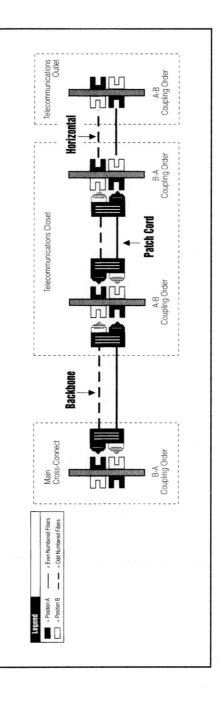

Connectors Recognized by TIA-568A

Two connector systems are most prevalent today: the SC (Figure 7.13) and the ST (Figure 7.14). Both are excellent optical fiber connector systems. The TIA committees on building wiring have called out the SC as the connector of choice for new installations. But, because the ST has a very large installed base in North America and Europe, it is allowed to remain in, or be added to, buildings and organizations that have an installed base of ST connectors.

The SC is a push-pull connector standardized by the ISO/IEC in 1992, when it was recognized as the preferred optical fiber connector for new installations in commercial buildings. The SC connector has since been recognized as the connector standard in the ANSIX3T9.3 Fiber Channel, ANSIX3T9.5 Low-Cost FDDI, Broadband ISDN T1E1 (ATM), and by the TIA-568A standards.

The SC is available as a simplex or duplex connector. It is keyed to prevent cross-mating, and uses a push-pull design to mate and unmate. The standard recommends it be color-coded beige for multimode, and blue for single-mode, but note that having those colors is not necessary for compliance.

ST connectors (BFOC/2.5) are the de facto standard connector of the late 1980s and early 1990s. This bayonet connector style (using push/turn motion to attach to its mating sleeve) has been officially recognized by only one standard: IEEE 10BaseF. It is available in two versions, the ST and STII; both use a 2.5-mm ferrule, which is keyed.

With the SC connector now preferred by most new standards, including the TIA-568A Commercial Building Wiring Standard, the use of the

FIGURE 7.13 SC fiber connectors.

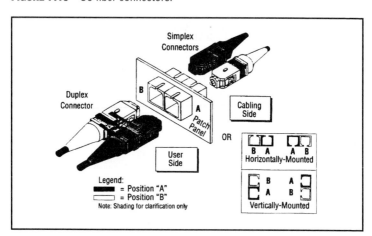

FIGURE 7.14   ST fiber connectors.

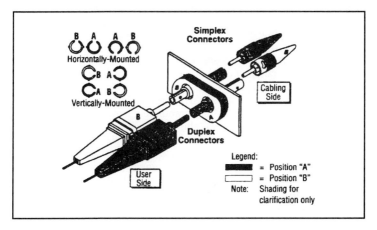

ST may decline except in preexisting installations in which its use is "grandfathered".

Hybrid Connectors

Because of the evolution from ST to SC connectors, hybrid solutions (jumpers and adapters) are on the market. These provide a path for current users (and manufacturers of network equipment) to migrate from the ST to the SC.

Fiber-Optic Closet Hardware

For fiber-optic cable, connecting hardware in the TC is primarily enclosed in patch panels. The horizontal cables and backbone cables are terminated with connectors, and connected to the patch panel couplers on the back of the panels. The cross-connection between the horizontal cables and equipment or backbone cables is done with simplex or duplex patch cords.

Distribution panels may also be used. A distribution panel differs from a patch panel in that the interconnections on a distribution panel are intended to be permanent, whereas patch panels are constructed to provide movable connections that allow the system to be modified.

In addition, a device called an *attenuator* may be needed to reduce the optical power levels into the detector, which can be oversaturated with optical power and thus cause an increase in the bit error rate. A fixed attenuator intentionally introduces loss into a system, usually in increments of 5, 10, 15, or 20 dB. It is typically mounted on the receiver side of the link, in a patch panel.

Fiber-Optic Patch Panels

Patch panels are used as intrabuilding interconnects or main cross-connects. They provide a central location for patching, testing, monitoring, attenuating, and restoring service to the fiber-optic transmission lines. The patch panel receives the fiber-optic patch cords or jumpers from the splice or panel or from equipment, and properly routes them to other pieces of equipment.

For LANs, the patch panel can accommodate main cross-connects or intermediate cross-connects to provide a location in which personnel can access backbone cabling in intrabuilding links and transmission equipment. Patch panels are available with various connector styles.

Patch panels come in wall- or rack-mounting styles; the deciding factor on which to use usually comes down to available floor space, and the existence of any other rack-mounted equipment. Wall-mount panels are usually used where floor space is limited, and there is no rack-mount equipment. Rack-mount panels are generally used where there is sufficient floor space to accommodate racks, and where there will be other rack-mounted equipment, such as hubs, routers, multiplexers, fiber-optic electronics, copper patch panels, and wire management products. Both mounting styles can usually be ordered with key-locked doors, which provide security by limiting user access.

Fiber-Optic Patch Cords

At the cross-connect, circuits are joined with patch cords—lengths of cable with connectors on each end (Figure 7.15). They may be simplex (one fiber per cable) or, more commonly, duplex (two fibers per cable) form.

FIGURE 7.15 Fiber-optic patch cord.

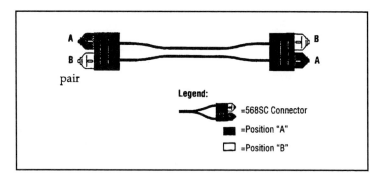

7.4 BACKBONE DISTRIBUTION PRACTICES

Backbone cabling for the MDF, IDF, equipment room, and TC is very application- and building-specific, and dependent upon the type of communications, type of building, and many other issues. Notwithstanding, the general specifications for backbone wiring are:

• Maximum of two levels
• Star topology
• Limited to 90 m (copper only) for category-rated applications, and 800 m for voice applications

Backbone Cabling Scope

The backbone cabling extends from the main cross-connect to the horizontal cross-connects and includes the main cross-connect, the intermediate cross-connect, the connecting hardware dedicated to the backbone cabling, and the cables that join them. The backbone cabling also includes the mechanical terminations and jumper/patch cords used for backbone-to-backbone connectivity. The backbone cabling does not include cables in the TCs, equipment rooms, or entrance facilities that connect directly to customer premises equipment (CPE).

The following requirements and recommendations are intended to assure that the backbone cabling system can accommodate present and future information technologies that are introduced over its planning period.

Choosing Backbone Cable Media

Considerations that factor into the selection and quantity of backbone cable pairs and fiber include:

• The planning life of the backbone system
• The bandwidth requirements for planned applications
• The number of work areas served by a given backbone segment

The first two criteria are unique to each installation and are left to the discretion of the electrical design professional.

Whenever possible, the different service requirements should first be determined. It is often convenient to group similar services together in a few categories such as voice, display terminal, Local Area Networks (LANs), and other digital connections. Within each group, individual types should be identified and required quantities projected.

When uncertain, use "worst case" scenarios when evaluating different backbone cabling alternatives. The higher the uncertainty, the more flexible the backbone cabling system needs to be.

Each recognized cable has individual characteristics that make it useful in a variety of situations. A single-cable type may not satisfy all the user requirements at the site. It is then necessary to use more than one medium in the backbone cabling.

Intrabuilding Backbone Cabling

The following requirements and recommendations are provided for backbone cabling runs within the building.

Optical fiber cabling

- For each intrabuilding backbone run that is greater than 90 m (295 ft) in length, at least one optical fiber cable should be provided. Two strands are required (1-TX, 1-RX).

This recommendation assures that high-speed data applications can be supported between the main and horizontal cross-connects. Fiber is desirable for intrabuilding backbone runs because of its bandwidth capability, grounding isolation, and the fact that it occupies less pathway space (per delivered bandwidth) than UTP/ScTP.

- It is recommended that at least two optical fiber cores/strands be provided for every known application to be served by the intrabuilding backbone over its planning period. A growth factor of 100 percent should be provided. For example:

Application	Fiber Count
Voice	2
Video (Security)	2
LAN (10Base-F)	2
Growth	6
Total	12

For this example, a 12-fiber intrabuilding backbone cable is recommended.

Unshielded/screened twisted-pair cabling

- For each intrabuilding backbone run, at least one UTP/ScTP cable should be provided.
- If the intrabuilding backbone run is less than or equal to 90 m (295 ft) in length, and optical fiber is not installed, at least one Category 5 cable should be provided.

This assures that both voice and high-speed data applications can be supported between main and horizontal cross-connects.

- It is recommended that at least two cable pairs (Category 3 or higher) be provided for each work area served by the intrabuilding backbone segment.

This recommendation provides for a minimum of one pair per work area to support low-speed data or voice applications, and at least one additional pair for growth or for applications that may require multiple pairs.

INTERBUILDING BACKBONE CABLING

The following recommendations are provided for all backbone cabling runs that extend between buildings.

Optical fiber cabling
- For each interbuilding backbone run, at least one optical fiber cable should be provided (two fibers: 1-TX and 1-RX).

Fiber is the preferred media for interbuilding runs because it typically provides more bandwidth than twisted-pair cabling and is immune to ground loop problems that may occur between buildings for some types of metallic media. It is also immune to the effects of lighting if a non-metallic-strength fiber is used.

- It is recommended that at least two optical fiber cores/strands be provided for every known application to be served by the interbuilding backbone over its planning period. A growth factor of 100 percent should be provided. For example:

Application	Fiber Count
Voice	2
Video (Security)	2
Video (Interactive)	2
LAN (10Base-F)	2
LAN (FDDI)	4
Data Mux (3 applications)	6
Growth	18
Total	36

- For this example, a 36-fiber interbuilding backbone cable is recommended.

Unshielded/screened twisted-pair
- For each interbuilding backbone run, at least one UTP/ScTP cable should be provided.
- It is recommended that at least two cable pairs (Category 3 or higher) be provided for each work area served by the interbuilding backbone segment. Note that this is a large quantity; it depends on the location of the PBX.

This recommendation provides for a minimum of one pair per work area to support low-speed data or voice applications between buildings,

TABLE 7.10 Distribution Backbone Maximum Cabling Distances

	Multimode Fiber	Singlemode Fiber	UTP/ScTP
MC to HC	2,000 m (6,560 ft)	3,000 m (9,840 ft)	800 m (2,624 ft)
MC to IC	1,500 m (4,920 ft)	2,500 m (8,200 ft)	300 m (984 ft)
IC to HC	500 m (1,640 ft)	500 m (1,640 ft)	500 m (1,640 ft)
HC to WA	90 m (295 ft)	Not Recognized	90 m (295 ft)

plus at least one additional pair for growth and for applications that may require multiple pairs.

Backbone Cabling Distances

The following requirements and recommendations apply to all portions of the backbone cabling. The connection of multiple campuses or buildings that extend beyond the distance limits specified in this section are outside the scope of this book. Table 7.10 summarizes the maximum cable length limits.

Backbone Cabling Links and Channels

For the purposes of this book, two terms are used to distinguish between backbone cabling subsystems with and without equipment cords. These terms are *backbone basic link* and *backbone channel.*

BACKBONE LINK MODEL

A backbone link model is considered to encompass all components of the backbone cabling subsystem covered within the scope of this book. These components include the backbone patch panel or connecting block, and the cables that extend between these connecting hardware fields.

BACKBONE CHANNEL MODEL

A backbone channel model encompasses all components of the backbone basic link model elements, plus any equipment cables that connect to the main, intermediate, and horizontal cross-connects and includes the patch cords and (equipment) connecting hardware fields. The backbone channel includes all of the physical cabling elements that are required to provide a continuous transmission path between active equipment on both ends of a backbone link model.

7.5 PATHWAYS—HORIZONTAL AND BACKBONE

Horizontal Pathways

Horizontal pathways provide the means for placement of telecommunications cable from the TC to the work area telecommunications outlet. Horizontal pathway types include: underfloor ducts, conduit, cable tray and wireway, ceiling, and perimeter (surface, recessed, moulding, modular furniture) raceway.

The design and layout of various types of pathways are major topics in themselves. It is assumed, however, that you are already familiar with the details of these topics as they are the fodder of electrical design professionals. The focus of the following guidelines, therefore, is on the proper sizing and other requirements necessary to accommodate telecommunications cabling.

Design Guidelines
GENERAL

Sizing

- Horizontal pathways should be capable of delivering a minimum of three four-pair UTP/ScTP cables, each with a diameter of 6 mm (0.25 in) to each individual work area served. For open office space, the pathways should be sized, based on 10 m^2 (100 ft^2) of usable floor space per work area.

Bend radii

- Minimum inside bend radius of horizontal pathways should not be less than 10 times the maximum cable diameter to be installed. Actually, 6 times up to 2 in C, 10 times over 2 in C and all fiber-optic cable.

UNDERFLOOR DUCT SYSTEMS

Duct sizing

- The practice for general office space, based on the assumption of three devices per work area and one work area per 10 m^2 (100 ft^2), is to provide 650 mm^2 (1 in^2) of cross-sectional underfloor duct area per 10 m^2 (100 ft^2) of usable floor space. This practice applies to both feeder and distribution duct.

Duct spacing

- Distribution duct runs at 1520 to 1825 mm (5 to 6 ft) apart at the midpoint of the building module.
- Runs adjacent to exterior building walls should be located 450 to 600 mm (1.8 to 2.4 ft) from the walls or column lines.

- In general, 18,000 mm (60 ft) spacing for access units is adequate.
- Provisions should be made to connect the system to the TC by a number of home runs.

Layout
- The layout of the distribution and feeder ducts is dependent upon the building module (structural system) and the work area (furniture systems) layout. Thus, the duct layout will have to be adjusted to these conditions, which may be less than ideal design conditions for the telecommunications distribution layout. Engineering judgement will have to be exercised in this case.

CELLULAR FLOOR SYSTEMS

Floor cell sizing
- The same as above for underfloor duct.
- Additionally, where the cellular unit profile provides different cross-sectional areas for the cells, the largest cell should be allocated for telecommunications and the smallest for electrical power.

Floor cell and insert spacing
- Cellular members are generally 610 mm (2 ft) wide and noncellular members are 610 to 915 mm (2 to 3 ft) wide.
- In office areas, in-floor service to work areas should be provided by locating cellular units on one 1220-mm or one 1525-mm (4- or 5-ft) center. Preset inserts, when installed, should be a minimum of 600-mm (2-ft) centers.

Cellular floor layout
- Distribution cells should run the length of the building to minimize the length of the feeder runs.
- An advantage of cellular floors over distribution ducts is that they typically provide a higher degree of flexibility for future changes and rearrangements of office furniture.

ACCESS FLOOR SYSTEMS

Loading performance
- The access floor system should be certified by the manufacturer to meet at least a medium-duty rating for the area it is intended to serve. Please refer to Annex A of ANSI/TIA/EIA-569A for access floor ratings.

Fire rating

- The floor panels exclusive of covering shall comply with the minimum flame spread rating as specified by applicable codes and regulations.
- Access floor distribution in spaces used for handling environmental air shall meet all applicable building and electrical codes.

Building structure

- When access floors are planned, the recommended building structure is "depressed slab" to avoid ramps and steps.

Floor height

- Where access flooring is used, the minimum access space between the subfloor and the underside of the floor tile shall be a minimum of 150 mm (6 in). Better would be 200 mm (8 in), to provide adequate cable management for high-performance twisted-pair and optical fiber cabling. If in a computer or control room environment where the space between the subfloor and raised access floor is used for environmental air, the minimum height should be 300 mm (1 ft).

Cable management

- Consideration should be given to using some method of containment for major runs of cabling. Defined telecommunications pathways for containment of major cabling runs include dedicated routes, cable tray, raceway distribution, or a zone distribution system.

Layout

- It is recommended that the access floor layout be complete before the floor plan is started. Once the floor plan is complete, the access floor cabling can be further optimized. This is offered because the accessibility of access floor cabling may be affected by the presence of equipment or furniture.

Bonding

- To provide electrostatic protection, it is recommended that the panel surfaces be conductive materials and the metallic parts to which they connect be bonded to ground. The resistance between the bare top surface of the panel and the pedestal should not exceed 1 Ω.

CONDUIT

General

- Conduit distribution has many limitations (limited access, fixed routes, fixed capacity) and should only be considered for use as a dedicated run to a single remote location or when horizontal runs are external to the building or when conduit is required to satisfy applicable codes and/or regulations.

- Magnetic conduit is generally preferred over nonmagnetic conduit because of its ability to block EMI. Metal flex conduit is not recommended due to its cable abrasion problems.
- A nylon pull cord should be installed in a conduit.
- A single conduit run extending from a TC should not serve more than three jacks in the work area.

Conduit sizing

- It is recommended that the conduit sizing requirements specified in Table 7.11 be observed.

Conduit termination

- A conduit protruding through the floor of a TC should be terminated 25 mm (1 in) to 100 mm (4 in) above the finished floor. Conduit protruding through the floor in other areas of the building should be terminated 25 mm (1 in) to 50 mm (2 in) above the finished floor.

Pull boxes

- Pull or splice boxes are used for fishing conduit runs (inserting a pull line or snake, then retracting it with a cable attached) and as interme-

TABLE 7.11 Conduit Sizing

Conduit		Number of Cables									
INTERNAL DIAMETER	TRADE SIZE	Wire O.D. mm (in)									
mm (in)		3.3 (.13)	4.6 (.18)	5.6 (.22)	6.1 (.24)	7.4 (.29)	7.9 (.31)	9.4 (.37)	13.5 (.53)	15.8 (.62)	17.8 (.70)
15.8 0.62	1/2	1	1	0	0	0	0	0	0	0	0
20.9 0.82	3/4	6	5	4	3	2	2	1	0	0	0
26.6 1.05	1	8	8	7	6	3	3	2	1	0	0
35.1 1.38	1 1/4	16	14	12	10	6	4	3	1	1	1
40.9 1.61	1 1/2	20	18	16	15	7	6	4	2	1	1
52.5 2.07	2	30	26	22	20	14	12	7	4	3	2
62.7 2.47	2 1/2	45	40	36	30	17	14	12	6	3	3
77.9 3.07	3	70	60	50	40	20	20	17	7	6	6
90.1 3.55	3 1/2	-	-	-	-	-	-	22	12	7	6
102.3 4.02	4	-	-	-	-	-	-	30	14	12	7

diate pull points (i.e., pulling the cable to the box and then looping the cable to be pulled into the next length of conduit).

- Copper cable splices are not allowed for horizontal cabling and they are not recommended for fiber (between the outlet and the closet).
- No section of conduit should be longer than 30 m (100 ft) or contain more than two 90° bends, or equivalent, between pull points.

CABLE TRAYS AND WIREWAYS

Location
- In addition to overhead mounting, cable trays may also be used in installations with access flooring.

Sizing of Cable Trays and Wireways
- The guideline of 650 mm^2 (1 in^2) of cross-sectional area of the tray per 10 m^2 (100 ft^2) of usable floor space should apply.

Cable tray/wireway bend radii
- The inside bend radii of cable trays and wireways for horizontal cable distribution should not be less than 150 mm (6 in). Cable trays that converge in a T formation should be provided with a transitional bend radius of 150 mm (6 in) in both directions.

CEILING PATHWAYS

Ceiling distribution systems primarily consist of the pathway types described previously in this section or may consist of discrete means of support such as open-ceiling cable supports (J hooks, for example). Cable delivery to WAOs is generally provided by means of conduits or interstud routing to wall outlets. In cases where wall outlets are not accessible or where they are located more than 3 m (10 ft) from the work area, other floor-to-ceiling fixtures such as utility columns can be used.

General
- Ceiling distribution systems shall provide full accessibility to distribution pathways. Lock-in-type ceiling tiles, drywall, or plaster ceilings shall not be used unless they meet one or more of the following criteria:

  An enclosed pathway is preexisting or is provided.

  The lock-in tiles have been modified to allow easy removal of tiles.

  A safe crawl or walk space exists, providing full access to the cabling.
- For open-ceiling cabling pathways, cable support systems shall be designed and installed to be a minimum of 75 mm (3 in) above the ceiling grid, which supports the suspended ceiling tiles.

Types of ceiling distribution
- Conduit, cable trays, wireways, discrete hangers, and catenary wire.

Cabling support
- Cable support shall be provided by means that are structurally independent of the suspended ceiling, its framework, or supports.
- Cable supports shall not be spaced more than 1.5 m (5 ft) apart. They should be located with an on-center spacing of 1 m (3 ft) apart.

Cabling pathway capacity
- The number of horizontal cables placed in a cable support or pathway (hooks, rings, and so on) shall be limited to a number of horizontal cables (UTP/ScTP or optical fiber cable), which will not cause a change to the geometric shape of the cables.
- Open-type cable supports (hooks, rings, and so forth) should be limited to no more than 30 horizontal cables (UTP/ScTP or optical fiber).

Environmental air spaces
- Ceiling distribution in spaces used for handling environmental air shall meet all applicable codes and regulations. In general, this means that if cables are not in an enclosed raceway such as conduit, they should be plenum-rated cable. All supports, including ty-wrap, shall also be plenum-rated.

PERIMETER PATHWAYS

Perimeter pathways are most commonly used in conjunction with other types of horizontal distribution systems for vertical runs from ceiling or underfloor pathways, when walls are not suitable for concealed cabling. Examples of cases when perimeter pathways are used include instances when walls are made of solid brick or poured concrete.

The practical capacity of telecommunications cabling in perimeter raceways ranges from 30 to 60 percent fill depending on the cable-bending radius, but is limited to 40 percent by the NEC. Also note that a minimum bending radius of 1 in should be maintained.

MISCELLANEOUS PATHWAYS

- *Interstud pathways:* Shall not have exposed sharp edges and shall be provided with bushings where required.
- *Exposed cabling:* Cables that extend between TCs and telecommunications outlet/connectors shall not be exposed in the work area or other locations with public access.

- *Poke-through floor pathways:* Not recommended because they reduce the structural strength of the floor slab and require fire stopping.

Backbone Pathways

Backbone pathways consist of intra- and interbuilding pathways. They may be either vertical or horizontal. Interbuilding backbone pathways extend between buildings. Intrabuilding backbone pathways are contained within a building.

- In general, the same design guidelines that apply to horizontal pathways also apply to backbone pathways, with a few additional considerations that follow.

INTRABUILDING PATHWAYS

The ideal vertical backbone pathway consists of TCs located on each floor, which are vertically stacked one above the other, and tied together by sleeves or slots.

In this context, the term *sleeve* refers to a circular opening in the wall, ceiling, or floor to permit the passage of cables between spaces. This is usually in the form of several rigid-metal or PVC conduit sleeves. A *slot* is the same as a sleeve, except that the shape of the opening is usually rectangular and indeed is a rectangular hole, either cut, framed out of a concrete pour, or formed by the omission of masonry brick or block.

Layout
- Intrabuilding pathways shall provide access to all entrance facility(ies), equipment room(s) and TCs located in the same building.
- Intrabuilding pathways shall be configured to support a star cabling topology.

Location
- Pathways shall not be located in or adjacent to elevator (lift) shafts, or power feeders. Table 7.12 shows the minimum separation distances between pathways and power wiring of 480 V or less.

Size
- Sizing of backbone pathways between building spaces containing cross-connect facilities should be sized based on a minimum of three (3), 100-mm (4-in) conduits. One for copper, one for fiber, and one spare.

Provisions for cable installation
- When closets are aligned with a common vertical pathway, some means for cable pulling, such as a steel anchor pulling iron or eye

TABLE 7.12 Separation of Telecommunications Pathways from 480-Volt or Less Power Lines

Condition	Minimum Separation Distance		
	< 2 kVA	2-5 kVA	> 5 kVA
Unshielded power lines or electrical equipment in proximity to open or nonmetal pathways.	127 mm (5 in)	305 mm (12 in)	610 mm (24 in)
Unshielded power lines or electrical equipment in proximity to a grounded metal conduit pathway	64 mm (2.5 in)	152 mm (6 in)	305 mm (12 in)
Power lines enclosed in a grounded metal conduit (or equivalent shielding) in proximity to a grounded metal conduit pathway.	- -	76 mm (3 in)	152 mm (6 in)

embedded in the concrete, should be provided above and in-line with the sleeves or slots at the uppermost closet of each vertical stack. Similar techniques may be required for long horizontal pathways.

INTERBUILDING PATHWAYS

Interbuilding pathways interconnect separate buildings such as in campus environments. They consist of underground, buried, aerial, and tunnel pathways.

- It is recommended that a telecommunications design plan be developed for all buildings, and the pathways between buildings, identified on the initial plot plan.

- Interbuilding pathways shall be provided between buildings served by the same main cross-connect.

- Interbuilding pathway design is very site specific and therefore beyond the scope of this book, however, requirements involving the media choices and length limitations shall be met.

7.6 TELECOMMUNICATIONS CLOSETS

General

The telecommunications closet (TC) is the hub of the horizontal cabling system and is the key to a well-organized horizontal cabling system. Because all telecommunication transmissions to and from the work area end up in the TC, the organization of the TC is critical to adds, moves, and changes. See Figure 7.16 for general closet layout.

First, the TC must be sufficient in size to handle the cross-connect field, the associated electronic equipment, the backbone and horizontal cables, and the pathways for the cables—and still have enough room for a craftsperson to work, without disrupting services. It must also meet NEC working clearances. The following specifications are from the Pathways Standard, TIA-569.

There should be one TC per floor, dedicated to telecommunications. If there are multiple TCs on a floor, they shall be interconnected by a minimum of one trade size 3-in conduit.

- The TC serving an office area should be of a specific minimum size to accommodate the current and future services in the area served (see Table 7.13). These sizes may seem a little large, but the TC needs to have enough room to allow electronic equipment to be added for voice, data, video, security, and so on. Please note that the following minimum sizes do not actually include space for the electronic equipment such as computers, servers, PBX, and so forth.

Lighting shall be adequate (minimum of 50 fc) so that the craftspersons can distinguish small lettering and work with the small color-coded wires.

- *Enough electrical service and outlets:* To provide power for the installed electronic equipment and the craftsperson's equipment.

TABLE 7.13 Minimum Recommended TC Sizes

Serving Area:		Closet Size:	
ft²	(m²)	ft. x ft.	(m x m)
10,000	(1000)	10 x 11	(3 x 3.4)
8,000	(800)	10 x 9	(3 x 2.8)
5,000	(500)	10 x 7	(3 x 2.2)

FIGURE 7.16 Typical telecommunications closet.

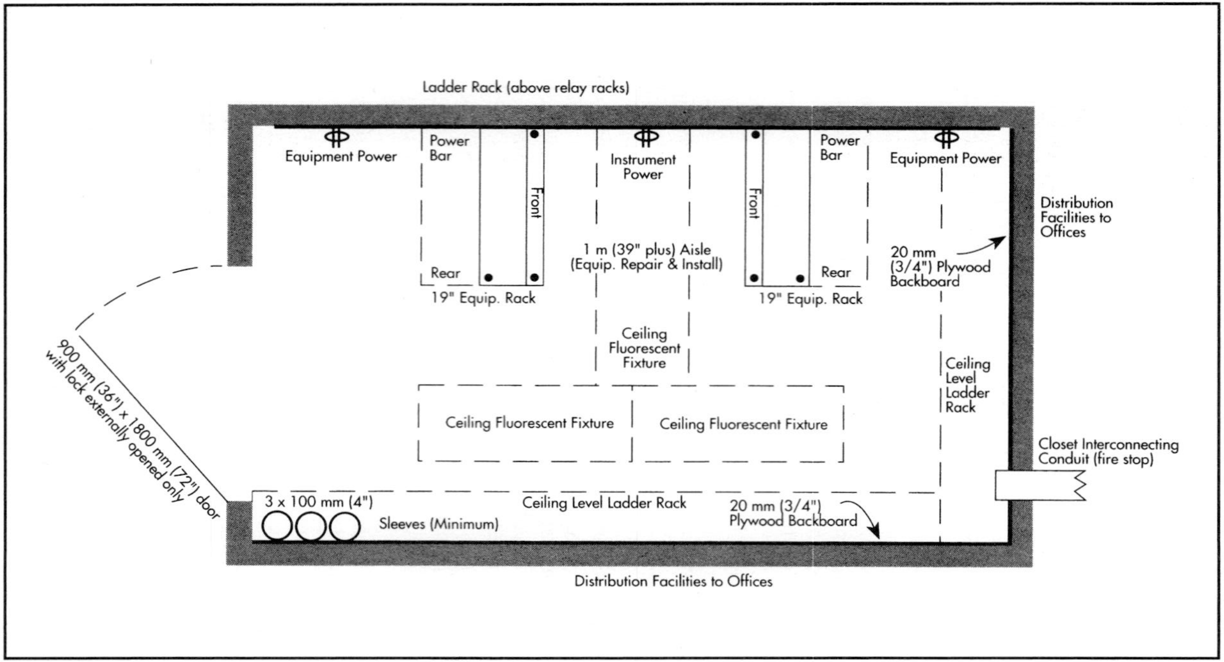

Ladder Rack (above relay racks)

Equipment Power

Power Bar

Front

Rear

19" Equip. Rack

Instrument Power

1 m (39" plus) Aisle
(Equip. Repair & Install)

Ceiling Fluorescent Fixture

Power Bar

Front

Rear

19" Equip. Rack

Equipment Power

20 mm (3/4") Plywood Backboard

Distribution Facilities to Offices

Ceiling Level Ladder Rack

Closet Interconnecting Conduit (fire stop)

Ceiling Fluorescent Fixture

Ceiling Fluorescent Fixture

Ceiling Fluorescent Fixture

Ceiling Level Ladder Rack

20 mm (3/4") Plywood Backboard

Distribution Facilities to Offices

Sleeves (Minimum)

3 x 100 mm (4")

900 mm (36") x 1800 mm (72") door
with lock externally opened only

498

- *Clean and free of clutter.*
- *Dedicated to telecommunications:* The TC should not be a storage room.
- *Climate-controlled:* Most electronic equipment designed for TCs requires a limited-temperature environment. Separately controlled air conditioning for the TC is almost always required.
- *Secure:* Businesses rely heavily on their communications systems, so access to the TC must be limited to authorized personnel. The door should be lockable.
- *Located in a room other than the power distribution or mechanical equipment.*
- *Door width shall be 36 in:* It should be lockable and open outward.
- *The TC should be firestopped:* Sleeves, slots, penetrations, and so on.

TC Cross-Connect Fields

TC cross-connect fields must be well organized to facilitate installation and changes, as this is the key to an easy-to-administer system.

- Cross-connects are usually mounted on a plywood backboard mounted to the walls of the TC (see Figure 7.17). The cross-connects are usually organized first by cable type (backbone, horizontal, equipment). Then they are often color-coded and organized into cabling system types and services due to the cross-connects required:

  Category 3 IDCs for voice (Category 3 100-Ω UTP)

  Category 5 IDCs or patch panel applications using Category 5 100-Ω UTP

  150-Ω STP patch panels

  Optical fiber patch panels
- EIA/TIA-606 suggests that TC cross-connects be organized into color-coded fields. Most cross-connect devices can have colored labels or markers added for identification (Table 7.14).
- Sufficient space and hardware must be provided to handle the size and weight of the backbone cables, horizontal cables, patchcords, and jumper cables.
- Provide a service loop at each termination for future adds, moves, and changes.
- For 100-Ω UTP and 150-Ω STP, only use jumper cables, patch cords, and cross-connect devices that comply with the category or extended requirements of the cabling system. Use of lower-performance components will cause performance degradation, poor-quality signals, and possible data loss.

FIGURE 7.17   Voice and data backboards in the telecommunications closet.

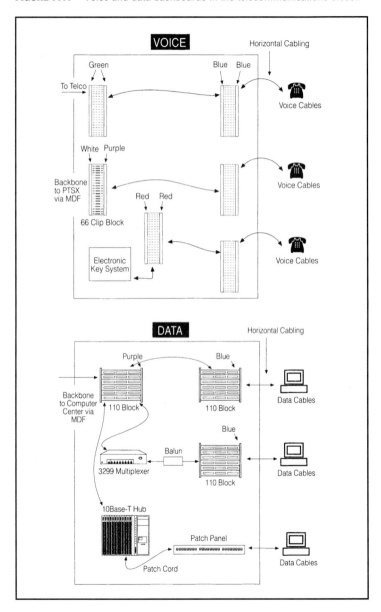

TABLE 7.14 Cross-Connect Field Color Codes

Crossconnect Field	Color
Backbone cable from main crossconnect	White
Backbone cable from intermediate crossconnect	Grey
Customer side of demarcation point	Green
Common equipment (PBXs, LANs)	Purple
Horizontal cable	Blue
Auxiliary circuits (alarms, etc.)	Yellow
Key Telephone	Red
Central Office Cable	Orange
Campus Cable	Brown

TC Electronic Equipment

With increasing use of communications by each person in a commercial environment, the TCs are being equipped with more electronic equipment. Thus, it is important to maintain sufficient space for cooling, servicing, and cable management. A reminder: ANSI/EIA/TIA-568A does not include space for this equipment in its recommended sizes.

The following electronic equipment is often housed in the TC:

1. Key system
2. Small PBX
3. Multiplexer or hub
4. Security system
5. File or print servers

TC Pathways

- The TC will have conduits or raceways entering the TC for backbone and horizontal cables. There will also generally be a cable tray pathway in the TC to distribute the horizontal and backbone cabling to the equipment.

Cabling System Management

To facilitate ongoing wiring system management and changes, wiring installations should be documented per the requirements of TIA-606. Proper wiring administration is a requirement of TIA-568A.

7.7 EQUIPMENT ROOMS

General

Equipment rooms typically contain a vast portion of the telecommunications equipment, cable terminations, and cross-connects. They may be thought of as serving the entire building or campus, whereas TCs are thought of as serving individual floors.

Any of the functions of TCs may alternatively be provided by an equipment room. Formerly known as apparatus closets, equipment rooms are mainly used for backbone and equipment terminations.

In addition to housing facilities for backbone terminations, main and intermediate cross-connects, the equipment room may also house equipment terminations and some cross-connects for horizontal and demarcation terminations for a portion of the building.

- In general, the requirements that apply to a TC also apply to an equipment room, with additional considerations that follow.

Location

- When practical, in a multistory building it is recommended that the equipment room be located on the middle floor of the building it is serving and in a location that provides easy access for the cabling pathways to TCs on other floors.

Size

- The size of the equipment room should take into account all types of equipment required as well as any connecting hardware requirements, including present and future needs.

Unknown Equipment

Where the specific equipment is not known, the following guidelines should be used:

- The recommended equipment room size should be a minimum of 14 m² (150 ft²) or 0.07 m² (0.75 ft²) of equipment room space for every 10 m² (100 ft²) work area space. In an environment in which the density of work areas is higher than usual, the equipment room size should be increased accordingly.

Multitenant Buildings

• In multitenant buildings, an equipment room may be located in a space that is common to all tenants, or each tenant may have their own equipment room.

Special-Use Buildings (e.g., Hotels, Hospitals, Universities)

• The size of the equipment room for special-use buildings should be based on the known number of work areas as opposed to usable floor area as shown below in Table 7.15.

Protective Finishes

• The floor, walls, and ceiling should be sealed to reduce dust. Finishes should be light in color to enhance room lighting. Flooring materials having antistatic properties should be selected.

Protective Measures

• To protect electronic equipment, it is recommended that secondary protection (e.g., overvoltage and/or sneak-current protection) be installed on any telecommunications cabling that extends outside of the building or is susceptible to extraneous voltages or currents beyond the rating of the equipment.

7.8   ENTRANCE FACILITY

The entrance facility (EF) typically contains the network demarcation point as well as inter- and intrabuilding backbone facilities.

The network demarcation point is the location within a building in which the local service provider installs an interface device for the cus-

TABLE 7.15   Equipment Room Floor Space (Special-Use Buildings)

| Work Areas | AREA | |
	(m²)	(ft²)
Up to 100	14	150
101 to 400	37	400
401 to 800	74	800
801 to 1,200	111	1,200

tomer premises cabling. This is the point at which the local exchange carrier (LEC) is released from liability as far as transmission and/or circuit integrity. This service is regulated by applicable codes and local regulatory agencies.

General

Because EFs are spaces used to house backbone cabling, the requirements that cover design and other functional requirements of this space are found in ANSI/TIA/EIA-569A.

In addition to housing facilities for the network demarcation terminations and associated building backbone cross-connects, the EF may also contain some cross-connects for interbuilding backbone facilities.

- In general, the requirements that apply to a TC also apply to an EF, with the additional considerations that follow.

Entrance Point

- An entrance point is the point of penetration of the foundation wall. An alternate EF should be provided in which security, continuity of service, or other special needs exist.

Layout (Access)

The layout for a typical underground access to an EF is shown in Figure 7.18.

FIGURE 7.18 Typical underground entrance facility layout.

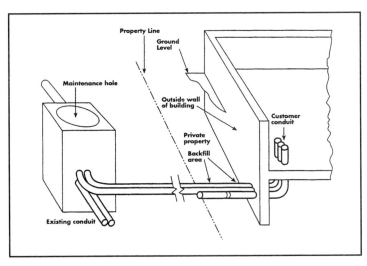

Location

- The entrance room or space shall be located in a dry area not subject to flooding.

- The location of other utilities, such as electrical, water, gas, and sewer, should be considered in the site selection of the telecommunications EF.

Size

The following tables specify space for all telecommunications equipment and associated cross-connections. Table 7.16 is based upon terminations and equipment mounted on a 2.5-m- (8-ft)-high wall. Table 7.17 is based upon terminations and equipment mounted on free-standing racks.

TABLE 7.16   Entrance Facility Wall Space (Minimum Equipment and Termination Wall Space)

GROSS FLOOR SPACE		WALL LENGTH	
m²	ft²	mm	in
500	5,000	990	39
1,000	10,000	990	39
2,000	20,000	1,060	42
4,000	40,000	1,725	68
5,000	50,000	2,295	90
6,000	60,000	2,400	96
8,000	80,000	3,015	120
10,000	100,000	3,630	144

TABLE 7.17   Entrance Facility Floor Space (Minimum Equipment and Termination Floor Space)

GROSS FLOOR SPACE		ROOM DIMENSIONS	
m²	ft²	mm	ft
7,000	70,000	3,660 × 1,930	12 × 6.3
10,000	100,000	3,660 × 1,930	12 × 6.3
20,000	200,000	3,660 × 2,750	12 × 9
40,000	400,000	3,660 × 3,970	12 × 13
50,000	500,000	3,660 × 4,775	12 × 15
60,000	600,000	3,660 × 5,588	12 × 18.3
80,000	800,000	3,660 × 6,810	12 × 22.3
100,000	1,000,000	3,660 × 8,440	12 × 27.7

In buildings with up to 10,000 m² (100,000 ft²) of usable floor space, wall-mounted terminating hardware may be suitable. Buildings of larger floor area may require free-standing frames for cable terminations.

EF Pathways

The methods of provisioning a service entrance pathway are underground conduit or duct, direct buried, aerial, and tunnel pathways. To determine the total number of pathways, consider the following:

1. Type and use of building
2. Growth
3. Difficulty of adding pathways in the future
4. Alternate entrance
5. Type and size of cables likely to be installed

7.9 TESTING CATEGORY 5 AND FIBER-OPTIC CABLING

Category 5 Cable Testing

Field-testing requirements for Category 5 cabling systems have been defined by the TIA in their Technical Service Bulletin TSB-67, *Link Performance Testing of Unshielded Twisted-Pair Cabling Systems.* The document defines pass/fail criteria to ensure that the installed cabling is capable of supporting high-speed LAN equipment. Please refer to this

FIGURE 7.19 Channel and link test configurations.

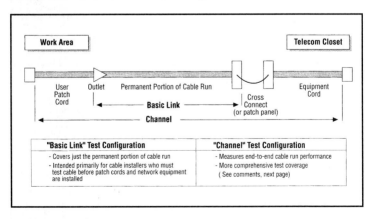

bulletin for details and the test equipment manufacturer's instructions. Figure 7.19 shows the channel and link test configurations.

Fiber-Optic Cable Testing

Testing the optical properties of fiber-optic cable involves measuring two characteristics: attenuation and bandwidth. Please refer to EIA/TIA-526-14, Method B: "Optical Power Loss Measurements of Installed Multimode Fiber Cable Plant" and the test equipment manufacturer's instructions.

NOTES

CHAPTER EIGHT
Miscellaneous
Special Applications

8.0 GENERAL

Introduction

It is the intent of this chapter to provide information and data that is often needed, but perhaps is a little bit outside of the mainstream day-to-day information required by the electrical design professional. In some cases, it represents emerging practices resulting from technological, code, or regulatory changes. In other cases, it represents popular *mis*application of established codes or other requirements that are sometimes misunderstood. And finally, it may simply be information that is needed but less frequently encountered.

8.1 MINERAL-INSULATED CABLE APPLICATIONS

Mineral-insulated (MI) cable has been around for a long time and is a cable of the highest thermal capacity and integrity. Historically, because of these qualities, and the premium cost associated with these qualities, its applications have been limited. This has bred a lack of familiarity and reluctance to use this cable in many applications.

The 1996 National Electrical Code and many state and local code and regulatory requirements are changing this. Because this type of cable has a 2-h fire-resistive rating as approved by the Underwriters Laboratories (UL), this type of cable is gaining popularity in meeting the latest code mandates.

When reviewed at a microscopic level, as compared with conventional construction, using this type of cable for 1-h and 2-h fire-resistive construction, it becomes a cost-effective solution in complying with these code mandates. It also requires considerably less space (in the order of *97 percent less space*) in meeting these requirements, which makes it particularly amenable to renovation/retrofit projects.

Fire Pump and Other MI Cable Applications

Independent tests have shown 90°C wire in conduit fails to ground in less than 3 min when exposed to temperatures of less than 500°F. Because a fire in a typical commercial building generates temperatures in the range of 1200°F to 1500°F, conduit and wire provides unacceptable reliability during a fire.

High-rise buildings frequently have thousands of feet of emergency system wiring routed throughout a building. The potential for some portion of this system being exposed to high temperatures during a fire is high. Loss of critical feeder and branch circuits from a fire will disable equipment long before it has served its intended purpose, impeding evacuation and jeopardizing lives.

The 1996 National Electrical Code has addressed this with two new sections. Section 700-9(c)(1) Fire Protection states: "Feeder-circuit wiring shall be installed either in spaces fully protected by approved automatic fire suppression systems (sprinklers, carbon dioxide systems, etc.) or shall be a listed electrical circuit protective system with a 1-hour fire rating."

The new Article 695 of the 1996 NEC details the installation requirements of the electrical power sources and interconnecting circuits of centrifugal fire pumps. Section 695-8(a) Supply Conductors, Exception No. 1 states: "Fire pump supply conductors on the load side of the disconnecting means as permitted by Section 695-3(c), Exception No. 1, shall be permitted to be routed through buildings using listed electrical circuit protective systems with a minimum of 1-h fire resistance. The installation shall comply with the restrictions provided for in the listing of such systems."

With a 2-h fire-resistive rating approved by UL, MI-type cable provides a technological and cost-effective solution to this requirement. The Commonwealth of Massachusetts and other states now require a 2-h fire rating for emergency feeders.

The following data in Tables 8.1, 8.2, and 8.3 will assist in the application of MI cable.

Classified Wiring (Hazardous) Locations

With approved terminations installed, MI cable meets the requirements of the NEC for wiring in areas classified as hazardous. The cable can be run in Classes I, II, and III, Divisions 1 and 2. Figure 8.1 shows a comparison between MI cable and conventional conduit/wire with accessories required for areas classified as hazardous. It has economic and technical merit.

TABLE 8.1 600-Volt MI Power Cable—Size and Ampacities

CURRENT RATING (75°C/90°C)* / TERMINATION SIZE	–/24 1/2"	–/18 1/2"	–/18 1/2"	–/14.4 1/2"	–/12.6 3/4"
16 AWG					
CABLE REFERENCE	1850/215/1	1850/340/2	1850/355/3	1850/387/4	1850/449/7
CURRENT RATING (75°C/90°C)* / TERMINATION SIZE	30/35 1/2"	20/25 1/2"	20/25 1/2"	16/20 3/4"	14/17.5 3/4"
14 AWG					
CABLE REFERENCE	1850/230/1	1850/371/2	1850/387/3	1850/418/4	1850/496/7
CURRENT RATING (75°C/90°C)* / TERMINATION SIZE	35/40 1/2"	25/30 1/2"	25/30 1/2"	20/24 3/4"	17.5/21 3/4"
12 AWG					
CABLE REFERENCE	1850/246/1	1850/402/2	1850/434/3	1850/465/4	1850/543/7
CURRENT RATING (75°C/90°C)* / TERMINATION SIZE	50/55 1/2"	35/40 3/4"	35/40 3/4"	28/32 3/4"	24.5/28 1"
10 AWG					
CABLE REFERENCE	1850/277/1	1850/449/2	1850/480/3	1850/527/4	1850/621/7
CURRENT RATING (75°C/90°C)* / TERMINATION SIZE	70/80 1/2"	50/55 3/4"	50/55 3/4"	40/44 3/4"	
8 AWG					
CABLE REFERENCE	1850/309/1	1850/512/2	1850/543/3	1850/590/4	
CURRENT RATING (75°C/90°C)* / TERMINATION SIZE	95/105 1/2"	65/75 3/4"	65/75 3/4"	52/60 1"	
6 AWG					
CABLE REFERENCE	1850/340/1	1850/590/2	1850/621/3	1850/684/4	
CURRENT RATING (75°C/90°C)* / TERMINATION SIZE	125/140 1/2"	85/95 1"	85/95 1"		
4 AWG					
CABLE REFERENCE	1850/402/1	1850/684/2	1850/730/3		

	3 AWG	2 AWG	1 AWG	1/0 AWG	2/0 AWG
CURRENT RATING (75°C/90°C)* / TERMINATION SIZE	145/165 1/2"	170/190 3/4"	195/220 3/4"	230/260 3/4"	265/300 3/4"
CABLE REFERENCE	1850/434/1	1850/465/1	1850/496/1	1850/543/1	1850/590/1

	3/0 AWG	4/0 AWG	250 kcmil	350 kcmil	500 kcmil
CURRENT RATING (75°C/90°C)* / TERMINATION SIZE	310/350 3/4"	360/405 1"	405/455 1"	505/570 1 1/4"	620/700 1 1/4"
CABLE REFERENCE	1850/637/1	1850/699/1	1850/746/1	1850/834/1	1850/1000/1

* Based on ampacities in the
National Electrical Code® (NEC).

TABLE 8.2 300-Volt MI Twisted-Pair and Shielded Twisted-Pair Cable Sizes

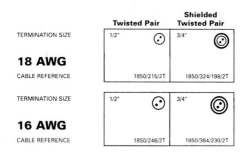

	Twisted Pair	Shielded Twisted Pair
18 AWG		
TERMINATION SIZE	1/2"	3/4"
CABLE REFERENCE	1850/215/2T	1850/324/198/2T
16 AWG		
TERMINATION SIZE	1/2"	3/4"
CABLE REFERENCE	1850/246/2T	1850/364/230/2T

FIGURE 8.1 MI cable versus conventional construction in hazardous (classified) locations.

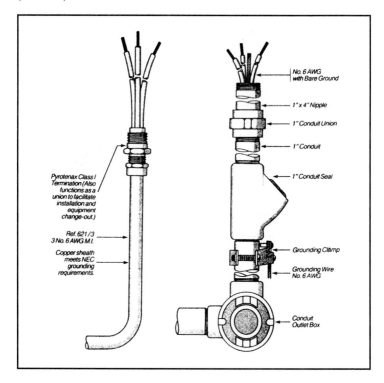

TABLE 8.3 Engineering Data—Calculating Voltage Drop and Feeder Sizing

Step I Determine Feeder Size

Estimate feeder size using the Voltage Drop Chart at right as in the following example:

       Run Length = 100'

       Circuit Voltage = 208 volts

       Circuit Amps = 400 amps

       Required Voltage Drop = 2% or 4.16 volts

Step II Verify Feeder Size

Using the formula and tables below, verify choice from Step I.

   1. Voltage Drop = $\dfrac{\text{(Run Length) X (Circuit Current) X (Temperature Constant) X (Factor from Voltage Drop Calculations Chart) X .87*}}{1000}$

     * .87 is multiplyer for 3-phase. Omit if making single phase calculation.

   2. Using the values of the example:

     $\dfrac{100' \text{ X } 400 \text{ X } 1.0 \text{ X } .1112 \text{ X } .87}{1000}$ = 3.87 Volts Voltage Drop

   3. Percentage Voltage Drop = $\dfrac{\text{Voltage Drop}}{\text{Circuit Voltage}}$ X 100%

   4. Values from example:

     $\dfrac{3.87}{208}$ X 100% = 1.86% Percent Voltage Drop

   5. Conclusion: Since 1.86% is better than the 2% voltage drop required, the choice of 250 MCM Pyrotenax MI Cable (746/1) is confirmed.

Temperature Constant Chart

Cable at full rated current	1.00
Cable at 3/4 rated current	0.95
Cable at 1/2 rated current	0.91
Cable at 1/4 rated current	0.88

Factors For Calculating Voltage Drop Using Pyrotenax MI Cable

AWG	Single Conductor	2 Conductor	3 Conductor	4 Conductor	7 Conductor
18		15.06	15.57	15.16	15.60
16	9.2	9.40	9.48	9.63	9.63
14	5.7	5.46	5.67	5.50	5.86
12	3.46	3.43	3.49	3.49	3.62
10	2.24	2.20	2.24	2.20	2.32
8	1.492	1.470	1.512	1.480	
6	.954	.928	.968	.944	
4	.602	.580	.608		
3	.478				
2	.406				
1	.314				
1/0	.254				
2/0	.202				
3/0	.1626				
4/0	.1296				
250 MCM	.1112				
350 MCM	.086				
500 MCM	.064				

Shaded area figures include an allowance for the effect of sheath loss (assuming the cables are run close together).

8.2 FIRE PUMP APPLICATIONS

The electrical requirements for electric-drive fire pumps are discussed in detail in Chapters 6 and 7 and Appendix A of NFPA 20. These requirements are supplemented by NFPA 70 (NEC), in particular, Articles 230, 430, and 700. The following guideline items are design highlights (based on Connecticut's and Massachusetts' requirements). Please refer to any different or additional codes or requirements that may be applicable in your state; however, the following should generally be applicable.

1. All electric fire pumps shall be provided with emergency power in accordance with Article 700 of NFPA 70. State of Connecticut requirement (add to Chapter 7, C.L.S.).

2. State of Massachusetts (add to 780 CMR, item 924.3): electrical fire pumps in many occupancies require emergency power per NFPA 20, and NEC Articles 695 and 700.

3. State of Massachusetts (add to 527 CMR, NEC, Article 700): emergency system feeders, generation and distribution equipment, including fire pumps, shall have a 2-h fire separation from all other spaces and equipment.

4. The fire pump feeder conductors shall be physically routed outside the building or enclosed in 2 in of concrete (1-h equivalent fire resistance) except in the electrical switchgear or fire pump rooms. NFPA 20, 6-3.1.1.

5. All pump room wiring shall be in rigid, intermediate, or liquid-tight flexible metal conduit. NFPA 20, 6-3.1.2 (MI cable is added to this in the 1993 version).

6. Maximum permissible voltage drop at the fire pump input terminals is 15 percent. NFPA 20, 6-3.1.4.

7. Protective devices (fuses or circuit breakers) ahead of the fire pump shall not open at the sum of the locked rotor currents of the facility or the fire pump auxiliaries. NFPA 20, 6-3.4.

8. The pump room feeder minimum size shall be 125 percent of the sum of the fire pump(s), jockey pump, and pump auxiliary full-load currents. NFPA 20, 6-3.5.

9. Automatic load shed and sequencing of fire pumps is permitted. NFPA 20, 6-7.

10. Remote annunciation of the fire pump controller is permitted per NFPA 20, 7-4.6 and 7-4.7. Note: A good practice is to assume this will happen and make provisions for it (i.e., fire alarm connections or wiring to the appropriate location).

11. When necessary, an automatic transfer switch may be used. It must be listed for fire pump use. It may be a separate unit or integrated with the fire pump controller in a barriered compartment. NFPA 20, 7-8.2.

12. A jockey pump is not required to be on emergency power.

13. Step-loading the fire pump onto an emergency generator can help control the generator size. A time-delay relay (0 to 60 s) to start or restart a fire pump when on generator power will help coordinate generator loading. The relay should be a part of the fire pump controller (see Item 9 above).

14. Reduced-voltage starters (i.e., autotransformer or wye-delta) for fire pumps are recommended.

FIGURE 8.2 Typical one-line diagram of fire pump system with separate ATS.

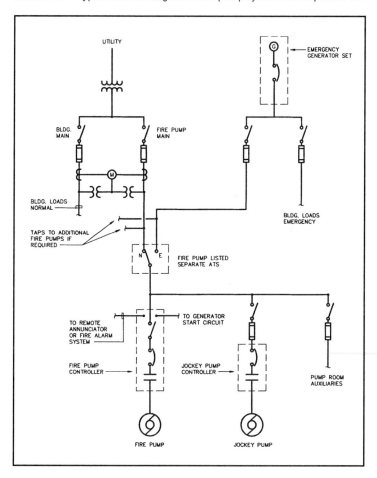

15. Fire pumps, fire pump controllers, and fire pump–listed automatic transfer switches are generally provided under Division 15. Division 16 is responsible for powering, wiring, and connecting this equipment.

Figures 8.2 and 8.3 are typical one-line diagrams showing fire pump systems; Figure 8.2 is with a separate ATS, and Figure 8.3 is with an ATS integrated with the fire pump controller.

FIGURE 8.3   Typical one-line diagram of fire pump system with ATS integrated with the fire pump controller.

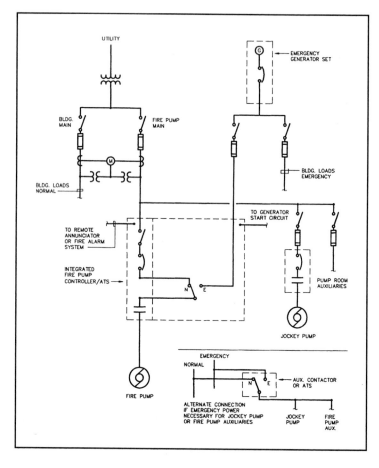

8.3 WIRING FOR PACKAGED ROOFTOP AHUS WITH REMOTE VFDS

An emerging trend in HVAC design is the use of packaged rooftop air-handling units (AHUs) with remote mounted variable-frequency drives (VFDs). In this circumstance, multiple electrical connections and significant additional wiring are required: not the traditional single point of connection previously needed. It is therefore critically important to coordinate closely with the mechanical design professionals to ensure that complete and proper wiring is provided.

Figure 8.4 shows an example of this situation with all of the additional wiring and connections required.

8.4 WYE-DELTA MOTOR STARTER WIRING

A common *mis*application that is encountered is the improper sizing of the six motor leads between the still very popular wye-delta reduced-voltage motor starter and the motor. This is best demonstrated by an example.

Assume that you have a 500-ton electrical centrifugal chiller operating at 460 V, three-phase, 60 Hz, with a nameplate rating of 588 full-load amps (FLA). You would normally apply the correct factor of 125 percent required by NEC Article 440, to arrive at the required conductor ampacity: $588 \times 1.25 = 735$ ampacity for each of the three conductors. Because there will be six conductors between the load side of the starter and the compressor motor terminals, the 735 ampacity is divided by two; you would select six conductors, each having an ampacity of not less than 368 A. Referring to NEC Article 310, Table 310-16 for insulated copper conductors at 75°C would result in the selection of 500-kcmil conductors.

This wire size is incorrect when used between the wye-delta starter and motor terminals. The problem is caused by a common failure to recognize that the motor may consist of a series of single-phase windings.

To permit the transition from wye-start to delta-run configuration, the motor is wound without internal connections. Each end of the three internal motor windings is brought out to a terminal, as shown in Figure 8.5.

The motor windings are configured as required for either starting or running at the starter as shown in Figure 8.6, panels *a* and *b,* respectively.

In the running-delta configuration, the field wiring from the load side of the starter to the compressor motor terminals consists of six conductors, electrically balancing the phases to each of the internal motor windings as described below in Figure 8.7.

Note, for example, that motor winding $T_1 - T_4$ is connected to the line voltage across phase $L_1 - L_2$.

FIGURE 8.4 Wiring of packaged rooftop AHUs with remote VFDs.

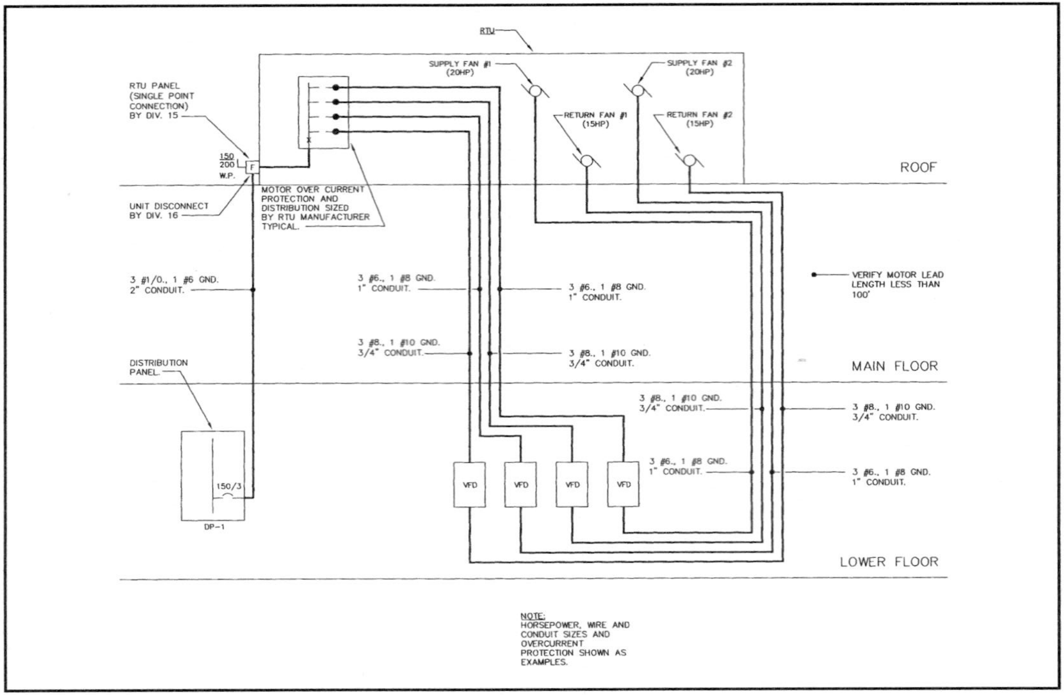

518

FIGURE 8.5 Wye-to-delta internal motor windings brought out to terminals.

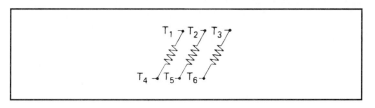

FIGURE 8.6 Wye-start, delta-run motor winding configuration.

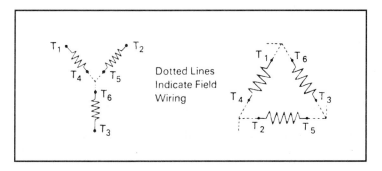

FIGURE 8.7 Field wiring between starter and motor in wye-start, delta-run configuration.

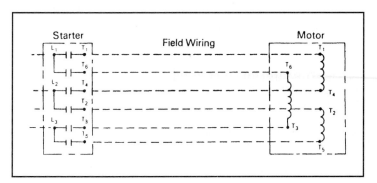

It should be apparent that the windings within the motor are single-phase-connected to the load side of the starter. Thus, the interconnecting field wiring between the starter and motor must be sized as though the motor were single-phase. Electrical terminology simply describes this motor as being phase-connected, and the current carried by the interconnecting conductors as phase amps.

To correctly size the conductors between the motor starter and the motor, therefore, it is necessary to calculate the ampacity with the 125 percent feeder-sizing factor required by the NEC on a single-phase basis as follows:

$$\text{Ampacity per terminal conductor} = \text{three-phase FLA} \times 1.25/1.73$$

For the example given:

$$\text{Ampacity per terminal conductor} = 588 \times 1.25/1.73 = 424$$

Thus, it is clear that the current in the conductors between the starter and the motor on a single-phase basis is 58 percent of the three-phase value, not 50 percent as originally assumed, because the current in one phase of a three-phase system in the delta-connected winding is one divided by the square root of three due to the vector relationship.

In the original example, the conductors were sized for a minimum ampacity of 368 A. From the NEC, 500-kcmil copper conductors at 75°C have a maximum allowable ampacity of 380. The preceding calculation discloses that the conductors should be selected for not less than 424 ampacity. Referring to the NEC again, 600-kcmil conductors have a maximum allowable ampacity of 420. In many cases, depending upon the interpretation of the local electrical inspector, 600 kcmil would be acceptable (usually within 3 percent is acceptable). Five-hundred–kilocircular mil wire would not be.

Almost needless to say, the conductors supplying the line side of the wye-delta starter are sized as conventional three-phase motor conductors.

8.5 MOTOR CONTROL DIAGRAMS

The following provides some basic motor control elementary and wiring diagrams of the most commonly encountered motor control requirements for convenient reference. The reader should refer to various motor control manufacturers for more extensive and detailed information that may be required for specific applications. The following diagrams (Figures 8.8 through 8.17) are courtesy of Square D Company.

FIGURE 8.8 Standard elementary diagram symbols.

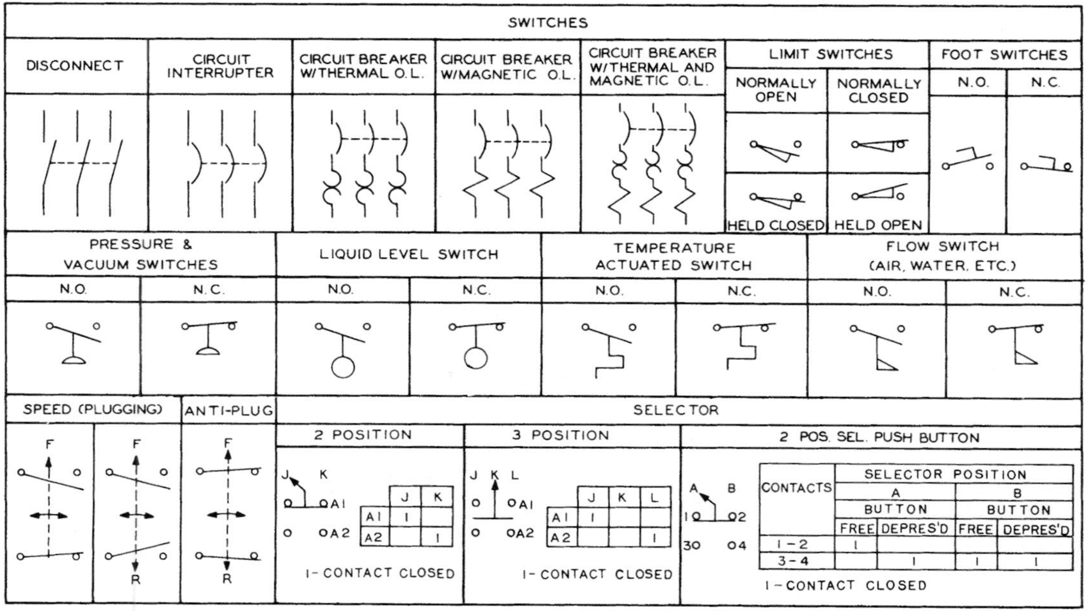

	J	K
A1	1	
A2		1

	J	K	L
A1	1		
A2			1

CONTACTS	SELECTOR POSITION			
	A		B	
	BUTTON		BUTTON	
	FREE	DEPRES'D	FREE	DEPRES'D
1-2	1			
3-4		1	1	1

FIGURE 8.8 Standard elementary diagram symbols. (*Continued*)

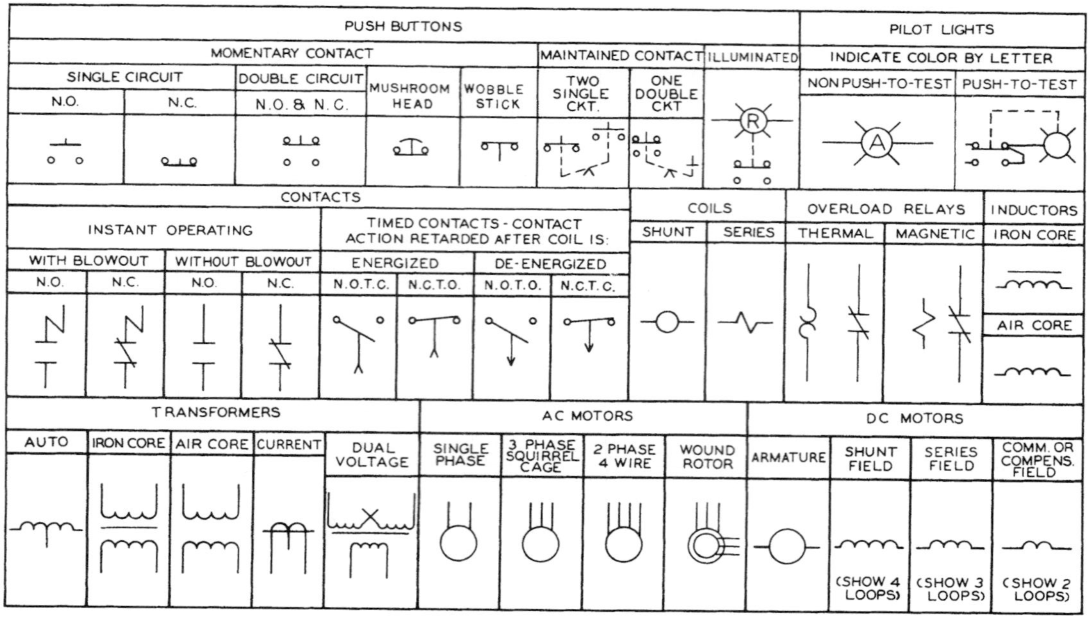

FIGURE 8.8 Standard elementary diagram symbols. (*Continued*)

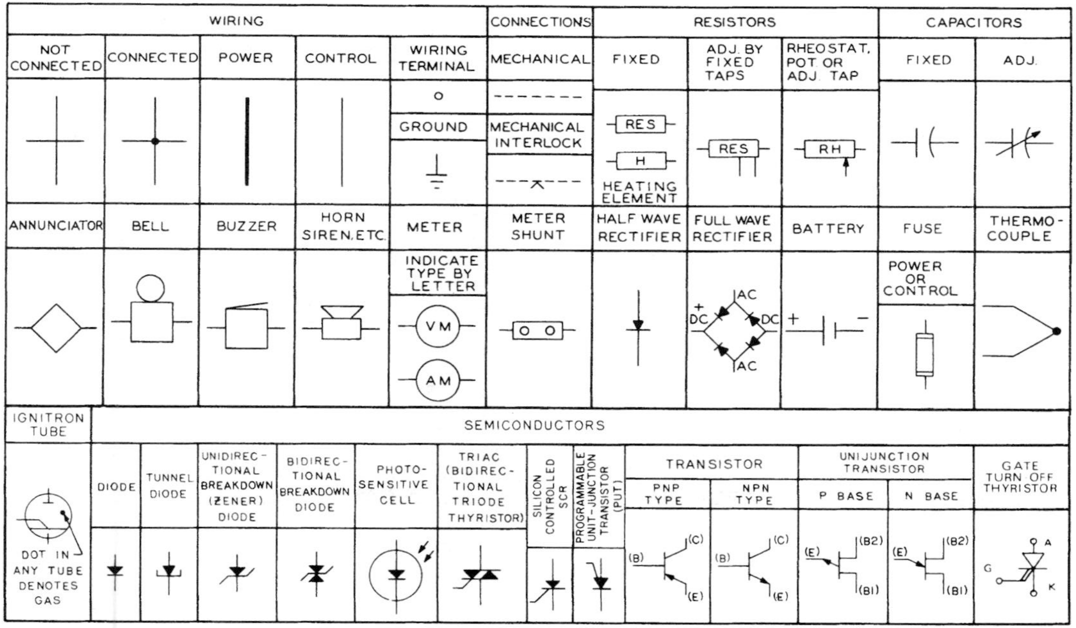

FIGURE 8.9 Supplementary contact symbols.

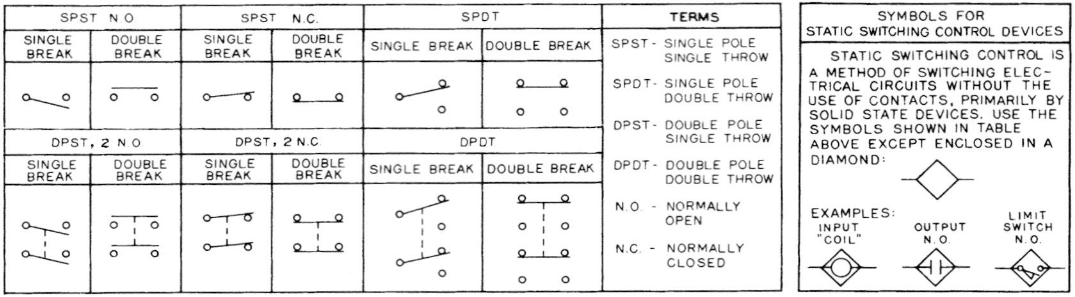
SPST N.O.
SINGLE BREAK
DOUBLE BREAK
SPST N.C.
SINGLE BREAK
DOUBLE BREAK
SPDT
SINGLE BREAK
DOUBLE BREAK
DPST, 2 N.O.
SINGLE BREAK
DOUBLE BREAK
DPST, 2 N.C.
SINGLE BREAK
DOUBLE BREAK
DPDT
SINGLE BREAK
DOUBLE BREAK
TERMS
SPST - SINGLE POLE SINGLE THROW
SPDT - SINGLE POLE DOUBLE THROW
DPST - DOUBLE POLE SINGLE THROW
DPDT - DOUBLE POLE DOUBLE THROW
N.O. - NORMALLY OPEN
N.C. - NORMALLY CLOSED
SYMBOLS FOR STATIC SWITCHING CONTROL DEVICES
STATIC SWITCHING CONTROL IS A METHOD OF SWITCHING ELECTRICAL CIRCUITS WITHOUT THE USE OF CONTACTS, PRIMARILY BY SOLID STATE DEVICES. USE THE SYMBOLS SHOWN IN TABLE ABOVE EXCEPT ENCLOSED IN A DIAMOND:
EXAMPLES:
INPUT "COIL"
OUTPUT N.O.
LIMIT SWITCH N.O.

FIGURE 8.10 Control and power connections—600 V or less, across-the-line starters (From NEMA Standard ICS 2-321A.60).

		1 PHASE	2 PHASE 4 WIRE	3 PHASE
LINE MARKINGS		L1, L2	L1, L3 - PHASE 1 L2, L4 - PHASE 2	L1, L2, L3
GROUND WHEN USED		L1 IS ALWAYS UNGROUNDED	—	L2
MOTOR RUNNING OVERCURRENT UNITS IN	1 ELEMENT	L1	—	—
	2 ELEMENT	—	L1, L4	—
	3 ELEMENT	—	—	L1, L2, L3
CONTROL CIRCUIT CONNECTED TO		L1, L2	L1, L3	L1, L2
FOR REVERSING INTERCHANGE LINES		—	L1, L3	L1, L3

FIGURE 8.11 Terminology.

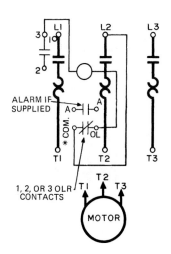

* Marked as "OL" if alarm contact is supplied.

WIRING DIAGRAM

A WIRING DIAGRAM shows, as closely as possible, the actual location of all of the component parts of the device. The open terminals (marked by an open circle) and arrows represent connections made by the user.

Since wiring connections and terminal markings are shown, this type of diagram is helpful when wiring the device, or tracing wires when troubleshooting. Note that bold lines denote the power circuit, and thin lines are used to show the control circuit. Conventionally, in ac magnetic equipment, black wires are used in power circuits and red wiring is used for control circuits.

A wiring diagram, however, is limited in its ability to convey a clear picture of the sequence of operation of a controller. Where an illustration of the circuit in its simplest form is desired, the elementary diagram is used.

FIGURE 8.11 Terminology. (*Continued*)

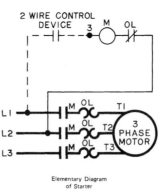

Elementary Diagram
of Starter
(2-wire control)

ELEMENTARY DIAGRAM

The elementary diagram gives a fast, easily understood picture of the circuit. The devices and components are not shown in their actual positions. All the control circuit components are shown as directly as possible, between a pair of vertical lines, representing the control power supply. The arrangement of the components is designed to show the sequence of operation of the devices, and helps in understanding how the circuit operates. The effect of operating various auxiliary contacts, control devices etc. can be readily seen — this helps in trouble shooting, particularly with the more complex controllers. This form of electrical diagram is sometimes referred to as a "schematic" or "line" diagram.

FIGURE 8.12 Examples of control circuits—elementary diagrams.

Low Voltage Release is a "two wire" control scheme using a maintained contact pilot device in series with the starter coil. This scheme is used when a starter is required to function automatically without the attention of an operator. If a power failure occurs while the contacts of the pilot device are closed, the starter will drop out. When the power is restored, the starter will pickup automatically through the closed contacts of the pilot device. The term "two wire" control arises from the fact that in the basic circuit, only two wires are required to connect the pilot device to the starter.

Low Voltage Protection is a "3 wire" control scheme using momentary contact push buttons or similar pilot devices to energize the starter coil. This scheme is used to prevent the unexpected starting of motors which could result in possible injury to machine operators or damage to driven machinery. The starter is energized by pressing the start button. An auxiliary "holding circuit" contact on the starter forms a parallel circuit around the start button contacts holding the starter in after the button is released. If a power failure occurs, the starter will drop out and will open the holding circuit contact. Upon resumption of power, the start button **must** be operated again before the motor will restart. The term "3 wire" control arises from the fact that in the basic circuit at least three wires are required to connect the pilot devices to the starter.

2 WIRE CONTROL

PILOT DEVICE SUCH AS
LIMIT SWITCH, PRESSURE SWITCH, ETC.

3 WIRE CONTROL

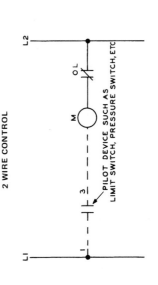

FIGURE 8.12 Examples of control circuits—elementary diagrams. (*Continued*)

3 WIRE CONTROL — MOMENTARY CONTACT MULTIPLE PUSH BUTTON STATION

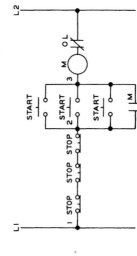

When a motor must be started and stopped from more than one location, any number of "Start" and "Stop" push buttons may be wired together as required. It is also possible to use only one "Start-Stop" station and have several "Stop" buttons at different locations to serve as emergency stop.

2 WIRE CONTROL — WITH MAINTAINED CONTACT HAND-OFF-AUTO SELECTOR SWITCH

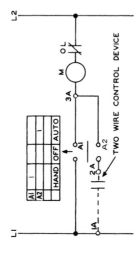

A "Hand-Off-Auto" selector switch is used on two wire control applications where it is desirable to operate the starter manually as well as automatically. The starter coil is energized manually when the switch is turned to the "Hand" position, and is energized automatically by the pilot device when the switch is in the "Auto" position.

FIGURE 8.12 Examples of control circuits—elementary diagrams. (*Continued*)

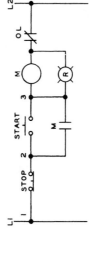

3 WIRE CONTROL WITH PILOT LIGHT TO INDICATE WHEN MOTOR IS STOPPED

A pilot light may be required to indicate when the motor is stopped. This can be done by wiring a normally closed auxiliary contact on the starter in series with the pilot light as shown. When the starter is deenergized, the pilot light is on. When the starter picks up, the auxiliary contact opens, turning off the light.

3 WIRE CONTROL WITH PILOT LIGHT TO INDICATE WHEN MOTOR IS RUNNING

A pilot light can be wired in parallel with the starter coil to indicate when the starter is energized and thus show that the motor is running.

FIGURE 8.12 Examples of control circuits—elementary diagrams. *(Continued)*

3 WIRE CONTROL WITH PUSH-TO-TEST PILOT LIGHT TO INDICATE WHEN MOTOR IS RUNNING

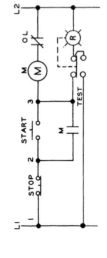

When the motor running pilot light is not lit, there may be doubt as to whether the circuit is open or whether the pilot light bulb is burned out. The push-to-test pilot light enables the testing of the bulb simply by pushing on the color cap.

3 WIRE CONTROL WITH ILLUMINATED PUSH BUTTON TO INDICATE WHEN MOTOR IS RUNNING

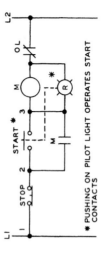

* PUSHING ON PILOT LIGHT OPERATES START CONTACTS

The illuminated push button combines a start button and a pilot light in one unit. Pressing the pilot light lens operates the start contacts. Space is saved by requiring only a two unit push button station instead of three.

FIGURE 8.12 Examples of control circuits—elementary diagrams. (*Continued*)

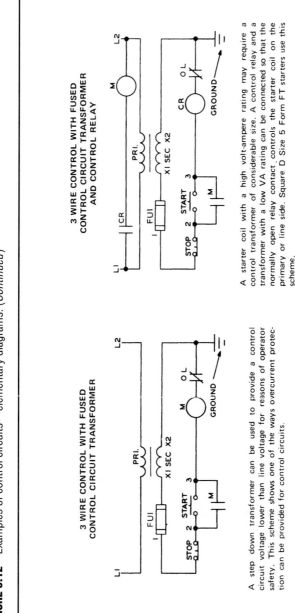

**3 WIRE CONTROL WITH FUSED
CONTROL CIRCUIT TRANSFORMER**

A step down transformer can be used to provide a control circuit voltage lower than line voltage for reasons of operator safety. This scheme shows one of the ways overcurrent protection can be provided for control circuits.

**3 WIRE CONTROL WITH FUSED
CONTROL CIRCUIT TRANSFORMER
AND CONTROL RELAY**

A starter coil with a high volt-ampere rating may require a control transformer of considerable size. A control relay and a transformer with a low VA rating can be connected so that the normally open relay contact controls the starter coil on the primary or line side. Square D Size 5 Form FT starters use this scheme.

531

FIGURE 8.12 Examples of control circuits—elementary diagrams. (*Continued*)

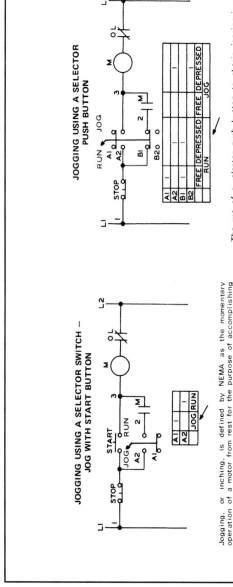

JOGGING USING A SELECTOR SWITCH – JOG WITH START BUTTON

		JOG	RUN
A1			—
A2		—	

Jogging, or inching, is defined by NEMA as the momentary operation of a motor from rest for the purpose of accomplishing small movements of the driven machine. One method of jogging is shown above. The selector switch disconnects the holding circuit contact and jogging may be accomplished by pressing the "Start" button.

JOGGING USING A SELECTOR PUSH BUTTON

	RUN		JOG	
	FREE	DEPRESSED	FREE	DEPRESSED
A1	—			—
A2	—	—		
B1	—	—		
B2				—

The use of a selector push button to obtain jogging is shown above. In the "Run" position the selector-push button gives normal 3 wire control. In the "Jog" position, the holding circuit is broken and jogging is accomplished by depressing the button.

532

FIGURE 8.12 Examples of control circuits—elementary diagrams. *(Continued)*

JOGGING USING A CONTROL RELAY

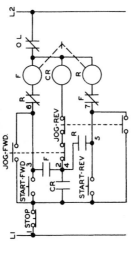

Pressing the "Start" button energizes the control relay which in turn energizes the starter coil. The normally open starter auxiliary contact, and relay contact then, form, a holding circuit around the "Start" button. Pressing the "Jog" button energizes the starter coil independent of the relay and no holding circuit forms, thus jogging can be obtained.

JOGGING USING A CONTROL RELAY FOR REVERSING STARTER

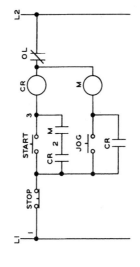

This control scheme permits jogging the motor either in the forward or reverse direction whether the motor is at standstill or is rotating in either direction. Pressing the "Start-Forward" or "Start-Reverse" buttons energizes the corresponding starter coil which closes the circuit to the control relay. The relay picks up and completes the holding circuit around the "Start" button. As long as the relay is energized either the forward or reverse contactor will remain energized. Pressing either the "Jog" button will deenergize the relay releasing the closed contactor. Further pressing of the "Jog" button permits jogging in the desired direction.

FIGURE 8.12 Examples of control circuits—elementary diagrams. (*Continued*)

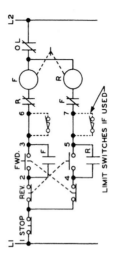

3 WIRE CONTROL - REVERSING STARTER

LIMIT SWITCHES IF USED

3 wire control of a reversing starter can be accomplished with a "Forward-Reverse-Stop" push button station as shown above. Limit switches can be added to stop the motor at a certain point in either direction. Jumpers 6 to 3 and 7 to 5 must then be removed.

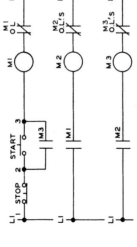

3 WIRE CONTROL - MORE THAN ONE STARTER, ONE PUSH BUTTON STATION CONTROLS ALL

When one "Start-Stop" station is required to control more than one starter, the scheme above can be used. A maintained overload on any one of the motors will drop out all three starters.

FIGURE 8.12 Examples of control circuits—elementary diagrams. (*Continued*)

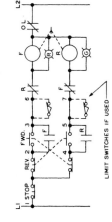

**3 WIRE CONTROL - REVERSING STARTER
WITH PILOT LIGHTS TO INDICATE
DIRECTION MOTOR IS RUNNING**

LIMIT SWITCHES IF USED

Pilot lights can be connected in parallel with the forward and reverse contactor coils to indicate which contactor is energized and thus which direction the motor is running.

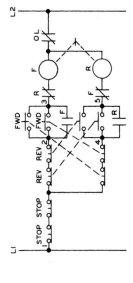

**3 WIRE CONTROL - REVERSING STARTER
MULTIPLE PUSH BUTTON STATION**

More than one "Forward-Reverse-Stop" push button station may be required and can be connected in the manner shown above.

535

FIGURE 8.12 Examples of control circuits—elementary diagrams. (*Continued*)

3 WIRE CONTROL - TWO SPEED STARTER WITH ONE PILOT LIGHT TO INDICATE MOTOR OPERATION AT EACH SPEED

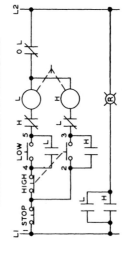

One pilot light can be used to indicate operation at both low and high speeds. One extra normally open auxiliary contact on each contactor is required. Two pilot lights, one for each speed, could be used by connecting pilot lights in parallel with high and low coils. (See Reversing Starter diagram above.)

3 WIRE CONTROL - TWO SPEED STARTER

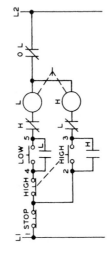

3 wire control of a two speed starter with a "High-Low-Stop" push button station is shown above. This scheme allows the operator to start the motor from rest at either speed or to change from low to high speed. The "Stop" button must be operated before it is possible to change from high to low speed. This arrangement is intended to prevent excessive line current and shock to motor and driven machinery which results when motors running at high speed are reconnected for a lower speed.

536

FIGURE 8.12 Examples of control circuits—elementary diagrams. (*Continued*)

PLUGGING A MOTOR TO A STOP FROM ONE DIRECTION ONLY

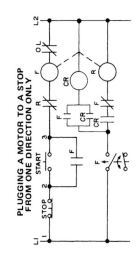

Plugging is defined by NEMA as a system of braking in which the motor connections are reversed so that the motor develops a counter torque, thus exerting a retarding force. In the above scheme the forward rotation of the motor closes the normally open plugging switch contact and energizing control relay CR. When the "Stop" button is operated the forward contactor drops out, the reverse contactor is energized through the plugging switch, the control relay contact as well as the normally closed forward auxiliary contact. This reverses the motor connections and the motor is braked to a stop. The plugging switch then opens and disconnects the reverse contactor, the control relay drops out as well. The control relay makes it impossible for the motor to be plugged in reverse by rotating the motor rotor clockwise. This type of control is used for plugging and not for running in reverse.

ANTI-PLUGGING - MOTOR TO BE REVERSED BUT MUST NOT BE PLUGGED

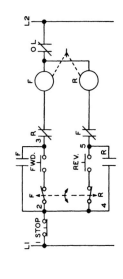

Anti-plugging protection is defined by NEMA as the effect of a device which operates to prevent application of counter-torque by the motor until the motor speed has been reduced to an acceptable value. In the scheme above, with the motor operating in one direction, a contact on the anti-plugging switch opens the control circuit of the contactor used for the opposite direction. This contact will not close until the motor has slowed down, after which the other contactor can be energized.

537

FIGURE 8.13 Examples of overcurrent protection for control circuits.

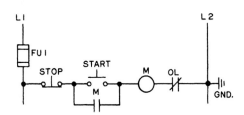

Common control with fusing in one line only and with both lines ungrounded or, if user's conditions permit, with one line grounded.

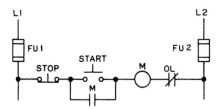

Common control with fusing in both lines and with both lines ungrounded.

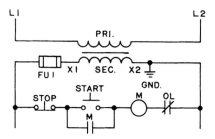

Control circuit transformer with fusing in one secondary line and with both secondary lines ungrounded or, if user's conditions permit, with one line grounded.

FIGURE 8.13   Examples of overcurrent protection for control circuits. (*Continued*)

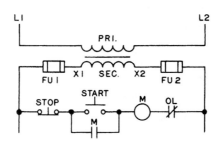

Control circuit transformer with fusing in both secondary lines and with both secondary lines ungrounded.

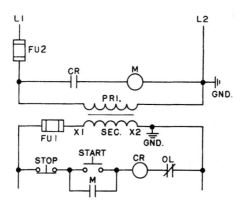

Control circuit transformer with fusing in one primary and one secondary line, and with all lines ungrounded, or, if user's conditions permit, with one primary and one secondary line grounded.

FIGURE 8.13 Examples of overcurrent protection for control circuits. (*Continued*)

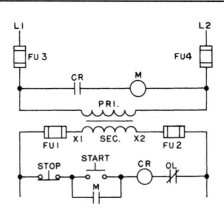

Control circuit transformer with fusing in both primary lines and both secondary lines and with all lines ungrounded.

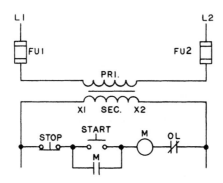

Control circuit transformer with fusing in both primary lines, with no secondary fusing and with all lines ungrounded.

FIGURE 8.14 AC manual starters and manual motor starting switches.

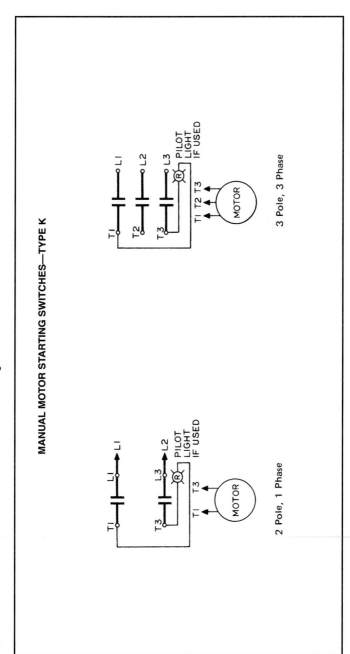

FIGURE 8.14 AC manual starters and manual motor starting switches. (*Continued*)

FRACTIONAL HORSEPOWER MANUAL STARTERS—TYPE F

1 Pole

2 Pole

2 Pole with Selector Switch

FIGURE 8.14 AC manual starters and manual motor starting switches. (*Continued*)

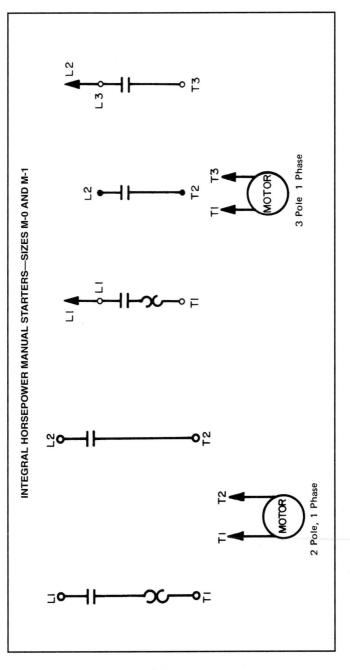

INTEGRAL HORSEPOWER MANUAL STARTERS—SIZES M-0 AND M-1

FIGURE 8.14 AC manual starters and manual motor starting switches. (*Continued*)

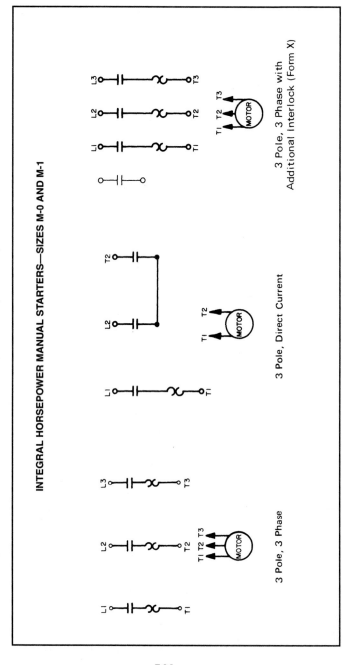

INTEGRAL HORSEPOWER MANUAL STARTERS—SIZES M-0 AND M-1

3 Pole, 3 Phase

3 Pole, Direct Current

3 Pole, 3 Phase with
Additional Interlock (Form X)

FIGURE 8.14 AC manual starters and manual motor starting switches. (*Continued*)

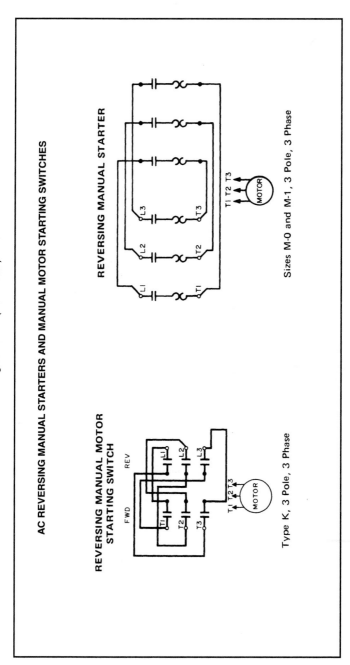

FIGURE 8.14 AC manual starters and manual motor starting switches. (*Continued*)

AC TWO SPEED MANUAL STARTERS AND MANUAL MOTOR STARTING SWITCHES

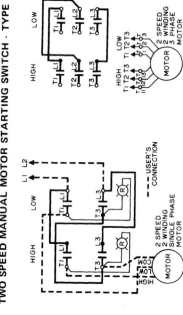

TWO SPEED MANUAL MOTOR STARTING SWITCH - TYPE K

2 Pole, Single Phase with Pilot Light

3 Pole, 3 Phase

FIGURE 8.14 AC manual starters and manual motor starting switches. (*Continued*)

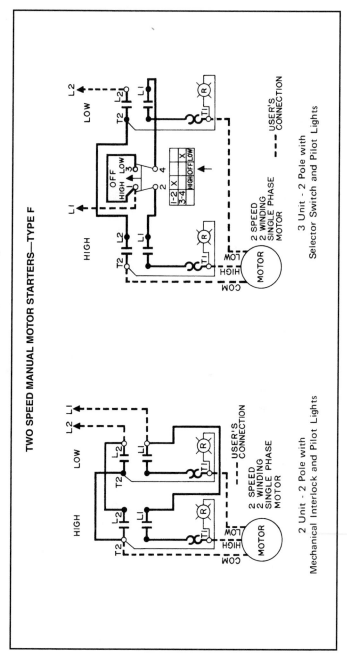

TWO SPEED MANUAL MOTOR STARTERS—TYPE F

2 Unit - 2 Pole with
Mechanical Interlock and Pilot Lights

3 Unit - 2 Pole with
Selector Switch and Pilot Lights

FIGURE 8.14 AC manual starters and manual motor starting switches. (*Continued*)

SIZES M-O AND M-1 - TWO SPEED MANUAL STARTERS

Two Speed Starter For Wye Connected Separate Winding Motor

FIGURE 8.15 Medium-voltage motor controllers.

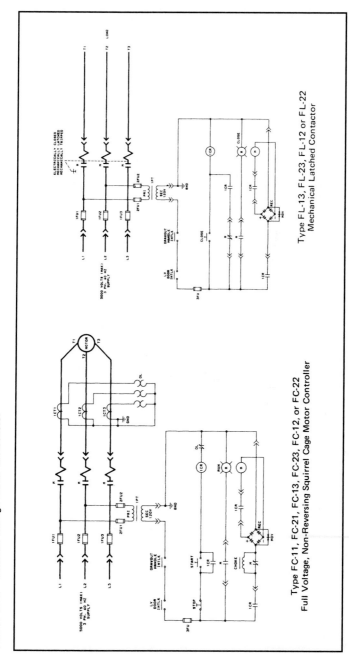

Type FC-11, FC-21, FC-13, FC-23, FC-12, or FC-22
Full Voltage, Non-Reversing Squirrel Cage Motor Controller

Type FL-13, FL-23, FL-12 or FL-22
Mechanical Latched Contactor

FIGURE 8.15 Medium-voltage motor controllers. (*Continued*)

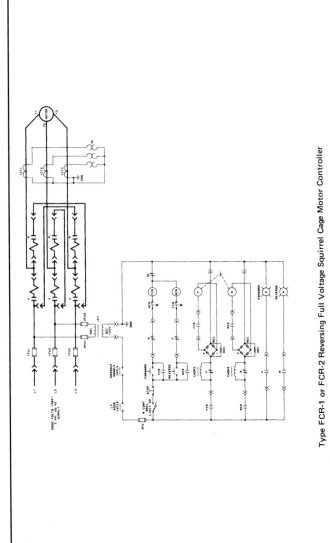

Type FCR-1 or FCR-2 Reversing Full Voltage Squirrel Cage Motor Controller

FIGURE 8.15 Medium-voltage motor controllers. (*Continued*)

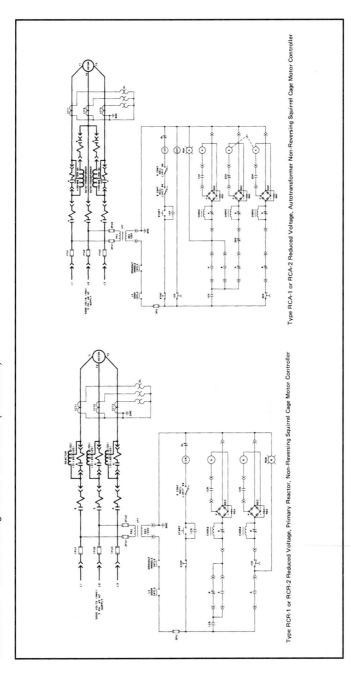

Type RCR-1 or RCR-2 Reduced Voltage, Primary Reactor, Non-Reversing Squirrel Cage Motor Controller

Type RCA-1 or RCA-2 Reduced Voltage, Autotransformer Non-Reversing Squirrel Cage Motor Controller

FIGURE 8.15 Medium-voltage motor controllers. (*Continued*)

Type FS-1 or FS-2 Full Voltage, Non-Reversing Synchronous Motor Controller

Type RSR-1 or RSR-2 Reduced Voltage Primary Reactor Non-Reversing Synchronous Motor Controller

552

FIGURE 8.15 Medium-voltage motor controllers. (*Continued*)

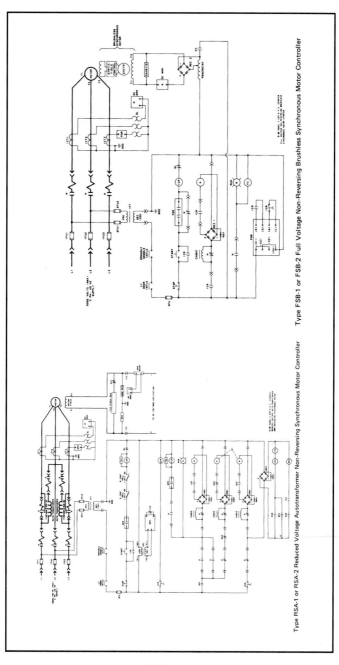

Type RSA-1 or RSA-2 Reduced Voltage Autotransformer Non-Reversing Synchronous Motor Controller

Type FSB-1 or FSB-2 Full Voltage Non-Reversing Brushless Synchronous Motor Controller

FIGURE 8.16 Reduced-voltage controllers.

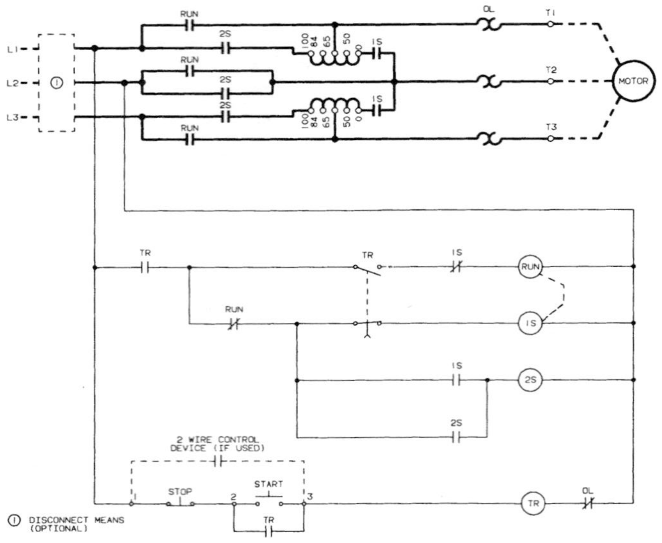

Sizes 2-5 Reduced Voltage Autotransformer Controllers
with Closed Transition Starting

FIGURE 8.16 Reduced-voltage controllers. (*Continued*)

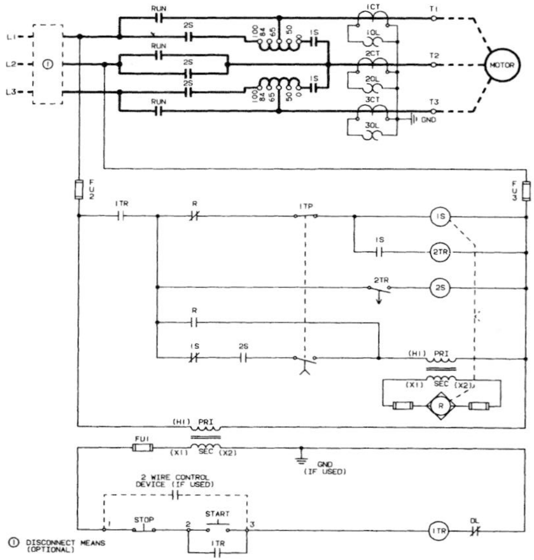

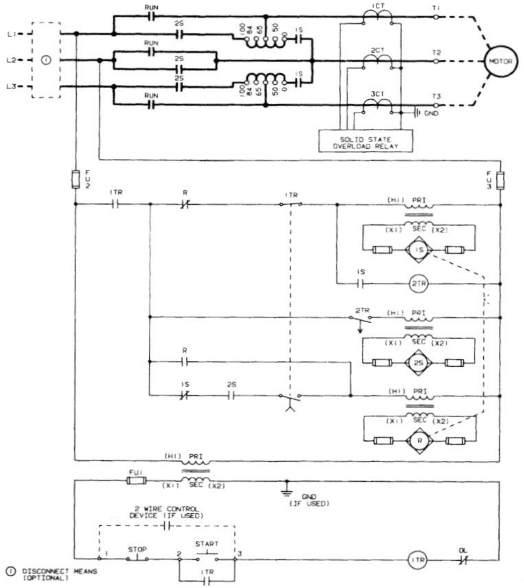

Size 7 Reduced Voltage Autotransformer Controller with Closed Transition Starting

Size 6 Reduced Voltage Autotransformer Controller with Closed Transition Starting

FIGURE 8.16 Reduced-voltage controllers. (*Continued*)

Sizes 1YΔ–5YΔ Controllers with Open Transition Starting

Sizes 1YΔ–5YΔ Controllers with Closed Transition Starting

FIGURE 8.16 Reduced-voltage controllers. (*Continued*)

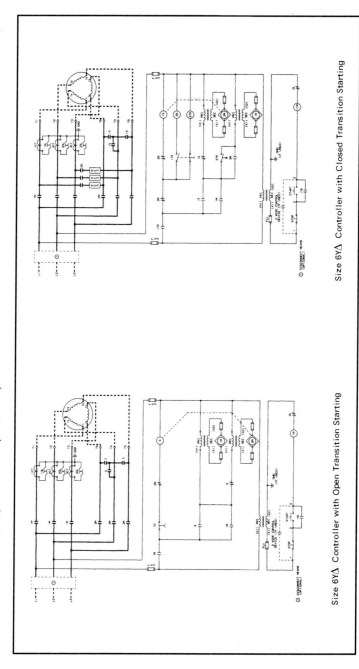

Size 6YΔ Controller with Closed Transition Starting

Size 6YΔ Controller with Open Transition Starting

FIGURE 8.16 Reduced-voltage controllers. (*Continued*)

Size 7, 3 Phase Primary Resistor Controller

Size 6, 3 Phase Primary Resistor Controller

FIGURE 8.16 Reduced-voltage controllers. *(Continued)*

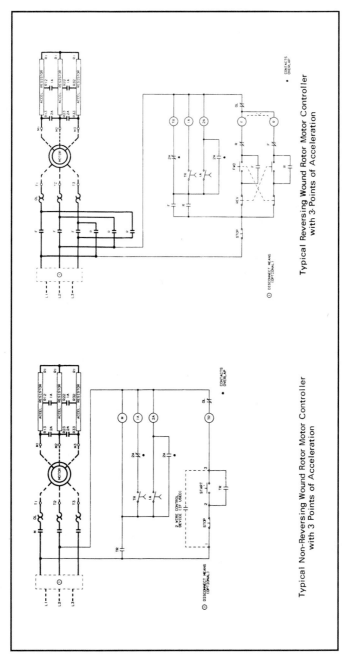

Typical Non-Reversing Wound Rotor Motor Controller
with 3 Points of Acceleration

Typical Reversing Wound Rotor Motor Controller
with 3 Points of Acceleration

FIGURE 8.17 Solid-state reduced-voltage controllers.

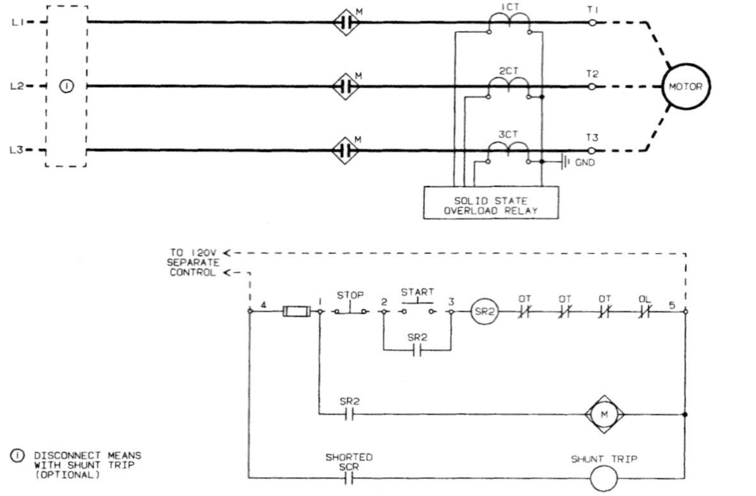

Solid State Reduced Voltage Controller
MH(200A), MJ(320A), MK(500A) and MM(750A) Devices

560

FIGURE 8.17 Solid-state reduced-voltage controllers. (*Continued*)

Solid State Reduced Voltage Controllers
with a Shorting Contactor MH(200A) Device

Solid State Reduced Voltage Controllers with a Shorting
Contactor MJ(320A), MK(500A) and MM(750A) Devices

FIGURE 8.17 Solid-state reduced-voltage controllers. (*Continued*)

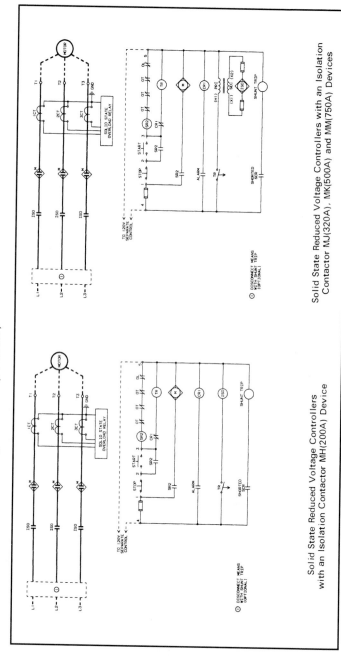

Solid State Reduced Voltage Controllers
with an Isolation Contactor MH(200A) Device

Solid State Reduced Voltage Controllers with an Isolation
Contactor MJ(320A), MK(500A) and MM(750A) Devices

FIGURE 8.17 Solid-state reduced-voltage controllers. (*Continued*)

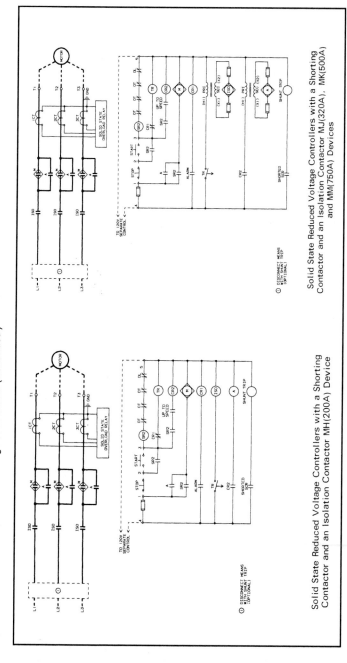

Solid State Reduced Voltage Controllers with a Shorting Contactor and an Isolation Contactor MH(200A) Device

Solid State Reduced Voltage Controllers with a Shorting Contactor and an Isolation Contactor MJ(320A), MK(500A) and MM(750A) Devices

8.6 ELEVATOR RECALL SYSTEMS

Elevator recall systems are discussed here rather than under Fire Alarm Systems in Chapter 7 because they can be installed as a stand-alone system, even though they are generally a part of a fire alarm system. Also, several codes are applicable to the installation of these systems, specifically ANSI/ASME A17.1, Safety Code for Elevators and Escalators; NFPA 72, National Fire Alarm Code; NFPA 13, Standard for Installation of Sprinklers; and NFPA 101, Life Safety Code—to which the reader is referred for complete details.

Further, applying these codes properly in combination can be problematic (for example, whether sprinklers are present), coupled with the requirements of the authority having jurisdiction (which are generally more stringent).

Briefly stated, ANSI/ASME A17.1 is written so as to ensure that an elevator car will not stop and open the door on a fire-involved floor by requiring elevators to be recalled nonstop to a designated safe floor when smoke detectors located in elevator lobbies, other than the designated level, are actuated. When the smoke detector at the designated level is activated, the cars return to an alternate level approved by the enforcing authority.

If the elevator is equipped with front and rear doors, it is necessary to have smoke detectors in both lobbies at the designated level.

Activation of a smoke detector in any elevator machine room, except a machine room at the designated level, shall cause all elevators having any equipment located in that machine room, and any associated elevators of a group automatic operation, to return nonstop to the designated level. When a smoke detector in an elevator machine room is activated that is at the designated level, with the other conditions being the same as above, the elevators shall return nonstop to the alternate level, or the appointed level when approved by the authority having jurisdiction.

NFPA 72 requires that in facilities without a building fire alarm system, these smoke detectors shall be connected to a dedicated fire alarm system control unit that shall be designated as "elevator recall control and supervisory panel." Thus, the stand-alone operation noted previously.

As noted, the foregoing is by no means complete, but captures the intent and basic cause-and-effect relationship between an elevator recall system's smoke detectors and elevator operation under the various stated conditions.

Figure 8.18 shows a typical elevator recall/emergency shutdown schematic. Please note that the authority having jurisdiction required that the elevator recall smoke detectors in this application be independent of the building fire alarm system smoke detectors. Figure 8.19 shows a typical elevator hoistway/machine room device installation detail for the same project application shown in Figure 8.18. Note that

FIGURE 8.18   Typical elevator recall/emergency shutdown schematic.

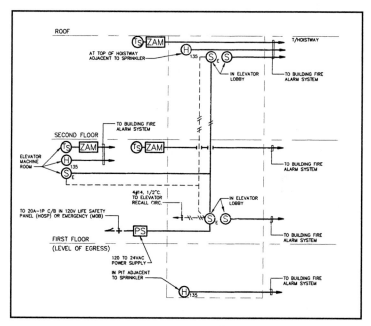

the fire alarm system is fully addressable and that the elevator machine rooms are at the designated level for egress.

8.7   MEDIUM-VOLTAGE CABLE AND ENGINEERING DATA

The following provides data on medium-voltage cable and engineering data. Although it would be nice to provide data for virtually every requirement, it is not the intent of this handbook. It would be impossible to show all such data. What is provided is most likely to be required in most situations. You might consider it a more narrow "bell curve" of data.

Ampacities

Experience has shown that most applications, usually college/university, hospital, or similar campus situations, involve underground distribution (conductors in duct bank or direct-buried). The most widely used con-

FIGURE 8.19 Typical elevator hoistway/machine room device installation detail.

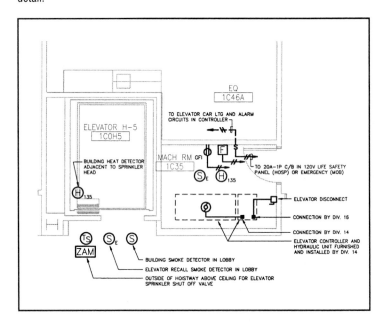

ductors are EPR-insulted, single conductors paralleled or triplexed, in conduit or duct bank. They may also be direct-buried or in air. The voltage class is usually 15 kV, although it may typically be 5 to 25 kV. With these parameters in mind, the following ampacity tables (Table 8.4 and Figures 8.20 and 8.21) are provided with the installation details upon which they are based.

Allowable Short-Circuit Currents

As indicated in Chapter 3, short-circuit currents for low-voltage cables (600 V and below) are not of significant concern for the cable withstand capability; however, for medium-voltage cable, it is of much greater concern. With this in mind, the following is provided in Figure 8.22.

DC Field Acceptance Testing

It is general practice, and obviously empirical, to relate the field test voltage upon installation to the final factory-applied DC potentials by using a factor of 80 percent. Table 8.5 shows these values.

FIGURE 8.20 Typical installations—underground in ducts.

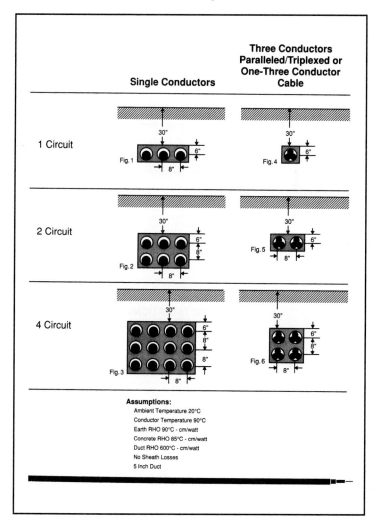

Three Conductors Paralleled/Triplexed or One-Three Conductor Cable

Single Conductors

1 Circuit

30"

6"

Fig. 1

8"

30"

6"

Fig. 4

2 Circuit

30"

6"

8"

Fig. 2

8"

30"

6"

Fig. 5

8"

4 Circuit

30"

6"

8"

8"

Fig. 3

8"

30"

6"

8"

Fig. 6

8"

Assumptions:
Ambient Temperature 20°C
Conductor Temperature 90°C
Earth RHO 90°C - cm/watt
Concrete RHO 85°C - cm/watt
Duct RHO 600°C - cm/watt
No Sheath Losses
5 Inch Duct

FIGURE 8.21 Typical installations—direct-buried and in-air.

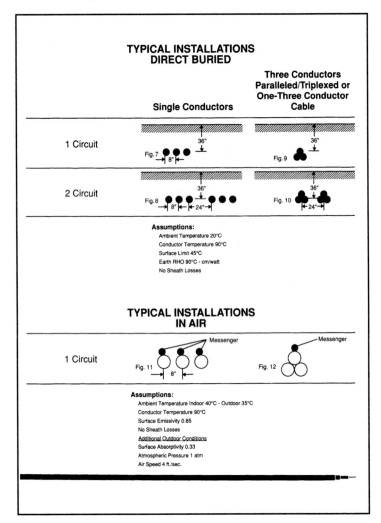

FIGURE 8.22 Allowable short-circuit currents for insulated copper conductors.

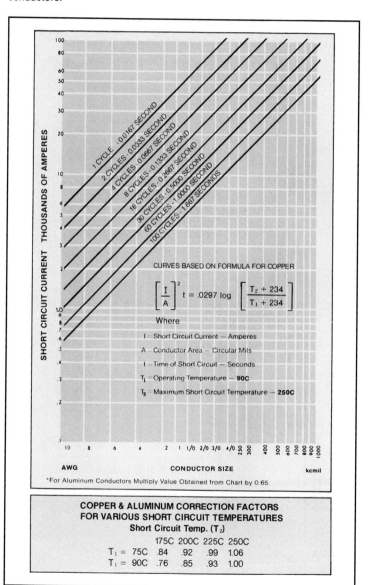

CURVES BASED ON FORMULA FOR COPPER

$$\left[\frac{I}{A}\right]^2 t = .0297 \log \left[\frac{T_2 + 234}{T_1 + 234}\right]$$

Where

I = Short Circuit Current — Amperes

A = Conductor Area — Circular Mils

t = Time of Short Circuit — Seconds

T_1 = Operating Temperature — **90C**

T_2 = Maximum Short Circuit Temperature — **250C**

*For Aluminum Conductors Multiply Value Obtained from Chart by 0.65.

**COPPER & ALUMINUM CORRECTION FACTORS
FOR VARIOUS SHORT CIRCUIT TEMPERATURES**
Short Circuit Temp. (T_2)

	175C	200C	225C	250C
T_1 = 75C	.84	.92	.99	1.06
T_1 = 90C	.76	.85	.93	1.00

TABLE 8.4 Triplexed or Paralleled Cable Ampacities, Single Conductors, Copper and Aluminum, EPR Insulated, 5 to 35 kV

Copper Conductors

Conductor Size AWG/kcmil	Underground In Ducts - Three 1/C Cable Per Duct									Direct Buried Three 1/C Cable Per Circuit						In Air Three Singles	
	1 Circuit Fig. 4 Load Factor			2 Circuits Fig. 5 Load Factor			4 Circuits Fig. 6 Load Factor			1 Circuit Fig. 9 Load Factor			2 Circuits Fig. 10 Load Factor			Fig. 12 Indoor	Fig. 12 Outdoor
	50	75	100	50	75	100	50	75	100	50	75	100	50	75	100		
6	101	97	92	97	91	84	91	81	73	139	114	92	135	104	83	101	130
4	131	125	119	126	117	108	117	105	93	182	146	117	175	132	105	133	171
2	174	166	156	167	154	141	154	136	120	234	189	152	224	170	136	179	219
1	199	189	178	190	175	160	175	154	135	268	214	172	254	192	153	205	252
1/0	227	215	202	216	199	181	198	174	153	306	242	194	287	216	173	235	289
2/0	259	245	230	246	226	205	225	197	173	351	273	219	324	244	195	270	332
3/0	295	279	261	280	256	233	256	223	195	402	308	247	366	275	220	311	382
4/0	337	317	297	319	291	264	290	253	221	460	349	279	413	310	248	358	439
250	372	350	326	352	320	289	319	277	241	504	382	306	452	339	271	398	485
350	450	422	392	424	384	346	383	331	287	603	455	364	539	404	322	488	594
500	549	513	475	516	465	417	463	398	344	727	547	437	647	483	385	605	735
750	680	633	584	636	571	510	568	485	418	892	671	536	791	590	470	760	905
1000	786	728	670	733	654	582	651	533	474	1023	767	612	903	672	535	893	1056

TABLE 8.4 Triplexed or Paralleled Cable Ampacities, Single Conductors, Copper and Aluminum, EPR Insulated, 5 to 35 KV (*Continued*)

Aluminum Conductors

| Conductor Size AWG/kcmil | Underground In Ducts - Three 1/C Cable Per Duct | | | | | | | | | Direct Buried Three 1/C Cable Per Circuit | | | | | | In Air Three Singles | |
| | 1 Circuit Fig. 4 Load Factor | | | 2 Circuits Fig. 5 Load Factor | | | 4 Circuits Fig. 6 Load Factor | | | 1 Circuit Fig. 9 Load Factor | | | 2 Circuits Fig. 10 Load Factor | | | Fig. 12 Indoor | Fig. 12 Outdoor |
	50	75	100	50	75	100	50	75	100	50	75	100	50	75	100		
6	76	73	70	74	69	64	69	62	56	109	88	71	105	80	64	77	101
4	99	95	90	96	89	83	89	80	71	142	112	90	135	102	81	101	133
2	134	127	120	128	118	109	118	105	93	182	146	118	173	131	105	137	171
1	153	145	137	146	135	123	134	119	105	208	165	133	196	148	119	157	197
1/0	174	165	156	166	153	140	153	134	118	239	187	150	222	167	134	181	226
2/0	199	188	177	189	174	158	173	152	134	273	211	169	250	189	151	208	260
3/0	227	215	201	216	198	180	197	173	151	313	238	191	283	213	170	239	299
4/0	260	245	229	246	225	204	224	196	171	356	270	216	320	241	192	276	344
250	286	270	252	271	247	224	246	214	187	390	295	236	350	263	210	307	381
350	349	327	304	329	298	269	297	257	223	468	353	283	419	314	250	378	467
500	428	400	371	402	363	326	361	311	269	567	427	341	506	378	301	472	581
750	539	501	463	504	452	404	449	384	330	704	528	421	624	465	370	608	743
1000	629	584	537	587	524	466	520	442	379	819	614	490	724	539	429	717	855

TABLE 8.5 High-Voltage Field Acceptance Test Prior to Being Placed in Service

Rated Voltage Phase to Phase	dc Hi-Pot Test (15 Minutes)		dc Hi-Pot Test	
	Wall - mils	kV	Wall - mils	kV
5000	90	25	115	35
8000	115	35	140	45
15000	175	55	220	65
25000	260	80	320	95
28000	280	85	345	100
35000	345	100	420	125
46000	445	130	580	170
69000	650	195	650	195

Note: *If the leakage current quickly stabilizes, the duration may be reduced to 10 minutes.

Installation Practices

Conduits or ducts should be properly constructed having smooth walls and of adequate size as determined by the overall cable diameter and recommended percentage fill of conduit area.

For groups or combinations of cables it is recommended that the conduit or tubing be of such size that the sum of the cross-sectional areas of the individual cables will not be more than the percentage of the interior cross-sectional area of the conduit or tubing as shown in Tables 8.7 through 8.10.

Clearance

Clearance refers to the distance between the uppermost cable in the conduit and the inner top of the conduit. Clearance should be ¼ in at minimum and up to 1 in for large-cable installations or installations involving numerous bends. Figure 8.23 shows how it is calculated.

When calculating clearance, ensure all cable diameters are equal. Use triplexed configuration formula if you are in doubt. Again, the cables may be of single- or multiple-conductor construction.

Jam Ratio

Jamming is the wedging of three cables lying side by side in a conduit. This usually occurs when cables are being pulled around bends or when cables twist.

Jam ratio is calculated by slightly modifying the ratio used to measure

Jacket Materials—Relative Performance

TABLE 8.6 Jacket Materials Selection Chart—Relative Performance Data

Mechanical	PVC	Polyethylene	Neoprene	Chlorosulphonated Polyethylene	Thermoplastic CPE
Abrasion Resistance	Good	Excellent	Good	Good	Excellent
Tensile Strength	Excellent	Excellent	Excellent	Excellent	Good
Elongation	Good	Excellent	Excellent	Excellent	Good
Compression Resistance	Good	Excellent	Excellent	Excellent	Good
Flexibility	Good	Fair	Excellent	Excellent	Fair
Environmental					
Flame	Good	Poor	Excellent	Excellent	Good
Moisture					
Fresh or salt water	Good	Exceptional	Good	Excellent	Excellent
Petroleum oils					
Motor oil		Excellent			Good
Fuel oil	Good	(Slight swelling	Good	Good	(Poor above 110°C)
Crude oil		above 60C)			
Creosote	Poor	Good	Fair	Fair	Good
Paraffinic Hydrocarbons					
Gasoline	Good	Excellent	Poor	Poor	Excellent
Kerosene		(Slight swelling at higher temperatures)			(Slight swelling at higher temperatures)
Alcohols					
Isopropyl					
Wood	Fair	Good	Fair	Good	Good
Grain					
Mineral Acids					
Sulfuric					
Nitric	Excellent	Excellent	Excellent	Excellent	Excellent
Hydrochloric					
Fixed Alkalies					
Sodium hydroxide (lye)					
Potassium hydroxide (potash)	Good	Excellent	Good	Excellent	Excellent
Calcium hydroxide (lime)					
Ketones					
Acetone	Poor	Good	Poor	Fair	Good
Methyl ethyl ketone (MEK)					
Esters					
Ethyl Acetate	Poor	Good	Poor	Fair	Good
Most lacquer thinners					
Halogenated Hydrocarbons					
Chloroform					
Carbon Tetrachloride	Poor	Poor	Poor	Poor	Poor
Methyl chloride					
General					
Leaves protective residue after combustion	Yes	No	Yes	Yes	Yes
Ozygen Index (ASTM D-2863)	23-30%	17-18%	31-39%	30-36%	30-34%
Halogen content - % Wt.	26	0	18	14	18-20
Minimum installation temperature	14F (-10C)	-40F (-40C)	-4F(-20C)	-4F (-20C)	−40°F (−40°C)
Dimensional stability under heat	Fair	Fair	Excellent	Excellent	Fair
Maximum operating temperature	75C (167F)	75C (167F)	90C (194F)	90C (194F)	75 C (167F)

NOTE: When cables are to be installed in cold weather. they should be kept in heated storage for at least 24 hrs before installation

configuration (*D/d*). A value of 1.05*D* is used for the inner diameter of the conduit, because bending a cylinder creates an oval cross-section in the bend (1.05*D/d*).

- If 1.05*D/d* is larger than 3.0, jamming is impossible.
- If 1.05*D/d* is between 2.8 and 3.0, serious jamming is probable.
- If 1.05*D/d* is less than 2.5, jamming is impossible but clearance should be checked.

TABLE 8.7 Dimensions of Conduit

Nominal size conduit inches	Internal diameter inches	Area square inches
1	1.049	0.86
1 1/4	1.380	1.50
1 1/2	1.610	2.04
2	2.067	3.36
2 1/2	2.469	4.79
3	3.068	7.38
3 1/2	3.548	9.90
4	4.026	12.72
5	5.047	20.00
6	6.065	28.89

TABLE 8.8 Maximum Percent Internal Area of Conduit or Tubing

	Number of cables				
	1	2	3	4	Over 4
Cables (not lead-covered)	53	31	40	40	40
Lead-covered cables	55	30	40	38	35

*This section summarizes procedures, calculations, and recommendations required for proper installation practices.

TABLE 8.9 Maximum Percent Internal Diameter of Conduit or Tubing

	Number of cables			
	1	2	3	4
Cables (not lead-covered)	72.8	39.3	36.5	31.6
Lead-covered cables	74.2	38.7	36.5	30.8

TABLE 8.10 Maximum Allowable Diameter (in Inches) of Individual Cables in Given Size Conduit

Non-metallic jacketed cable— all cables of same outside diameter				
Nominal size conduit	Number of cables having same O.D.			
	1	2*	3*	4*
1/2	0.453	0.244	0.227	0.197
3/4	0.600	0.324	0.301	0.260
1	0.763	0.412	0.383	0.332
1 1/4	1.010	0.542	0.504	0.436
1 1/2	1.173	0.633	0.588	0.509
2	1.505	0.812	0.754	0.653
2 1/2	1.797	0.970	0.901	0.780
3	2.234	1.206	1.120	0.970
3 1/2	2.583	1.395	1.296	1.121
4	2.930	1.583	1.470	1.273
5	3.675	1.985	1.844	1.595
6	4.416	2.385	2.215	1.916

NOTE: To determine the size conduit required for any number (n) of equal diameter cables in excess of four, multiply the diameter of one cable by $\sqrt{\dfrac{n}{4}}$

This will give the "equivalent" diameter of four such cables and the conduit size required for (n) cables may then be found by using the column for four cables.

*These diameters are based on percent fill only. The Jam Ratio, Conduit ID to Cable O.D., should be checked to avoid possible jamming.

Because there are manufacturing tolerances on cable, the actual overall diameter should be measured prior to computing the jam ratio.

Pulling Tensions

Most major cable manufacturers provide examples of pulling tension calculations in their catalogs and the reader should refer to these for preliminary calculations. It is recommended, however, that you provide to the cable manufacturer that you plan to use the necessary application data for calculations by them.

Minimum Bending Radii

Refer to Table 8.11 for information on minimum bending radii.

FIGURE 8.23 Clearance of cables in conduit.

# Of Conductors/Cables	Configuration	Formula
1	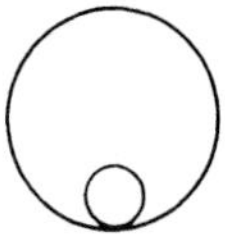	$D-d$
3	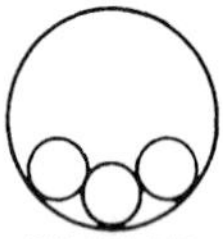 CRADLED	$\dfrac{D}{2} - 1.366d + \dfrac{D-d}{2} \sqrt{1 - \left(\dfrac{d}{D-d}\right)^2}$
3	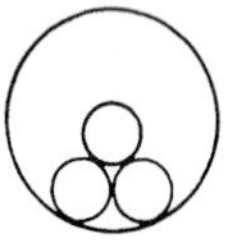 TRIPLEXED	$\dfrac{D}{2} - \dfrac{d}{2} + \dfrac{D-d}{2} \sqrt{1 - \left[\dfrac{d}{2(D-d)}\right]^2}$

TABLE 8.11 Minimum Bending Radii—Power and Control Cables
with Metallic Shielding or Armor

Type of Cable	Minimum Bending Radius as a Multiple of Cable Diameter	
	Power	Control
Armored, flat tape or wire type	12	12
Armored, smooth aluminum sheath, up to		
0.75 inches cable diameter.	10*	10*
0.76 to 1.5 inches cable diameter.	12	12
over 1.5 inches cable diameter	15	15
Armored, corrugated sheath or		
interlocked type	7	7
with shielded single conductor	12	12
with shielded multi-conductor.	**	**
Non-armored, flat or corrugated		
tape shielded single conductor	12	12
tape shielded multi-conductor.	**	**
multi-conductor overall tape shield.	12	12
LCS with PVC jacket.	15	15
Non-armored, flat strap shielded.	8	—

   * with shielded conductors 12
  ** 12 times single conductor diameter or 7 times overall
      cable diameter — whichever is greater

LCS = longitudinally applied corrugated shield

8.8   HARMONIC EFFECTS AND MITIGATION

Introduction

Harmonics are the result of nonlinear loads so prevalent with late-twentieth-century technology. Personal computers, adjustable speed drives, uninterruptible power supplies, to name a few, all have nonlinear load characteristics. What all nonlinear loads have in common is that they convert AC to DC and contain some kind of rectifier.

A sinusoidal system can supply nonsinusoidal current demands because any nonsinusoidal waveform can be generated by the proper combination of harmonics of the fundamental frequency. Each harmonic in the combination has a specific amplitude and phase relative to the fundamental. The particular harmonics drawn by a nonlinear load are a function of the rectifier circuit and are not affected by the type of load.

Harmonic Origins

Harmonics have two basic origins—current wave distortion and voltage wave distortion.

HARMONICS-PRODUCING EQUIPMENT (VOLTAGE DISTORTION)

Uninterruptible power supplies

Variable-frequency drives

Large battery chargers

Elevators

Synchronous clock systems

Radiology equipment

Large electronic dimming systems

Arc heating devices

HARMONICS-PRODUCING EQUIPMENT (CURRENT DISTORTION)

Personal computers

Desktop printers

Small battery chargers

Electric-discharge lighting

Electronic/electromagnetic ballasts

Small electronic dimming systems

It should be noted that voltage distortion is more difficult to deal with because it is system-wide.

Harmonic Characteristics

- Harmonics are integer multiples of the fundamental frequency.
- First order is the fundamental frequency (e.g., 60 Hz); the second order is $2 \times 60 = 120$ Hz; the third order is $3 \times 60 = 180$ Hz; and so on.
- In three-phase systems, even harmonics cancel; odd harmonics are additive in the neutral and ground paths.
- Harmonics that are multiples of three are called triplens (i.e., 3rd, 9th, 15th, and so forth).
- Triplen harmonics, particularly the third, cause major problems in electrical distribution systems.

Problems with Harmonics

- Harmonics do no work, but contribute to the rms current that the system must carry.
- Triplen harmonics are additive in the system neutral.
- These currents return to the transformer source over the neutral and are dissipated as heat in the transformer, cables, and load devices.

Symptoms of Harmonic Problems

- Overheated neutral conductors, panels, and transformers
- Premature failure of transformers, generators, and UPS systems
- Lost computer data
- Interference on communication lines
- Operation of protective devices without overload or short circuit
- Random component failure in electronic devices
- Operating problems with electronic devices not traceable to component problems
- Interaction between multiple VFDs throwing off set points
- Interaction between UPSs and their supplying generators
- System power factor reduction and related system capacity loss
- Problems with capacitor operation and life

Harmonic Mitigation

Currently there are no devices that completely eliminate harmonics, and thus their effects; however, they can be mitigated substantially to control their deleterious consequences. Essentially, current techniques consist of accommodating harmonics, and include the following:

- Increasing neutral sizes, usually doubling feeder neutral sizes and installing a separate neutral with each single-phase branch circuit of a three-phase system, effectively a triple-neutral, rather than a single common neutral of the same size as the phase conductor.
- K-rated transformers.
- Harmonic-rated distribution equipment such as panelboards.
- Passive filters such as phase shifters, phase cancellers, zigzag transformers, and zero-sequence transformers.
- Active filters, electronic, primarily protects upstream equipment/devices.
- Proper grounding.
- Isolation transformers (electrostatically shielded).
- Motor-generator sets.
- Oversizing equipment.

Most of the above involve "beefing up" to accommodate harmonics.

ACTIVE VERSUS PASSIVE DEVICES

Active Devices

PROS

Works well for mitigation of harmonics upstream of the device.
Protects the transformer.

CONS

Expensive.
High maintenance costs.
Uses power.
Works only upstream.

Passive Devices

PROS

No electronic circuitry.
Very reliable.

CONS

Work only upstream to accommodate harmonics.
Location is critical.
Phase loads must be balanced.
Can be overloaded.
Dissipate heat.
Require fused disconnect.

Ultimate/Ideal Solution

The ultimate ideal solution would be:

- Eliminate the production of harmonics at the source (not just accommodate them).
- Be passive and therefore cost-effective, reliable, and efficient.
- Be easily installed and not require protection.
- Handle any load on the distribution system (not require load balancing to be effective).
- Resist overloading (not become a harmonic sink for the rest of the distribution system).

NOTES

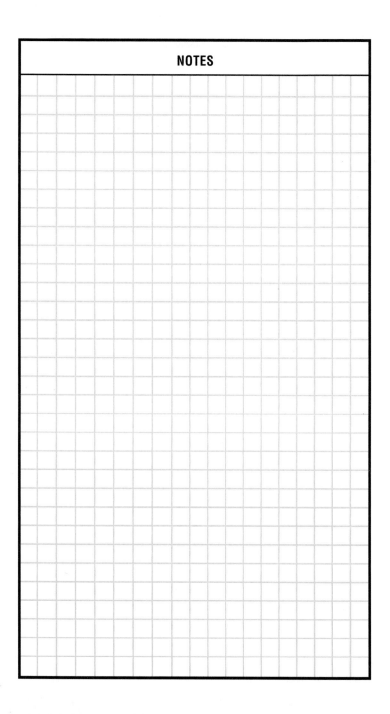

Index

ABOUT THE AUTHOR

Robert B. (Bob) Hickey, P.E., is a licensed professional engineer in four states and is president and chief exectuvie officer of vanZelm, Heywood & Shadford, Inc., a leading northeast U.S. mechanical and electrical consulting engineering firm based in West Hartford, Connecticut. His 35 years of experience spans the electric utility, contracting, and consulting engineering areas of the industry spectrum. He has taught electrical engineering technology as an adjunct faculty member over a two-year period in Connecticut's Community-Technical College system. He also coauthored the electrical chapter of McGraw-Hill's *Field Inspection Handbook*, Second Edition.

NOTES

NOTES

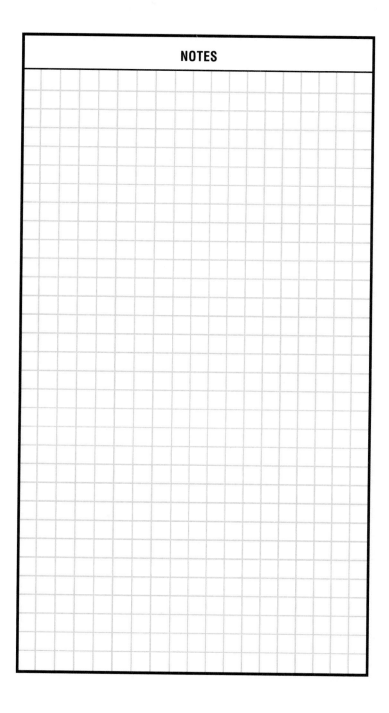

NOTES

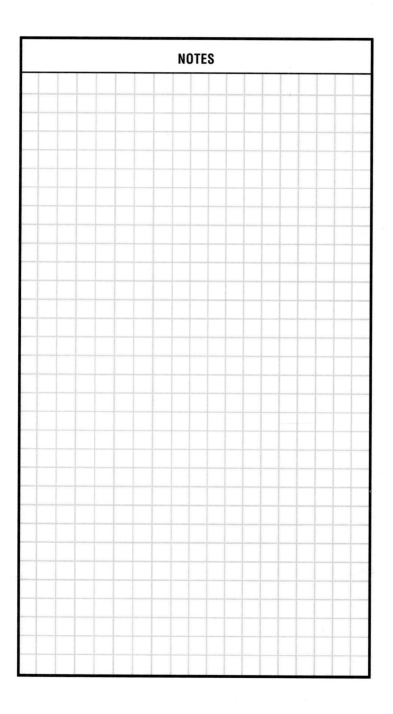

NOTES

NOTES

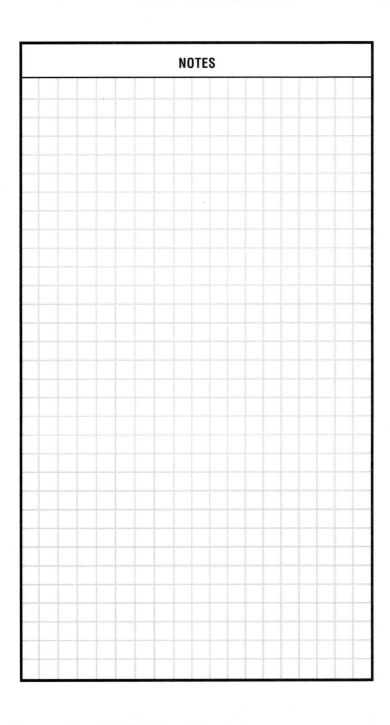